Mathematical Foundations of Quantum Computing

A Scaffolding Approach

Peter Y. Lee

James M. Yu

Ran Cheng

POLARIS
QCI PUBLISHING

ISBN 978-1-961880-08-5 (ebook)

ISBN 978-1-961880-09-2 (paperback)

ISBN 978-1-961880-10-8 (hardcover)

Library of Congress Control Number (LCCN): 2024947285

First edition, March 2025

This document is typeset using LaTeX.

Design and layout of this book are based on the Legrand Orange Book template from https://latextemplates.com/template/the-legrand-orange-book.

Publisher Website: https://polarisqci.com

Synopsis

In an era where quantum computing is poised to revolutionize technology and science, *Mathematical Foundations of Quantum Computing: A Scaffolding Approach* provides the essential mathematical tools needed to navigate this rapidly evolving field. This book is part of an educational series designed to systematically guide beginning graduate students, senior undergraduates, and all those intrigued by the mathematical underpinnings of quantum computation.

The book adopts a "scaffolding approach," offering a step-by-step progression from foundational mathematics to more advanced concepts. Drawing inspiration from educational theories like Vygotsky's Zone of Proximal Development, this approach ensures that readers build confidence and understanding as they move through increasingly sophisticated topics. Key mathematical concepts—such as linear algebra, matrix theory, and probability—are introduced gradually, with frequent revisiting and reinforcement of core ideas. The use of diagrams, tables, and boxed highlights enhances learning while avoiding cognitive overload.

Structured into four parts, the book begins with a review of essential mathematical preliminaries, followed by an in-depth study of vectors, matrices, and linear spaces tailored for quantum computing. From there, it delves into advanced matrix analysis, introducing the indispensable tools of tensor products, matrix decompositions, and quantum operators. The final section focuses on the probability foundations necessary for quantum algorithms, exploring stochastic processes, Markov chains, and Monte Carlo methods.

Whether you are new to quantum computing or wish to solidify your understanding of the mathematical framework, this book offers a clear, structured pathway to mastering the subject. Prepare to engage deeply with the mathematical principles that form the backbone of quantum computing, and embark on your journey into this exciting and complex field.

About the Authors

Dr. Peter Y. Lee: Holds a Ph.D. in Electrical Engineering from Princeton University. His research at Princeton focused on quantum nanostructures, the fractional quantum Hall effect, and Wigner crystals. Following his academic tenure, he joined Bell Labs, making significant contributions to the fields of photonics and optical communications and securing over 20 patents. Dr. Lee's multifaceted expertise extends to educational settings; he has a rich history of teaching, academic program oversight, and computer programming.

Dr. James M. Yu: Earned his Ph.D. in Mechanical Engineering from Rutgers University in New Jersey, specialized in mathematical modeling and simulation of biophysical phenomena. Following his doctorate studies, He continued to conduct research as a postdoctoral associate at Rutgers University. Currently, he is a faculty member at Fei Tian College, Middletown where he dedicates to teaching mathematics, statistics, and computer science.

Dr. Ran Cheng: Earned his Ph.D. in Physics from the University of Texas at Austin, with a specialization in condensed matter theory, particularly in spintronics and magnetism. Following a postdoctoral position at Carnegie Mellon University, he joined the faculty at the University of California, Riverside, where he was honored with the NSF CAREER and DoD MURI awards.

Contents

Level Indicators

Unmarked content: Foundational, appropriate for all readers, including newcomers to linear algebra.

✳: Typically an initial conceptual overview accessible to general readers, followed by math-intensive segments for a senior undergraduate or early graduate-level audience with a background in linear algebra. Assumes familiarity with unmarked content.

✳✳: Advanced exploration, extending beyond ✳-level topics.

III Matrix Methods for Quantum Computing

Preface

The *Scaffolding* Series

This book is part of a series of textbooks designed to provide a pedagogically sound and systematic approach to teaching and learning quantum computing. The series currently includes the following titles:

- *Mathematical Foundations of Quantum Computing: A Scaffolding Approach*

- *Quantum Computing and Information: A Scaffolding Approach*

- *Quantum Algorithms and Applications: A Scaffolding Approach*

While each book serves as a standalone guide to its topic, together they form a comprehensive resource on quantum computing.

This book is aimed at beginning graduate students and advanced undergraduates, with helpful markers for readers at both entry-level and more advanced stages.

Why This Book

Quantum Computing and Information (QCI) introduces a paradigm shift not only in computation but also in the mathematics required to advance in the field. While linear algebra remains central, its application here extends beyond traditional uses. In quantum computing, matrices act dynamically as operators and transformations, and tools like tensor products, trace operations, decompositions, and matrix functions become essential for the mathematical precision and flexibility that quantum mechanics demands.

Dirac notation serves as the primary language for expressing vectors, operators, and their interactions, helping students transition smoothly into quantum mechanics. Special matrices such as Hermitian, unitary, and Pauli matrices are introduced not as abstract concepts but as practical building blocks that encode quantum states and drive quantum transformations, directly supporting algorithms and quantum error correction.

Because the probabilistic structure of quantum mechanics departs from classical probability, this book equips students with a foundation in probability theory, sampling, and key stochastic methods like Markov chains and MCMC, preparing them for probabilistic quantum algorithms and future topics in quantum probability.

Recognizing the wide-ranging prerequisites for quantum computing, we start with a focused review of complex numbers, trigonometry, and summation rules, tailored for quantum applications. By omitting subjects like differential equations and complex analysis—valuable but not essential for QC—this book emphasizes efficiency and accessibility, guiding students toward mastering topics directly aligned with quantum studies.

Rather than offering a comprehensive mathematical reference, *Mathematical Foundations of Quantum Computing* serves as a streamlined guide, bridging traditional mathematical training and the specialized demands of quantum computing and quantum algorithms.

Parts of the Book

The book is organized into four main parts:

- Part I: Preliminaries. A concise review of complex numbers, trigonometry, summation rules, and sets, groups, and functions, laying the groundwork for the following sections.

- Part II: Vectors, Matrices, and Linear Spaces. Covers topics traditionally studied in a first-year linear algebra course, but with an early introduction to complex numbers and Dirac notation, which are used throughout.

- Part III: Matrix Methods for Quantum Systems. Focuses on advanced linear algebra concepts essential for quantum computing, including tensor products, matrix functions, Pauli matrices and strings, and singular value decomposition.

- Part IV: Probability Foundations for Computation. Explores topics such as probability, stochastic processes, Markov chains, and Monte Carlo methods, with an emphasis on their relevance to quantum computing.

Recommended Use

This book provides the foundational mathematics necessary for further study in quantum computing and quantum algorithms. It serves both as a stepping stone to the second and third books in this series and as a standalone reference.

To fully engage with the material, students are encouraged to complete the exercises and problems at the end of each chapter. For a two-semester course, this approach allows for a comprehensive exploration of the content. However, students already familiar with basic linear algebra may complete the book in a single semester by focusing primarily on Parts III and IV.

Furthermore, while not all topics covered here are essential for the second book, *Quantum Computing and Information*, they will be important for the third book, *Quantum Algorithms and Applications*. These topics can be deferred until studying the third book or specific quantum algorithms that require them. They include:

- Discrete Fourier Transform (§ 11.2.5)

- Pauli String Basis (§ 12.4) and Pauli Groups (§ 12.5)

- Jordan, Singular Value, and Schmidt Decompositions (Chapter 13)

- Markov Chains (Chapter 16) and Monte Carlo Methods (Chapter 17)

The Scaffolding Approach

Quantum Computing and Information (QCI) is a complex field that blends advanced mathematics, quantum mechanics, and intricate algorithms. To facilitate learning, this book employs a scaffolding approach that builds understanding progressively from fundamental concepts to more advanced ideas.

Inspired by educational theories like Vygotsky's Zone of Proximal Development, this approach ensures that learners are well-prepared for each new topic. Key strategies include:

- Progressive Learning: Concepts are introduced incrementally, beginning with intuitive examples and moving toward more abstract ideas.

- Spiral Curriculum: Core ideas are revisited from different perspectives throughout the text, deepening understanding with each iteration.

- Active Engagement: Exercises and problems are integrated into each chapter, encouraging learners to apply concepts as they go.

- Cognitive Load Management: Clear explanations, diagrams, and tables are used to reduce cognitive overload, making complex topics more approachable.

By following this method, readers will gain both knowledge and the skills to navigate the interconnected areas of quantum computing and information.

Navigation Aids

To help readers navigate the book, we use several special features:

Concept Box

This box is reserved for key concepts, postulates, and theorems.

Highlight Box

This box highlights corollaries, summaries, implications, and other key points.

Exercise 0.1 Exercises are interspersed throughout the text to sharpen your skills and reinforce your understanding of the material.

Each chapter ends with a Problem Set, providing more challenging problems to encourage deeper engagement with the content.

(i) Used for tips, alerts, connections between concepts, and practical advice.

Info Box
Provides additional context, related concepts, or supplementary information.

Level Indicators

To accommodate learners from various academic backgrounds, some sections are marked with level indicators.

Unmarked content is fundamental and appropriate for all readers, including those new to quantum computing.

∗ marks more advanced sections. These typically start with an overview accessible to all readers, followed by more math-intensive content aimed at senior undergraduates or graduate students with a background in linear algebra.

✳ indicates content for further exploration, going beyond the scope of ✳.

Acknowledgements

We extend our deep thanks to Ms. Liang Zhou and Ms. Elsie He for their creative talents in designing the book cover, and for their thoughtful and visually compelling interpretation of the subject matter.

We are also profoundly grateful to Ms. Vivian Liu and Ms. Nathalie Chiao, whose editorial insights greatly improved the manuscript's readability, coherence, and clarity.

To our colleagues and students—Mr. John Hurst, Mr. Yiyong Huang, Dr. Jason Wang, Dr. Joseph Zhao, and Mr. Jacky Guan—your intellectual curiosity, support, and feedback have been vital in shaping this work into a more practical and educational resource.

Our sincere thanks to our expert reviewers, whose feedback and suggestions were instrumental in enhancing the rigor and accessibility of this text:

- Leonard Kahn, Professor and Chair, Department of Physics, University of Rhode Island

- Andrew Kent, Professor of Physics, The Center for Quantum Phenomena, New York University

- Ying Nian Wu, Professor, Department of Statistics and Data Science, University of California in Los Angeles

- Steven Frankel, Rosenblatt Professor, Faculty of Mechanical Engineering, Technion - Israel Institute of Technology

- Yamamoto Fujio, Professor Emeritus, Kanagawa Institute of Technology, Japan

- Tony Holdroyd, Retired Senior Lecturer in Computer Science and Mathematics

- John Hurst, Veteran Software Developer and Mathematics Enthusiast

Finally, we express our deep appreciation to the broader quantum computing community, whose advancements continue to inspire and guide us. Our goal is to distill and share this knowledge, making quantum computing accessible and exciting for students.

Dedication

This book is dedicated to all inquisitive minds and relentless spirits who believe in the power of learning and the boundless potential of human intellect.

Reviews

❝ With the move toward introducing quantum computing as a first-year course, the structure of *Mathematical Foundations of Quantum Computing* makes it a strong contender as a text that can be used throughout an academic career. The authors have successfully designed a text that can be used at multiple stages of development, from introductory, through intermediate and graduate levels, as well as a useful reference work. From the introduction of vectors and matrices, each topic is revisited with increasing complexity, an ideal implementation of the scaffolding approach. The layout of the text, accompanied by a variety of exercises, examples, and clear graphics, advances the authors' goal of creating a valuable learning and teaching aid. The text, along with its companion *Quantum Computing and Information*, deserves serious consideration by those who are designing a full-range quantum computing curriculum. ❞

— Leonard Kahn, Professor and Chair
Department of Physics, University of Rhode Island

❝ The QCI book (*Quantum Computing and Information: A Scaffolding Approach*) presents quantum computing in a wonderfully friendly manner, making this complex field accessible to anyone with basic undergraduate math preparation. The companion text (*Mathematical Foundations of Quantum Computing: A Scaffolding Approach*), with its comprehensive coverage of mathematical foundations, provides all the essential tools needed to dive into quantum concepts with confidence. I found the chapters on probability to be expertly written, offering a clear, engaging, and quantum-relevant introduction. Together, these books form an inviting and masterful gateway for learners eager to explore quantum computing.❞

— Ying Nian Wu, Professor
Department of Statistics and Data Science, University of California in Los Angeles

❝ This comprehensive and accessible text presents, in a single volume, the mathematical foundation of quantum information. Beginning with the essentials—linear algebra, probability, and matrix analysis—and advancing to topics like tensor products, spectral decompositions, and Markov Chain Monte Carlo simulations, the authors guide the reader with clarity and rigor. Rarely is so much mathematical depth presented in such a student-friendly way. This volume will serve both newcomers and experts alike, providing a strong foundation for gaining facility with the mathematics required to understand quantum systems.❞

— Andrew Kent, Professor of Physics
The Center for Quantum Phenomena, New York University

" A beautiful, colorfully crystal clear, and veritable one-stop-shop, this resource offers everything mathematical essential to quantum computing. Covering vector spaces, matrix methods including tensor products, and probability theory, it is a must-read for quantum computing researchers and practitioners alike."

— *Steven Frankel, Rosenblatt Professor*
Faculty of Mechanical Engineering, Technion - Israel Institute of Technology

" This book is a learned and thorough exposition of the mathematics that supports quantum computing. The authors have gone to great lengths to make it both learner-friendly and detailed while maintaining rigor. It covers topics ranging from the fundamentals of quantum mathematics to the complexities of vector and matrix algebra, as well as the probabilities central to quantum computing. The text is complemented by numerous supporting figures that effectively illustrate key concepts. Applications of quantum computing are introduced and seamlessly integrated throughout the book. This volume, along with its companion, *Quantum Computing and Information - a Scaffolding Approach*, is an essential addition to the bookshelf of anyone seeking a deeper understanding of quantum computing and its mathematical foundations."

— *Tony Holdroyd*
Retired Senior Lecturer in Computer Science and Mathematics

" This book provides a thorough explanation of the mathematics underlying quantum computing. Dirac (bra-ket) notation is introduced right at the beginning of Part II. Part III then offers a detailed treatment of matrix operations fundamental to quantum computing, with particular emphasis on tensor products. The text also gives careful attention to change of basis—crucial in applications such as quantum key distribution—and to the Kronecker product, which is central to describing composite quantum systems. Equally significant, Part IV presents an in-depth discussion of probability, an essential tool for understanding quantum computing in contrast to classical computing."

— *Yamamoto Fujio*
Professor Emeritus, Kanagawa Institute of Technology, Japan

Preliminaries

The most renowned algorithm in quantum computing is, arguably, Shor's algorithm. To understand its significance and functionality, one must be comfortable working with equations such as the following:

$$\frac{1}{\sqrt{N}} \sum_{j=0}^{N-1} |\psi_j\rangle = \frac{1}{N} \sum_{j=0}^{N-1} \sum_{k=0}^{N-1} \omega_N^{jk} |a^k\rangle$$

$$= |1\rangle + \frac{1}{N} \sum_{k=1}^{N-1} \left(\sum_{j=0}^{N-1} \omega_N^{jk} \right) |a^k\rangle = |1\rangle,$$

where N is a large integer, and $\omega_N = e^{\frac{2\pi i}{N}}$ represents the primitive Nth root of unity.

In the near term, one of the most promising applications of quantum computing is expected to be the simulation of chemical processes. In such simulations, a molecule consisting of K nuclei and n electrons is modeled by a Hamiltonian of the form:

$$H = -\sum_{i=1}^{n} \frac{(\nabla_i)^2}{2} - \sum_{I=1}^{K} \frac{(\nabla_I)^2}{2M_I} - \sum_{i,I} \frac{Z_I}{|r_i - R_I|} + \frac{1}{2} \sum_{i \neq j} \frac{1}{|r_i - r_j|} + \frac{1}{2} \sum_{I \neq J} \frac{Z_I Z_J}{|R_I - R_J|},$$

where ∇ denotes the gradient operator, r_i is the position of the ith electron, and R_I and Z_I represent the position and charge of the Ith nucleus, respectively.

These examples highlight a fundamental requirement for mastering quantum computing: a solid understanding of summation and product notations (i.e., $\sum$ and $\prod$), complex numbers (e.g., $\omega_N = e^{\frac{2\pi i}{N}}$), trigonometry, and the mathematics of sets, groups, and functions. Familiarity with these preliminary concepts will ease the transition into the computational and physical aspects of quantum computing. The purpose of Part I is to help readers review and refresh their knowledge of these preliminaries, ensuring they are well-prepared for the advanced topics that follow.

Another important branch of mathematics relevant to quantum computing is advanced calculus and differential equations, as illustrated by the gradient operator ∇ in the Hamiltonian equation above. While this book does not cover advanced calculus and differential equations in depth, the topics included here, combined with a basic understanding of calculus, provide a sufficient foundation for many areas of quantum computing, including algorithms and programming.

1. Summation and Product Notations

Contents

Summation (Sigma, $\sum$) and Product (Pi, $\prod$) notations are essential tools in mathematics for simplifying the operations of sums and products over sequences. These notations are ubiquitous in quantum computing as well as across various mathematical disciplines. Given their complexity and wide-ranging applications, a focused study of these notations is invaluable for anyone delving into mathematical and scientific fields, where they are frequently used to succinctly express and manipulate long series and products.

1.1 Summation over a Single Variable

1.1.1 The Sigma Notation

$$n \longleftarrow \text{Ending Index}$$

$$\sum f(i) \longleftarrow \text{Quantity to Sum}$$

$$\text{Index Variable} \longrightarrow i{=}1 \longleftarrow \text{Starting Index}$$

1 Definition

As shown in the figure above, we use $\sum$ (Sigma) to represent the sum of a series:

$$\sum_{i=m}^{n} a_i = a_m + a_{m+1} + a_{m+2} + \cdots + a_{n-1} + a_n. \tag{1.1}$$

where i is the index variable, index in short; m is the starting index, and n is the ending index. By convention, the step size of i is 1.

For example,

$$\sum_{i=1}^{100} i = 1 + 2 + 3 + 4 + \cdots + 99 + 100,$$

$$\sum_{i=3}^{7} i^2 = 3^2 + 4^2 + 5^2 + 6^2 + 7^2.$$

2 Infinite Sums

Sums can contain an infinite number of terms. A sum is said to converge if the sequence of its partial sums approaches a specific finite value as the number of terms increases indefinitely. Conversely, a sum diverges if the partial sums do not approach any finite limit. Here are two famous converging infinite sums:

$$\sum_{i=0}^{\infty} \frac{1}{2^i} = 1 + \frac{1}{2} + \frac{1}{4} + \frac{1}{8} + \frac{1}{16} + \cdots = 2, \tag{1.2a}$$

$$\sum_{i=1}^{\infty} \frac{1}{i^2} = \frac{1}{1^2} + \frac{1}{2^2} + \frac{1}{3^2} + \frac{1}{4^2} + \cdots = \frac{\pi^2}{6}. \tag{1.2b}$$

The first example is an infinite geometric series, where each term is half the previous term. The sum of the first n terms, called the partial sum, is given by $S_n = \sum_{i=0}^{n} \frac{1}{2^i} = \frac{1 - \frac{1}{2^n}}{1 - \frac{1}{2}}$. As $n \to \infty$, $\frac{1}{2^n} \to 0$. Consequently, $S_n \to \frac{1}{1 - \frac{1}{2}}$, which simplifies to 2.

The second example involves the sum of the reciprocals of the squares of positive integers, and its convergence to $\frac{\pi^2}{6}$ is a result that is not immediately obvious and typically requires tools from calculus to prove rigorously.

3 General Notation

The index in a summation (commonly i, as in the examples above) can be represented by any algebraic symbol without changing the meaning of the sum. However, in contexts involving complex numbers, where i conventionally denotes $\sqrt{-1}$, it is advisable to use alternative symbols for the index to avoid ambiguity.

Moreover, sums can be specified using descriptions other than starting and ending values for the index. For example,

$$\sum_{p \in P} f(p) \quad \text{where } f(p) = \frac{1}{2^p} \quad \text{and } P = \{2, 3, 5, 7, 11, \ldots\} \text{ (the prime numbers)}$$

represents the sum $\dfrac{1}{2^2} + \dfrac{1}{2^3} + \dfrac{1}{2^5} + \dfrac{1}{2^7} + \dfrac{1}{2^{11}} + \cdots$.

Exercise 1.1 Calculate $f(x) = \displaystyle\sum_{n=0}^{\infty} a_n$, where $a_0 = 1$ and $a_{k+1} = \frac{a_k}{2} + \frac{1}{2^k}$ for $k = 0, 1, 2, \ldots$.

4 Sums as Functions

Sums can contain parameters other than the index, resulting in functions of those parameters. For example, the discrete Fourier transform (DFT) is given by,

$$\tilde{x}_k = \frac{1}{\sqrt{N}} \sum_{n=0}^{N-1} x_n e^{-\frac{2\pi i}{N} kn}, \quad k = 0, 1, \ldots, N - 1, \tag{1.3}$$

where x_n represents N values indexed by n, and $\tilde{x}_k$ are the Fourier coefficients. Here, i is the imaginary unit, and N is a positive integer representing the dimension of the DFT. (If you are not familar with complex numbers or DFT, don't worry. We will delve into complex numbers in Chapter 3, and DFT in § 11.2.5.)

Exercise 1.2 Interpret $f(x) = \displaystyle\sum_{n=0}^{\infty} \frac{x^{\frac{n}{2}}}{n!}$. (By definition, $0! = 1$.)

1.1.2 Useful Sums

The following are some summation formulas commonly encountered in quantum computing, as well as in various mathematical and engineering disciplines:

$$\sum_{i=1}^{n} i = \frac{n(n+1)}{2} \tag{1.4}$$

$$\sum_{i=1}^{n} i^2 = \frac{n(n+1)(2n+1)}{6} \tag{1.5}$$

$$\sum_{i=1}^{n} i^3 = \left(\frac{n(n+1)}{2}\right)^2 \tag{1.6}$$

$$\sum_{i=0}^{n} (a_0 + id) = (n+1)\left(a_0 + \frac{nd}{2}\right) \quad \text{(arithmetic series)} \tag{1.7}$$

$$\sum_{i=0}^{n} a^i = \frac{1 - a^{n+1}}{1 - a} \quad \text{(geometric series)} \tag{1.8}$$

$$(a + b)^n = \sum_{i=0}^{n} \binom{n}{i} a^{n-i} b^i \quad \text{(binomial theorem)} \tag{1.9}$$

$$\frac{1}{1 - x} = \sum_{n=0}^{\infty} x^n = 1 + x + x^2 + x^3 + \cdots \quad \text{(for } |x| < 1) \tag{1.10}$$

$$\frac{1}{(1 - x)^2} = \sum_{n=1}^{\infty} n x^{n-1} = 1 + 2x + 3x^2 + 4x^3 + \cdots \quad \text{(for } |x| < 1) \tag{1.11}$$

$$\ln(1 + x) = \sum_{n=1}^{\infty} \frac{(-1)^{n+1}}{n} x^n = x - \frac{x^2}{2} + \frac{x^3}{3} - \cdots \quad \text{(for } |x| < 1) \tag{1.12}$$

$$e^x = \sum_{n=0}^{\infty} \frac{x^n}{n!} = 1 + x + \frac{x^2}{2!} + \frac{x^3}{3!} + \cdots \tag{1.13}$$

$$\sin x = \sum_{n=0}^{\infty} \frac{(-1)^n}{(2n + 1)!} x^{2n+1} = x - \frac{x^3}{3!} + \frac{x^5}{5!} - \cdots \tag{1.14}$$

$$\cos x = \sum_{n=0}^{\infty} \frac{(-1)^n}{(2n)!} x^{2n} = 1 - \frac{x^2}{2!} + \frac{x^4}{4!} - \cdots \tag{1.15}$$

Exercise 1.3 Write $\dfrac{1}{1 + 2x^2}$ as an infinite sum in $\sum$ notation.

Exercise 1.4 Write $e^{\frac{1}{x}}$ as an infinite sum in $\sum$ notation.

1.1.3 Summation Rules

Presented below are several fundamental rules and properties of the $\sum$ summation, which are essential in simplifying and manipulating $\sum$ expressions:

$$\sum_{i=m}^{n} a_i = \sum_{j=m}^{n} a_j \quad \text{(change of index variable)} \tag{1.16}$$

$$\sum_{i=s}^{t} f(i) = \sum_{n=s}^{t} f(n) \quad \text{(change of index variable)} \tag{1.17}$$

$$\sum_{n=s}^{t} f(n) = \sum_{n=s}^{j} f(n) + \sum_{n=j+1}^{t} f(n) \quad \text{(splitting a sum)} \tag{1.18}$$

$$\sum_{n=s}^{t} f(n) = \sum_{n=0}^{t-s} f(t - n) \quad \text{(reverse order)} \tag{1.19}$$

$$\sum_{n=s}^{t} f(n) = \sum_{n=s+p}^{t+p} f(n - p) \quad \text{(index shift)} \tag{1.20}$$

$$\sum_{n=s}^{t} a \cdot f(n) = a \cdot \sum_{n=s}^{t} f(n) \quad \text{(distributivity)} \tag{1.21}$$

$$\sum_{n=s}^{t} f(n) \pm \sum_{n=s}^{t} g(n) = \sum_{n=s}^{t} (f(n) \pm g(n)) \quad \text{(commutativity)} \tag{1.22}$$

Exercise 1.5 Change the index in $\sum_{k=6}^{24}(5k + 8)$ so that it starts with 0.

1.2 Products and Other Notations

1.2.1 The Pi Notation

Similar to $\sum$ for sum, we use $\prod$ (Pi) to denote the product of a series of terms:

$$\prod_{i=m}^{n} a_i = a_m \cdot a_{m+1} \cdot a_{m+2} \cdot \cdots \cdot a_{n-1} \cdot a_n. \tag{1.23}$$

A familiar example is $\prod_{i=1}^{n} i = n!$, which represents the factorial of n.

Here are two useful equations relating $\sum$ and $\prod$:

$$b^{\sum_{n=s}^{t} f(n)} = \prod_{n=s}^{t} b^{f(n)}, \tag{1.24a}$$

$$\sum_{n=s}^{t} \log_b f(n) = \log_b \prod_{n=s}^{t} f(n). \tag{1.24b}$$

Exercise 1.6 Convert $\prod_{k=0}^{n-1}(k+1)e^{i\frac{2k\pi}{n}}$ to $\sum$ notation. (If you are not familar with the basics of complex numbers or exponential functions, come back to this exercise after Chapter 3.)

1.2.2 Other Notations

In quantum computing and linear algebra, similar notations are used, such as $\oplus$ for the modulo-2 sum (i.e., the bitwise XOR operation), or depending on the context, the direct sum of linear spaces, and $\otimes$ for tensor product. However, we will not delve into these notations at this time.

1.3 Summation over Multiple Variables

This section explores summation extended to multiple variables, beginning with the case of two variables.

1.3.1 Rectangular Sums

The double (two-dimensional) summation over a rectangular array is given by:

$$\sum_{i=1,j=1}^{n_1,n_2} a_{i,j} = \sum_{i=1}^{n_1}\sum_{j=1}^{n_2} a_{i,j} = \sum_{j=1}^{n_2}\sum_{i=1}^{n_1} a_{i,j}$$

$$
\begin{aligned}
&= a_{1,1} + a_{1,2} + a_{1,3} + a_{1,4} + \cdots + a_{1,n_2} \\
&+ a_{2,1} + a_{2,2} + a_{2,3} + a_{2,4} + \cdots + a_{2,n_2} \\
&+ a_{3,1} + a_{3,2} + a_{3,3} + a_{3,4} + \cdots + a_{3,n_2} \\
&+ a_{4,1} + a_{4,2} + a_{4,3} + a_{4,4} + \cdots + a_{4,n_2} \\
&+ \cdots \\
&+ a_{n_1,1} + a_{n_1,2} + a_{n_1,3} + a_{n_1,4} + \cdots + a_{n_1,n_2}
\end{aligned}
\tag{1.25}
$$

Here, $\displaystyle\sum_{i=1}^{n_1}\sum_{j=1}^{n_2} a_{i,j}$ represents summing over each row first and then summing the results, while $\displaystyle\sum_{j=1}^{n_2}\sum_{i=1}^{n_1} a_{i,j}$ represents summing over each column first and then summing those results. The notation $\displaystyle\sum_{i=1,j=1}^{n_1,n_2} a_{i,j}$ implies summation over all elements in a two-dimensional array, irrespective of the order. For rectangular arrays, the order of summation can be interchanged without affecting the final result.

1.3.2 Product of Two Sums

The product of two sums can be expanded into a double sum as follows:

$$\left(\sum_{i=1}^{m} a_i\right)\left(\sum_{j=1}^{n} b_j\right) = (a_1 + a_2 + \cdots + a_m)(b_1 + b_2 + \cdots + b_n)$$

$$
\begin{aligned}
&= a_1 b_1 + a_1 b_2 + a_1 b_3 + a_1 b_4 + \cdots + a_1 b_n \\
&+ a_2 b_1 + a_2 b_2 + a_2 b_3 + a_2 b_4 + \cdots + a_2 b_n \\
&+ a_3 b_1 + a_3 b_2 + a_3 b_3 + a_3 b_4 + \cdots + a_3 b_n \\
&+ \cdots \\
&+ a_m b_1 + a_m b_2 + a_m b_3 + a_m b_4 + \cdots + a_m b_n \\
&= \sum_{i=1}^{m}\sum_{j=1}^{n} a_i b_j = \sum_{i=1}^{m} a_i \sum_{j=1}^{n} b_j.
\end{aligned}
\tag{1.26}
$$

For $\left(\displaystyle\sum_{i=1}^{m} a_i\right)\left(\displaystyle\sum_{j=1}^{n} b_j\right)$, the left-hand side is a product of two sums, which expands to the same set of terms as the double sum on the right-hand side. This demonstrates the distributive property of multiplication over addition.

For the expression $\displaystyle\sum_{i=1}^{m}\sum_{j=1}^{n} a_i b_j$, the factor a_i is constant with respect to the inner summation over j, and can therefore be factored out, leading to the simplification

$\sum_{i=1}^{m} a_i \sum_{j=1}^{n} b_j$. This is an application of the distributive property, which allows for the rearrangement and combination of sums in this manner.

1.3.3 Triangular Sums

The triangular sum over a lower triangular array is given by:

$$\sum_{1 \le j \le i \le n} a_{i,j} = \sum_{i=1}^{n} \sum_{j=1}^{i} a_{i,j} = \sum_{j=1}^{n} \sum_{i=j}^{n} a_{i,j} = \sum_{j=0}^{n-1} \sum_{i=1}^{n-j} a_{i+j,i}$$

$$= a_{1,1}$$
$$+ a_{2,1} + a_{2,2} \tag{1.27}$$
$$+ a_{3,1} + a_{3,2} + a_{3,3}$$
$$+ a_{4,1} + a_{4,2} + a_{4,3} + a_{4,4}$$
$$+ \cdots$$
$$+ a_{n,1} + a_{n,2} + a_{n,3} + a_{n,4} + \cdots + a_{n,n}$$

Here, $\sum_{1 \le j \le i \le n} a_{i,j}$ denotes the sum over all elements in a lower triangular array including the diagonal. The notation $\sum_{i=1}^{n} \sum_{j=1}^{i} a_{i,j}$ sums each row up to the ith element, then aggregates these sums. Conversely, $\sum_{j=1}^{n} \sum_{i=j}^{n} a_{i,j}$ sums each column starting from the jth element downwards, then aggregates these sums.

The expression $\sum_{j=0}^{n-1} \sum_{i=1}^{n-j} a_{i+j,i}$ sums along the diagonals, where $j = 0$ corresponds to the main diagonal and $j = n - 1$ to the first off-diagonal, which is a single term.

Exercise 1.7 The sum in Eq. 1.27 is for a lower triangular array, including the diagonal elements. Modify Eq. 1.27 to represent the sum over the lower triangular array excluding the diagonal elements.

Exercise 1.8 Eq. 1.27 sums over a lower triangular array. In contrast, $\sum_{1 \le i \le j \le n} a_{i,j}$ sums over an upper triangular array. Express this sum in a row-first, column-first, and diagonal-first manner, analogous to Eq. 1.27.

1.3.4 Multiple Variables

The techniques and principles established for two-variable sums extend naturally to multivariate sums and products, providing a foundation for more complex scenarios.

■ **Example 1.1 — Expansion of a Product of Sums.**

Let's try to understand the equation below, which represents the expansion of a product of sums:

$$\prod_{i=1}^{n}(1-x_i) = 1 + \sum_{k=1}^{n}\left(\sum_{1 \le i_1 < i_2 < \cdots < i_k \le n}\prod_{j=1}^{k}x_{i_j}\right). \tag{1.28}$$

The left-hand side represents the product $(1+x_1)(1+x_2)(1+x_3)\cdots(1+x_n)$. The right-hand side represents its expansion, which includes 1 and the sum of all unique product terms of the form $\prod_{j=1}^{k}x_{i_j} = x_{i_1}x_{i_2}\cdots x_{i_k}$. The condition $1 \le i_1 < i_2 < \cdots < i_k \le n$ ensures the uniqueness of each term, where k ranges from 1 to n.

The inner sum sums all products $\prod_{j=1}^{k}x_{i_j}$ over all possible k-element subsets of $1, 2, \ldots, n$. The outer sum then sums these over all possible values of k from 1 to n.

If all the x_i are equal, then the equation simplifies to the binomial theorem:

$$(1+x)^n = \sum_{k=0}^{n}\binom{n}{k}x^k, \tag{1.29}$$

where $\binom{n}{k}$ is the binomial coefficient representing the number of ways to choose k elements from a set of n distinct elements. ∎

Exercise 1.9 Interpret the expression $\sum_{i,j,k=0}^{1}c_{ijk}4^i 2^j 1^k$.

Exercise 1.10 Evaluate $\sum_{p=1}^{n}\sum_{k_1=0}^{1}\cdots\sum_{k_n=0}^{1}k_p 2^{-p}$ for $n = 4$.

Exercise 1.11 Write a $\sum$-$\prod$ notation for the expansion of $\prod_{i=1}^{n}(1+x_i+x_i^2)$, similar to Eq. 1.28.

Problem Set 1

1.1 Evaluate $\sum_{k=11}^{100}(2+3k+4k^2)$.

1.2 Write out the first few terms of the double series $\sum_{1 \le j \le i \le \infty}\frac{1}{2^{ij}}$ in row-first, column-first, and diagonal-first order.

1.3 Evaluate $\sum_{i=0}^{\infty}\sum_{j=0}^{\infty}\frac{1}{2^{i+j}}$.

1.4 Write a $\sum$-$\prod$ notation for the expansion of $\prod_{i=1}^{n}(1+x_i+y_i)$, similar to Eq. 1.28.

1.5 Evaluate

$$f_n = \sum_{\substack{m>0 \ k_1+k_2+\cdots+k_m=n \\ k_1,k_2,\ldots,k_m>0}} k_1 + k_2 + \cdots + k_m.$$

1.6 Suppose e^x is defined as $e^x = \sum_{n=0}^{\infty} \frac{1}{n!}x^n$. Show that $e^{x+y} = e^x e^y$.

1.7 $*$ Consider a set $\{x_{ij}\}$, where $i, j = 1, 2, \ldots, n$. Assume it satisfies the condition

$$\sum_{k=1}^{n} x_{ik}x_{jk} = \delta_{ij},$$

where δ_{ij} is the Kronecker δ function, defined as $\delta_{ij} = 1$ if $i = j$ and $\delta_{ij} = 0$ if $i \neq j$. Define

$$A_{ij} = \sum_{k=1}^{n} c_k x_{ik}x_{jk},$$

for any constants $c_1, c_2, \ldots, c_n$ and integers i, j. Prove that for any positve integer p,

$$\sum_{m_1,m_2,\ldots,m_p=1}^{n} A_{im_1}A_{m_2m_3}A_{m_3m_4}\cdots A_{m_pj} = \sum_{k=1}^{n} c_k^{p+1} x_{ik}x_{jk}.$$

1.8 $*$ (Consider numpy and itertools in Python, particularly np.sum.)

Part 1

(a) Write a Python function to generate a 2D array of size $n \times n$ filled with random integers in a specified range.

(b) Write four functions to compute the following equivalent sums using the 2D array and n as input:

$$\sum_{1 \leq j \leq i \leq n} a_{i,j} = \sum_{i=1}^{n}\sum_{j=1}^{i} a_{i,j} = \sum_{j=1}^{n}\sum_{i=j}^{n} a_{i,j} = \sum_{j=0}^{n-1}\sum_{i=1}^{n-j} a_{i+j,i}.$$

(c) Verify that the four sums computed in (b) are equal using the arrays generated in (a). Repeat this verification with different sets of random integers.

Part 2

(a) Write a Python function to generate an array of length n with random integers in a specified range.

(b) Implement a function to calculate $P_n = \prod_{i=1}^{n}(1 + x_i)$, where x_i are elements of the array from (a).

(c) Implement a function for $Q_n = 1 + \sum_{k=1}^{n} \left(\sum_{1 \leq i_1 < i_2 < \cdots < i_k \leq n} \prod_{j=1}^{k} x_{i_j} \right),$

where x_{i_j} are elements of the array from (a).

(d) Test whether P_n and Q_n produce equivalent results.

2. Trigonometry

Contents

Trigonometry is a fundamental aspect of quantum computing, crucial in the formulation and understanding of quantum algorithms. This section offers a succinct overview of trigonometric principles, serving as an essential primer for those venturing into the realm of quantum computation.

Angles are empirically measured in degrees. However, in trigonometry, we usually measure angles in radians. A full circle, or a complete turn, is $360°$, equivalent to 2π radians.

2.1 Definitions

2.1.1 From Triangle to Unit Circle

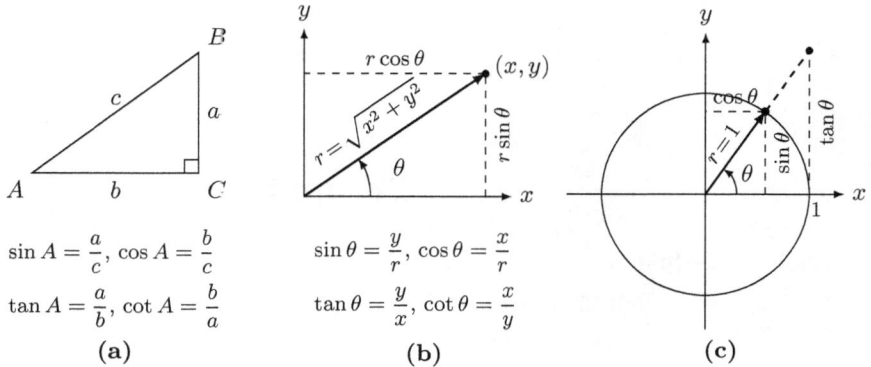

$$\sin A = \frac{a}{c}, \ \cos A = \frac{b}{c}$$

$$\tan A = \frac{a}{b}, \ \cot A = \frac{b}{a}$$

(a)

$$\sin \theta = \frac{y}{r}, \ \cos \theta = \frac{x}{r}$$

$$\tan \theta = \frac{y}{x}, \ \cot \theta = \frac{x}{y}$$

(b)

(c)

Figure 2.1: Definitions of the sin and cos Functions

As shown in Fig. 2.1(a), the trigonometric functions are initially defined as ratios between the sides of a right triangle. They are then generalized in terms of (x, y) coordinates, as shown in Fig. 2.1(b). These functions can also be visualized on the unit circle, as depicted in Fig. 2.1(c).

Exercise 2.1 Using the unit circle diagram below and in Fig. 2.1, describe the geometric interpretation of the sin, cos, tan, cot, sec, and csc functions. Reflect on how these interpretations relate to the etymology of their names: "sine" corresponds to a line segment within a circle, "tangent" to a line that touches the circle at a single point, and "secant" to a line that intersects the circle at two points.

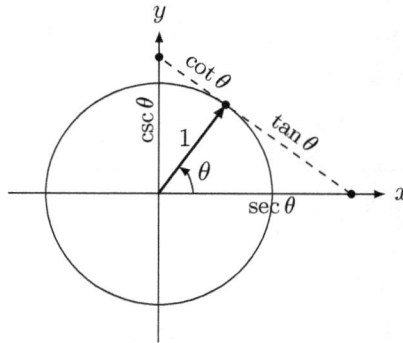

2.1.2 Trigonometric Curves

With both the coordinate and unit-circle representations, the angle is no longer limited to 0 to 90°. It can be extended to all real values, including negatives. Trigonometric ratios are thus treated as mathematical functions with a domain of $(-\infty, \infty)$, and their plots are illustrated in Fig. 2.2.

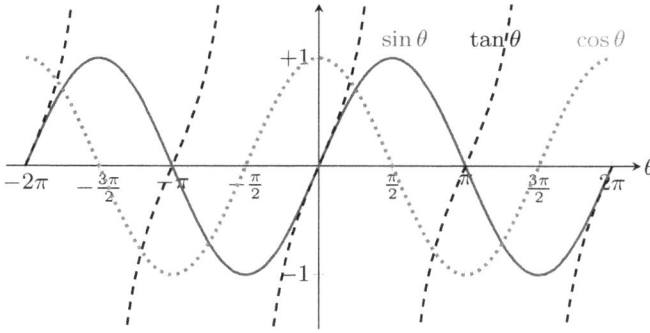

Figure 2.2: Graphs of the sin, cos, and tan Curves

> ⓘ The standard function notation $\sin(\theta)$ is often abbreviated to $\sin\theta$ when the argument is unambiguous. Thus, $\sin\frac{\pi}{4}$ is valid notation. However, parentheses are required in expressions like $\sin\left(\frac{\pi}{2} - \theta\right)$ to avoid ambiguity. This rule similarly applies to other trigonometric functions.
>
> Other standard notations in trigonometry, which may seem unconventional in regular algebra, include $\sin^2\theta$, which means $(\sin\theta)^2$ rather than $\sin(\sin\theta)$. Similarly, $\sin^{-1}x$ represents $\arcsin x$ rather than $1/\sin x$. These notational conventions are specific to trigonometry and are universally accepted in mathematical discourse.

2.2 Basic Properties and Inverse Functions

2.2.1 Periodic Properties

Trigonometric functions are periodic: $\sin\theta$, $\cos\theta$, $\sec\theta$, and $\csc\theta$ have a period of 2π, while $\tan\theta$ and $\cot\theta$ have a period of π. The functions $\sin\theta$ and $\cos\theta$ are bounded within $[-1, 1]$, whereas the others are unbounded. These properties are summarized in Table 2.1.

	$\sin\theta$	$\cos\theta$	$\tan\theta$	$\csc\theta$	$\sec\theta$	$\cot\theta$
Definition	y/r	x/r	y/x	r/y	r/x	x/y
Period	2π	2π	π	2π	2π	π
Range	$[-1, 1]$	$[-1, 1]$	$(-\infty, \infty)$	$(-\infty, -1] \cup [1, \infty)$		$(-\infty, \infty)$
Zeros	$n\pi$	$(n+\frac{1}{2})\pi$	$n\pi$			$(n+\frac{1}{2})\pi$
Poles			$(n+\frac{1}{2})\pi$	$n\pi$	$(n+\frac{1}{2})\pi$	$n\pi$

Note: n is an integer.

Table 2.1: Basic Properties of Trigonometric Functions

2.2.2 Symmetry Properties

Trigonometric functions are not only periodic but also exhibit symmetry within their period. Notably, $\cos\theta$ is symmetric around $\theta = 0$, while $\sin\theta$ and $\tan\theta$ are antisymmetric around $\theta = 0$.

$$\sin(-\theta) = -\sin\theta, \quad \sin(\pi - \theta) = \sin\theta, \quad \sin(\pi + \theta) = -\sin\theta.$$
$$\cos(-\theta) = \cos\theta, \quad \cos(\pi - \theta) = -\cos\theta, \quad \cos(\pi + \theta) = -\cos\theta. \quad (2.1)$$
$$\tan(-\theta) = -\tan\theta, \quad \tan(\pi - \theta) = -\tan\theta, \quad \tan(\pi + \theta) = \tan\theta.$$

Exercise 2.2 Determine the periodicity, range, and zeros of $y = 5\sin(\frac{2}{3}x + 1)$.

2.2.3 Inverse Functions

The inverse of sin is denoted as arcsin, or asin, or $\sin^{-1}$. This notation applies similarly to other trigonometric functions. Since trigonometric functions are periodic, their inverses are defined over specific ranges to ensure they are single-valued. These properties are summarized in Table 2.2.

Function	sin	cos	tan	csc	sec	cot
Inverse	$\sin^{-1}$ arcsin	$\cos^{-1}$ arccos	$\tan^{-1}$ arctan	$\csc^{-1}$	$\sec^{-1}$	$\cot^{-1}$
Domain	$[-1,1]$	$[-1,1]$	$(-\infty,\infty)$	$(-\infty,-1]\cup[1,\infty)$		$(-\infty,\infty)$
Range	$[-\frac{\pi}{2},\frac{\pi}{2}]$	$[0,\pi]$	$(-\frac{\pi}{2},\frac{\pi}{2})$	$[-\frac{\pi}{2},\frac{\pi}{2}]\setminus\{0\}$	$[0,\pi]\setminus\{\frac{\pi}{2}\}$	$(-\frac{\pi}{2},\frac{\pi}{2})$

Table 2.2: Inverse Trigonometric Functions

Exercise 2.3 Find the exact values of

(a) $\arcsin\left(\cos\frac{\pi}{6}\right)$

(b) $\tan\left(\arcsin\frac{5}{13}\right)$

2.2.4 ∗Extended Inverse Functions

There is also an extended version of inverse trigonometric functions. For example, the function arctan2(y,x) (or atan2) extends the range of the regular arctan function to $[-\pi,\pi]$ by considering the signs of x and y. This extension enables the function to distinguish angles in all four quadrants of the Cartesian plane, resolving the ambiguity present in the standard arctan function, which is limited to $(-\frac{\pi}{2},\frac{\pi}{2})$. The arctan2$(y,x)$ function is defined as:

$$\text{arctan2}(y,x) = \begin{cases} \arctan\frac{y}{x} & \text{if } x > 0, \\ \arctan\frac{y}{x} + \pi & \text{if } x < 0 \text{ and } y \geq 0, \\ \arctan\frac{y}{x} - \pi & \text{if } x < 0 \text{ and } y < 0, \\ +\frac{\pi}{2} & \text{if } x = 0 \text{ and } y > 0, \\ -\frac{\pi}{2} & \text{if } x = 0 \text{ and } y < 0, \\ 0 & \text{if } x = 0 \text{ and } y = 0. \end{cases} \quad (2.2)$$

Exercise 2.4 Construct a function arcsin2(y,x) that returns a unique angle θ in the range $[-\pi,\pi]$. This function should satisfy $\sin\theta = \dfrac{y}{\sqrt{x^2 + y^2}}$, ensuring that the angle corresponds correctly to the quadrant of the point (x, y), similar to the behavior of arctan2(y, x) as defined in Eq. 2.2.

2.3 Special Angles and Function Values

Trigonometric functions generally yield transcendental values. However, for certain special angles, their values can be expressed as finite radical expressions. These values are summarized in Fig. 2.3.

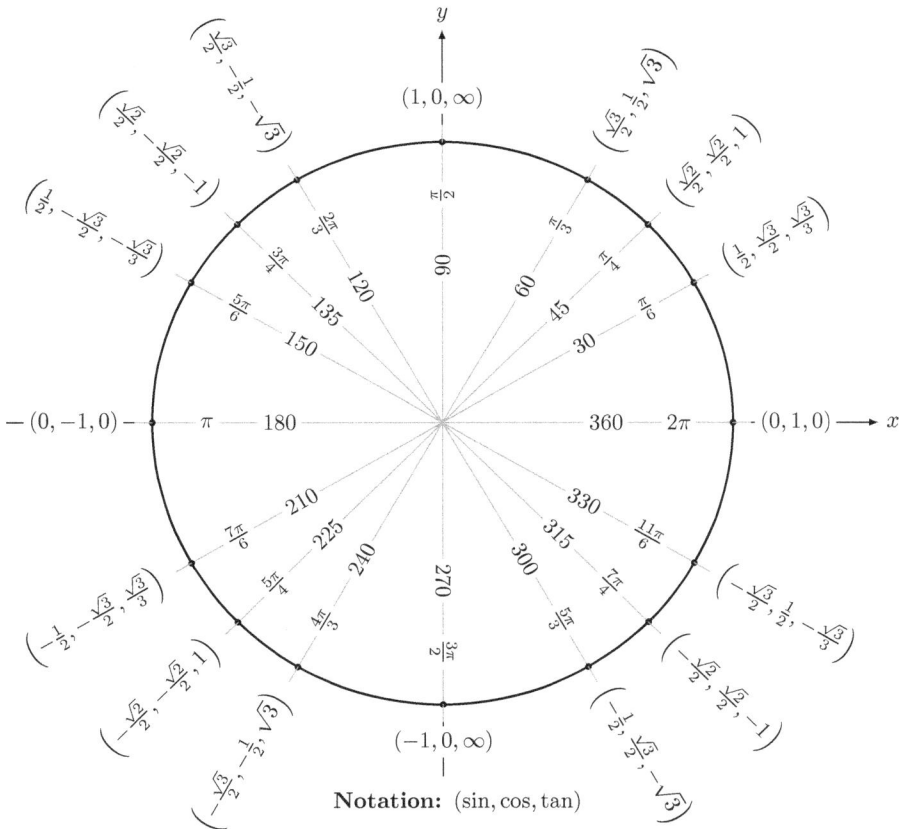

Figure 2.3: Special Angles and Their Corresponding sin, cos, tan Values

■ **Example 2.1** Determine all solutions to the equation $\sin\theta = \frac{1}{2}$.

Solution. The inverse sin function, arcsin, yields $\theta = \arcsin\frac{1}{2} = \frac{\pi}{6}$, which is the principal value within the range of $[-\frac{\pi}{2}, \frac{\pi}{2}]$. However, due to the symmetric nature of the sin function, there is another solution in the first period, given by $\sin(\pi - \theta) = \sin\theta$. Therefore, $\theta = \pi - \frac{\pi}{6} = \frac{5\pi}{6}$ is also a solution. Taking into account the periodic nature of the sin function, where $\sin(\theta + 2n\pi) = \sin\theta$ for any integer n, the complete set of solutions is:

$$\theta = \frac{\pi}{6} + 2n\pi \quad \text{or} \quad \frac{5\pi}{6} + 2n\pi, \quad \text{where } n \in \mathbb{Z}.$$

■

Exercise 2.5 Determine all solutions to the equation $\cos\theta = \frac{1}{2}$.

2.4 Trigonometric Identities

Trigonometric identities are indispensable tools in quantum computing, where they frequently facilitate the simplification of complex equations. This subsection categorizes these identities for easy reference. The sum and difference formulas are typically derived using geometric principles, while the double-angle and half-angle formulas are derived from these foundational formulas.

2.4.1 Single-Angle Identities

1 Interrelations

$$\tan\theta = \frac{\sin\theta}{\cos\theta}, \quad \cot\theta = \frac{1}{\tan\theta}, \quad \sec\theta = \frac{1}{\cos\theta}, \quad \csc\theta = \frac{1}{\sin\theta}. \tag{2.3}$$

2 Cofunction Formulas

$$\sin\left(\frac{\pi}{2}-\theta\right) = \cos\theta, \quad \cos\left(\frac{\pi}{2}-\theta\right) = \sin\theta, \quad \cot\left(\frac{\pi}{2}-\theta\right) = \tan\theta. \tag{2.4}$$

3 Pythagorean Identities

$$\sin^2\theta + \cos^2\theta = 1, \quad \tan^2\theta + 1 = \sec^2\theta, \quad 1 + \cot^2\theta = \csc^2\theta. \tag{2.5}$$

■ **Example 2.2**

$$\begin{aligned}
\frac{\sin\theta}{1-\cos\theta} &= \frac{\sin\theta}{(1-\cos\theta)} \cdot \frac{(1+\cos\theta)}{(1+\cos\theta)} \\
&= \frac{\sin\theta(1+\cos\theta)}{1-\cos^2\theta} \\
&= \frac{\sin\theta(1+\cos\theta)}{\sin^2\theta} \qquad \text{Pythagorean identity} \\
&= \frac{\sin\theta(1+\cos\theta)}{\sin\theta\sin\theta} \\
&= \frac{1+\cos\theta}{\sin\theta}
\end{aligned}$$

■

Exercise 2.6 Prove the following identities:

(a) $\cos^2\theta\tan^3\theta = \tan\theta - \sin\theta\cos\theta$

(b) $\dfrac{1}{\sec\theta - \tan\theta} = \sec\theta + \tan\theta$

(c) $\csc\theta - \cot\theta = \dfrac{\sin\theta}{1+\cos\theta}$

4 Double Angle Formulas

$$\sin 2\theta = 2 \sin \theta \cos \theta, \tag{2.6a}$$

$$\cos 2\theta = \cos^2 \theta - \sin^2 \theta = 2 \cos^2 \theta - 1 = 1 - 2 \sin^2 \theta, \tag{2.6b}$$

$$\tan 2\theta = \frac{2 \tan \theta}{1 - \tan^2 \theta}. \tag{2.6c}$$

These identities can be derived from the sum and difference of angles formulas in Equation 2.8.

5 Half Angle Formulas

$$\sin^2 \frac{\theta}{2} = \frac{1 - \cos \theta}{2}, \tag{2.7a}$$

$$\cos^2 \frac{\theta}{2} = \frac{1 + \cos \theta}{2}, \tag{2.7b}$$

$$\tan^2 \frac{\theta}{2} = \frac{1 - \cos \theta}{1 + \cos \theta}, \tag{2.7c}$$

$$\tan \frac{\theta}{2} = \frac{\sin \theta}{1 + \cos \theta} = \frac{1 - \cos \theta}{\sin \theta}. \tag{2.7d}$$

These identities can be derived from the double angle formula Eq. 2.6b.

2.4.2 Double-Angle Identities

1 Sum and Difference Formulas

$$\sin(\alpha \pm \beta) = \sin \alpha \cos \beta \pm \cos \alpha \sin \beta, \tag{2.8a}$$

$$\cos(\alpha \pm \beta) = \cos \alpha \cos \beta \mp \sin \alpha \sin \beta, \tag{2.8b}$$

$$\tan(\alpha \pm \beta) = \frac{\tan \alpha \pm \tan \beta}{1 \mp \tan \alpha \tan \beta}. \tag{2.8c}$$

Figure 2.4 illustrates the geometric relationships giving rise to the sum of angles formulas for $\sin(\alpha + \beta)$ and $\cos(\alpha + \beta)$.

Exercise 2.7 Illustrate the geometric relationships for the $\sin(\alpha - \beta)$ and $\cos(\alpha - \beta)$ formulas, similar to Fig. 2.4.

2 Product-to-Sum Formulas

$$2 \sin \alpha \sin \beta = \cos(\alpha - \beta) - \cos(\alpha + \beta), \tag{2.9a}$$

$$2 \cos \alpha \cos \beta = \cos(\alpha - \beta) + \cos(\alpha + \beta), \tag{2.9b}$$

$$2 \sin \alpha \cos \beta = \sin(\alpha + \beta) + \sin(\alpha - \beta), \tag{2.9c}$$

$$2 \cos \alpha \sin \beta = \sin(\alpha + \beta) - \sin(\alpha - \beta). \tag{2.9d}$$

These relationships can be derived from the sum and difference of angles formulas in Equation 2.8.

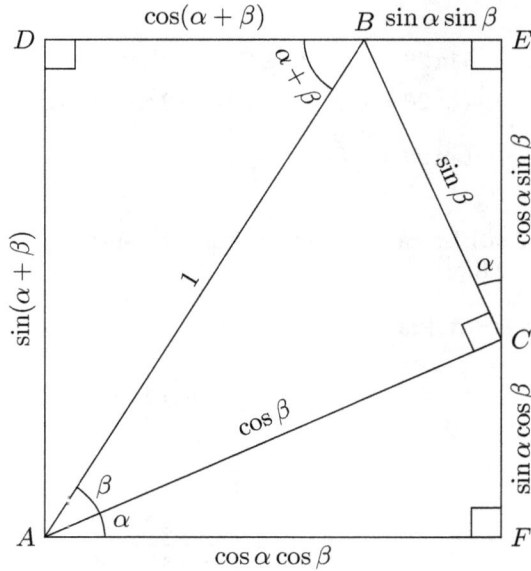

Figure 2.4: Geometric Relationships for $\sin(\alpha + \beta)$ and $\cos(\alpha + \beta)$

3   Sum-to-Product Formulas

$$\sin \alpha + \sin \beta = 2 \sin \frac{\alpha + \beta}{2} \cos \frac{\alpha - \beta}{2}, \tag{2.10a}$$

$$\sin \alpha - \sin \beta = 2 \cos \frac{\alpha + \beta}{2} \sin \frac{\alpha - \beta}{2}, \tag{2.10b}$$

$$\cos \alpha + \cos \beta = 2 \cos \frac{\alpha + \beta}{2} \cos \frac{\alpha - \beta}{2}, \tag{2.10c}$$

$$\cos \alpha - \cos \beta = -2 \sin \frac{\alpha + \beta}{2} \sin \frac{\alpha - \beta}{2}. \tag{2.10d}$$

These relationships can be derived from the product-to-sum formulas.

■ **Example 2.3** Find the exact value of $\sin \frac{19\pi}{12}$.

Solution. We first express $\frac{19\pi}{12}$ as a sum or difference of common angles. Then we use the sum or difference formula.

$$\sin \frac{19\pi}{12} = \sin \left(\frac{4\pi}{3} + \frac{\pi}{4} \right)$$

$$= \sin \frac{4\pi}{3} \cos \frac{\pi}{4} + \cos \frac{4\pi}{3} \sin \frac{\pi}{4}$$

$$= \left(-\frac{\sqrt{3}}{2} \right) \left(\frac{\sqrt{2}}{2} \right) + \left(-\frac{1}{2} \right) \left(\frac{\sqrt{2}}{2} \right)$$

$$= -\frac{\sqrt{6} + \sqrt{2}}{4}.$$

■

■ **Example 2.4** Derive the formula for $\cos 2\theta$ using the formula for $\cos(\alpha + \beta)$.

Solution.

$$
\begin{aligned}
\cos 2\theta &= \cos(\theta + \theta) \\
&= \cos\theta\cos\theta - \sin\theta\sin\theta \\
&= \cos^2\theta - \sin^2\theta \\
&= (1 - \sin^2\theta) - \sin^2\theta \\
&= 1 - 2\sin^2\theta.
\end{aligned}
$$

■

■ **Example 2.5** Derive the sum formula for tan using the sum formulas of sin and cos.

Solution.

$$
\begin{aligned}
\tan(\alpha + \beta) &= \frac{\sin(\alpha + \beta)}{\cos(\alpha + \beta)} && \text{quotient identity for tan} \\
&= \frac{\sin\alpha\cos\beta + \cos\alpha\sin\beta}{\cos\alpha\cos\beta - \sin\alpha\sin\beta} && \text{sum identities for cos and sin} \\
&= \frac{\sin\alpha\cos\beta + \cos\alpha\sin\beta}{\cos\alpha\cos\beta - \sin\alpha\sin\beta} \cdot \frac{\frac{1}{\cos\alpha\cos\beta}}{\frac{1}{\cos\alpha\cos\beta}} && \text{goal: } \frac{\sin\alpha}{\cos\alpha} \text{ and } \frac{\sin\beta}{\cos\beta} \\
&= \frac{\frac{\sin\alpha\cos\beta}{\cos\alpha\cos\beta} + \frac{\cos\alpha\sin\beta}{\cos\alpha\cos\beta}}{\frac{\cos\alpha\cos\beta}{\cos\alpha\cos\beta} - \frac{\sin\alpha\sin\beta}{\cos\alpha\cos\beta}} \\
&= \frac{\tan\alpha + \tan\beta}{1 - \tan\alpha\tan\beta}.
\end{aligned}
$$

■

■ **Example 2.6** Solve $\sin 2x = \sqrt{3}\cos x$ for $x \in [0, 2\pi)$.

Solution. We first transform the original equation into a product of simple expressions of $\sin x$ and $\cos x$ which equals 0:

$$
\begin{aligned}
\sin 2x &= \sqrt{3}\cos x \\
2\sin x\cos x &= \sqrt{3}\cos x \\
2\sin x\cos x - \sqrt{3}\cos x &= 0 \\
\cos x\left(2\sin x - \sqrt{3}\right) &= 0
\end{aligned}
$$

Thus, we must have $\cos x = 0$ or $\sin x = \frac{\sqrt{3}}{2}$.

From $\cos x = 0$, we obtain $x = \frac{\pi}{2} + \pi k$ for integers k.

From $\sin x = \frac{\sqrt{3}}{2}$, we get $x = \frac{\pi}{3} + 2\pi k$ or $x = -\frac{\pi}{3} + 2\pi k$ for integers k.

Finally, the solutions which lie in $[0, 2\pi)$ are $x = \frac{\pi}{2}, 3\frac{\pi}{2}, \frac{\pi}{3}$, and $\frac{5\pi}{3}$.

■

Exercise 2.8 Given $\sin\alpha = \frac{3}{5}$, where $0 < \alpha < \frac{\pi}{2}$, and $\cos\beta = \frac{12}{13}$, where $\frac{3\pi}{2} < \beta < 2\pi$, find $\sin(\alpha + \beta)$ and $\cos(\alpha + \beta)$.

Exercise 2.9 Prove the following identities:

(a) $\dfrac{\cos(\alpha + \beta)}{\cos(\alpha - \beta)} = \dfrac{1 - \tan\alpha\tan\beta}{1 + \tan\alpha\tan\beta}$

(b) $\tan(2\theta) = \dfrac{1}{1 - \tan(\theta)} - \dfrac{1}{1 + \tan(\theta)}$

Exercise 2.10 Express the following function in the forms $C(t) = A\cos(\omega t + \phi) + B$ and $S(t) = A\sin(\omega t + \phi) + B$ for $\omega > 0$ and $0 \le \phi < 2\pi$.

(a) $f(t) = -\sin t + \cos t - 2$

(b) $f(t) = \sqrt{3}\cos t - \sqrt{2}\sin t$

2.5 The Spherical Coordinate System

2.5.1 Coordinate Systems

In addition to the familiar Cartesian coordinate system, cylindrical and spherical coordinate systems provide alternative representations for points in three-dimensional space, as shown in Fig. 2.5.

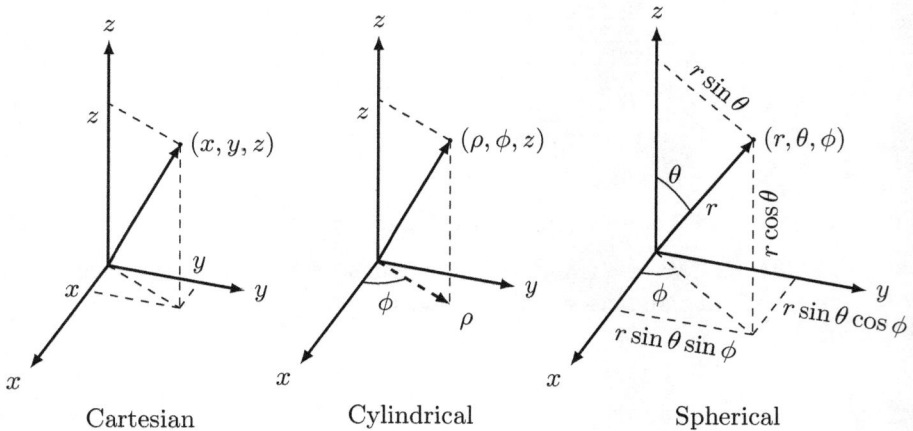

Figure 2.5: Cylindrical and Spherical Coordinate Systems

The cylindrical coordinate system, akin to the polar coordinate system but extended into three dimensions, is beneficial for describing structures like cylinders or helices.

The spherical coordinate system is particularly useful for describing rotations in three-dimensional space. It is notably used in quantum computing to represent qubit states on the Bloch sphere, a powerful visualization tool for the state space of qubits.

This system employs a radius (r), a polar angle (θ), and an azimuthal angle (ϕ) to specify the location of a point. The azimuthal angle ϕ is measured in the xy-plane from the positive x-axis towards the positive y-axis, with values ranging from $(-\pi, \pi]$ or $[0, 2\pi)$. The polar angle θ is measured from the positive z-axis towards the xy-plane, with values ranging from $[0, \pi]$.

2.5.2 Conversion Formulas

Conversion from spherical coordinates to Cartesian coordinates is given by:

$$x = r \sin \theta \cos \phi, \tag{2.11a}$$
$$y = r \sin \theta \sin \phi, \tag{2.11b}$$
$$z = r \cos \theta. \tag{2.11c}$$

Conversion from Cartesian coordinates to spherical coordinates is given by:

$$r = \sqrt{x^2 + y^2 + z^2}, \tag{2.12a}$$
$$\phi = \arctan2(y, x), \tag{2.12b}$$
$$\theta = \arccos \frac{z}{r}, \tag{2.12c}$$

where $\arctan2(y, x)$ is a variant of the arctan function that takes the coordinates (x, y) as input and returns a unique value in the range $[-\pi, \pi]$.

Exercise 2.11 Convert the following spherical coordinates to Cartesian coordinates.

(a) $(1, \frac{\pi}{2}, \frac{\pi}{4})$

(b) $(2, \frac{\pi}{3}, \frac{\pi}{4})$

Exercise 2.12 Convert the following Cartesian coordinates to spherical coordinates.

(a) $(1, 1, 0)$

(b) $(1, -1, -1)$

2.6 Laws of Sines, Cosines, and Tangents

The laws of sines, cosines, and tangents establish relationships between the sides (a, b, c) and angles (A, B, C) of a triangle. While an integral component of trigonometry, they are less relevant to quantum computing. So we only present the formulas of these laws below.

$$\text{The Law of Sines: } \frac{\sin A}{a} = \frac{\sin B}{b} = \frac{\sin C}{c}. \tag{2.13}$$

$$\text{The Law of Cosines: } a^2 = b^2 + c^2 - 2bc \cos A, \tag{2.14a}$$

$$\text{or, } \cos A = \frac{b^2 + c^2 - a^2}{2bc}. \tag{2.14b}$$

$$\text{The Law of Tangents: } \frac{a - b}{a + b} = \frac{\tan \frac{1}{2}(A - B)}{\tan \frac{1}{2}(A + B)}. \tag{2.15}$$

Problem Set 2

2.1 Determine all solutions to the equation $\cos\theta = 2\sin\theta$.

2.2 Find the exact values of

(a) $\arcsin\left(\sin\left(\frac{11\pi}{6}\right)\right)$

(b) $\arccos\left(\cos\left(\frac{2\pi}{3}\right)\right)$

(c) $\cos\left(\arcsin\left(-\frac{5}{13}\right)\right)$

2.3 Prove the following identities:

(a) $\sin\theta(\tan\theta + \cot\theta) = \sec\theta$

(b) $\dfrac{1}{\csc\theta + \cot\theta} = \csc\theta - \cot\theta$

(c) $\dfrac{1}{1 + \cos\theta} = \csc^2\theta - \csc\theta\cot\theta$

(d) $\dfrac{1}{1 + \sin\theta} = \sec^2\theta - \sec\theta\tan\theta$

(e) $\dfrac{\sin^2(-\theta) - \cos^2(-\theta)}{\sin(-\theta) - \cos(-\theta)} = \cos\theta - \sin\theta$

2.4 Illustrate the geometric relationships for the $\tan(\alpha+\beta)$ and $\tan(\alpha-\beta)$ formulas, similar to Fig. 2.4.

2.5 Given $\sin\alpha = \frac{4}{5}$, where $0 < \alpha < \frac{\pi}{2}$, and $\cos\beta = \frac{5}{13}$, where $\frac{3\pi}{2} < \beta < 2\pi$, find

(a) $\cos(\alpha - \beta)$

(b) $\tan(\alpha - \beta)$

2.6 Derive the half-angle formulas (Eq. 2.7) using the double-angle formulas (Eq. 2.6). Hint: substitute 2θ with θ in the double-angle formulas first.

2.7 Derive the product-to-sum formulas (Eq. 2.9) using the sum and difference formulas (Eq. 2.8).

2.8 Derive the sum-to-product formulas (Eq. 2.10) using the product-to-sum formulas (Eq. 2.9).

2.9 Derive the formulas for $\sin 3\theta$ and $\cos 3\theta$ in terms of $\sin\theta$ and $\cos\theta$.

2.10 Prove the following identities:

(a) $\dfrac{\tan(\alpha + \beta)}{\tan(\alpha - \beta)} = \dfrac{\sin\alpha\cos\alpha + \sin\beta\cos\beta}{\sin\alpha\cos\alpha - \sin\beta\cos\beta}$

(b) $\dfrac{1}{\cos\theta - \sin\theta} + \dfrac{1}{\cos\theta + \sin\theta} = \dfrac{2\cos\theta}{\cos 2\theta}$

2.11 Express the following function in the forms $C(t) = A\cos(\omega t + \phi) + B$ and $S(t) = A\sin(\omega t + \phi) + B$ for $\omega > 0$ and $0 \le \phi < 2\pi$.

(a) $f(t) = 2\sqrt{3}\cos t - 2\sin t$

(b) $f(t) = -\frac{1}{2}\cos(5t) - \frac{\sqrt{3}}{2}\sin(5t)$

2.12 ✳ Find the exact values of

(a) $\sin\left(\arcsin\left(-\dfrac{5}{13}\right) + \dfrac{\pi}{4}\right)$

(b) $\cos(\operatorname{arcsec}(3) + \arctan(2))$

(c) $\sec(\arctan(2x))\tan(\arctan(2x))$

2.13 ✳ Solve the following equations for $x \in [0, 2\pi)$.

(a) $\sin x + \cos x = 1$

(b) $\cos 2x \cos x + \sin 2x \sin x = 1$

(c) $\sin x + \sqrt{3}\cos x = 1$

2.14 ✴ Use Python or another programming language to graph the three-dimensional surfaces defined by the following equations in spherical coordinates. Analyze the resulting shapes to understand the symmetries and patterns that emerge and how they relate to the functions involved.

(a) $r = 1 + \cos 2\theta$,

(b) $r = \cos^2 \phi$,

(c) $r = 2 + \sin(3\phi + 5\theta)$.

3. Complex Numbers

Contents

Complex numbers, consisting of a real and an imaginary part, are a fundamental concept in mathematics and form the backbone of various advanced fields, including quantum computing. In quantum mechanics, quantum states are described by vectors composed of complex numbers, making an understanding of these numbers essential. By mastering complex numbers, one gains a necessary tool to navigate and comprehend the intricate behaviors of quantum systems.

3.1 Cartesian Form

3.1.1 Definition and the Number System

A complex number z is defined as $z = x + iy$, where x and y are real numbers, and i is the imaginary unit with the property that $i^2 = -1$. This is referred to as the

Cartesian form because z corresponds to a point or a vector in the two-dimensional coordinate system known as the complex plane, as shown in Fig. 3.1.

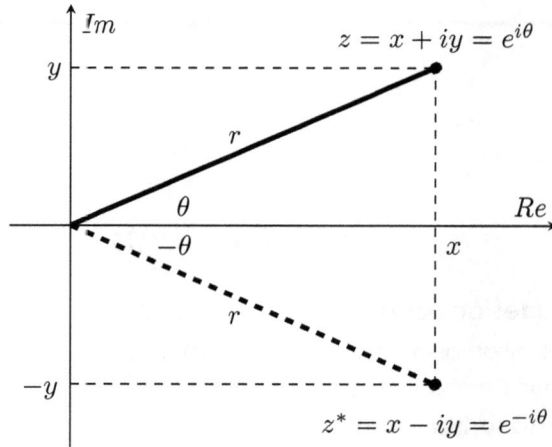

Figure 3.1: The Complex Plane

Real numbers are considered a subset of complex numbers where the imaginary part is zero. Thus, the set of all complex numbers, denoted by $\mathbb{C}$, is the superset of numbers, including the set of all real numbers $\mathbb{R}$, the set of integers $\mathbb{Z}$, and the set of natural numbers $\mathbb{N}$. Specifically, $\mathbb{C} \supset \mathbb{R} \supset \mathbb{Z} \supset \mathbb{N}$, as illustrated in Fig. 3.2.

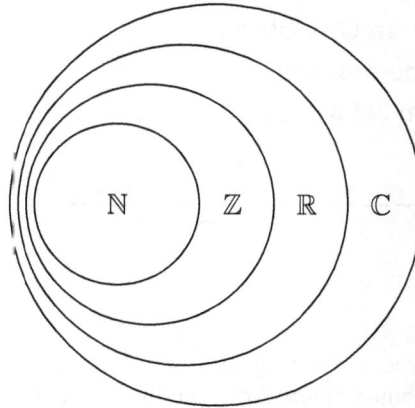

Figure 3.2: The Number System

3.1.2 Basic Relations

The following are key definitions pertaining to complex numbers and their properties, as illustrated in Fig. 3.1:

- Real part: $x = \text{Re}(z)$

- Imaginary part: $y = \text{Im}(z)$

- Conjugate: $z^* = x - iy$

- Modulus or absolute value: $r = |z|$
- Phase angle or argument: θ
- Imaginary unit: $i = \sqrt{-1}$ with $i^2 = -1$, $i^3 = -i$, $i^4 = 1$, and so on
- Modulus calculation: $r = \sqrt{zz^*} = \sqrt{x^2 + y^2}$
- Argument calculation: $\tan \theta = \frac{y}{x}$, hence $\theta = \arctan2(y, x)$. (See § 2.2.4.)
- Conversion from polar coordinates: $x = r \cos \theta$ and $y = r \sin \theta$
- Polar form of z: $z = r(\cos \theta + i \sin \theta)$

■ **Example 3.1** Given $z = 1 + \sqrt{3}i$, the following properties hold:

$$z^* = 1 - \sqrt{3}i,$$

$$|z| = \sqrt{1^2 + (\sqrt{3})^2} = 2,$$

$$zz^* = (1 + \sqrt{3}i)(1 - \sqrt{3}i) = 1 - (\sqrt{3}i)^2 = 1 + 3 = 4 = |z|^2,$$

$$\theta = \arctan \frac{\sqrt{3}}{1} = \frac{\pi}{3}.$$

■

Exercise 3.1 Calculate z^* and θ for $z = -1 - \sqrt{3}i$.

3.2 Exponential Form

3.2.1 From Euler's Formula to Exponential Form

Given a complex number in Cartesian form $z = r \cos \theta + ir \sin \theta$ and Euler's formula:

$$e^{i\theta} = \cos \theta + i \sin \theta, \tag{3.1}$$

we can express a complex number in its exponential form:

$$z = re^{i\theta}. \tag{3.2}$$

A Proof of the Euler's Formula

The Euler's formula can be proved using Taylor series expansions, which, for e^x, $\sin x$, and $\cos x$ around 0, are given by:

$$e^x = 1 + x + \frac{x^2}{2!} + \frac{x^3}{3!} + \frac{x^4}{4!} + \cdots, \tag{3.3a}$$

$$\sin x = x - \frac{x^3}{3!} + \frac{x^5}{5!} - \frac{x^7}{7!} + \cdots, \tag{3.3b}$$

$$\cos x = 1 - \frac{x^2}{2!} + \frac{x^4}{4!} - \frac{x^6}{6!} + \cdots. \tag{3.3c}$$

When x is replaced with ix, the Taylor series for e^{ix} becomes:

$$e^{ix} = 1 + ix + \frac{(ix)^2}{2!} + \frac{(ix)^3}{3!} + \frac{(ix)^4}{4!} + \cdots.$$

Simplifying this using the properties of i (notably that $i^2 = -1$, $i^3 = -i$, $i^4 = 1$, and so on), we get:

$$e^{ix} = 1 + ix - \frac{x^2}{2!} - i\frac{x^3}{3!} + \frac{x^4}{4!} + i\frac{x^5}{5!} - \frac{x^6}{6!} - i\frac{x^7}{7!} + \cdots.$$

This series can be split into its real and imaginary parts:

$$e^{ix} = \left(1 - \frac{x^2}{2!} + \frac{x^4}{4!} - \frac{x^6}{6!} + \cdots\right) + i\left(x - \frac{x^3}{3!} + \frac{x^5}{5!} - \frac{x^7}{7!} + \cdots\right).$$

Notice that the real part is the Taylor series for $\cos x$, and the imaginary part is the Taylor series for $\sin x$. Therefore, we arrive at the Euler's formula:

$$e^{ix} = \cos x + i \sin x.$$

3.2.2 Conversion Formulas

	Cartesian Form	Exponential Form
	$z = x + iy$	$z = re^{i\theta}$
Conjugate	$z^* = x - iy$	$z^* = re^{-i\theta}$
Modulus	$|z| = \sqrt{zz^*} = \sqrt{x^2 + y^2}$	$|z| = r$
Conversion	$x = r\cos\theta$	$r = \sqrt{x^2 + y^2}$
	$y = r\sin\theta$	$\theta = \arctan2(y, x)$

Table 3.1: Conversion Between Cartesian and Exponential Forms

■ **Example 3.2** To convert $z = 1 + \sqrt{3}i$ to its exponential form, we first find its modulus and phase angle, as in the previous example: $|z| = 2$, $\theta = \frac{\pi}{3}$. Then we can express z as $z = 2e^{i\frac{\pi}{3}}$.   ■

Exercise 3.2

(a) For $z = 3 - 4i$, find $|z|$ and θ.

(b) Convert $z = -2 + 2i$ to its exponential form.

(c) Convert $z = 2e^{i\frac{\pi}{6}}$ to its Cartesian form.

3.3 Basic Operations

3.3.1 Addition and Subtraction

Given two complex numbers $z_1 = x_1 + iy_1$ and $z_2 = x_2 + iy_2$, addition and subtraction are straightforward in Cartesian form:

$$z_1 + z_2 = (x_1 + x_2) + i(y_1 + y_2), \tag{3.4a}$$
$$z_1 - z_2 = (x_1 - x_2) + i(y_1 - y_2). \tag{3.4b}$$

What if they are given in exponential form, $z_1 = r_1 e^{i\theta_1}$ and $z_2 = r_2 e^{i\theta_2}$? To perform addition or subtraction, we first convert them to Cartesian form.

3.3.2 Multiplication and Division

Given $z_1 = x_1 + iy_1 = r_1 e^{i\theta_1}$ and $z_2 = x_2 + iy_2 = r_2 e^{i\theta_2}$, multiplication and division are more intuitive in exponential form:

$$z_1 \cdot z_2 = r_1 e^{i\theta_1} \cdot r_2 e^{i\theta_2} = r_1 r_2 e^{i(\theta_1 + \theta_2)}, \tag{3.5a}$$

$$\frac{z_1}{z_2} = \frac{r_1 e^{i\theta_1}}{r_2 e^{i\theta_2}} = \frac{r_1}{r_2} e^{i(\theta_1 - \theta_2)}. \tag{3.5b}$$

These operations can also be performed in Cartesian form:

$$z_1 \cdot z_2 = (x_1 + iy_1)(x_2 + iy_2)$$
$$= x_1 x_2 - y_1 y_2 + i(x_1 y_2 + x_2 y_1), \tag{3.6a}$$

$$\frac{z_1}{z_2} = \frac{(x_1 + iy_1)(x_2 - iy_2)}{x_2^2 + y_2^2}$$
$$= \frac{x_1 x_2 + y_1 y_2}{x_2^2 + y_2^2} + i \frac{x_2 y_1 - x_1 y_2}{x_2^2 + y_2^2}. \tag{3.6b}$$

■ **Example 3.3** Consider $z_1 = \frac{1}{2} + \frac{\sqrt{3}}{2}i = e^{i\frac{\pi}{3}}$ and $z_2 = \frac{\sqrt{2}}{2} + \frac{\sqrt{2}}{2}i = e^{i\frac{\pi}{4}}$.

In exponential form:

$$z_1 z_2 = e^{i\left(\frac{\pi}{3} + \frac{\pi}{4}\right)} = e^{i\frac{7\pi}{12}} = \cos\frac{7\pi}{12} + i\sin\frac{7\pi}{12},$$

$$\frac{z_1}{z_2} = e^{i\left(\frac{\pi}{3} - \frac{\pi}{4}\right)} = e^{i\frac{\pi}{12}} = \cos\frac{\pi}{12} + i\sin\frac{\pi}{12}.$$

In Cartesian form:

$$z_1 z_2 = \left(\frac{\sqrt{2}}{4} - \frac{\sqrt{6}}{4}\right) + \left(\frac{\sqrt{2}}{4} + \frac{\sqrt{6}}{4}\right)i,$$

$$\frac{z_1}{z_2} = \left(\frac{\sqrt{2}}{4} + \frac{\sqrt{6}}{4}\right) + \left(-\frac{\sqrt{2}}{4} + \frac{\sqrt{6}}{4}\right)i.$$

The two forms are equivalent, therefore:

$$\cos\frac{7\pi}{12} = \frac{\sqrt{2}}{4} - \frac{\sqrt{6}}{4},$$

$$\sin\frac{\pi}{12} = -\frac{\sqrt{2}}{4} + \frac{\sqrt{6}}{4}.$$

■

Exercise 3.3 For $z_1 = 1 + i$ and $z_2 = 1 + \sqrt{3}i$, evaluate in both Cartesian and exponential forms:

(a) $z_1 z_2$

(b) $\dfrac{z_1}{z_2}$

3.3.3 Conjugation

Complex number conjugation has the following properties:

$$|z^*| = |z|, \tag{3.7a}$$
$$(z_1 \pm z_2)^* = z_1^* \pm z_2^*, \tag{3.7b}$$
$$(z_1 \cdot z_2)^* = z_1^* \cdot z_2^*, \tag{3.7c}$$
$$(z_1/z_2)^* = z_1^*/z_2^*, \tag{3.7d}$$
$$(z^x)^* = (z^*)^x \quad (x \in \mathbb{R}), \tag{3.7e}$$
$$(x^z)^* = x^{z^*} \quad (x \in \mathbb{R}). \tag{3.7f}$$

Exercise 3.4 Prove $(x^z)^* = x^{z^*}$ (where $x \in \mathbb{R}$) from $(e^{i\theta})^* = e^{-i\theta}$.

3.3.4 Powers and Roots

Given a complex number $z = re^{i\theta}$ and a real number s, its power is given by:

$$z^s = r^s e^{is\theta}. \tag{3.8}$$

From this, we derive De Moivre's Theorem:

$$(\cos\theta + i\sin\theta)^s = \cos s\theta + i\sin s\theta. \tag{3.9}$$

Exercise 3.5 Use De Moivre's Theorem to derive the double angle formulas of $\sin 2\theta$ and $\cos 2\theta$.

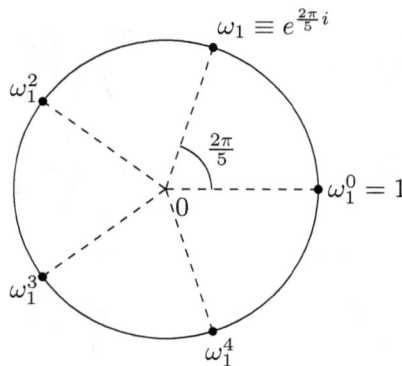

Figure 3.3: Roots of Unity

In particular, the nth roots of unity, where n is a positive integer (i.e., solutions to $\omega^n = 1$), are given by:

$$\omega_k = e^{\frac{2k\pi i}{n}} = \omega_1^k, \tag{3.10}$$

where $k = 0, 1, \ldots, n-1$, and $\omega_1 = e^{\frac{2\pi i}{n}}$.

▪ **Example 3.4** Demonstrate that

$$\sum_{k=0}^{n-1} \omega_k = \sum_{k=0}^{n-1} \omega_1^k = 0. \tag{3.11}$$

Proof. Applying the formula for the sum of a geometric series, we have

$$\sum_{k=0}^{n-1} \omega_1^k = \frac{1 - \omega_1^n}{1 - \omega_1} = 0,$$

since $\omega_1^n = 1$ by the definition of roots of unity.

Using the same argument, we can generalize Eq. 3.11 as

$$\sum_{k=0}^{n-1} \omega^k = 0. \tag{3.12}$$

where ω is any nth root of unity except 1, i.e., $\omega^n = 1$ and $\omega \neq 1$. ▪

▪ **Example 3.5 — ✳ DFT Orthonormality Condition.**

Let's evaluate the following sum, which depends on two parameters, k and l:

$$\frac{1}{N} \sum_{n=0}^{N-1} e^{-\frac{2\pi i}{N} kn} e^{\frac{2\pi i}{N} ln} = \delta_{k-l \bmod N}, \tag{3.13}$$

where $\delta_{k-l \bmod N} = 1$ if and only if $k \equiv l \pmod{N}$; otherwise, it is 0.

Proof. We define $\omega = e^{i\frac{2\pi}{N}}$, which is a primitive Nth root of unity, satisfying $\omega^N = 1$. Then, the left-hand side of the above equation can be transformed into the right-hand side as follows:

$$\frac{1}{N} \sum_{n=0}^{N-1} e^{-\frac{2\pi i}{N} kn} e^{\frac{2\pi i}{N} ln} = \frac{1}{N} \sum_{n=0}^{N-1} \omega^{-kn} \omega^{ln}$$

$$= \frac{1}{N} \sum_{n=0}^{N-1} \omega^{n(l-k)}$$

$$= \begin{cases} \dfrac{1}{N} \sum_{n=0}^{N-1} 1 = 1 & \text{if } l \equiv k \pmod{N}, \\[2mm] \dfrac{1}{N} \dfrac{1 - \omega^{(l-k)N}}{1 - \omega^{(l-k)}} = 0 & \text{if } l \not\equiv k \pmod{N} \end{cases}$$

$$= \delta_{k-l \bmod N}.$$

Here, we have used the fact that $\omega^{nN} = \left(\omega^N\right)^n = 1^n = 1$ for any integer n. This example can be considered an extension of Example 3.4. ▪

Kronecker Delta Function and Levi-Civita Permutation Symbol

The Kronecker delta function, δ_{ij}, is defined as:

$$\delta_{ij} = \begin{cases} 1 & \text{if } i = j, \\ 0 & \text{if } i \neq j. \end{cases} \tag{3.14}$$

The single-argument delta function is defined as $\delta_i \equiv \delta_{i0}$.

The Levi-Civita permutation symbol, ϵ_{ijk}, is defined as:

$$\epsilon_{ijk} = \begin{cases} +1 & \text{if } (i,j,k) \text{ is } (1,2,3), (2,3,1), \text{ or } (3,1,2), \\ -1 & \text{if } (i,j,k) \text{ is } (3,2,1), (1,3,2), \text{ or } (2,1,3), \\ 0 & \text{if any two indices are equal.} \end{cases} \tag{3.15}$$

Exercise 3.6 Evaluate $\prod_{k=0}^{n-1} \omega_k$.

3.4 Advanced Operations

We illustrate some more advanced operations on complex numbers in the example below. These skills are essential for working with quantum algorithms.

■ **Example 3.6**

(a) Evaluate $\sqrt{i}$, which is $\sqrt{\sqrt{-1}}$.

Solution.

$$\sqrt{i} = \left(e^{\frac{\pi i}{2}}\right)^{\frac{1}{2}} = e^{\frac{\pi i}{4}} = \cos\frac{\pi}{4} + i\sin\frac{\pi}{4} = \frac{1}{\sqrt{2}}(1+i),$$

$$\text{Inverse: } \frac{1}{2}(1+i)^2 = \frac{1}{2}(1+2i+i^2) = \frac{1}{2}(2i) = i.$$

(b) Evaluate $\left(\frac{1}{2} + \frac{\sqrt{3}}{2}i\right)^{50}$.

Solution.

$$\left(\frac{1}{2} + \frac{\sqrt{3}}{2}i\right)^{50} = \left(e^{\frac{\pi i}{3}}\right)^{50} = e^{\frac{50\pi i}{3}} = e^{(16+\frac{2}{3})\pi i} = e^{\frac{2\pi i}{3}} = -\frac{1}{2} + \frac{\sqrt{3}}{2}i.$$

Try to work out the inverse: $\left(-\frac{1}{2} + \frac{\sqrt{3}}{2}i\right)^{\frac{1}{50}}$.

(c) Given $z^5 = \frac{1}{2} + \frac{\sqrt{3}}{2}i$, solve for z.

Solution. In the real domain, this equation has no solution. In the complex domain, however, there are five unique solutions:

$$z_k = e^{\frac{\pi i}{15}} e^{\frac{2k\pi i}{5}}, \quad \text{where } k = 0, 1, 2, 3, 4.$$

(d) Evaluate 2^{3+4i}.

Solution.

$$2^{3+4i} = 2^3 \cdot 2^{4i} = 8 \cdot e^{4\ln(2)i} = 8\cos(4\ln(2)) + i8\sin(4\ln(2)).$$

(e) Evaluate $\cos(3+4i)$.

Solution. From Euler's formula $e^{i\theta} = \cos\theta + i\sin\theta$ and its conjugate $e^{-i\theta} = \cos\theta - i\sin\theta$, we obtain $\cos\theta = \dfrac{1}{2}\left(e^{i\theta} + e^{-i\theta}\right)$. (Yes, these formulas work even if θ is a complex number!) Therefore,

$$
\begin{aligned}
\cos(3+4i) &= \frac{1}{2}\left(e^{i(3+4i)} + e^{-i(3+4i)}\right)\\
&= \frac{1}{2}\left(e^{-4+3i} + e^{4-3i}\right)\\
&= \frac{1}{2}e^{-4}(\cos 3 + i\sin 3) + \frac{1}{2}e^{4}(\cos 3 - i\sin 3)\\
&= \frac{1}{2}\left(e^{-4} + e^{4}\right)\cos 3 + i\frac{1}{2}\left(e^{-4} - e^{4}\right)\sin 3.
\end{aligned}
$$

∎

Exercise 3.7

(a) Given $z^8 = i$, solve for z.

(b) For $z_1 = \frac{\sqrt{3}}{2} + \frac{1}{2}i$, and $z_2 = \frac{1}{2} + \frac{\sqrt{3}}{2}i$, find $z_1^\alpha z_2^\beta$, where $\alpha, \beta \in \mathbb{R}$.

3.5 ✳ Mandelbrot Set

As a simple yet fascinating application of complex numbers, we explore the Mandelbrot Set, one of the most iconic figures in mathematical visualization. It serves as a prime example of a complex structure arising from the application of simple mathematical rules.

The Mandelbrot Set is generated through an iterative process. For each point $c = x + iy$ on a two-dimensional grid representing the complex plane:

1. Begin with $z_0 = 0$.

2. Compute $z_1 = z_0^2 + c$.

3. Compute $z_2 = z_1^2 + c$.

4. Continue iteratively with $z_n = z_{n-1}^2 + c$.

This iteration continues until the absolute value of z_n, denoted $|z_n|$, exceeds a predetermined threshold. This threshold is set to determine when the sequence is likely to diverge to infinity, indicating that it is unbounded. The process also stops if a maximum number of iterations is reached, to ensure the computation completes.

Points c are then plotted and colored based on the iteration count when $|z_n|$ first exceeds the threshold. Points for which $|z_n|$ never exceeds the threshold are considered to be part of the Mandelbrot Set.

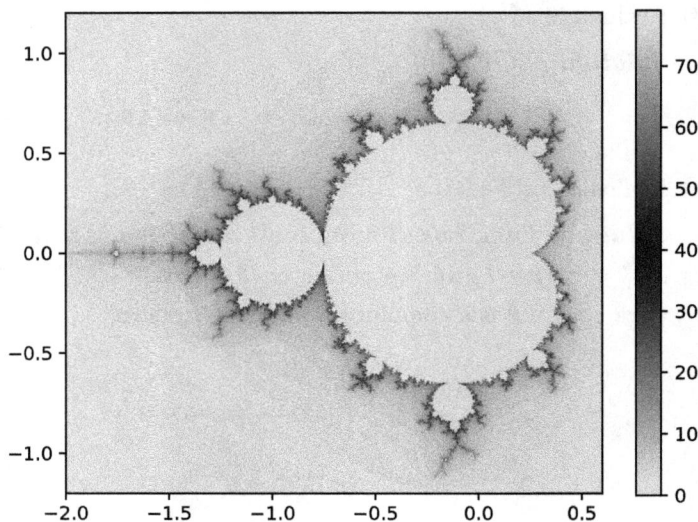

Figure 3.4: Mandelbrot Set

Note that the Mandelbrot Set specifically refers to the set generated by the function $z_{n+1} = z_n^2 + c$. A more general concept is *fractals*, which include images generated by other iterative relations in the complex domain.

Problem Set 3

3.1 Find z^*, zz^*, $|z|$, θ, and the exponential form of z, for the following complex numbers z:

  (a) $z = \frac{1}{2} - \frac{\sqrt{3}}{2}i$

  (b) $z = -\frac{1}{2} + \frac{\sqrt{3}}{2}i$

  (c) $z = -\frac{\sqrt{3}}{2} - \frac{1}{2}i$

3.2 Consider $z_1 = \frac{1}{2} - \frac{\sqrt{3}}{2}i$ and $z_2 = -\frac{\sqrt{2}}{2} + \frac{\sqrt{2}}{2}i$. Find $z_1 z_2$, z_1/z_2, $\sqrt{z_1}$, and z_2^9.

3.3 Consider $z_1 = 4e^{-i\frac{\pi}{4}}$ and $z_2 = 3e^{i\frac{5\pi}{3}}$. Find $z_1 + z_2$, $z_1 z_2$ and z_1/z_2.

3.4   (a) Given $z_1^9 = i$. $z_2^8 = -1$, solve for $z_1 z_2$ and $z_1^8 z_2^9$.

  (b) Evaluate $2^{e^{\frac{\pi}{4}i}}$.

  (c) Evaluate $\sin(4 + 3i)$.

  (d) Evaluate $\ln(8ei)$.

3.5 Let z_1 and z_2 be two complex numbers. Prove that $|z_1 z_2| = |z_1||z_2|$.

3.6 Let α and β be two complex numbers satisfying $|\alpha|^2 + |\beta|^2 = 1$. Prove that $1 - (|\alpha|^2 - |\beta|^2)^2 = 4|\alpha\beta|^2$.

3.7 Utilize the results from Euler's formula to derive the sum and difference formulas for sine and cosine. From the Euler's formula:

$$e^{i\alpha} = \cos\alpha + i\sin\alpha, \qquad\qquad e^{i\alpha}e^{i\beta} = e^{i(\alpha+\beta)},$$

$$e^{i\beta} = \cos\beta + i\sin\beta, \qquad\qquad \frac{e^{i\alpha}}{e^{i\beta}} = e^{i(\alpha-\beta)},$$

Using these identities, prove:

$$\sin(\alpha \pm \beta) = \sin\alpha\cos\beta \pm \cos\alpha\sin\beta,$$
$$\cos(\alpha \pm \beta) = \cos\alpha\cos\beta \mp \sin\alpha\sin\beta.$$

3.8 $*$ Let

$$x = \sqrt[3]{2 + \sqrt{-4}} + \sqrt[3]{2 - \sqrt{-4}}.$$

Show that

(a) $x^3 - 6x - 4 = 0$;

(b) x can take the three real values

$$1 + \sqrt{3}, \qquad -2, \qquad 1 - \sqrt{3}.$$

3.9 $*$ Create a fractal image with a different iterative function, similar to the Mandelbrot set ($z_{n+1} = z_n^2 + c$), using Python or another programming language.

3.10 $*$ Up for a math challenge?

The Gaussian sum identity is given below:

$$\frac{1}{\sqrt{N}}\sum_{n=0}^{N-1} e^{i\frac{2\pi}{N}n^2} = \begin{cases} 1+i & \text{if } N = 4m \\ 1 & \text{if } N = 4m+1 \\ 0 & \text{if } N = 4m+2 \\ i & \text{if } N = 4m+3 \end{cases}$$

The proof of this result took Gauss two years. Since the original proof by Gauss, it is an ongoing challenge among mathematicians to present new, possibly better, proofs of this result. Study some of these proofs, and better yet, come up with a new one!

4. Sets, Groups, and Functions

Contents

Sets, groups, and functions form the backbone of many mathematical concepts essential for quantum computing. This chapter provides a thorough introduction to these foundational topics, covering definitions, operations, and key properties. A solid understanding of these areas is crucial for grasping more advanced concepts in quantum algorithms and computation. Our goal is to equip readers with the necessary knowledge to build a strong mathematical foundation for their journey into quantum computing.

4.1 Sets

The concept of sets is fundamental to many areas of mathematics. A set is a well-defined collection of distinct objects, considered as an object in its own right. Sets can be used to group numbers, points, or any other elements, providing a foundation for defining more complex structures and operations. Understanding the properties and operations of sets is essential for progressing in mathematical studies.

4.1.1 Definitions

> **Definition 4.1 — Set.** A set is a (unordered) collection of objects. These objects are sometimes called elements or members of the set.

■ **Example 4.1**

- First nine prime numbers

$$A = \{2, 3, 5, 7, 11, 13, 17, 19, 23\}$$

- Vowels in the English alphabet

$$V = \{a, e, i, o, u\}$$

- All prime numbers

$$P = \{2, 3, 5, 7, 11, 13, 17, 19, \ldots\}$$

■

Note the order of elements in a set is inconsequential. That is,

$$\{2, 3, 5, 7, 11, 13, 17, 19, 23\} \equiv \{13, 5, 23, 11, 2, 17, 7, 3, 19\}$$
$$\{a, e, i, o, u\} \equiv \{e, i, a, u, o\}.$$

Notation on membership: $3 \in A$, $4 \notin A$, $e \in V$, and $b \notin V$.

Exercise 4.1 Write down the set containing the first 5 square numbers.

Besides listing all the set members in $\{\ldots\}$, we can also use formulas to construct sets. For example, the set of the first n square numbers can be represented as:

$$S_n = \{i^2 \mid i = 1, 2, \ldots, n\}.$$

> **Definition 4.2 — Tuple.** A tuple (or sequence) is an ordered list of elements.

■ **Example 4.2** $T = (a, b, c, d)$. Note that tuples are denoted using parentheses (...) instead of curly braces {...}, which are used for sets. ■

> **Definition 4.3 — Cardinality.** The cardinality of a set A, denoted $|A|$, is the number of elements in A.

Sets can be categorized based on their cardinality as follows:

- Finite: A set is finite if it contains a specific number of elements, i.e., $|A| < \infty$.

- Countably Infinite: A set is countably infinite if its elements can be put into a one-to-one correspondence with the set of natural numbers $\mathbb{N}$, such as the set of integers $\mathbb{Z}$.

- Uncountably Infinite: A set is uncountably infinite if it cannot be put into a one-to-one correspondence with $\mathbb{N}$, indicating a larger kind of infinity, such as the set of real numbers $\mathbb{R}$.

Definition 4.4 — Subset and Superset. B is a subset of A, denoted $B \subseteq A$, if all elements in B are also in A. In this case, A is a superset of B, denoted $A \supseteq B$. If B is a subset of A but not equal to A, then B is called a proper subset of A, denoted $B \subset A$.

Exercise 4.2

(a) Is $\{1,2\} \in \{\{1,2\}, \{3,4\}\}$?

(b) Is $3 \in \{\{1\}, \{2\}, \{3\}\}$?

(c) Is $\{1,2,1\} \subseteq \{1,2\}$?

(d) What is $|\{1,2,3,4,\{1,2\},\{3,4\}\}|$?

4.1.2 Set Operations

Definition 4.5 — Union. The union of two sets A and B, denoted by $A \cup B$, is the set containing all the elements that are in A, in B, or in both.

■ **Example 4.3** Given $A = \{1,2,3,6\}$ and $B = \{2,4,6,9\}$, their union is

$$A \cup B = \{1,2,3,4,6,9\}.$$

■

Definition 4.6 — Intersection. The intersection of two sets A and B, denoted by $A \cap B$, is the set containing all the elements that are both in A and B.

■ **Example 4.4** Given $A = \{1,2,3,6\}$ and $B = \{2,4,6,9\}$, their intersection is

$$A \cap B = \{2,6\}.$$

■

Definition 4.7 — Difference. The difference between two sets A and B, denoted by $A - B$ or $A \setminus B$, is the set containing all the elements that are in A but not in B.

■ **Example 4.5** Given $A = \{1,2,3,6\}$ and $B = \{2,4,6,9\}$, their differences are

$$A - B = \{1,3\},$$

$$B - A = \{4,9\}.$$

■

Definition 4.8 — Universal Set. The universal set, denoted by U, is the set that contains all the elements under consideration, usually for a particular discussion or problem. Every other set in that context is a subset of the universal set U.

Definition 4.9 — Complement. The complement of a set A, denoted by $\overline{A}$ or A^c, is the set of all elements in the universal set U that are not in A.

■ **Example 4.6** If the universal set is $U = \{1, 2, 3, 4, 5, 6, 7, 8, 9\}$ and $A = \{2, 4, 6, 9\}$, then the complement of A is

$$\overline{A} = \{1, 3, 5, 7, 8\}.$$

■

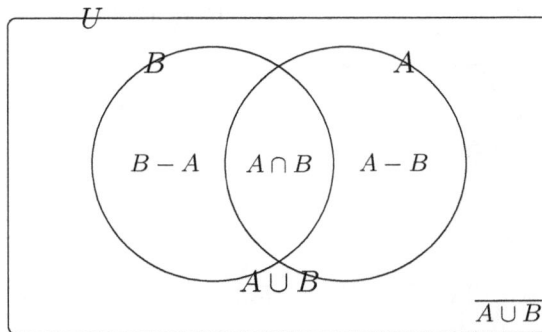

Figure 4.1: Venn Diagram for Set Relations

Theorem 4.1 — Inclusion-Exclusion Principle.

$$|A \cup B| = |A| + |B| - |A \cap B| \tag{4.1}$$

$$|A \cup B \cup C| = |A| + |B| + |C| - |A \cap B| - |A \cap C| - |B \cap C| + |A \cap B \cap C|$$

■ **Example 4.7** The first part of this theorem, Eq. 4.1, is illustrated using the Venn diagram in Fig. 4.1. We can verify it with the sets A and B used in Examples 4.3 to 4.6: $|A \cup B| = 6$, $|A| = 4$, $|B| = 4$, and $|A \cap B| = 2$.

■

Definition 4.10 — Cartesian Product. The Cartesian product of two sets A and B, denoted $A \times B$, is the set of all ordered pairs (a, b) such that the first element a is in A and the second element b is in B.

■ **Example 4.8** Given $A = \{a, b, c\}$ and $B = \{1, 2, 3\}$, their Cartesian product is:

$$A \times B = \{(a, 1), (a, 2), (a, 3), (b, 1), (b, 2), (b, 3), (c, 1), (c, 2), (c, 3)\}.$$

■

■ **Example 4.9** Consider the set $B = \{0, 1\}$, which is often referred to as the binary set. The Cartesian product B^2 is:

$$B^2 \equiv B \times B = \{(0, 0), (0, 1), (1, 0), (1, 1)\},$$

which is commonly written as

$$\{0, 1\}^2 = \{00, 01, 10, 11\}.$$

Similarly,
$$\{0,1\}^n = \{000\ldots00, 000\ldots01, \ldots, 111\ldots11\}$$
represents the set of 2^n all possible n-bit strings.   ■

Apparently,
$$|A \times B| = |A| \cdot |B|. \tag{4.2}$$

> **Definition 4.11 — Disjoint Sets.** Two or more sets are said to be disjoint if they have no elements in common, that is, their intersection is the empty set: $A \cap B = \emptyset$.

> **Definition 4.12 — Set Partitions.** A partition of a set A is a collection of disjoint subsets of A such that every element in A is included in exactly one subset. The subsets are called blocks of the partition.

■ **Example 4.10** The partitions of the set $\{1, 2, 3\}$, classified by the number of blocks, are as follows:

- 1 block: $\{\{1, 2, 3\}\}$
- 2 blocks: $\{\{1, 2\}, \{3\}\}, \{\{1, 3\}, \{2\}\}, \{\{2, 3\}, \{1\}\}$
- 3 blocks: $\{\{1\}, \{2\}, \{3\}\}$ ■

> **Theorem 4.2 — Set Partition.** If $\{A_1, A_2, \ldots A_n\}$ is a partition of set A, and $B \subseteq A$, then $\{A_1 \cap B, A_2 \cap B, \ldots A_n \cap B\}$ is a partition of set B.

> **Exercise 4.3** Use an example to illustrate that the Set Partition Theorem does not hold without the condition $B \subseteq A$.

> **Definition 4.13 — Power Set.** The power set of A is the set of all subsets of A.

■ **Example 4.11** Given $A = \{a, b, c\}$, the power set of A, denoted by $\mathcal{P}(A)$ or 2^A, is:

$$\mathcal{P}(A) = \{\emptyset, \{a\}, \{b\}, \{c\}, \{a, b\}, \{a, c\}, \{b, c\}, \{a, b, c\}\}.$$

Here, $\emptyset$ denotes the empty set.   ■

The power set has the following property:
$$|\mathcal{P}(A)| \equiv |2^A| = 2^{|A|}. \tag{4.3}$$

> **Exercise 4.4** If $A = \{a\}$, What is $\mathcal{P}(\mathcal{P}(A))$?

4.1.3 Set Identities

Identity Laws:

$$A \cup \emptyset = A, \tag{4.4a}$$
$$A \cap U = A. \tag{4.4b}$$

Domination Laws:

$$A \cup U = U, \tag{4.5a}$$
$$A \cap \emptyset = \emptyset. \tag{4.5b}$$

Idempotent Laws:

$$A \cup A = A, \tag{4.6a}$$
$$A \cap A = A. \tag{4.6b}$$

Absorption Laws:

$$A \cup (A \cap B) = A, \tag{4.7a}$$
$$A \cap (A \cup B) = A. \tag{4.7b}$$

Complement Laws:

$$A \cup \overline{A} = U, \tag{4.8a}$$
$$A \cap \overline{A} = \emptyset, \tag{4.8b}$$
$$\overline{\overline{A}} = A. \tag{4.8c}$$

Commutative Laws:

$$A \cup B = B \cup A, \tag{4.9a}$$
$$A \cap B = B \cap A. \tag{4.9b}$$

Associative Laws:

$$A \cup (B \cup C) = (A \cup B) \cup C, \tag{4.10a}$$
$$A \cap (B \cap C) = (A \cap B) \cap C. \tag{4.10b}$$

Distributive Laws:

$$A \cup (B \cap C) = (A \cup B) \cap (A \cup C), \tag{4.11a}$$
$$A \cap (B \cup C) = (A \cap B) \cup (A \cap C). \tag{4.11b}$$

De Morgan's Laws:

$$\overline{A \cap B} = \overline{A} \cup \overline{B}, \tag{4.12a}$$
$$\overline{A \cup B} = \overline{A} \cap \overline{B}. \tag{4.12b}$$

Exercise 4.5 Is each of the following statements true or false?

(a) $\{1, 2, 3\} \subseteq A$ implies that $2 \in A$ and $\{1, 3\} \subseteq A$.

(b) $\{2, 3, 4\} \in A$ and $\{2, 3\} \in B$ implies that $\{4\} \subseteq A - B$.

(c) $A \cap B \supseteq \{3, 4, 5\}$ implies that $\{3, 4\} \subseteq A$ and $\{4, 5\} \subseteq B$.

| (d) $A - B \supseteq \{3, 4\}$ and $\{1, 2\} \subseteq B$ implies that $\{1, 2, 3, 4\} \subseteq A \cup B$.

4.1.4 ∗ Supremum and Infimum

The supremum function is essentially the least upper bound, or the smallest value that is greater than or equal to all elements of a set. Similarly, the infimum function is the greatest lower bound, or the largest value that is smaller than or equal to all elements of a set. They are not termed 'max' or 'min' because not every set has a maximum or minimum element, but many sets still have a least upper bound and/or greatest lower bound. For example, the open interval $(0, 1)$ does not have a maximum element, but has a least upper bound 1; thus, $\sup(0, 1) = 1$.

Rigorous definitions of supremum and infimum involve the concept of partially ordered sets, as explained below.

> **Definition 4.14 — Totally Ordered Set.** In a totally ordered set, every pair of elements is comparable. This means for any two elements a and b, either $a \leq b$ or $b \leq a$.

Examples of totally ordered sets include the set of real numbers with the usual ordering and the set of integers.

> **Definition 4.15 — Partially Ordered Set (Poset).** In a poset, not every pair of elements needs to be comparable. There might be some incomparable elements where neither $a \leq b$ nor $b \leq a$ holds.

As an example, consider the power set of $\mathbb{Z}$ (the set of subsets of the set of integers), ordered by inclusion ($\subseteq$). Some subsets might not be comparable, e.g., $\{1, 2\}$ and $\{2, 3\}$ are not comparable.

Another example is the set of positive integers, ordered by divisibility ($a \leq b$ if a divides b). Some numbers (e.g., 5 and 7) might not be comparable.

> **Definition 4.16 — Supremum.** Let S be a non-empty subset of a partially ordered set P. An element $u \in P$ is the supremum of S, denoted by $u = \sup S$, if it satisfies the following conditions:
>
> 1. Upper Bound: Every element $s \in S$ satisfies $s \leq u$.
>
> 2. Least Upper Bound: If v is any upper bound of S, then $u \leq v$.

For example, $\sup\{-1, -2, -3, \ldots\} = -1$. And $\sup\{\sin x : x \in [0, \pi]\} = 1$.

> **Definition 4.17 — Infimum.** Let S be a non-empty subset of a partially ordered set P. An element $l \in P$ is the infimum of S, denoted by $l = \inf S$, if it satisfies the following conditions:
>
> 1. Lower Bound: Every element $s \in S$ satisfies $l \leq s$.
>
> 2. Greatest Lower Bound: If v is any lower bound of S, then $v \leq l$.

For example, $\inf\{e^{-x} : x > 0\} = 0$, even though 0 is not an element of the set.

Exercise 4.6 For each set A described below, identify $\sup A$ and $\inf A$ if they exist.

(a) $A = \left\{\frac{1}{n} : n \in \mathbb{N}\right\}$

(b) $A = \left\{\frac{e^x - e^{-x}}{e^x + e^{-x}} : x \in \mathbb{R}\right\}$

4.2 Groups

Groups, rings, and fields provide foundational structures in mathematics that endow sets with additional algebraic operations. These structures are not merely academic; they are ubiquitous in mathematics and physics, with sets such as the integers $\mathbb{Z}$ and the real numbers $\mathbb{R}$ forming familiar examples of fields. In the quantum realm, these abstract concepts become crucial for understanding symmetries and other fundamental properties.

More specifically, a group is a set equipped with a single binary operation that exhibits certain properties akin to standard addition or multiplication. A ring expands upon this by incorporating two operations, typically referred to as addition and multiplication, where the set forms a group under the operation of addition and adheres to some group-like properties under multiplication. A field is a more stringent structure where the set is a group under both operations, with multiplication also being commutative, and every non-zero element has a multiplicative inverse.

4.2.1 Groups

Definition 4.18 — Group. A group is a set G which is *closed* under an operation $*$ (that is, for any $x, y \in G$, $x * y \in G$) and satisfies the following properties:

1. Identity: There exists an element $e \in G$, such that for every $x \in G$, $x * e = x = e * x$, where e is called the identity element.

2. Inverse: For every $x \in G$, there exists an element $y \in G$ such that $x * y = e = y * x$, where e is the identity element.

3. Associativity: The operation $*$ is associative for every $x, y, z \in G$, i.e.,

$$x * (y * z) = (x * y) * z.$$

$\boxed{i}$ When unambiguous from context, it is common to omit the symbol $*$ for the group operation and write $x * y$ simply as xy.

■ **Example 4.12 — $\mathbb{Z}$, the group of integers.** The set of integers $\mathbb{Z}$ under addition forms a group, denoted by $(\mathbb{Z}, +)$. In this group, the identity element is 0, and the inverse of any integer x is its negation $-x$.

The set of integers under multiplication does not form a group since not all elements have a multiplicative inverse within the integers. For example, there is no integer y such that $2y = 1$, as $1/2$ is not an integer.   ■

■ **Example 4.13 — $\mathbb{Z}/n\mathbb{Z}$, the group of integers modulo n.** The set of integers modulo n, denoted by $\mathbb{Z}/n\mathbb{Z}$, forms a group under addition modulo n. It satisfies the group axioms as follows:

1. Closure: The sum of any two integers modulo n is also an integer modulo n.

2. Identity: The integer 0 modulo n serves as the identity element since $a + 0 \equiv a$ (mod n).

3. Inverse: For any integer a modulo n, the integer $n - a$ serves as its inverse since $a + (n - a) \equiv 0$ (mod n).

4. Associativity: The addition of integers modulo n is associative, as it inherits the associativity of integer addition:

$$a + (b + c) \equiv (a + b) + c \quad (\text{mod } n).$$

∎

Exercise 4.7 Give two examples of sets of numbers that form a group under regular multiplication. Ensure each example set is closed under multiplication, contains a multiplicative identity, and each element has a multiplicative inverse within the set.

Hint: Consider the sets of non-zero rational numbers ($\mathbb{Q}^*$), non-zero real numbers ($\mathbb{R}^*$), and non-zero complex numbers ($\mathbb{C}^*$).

Definition 4.19 — Abelian Group. A group is said to be *abelian* if the operation $*$ is commutative for every $x, y \in G$, that is, $x * y = y * x$.

The examples provided thus far are all abelian groups. In contrast, the set of symmetries of an equilateral triangle, known as the dihedral group D_3, forms a non-abelian group of size 6 when considering the composition of symmetries.

Popular Symmetry Groups

The following symmetry groups are fundamental in the study of symmetry and are key constructs in mathematics, physics, and particularly quantum computing:

SO(N): The special orthogonal group in N dimensions, SO(N), consists of all $N \times N$ orthogonal matrices with determinant 1, representing rotations in N-dimensional space. These rotations preserve distances and the orientation of objects. SO(2) is exemplified by the symmetries of a circle, while SO(3) the symmetries of a sphere.

SU(2): The special unitary group of degree 2, SU(2), comprises 2×2 complex unitary matrices with determinant 1. It is closely related to SO(3) and finds applications in the quantum realm, especially in describing spins and qubit states. SU(2) is a double cover of SO(3), meaning each rotation in SO(3) corresponds to two points in SU(2).

SU(N): SU(N) represents the special unitary group of degree N, extending the concepts of SU(2). These groups are useful in the study of quantum entanglement in quantum computing, where N-level quantum systems (quNits) are considered.

We will revisit these topics in § 11.2.2.

4.2.2 Subgroups

> **Definition 4.20 — Subgroup.** A subgroup H of a group G is a subset of G that is itself a group with the group operation inherited from G. That is, H is a subgroup of G if:
>
> 1. The identity element of G is in H.
>
> 2. If $h_1, h_2 \in H$, then $h_1 * h_2$ is also in H.
>
> 3. If $h \in H$, then the inverse of h is also in H.

■ **Example 4.14** In the group $\mathbb{Z}$ under addition, the set of all even integers $2\mathbb{Z} = \{2n \mid n \in \mathbb{Z}\}$ forms a subgroup.

The group $\mathbb{Z}/4\mathbb{Z} = \{0, 1, 2, 3\}$ under addition modulo 4 has a subgroup $\{0, 2\}$. ■

A key property of subgroups is:

> **Theorem 4.3 — Lagrange's Theorem.** For any finite group G and any subgroup H of G, the order of H (the number of elements in H) divides the order of G, i.e., $|G| \equiv 0 \mod |H|$.

■ **Example 4.15** Consider the group $G = \mathbb{Z}/6\mathbb{Z}$, the integers modulo 6 under addition, which is a finite group with 6 elements:

$$G = \{0, 1, 2, 3, 4, 5\}.$$

The order of G is $|G| = 6$.

Consider a subgroup $H = \{0, 3\}$. The order of H is $|H| = 2$.

According to Lagrange's Theorem, the order of H should divide the order of G. In this case, clearly, $|H| = 2$ divides $|G| = 6$. ■

4.2.3 ✳Cosets and Cyclic Groups

In the realm of quantum computing, Simon's and Shor's algorithms stand out as groundbreaking advancements, showcasing the power of quantum computation in solving specific problems more efficiently than classical methods. Interestingly, these algorithms can be generalized to cosets and cyclic groups, extending their applicability. In this subsection, we introduce the mathematical background of cosets and related concepts.

Cosets are used to partition a group into equivalence classes based on a subgroup. Given a subgroup H of a group G, the concept of cosets allows us to divide G into distinct subsets.

> **Definition 4.21 — Coset.** Given a group G and a subgroup H of G, the left coset of H with representative $g \in G$ is the set $gH = \{gh \mid h \in H\}$. Similarly, the right coset is $Hg = \{hg \mid h \in H\}$.

> ⓘ The notation hg is shorthand for $h * g$, where $*$ is the group operation, as is commonly used in mathematical texts.

■ **Example 4.16** Consider the group $G = \mathbb{Z}$ (the integers under addition) and the subgroup $H = 3\mathbb{Z}$, which consists of all multiples of 3. A coset of H in G is of the form $g + 3\mathbb{Z}$ for some $g \in \mathbb{Z}$. For example, the coset $1 + 3\mathbb{Z}$ is:

$$1 + 3\mathbb{Z} = \{1 + 3k \mid k \in \mathbb{Z}\} = \{\ldots, -5, -2, 1, 4, 7, \ldots\}.$$

Similarly, $0 + 3\mathbb{Z}$ and $2 + 3\mathbb{Z}$ are also cosets $3\mathbb{Z}$ in $\mathbb{Z}$. ■

> **Theorem 4.4 — Partition Theorem for Cosets.** The collection of all left cosets of a subgroup H forms a partition of the group G. This means:
>
> 1. Every element of G belongs to exactly one coset of H.
>
> 2. Cosets are disjoint (have no elements in common).

■ **Example 4.17** Consider the group $\mathbb{Z}$ and its subgroup $4\mathbb{Z}$. We can partition $\mathbb{Z}$ into cosets based on congruence modulo 4. This results in the partitioning of $\mathbb{Z}$ into $0 + 4\mathbb{Z}$, $1 + 4\mathbb{Z}$, $2 + 4\mathbb{Z}$, and $3 + 4\mathbb{Z}$. ■

> **Definition 4.22 — Normal Subgroup.** A subgroup N of a group G is called a normal subgroup if it is invariant under conjugation by elements of G. This means that for every $n \in N$ and every $g \in G$, the element gng^{-1} is also in N. In notation, $N \triangleleft G$ if:
> $$gNg^{-1} = \{gng^{-1} \mid n \in N\} \subseteq N \quad \text{for all} \quad g \in G.$$

Importantly, the left and right cosets of a normal subgroup N are the same, allowing the group operation on cosets to be well-defined, leading to the concept of quotient groups:

> **Definition 4.23 — Quotient Group.** Let G be a group, and let N be a normal subgroup of G. The quotient group G/N is the set of cosets of N in G, with the group operation defined by:
> $$(gN)(hN) = (gh)N,$$
> for all $g, h \in G$.

■ **Example 4.18 — $\mathbb{Z}/2\mathbb{Z}$.** Consider the group $\mathbb{Z}$ of integers under addition and the subgroup $2\mathbb{Z}$ consisting of all even integers. The quotient group $\mathbb{Z}/2\mathbb{Z}$ is the set of cosets of $2\mathbb{Z}$ in $\mathbb{Z}$:

- The coset $0 + 2\mathbb{Z} = \{\ldots, -4, -2, 0, 2, 4, \ldots\}$ represents the even integers.

- The coset $1 + 2\mathbb{Z} = \{\ldots, -3, -1, 1, 3, 5, \ldots\}$ represents the odd integers.

Thus, $\mathbb{Z}/2\mathbb{Z}$ has two elements: $0 + 2\mathbb{Z}$ and $1 + 2\mathbb{Z}$, corresponding to the even and odd integers, respectively. The group operation is addition modulo 2.

This quotient group is *isomorphic* to $\mathbb{Z}_2 = \{0, 1\}$ under addition modulo 2, denoted as $\mathbb{Z}/2\mathbb{Z} \cong \mathbb{Z}_2$. Two groups are isomorphic if they have the same structure, meaning there is a one-to-one correspondence between the elements of the two groups that preserves the group operation. ■

By common convention,

- $n\mathbb{Z}$ denotes the subgroup of integer multiples of n (i.e., $\{\ldots, -2n, -n, 0, n, 2n, \ldots\}$).

- $\mathbb{Z}_n$ denotes the group (or ring, depending on context) of integers modulo n (i.e., $\{0, 1, 2, \ldots n-1\}$).

- $\mathbb{Z}/n\mathbb{Z}$ is the *quotient group* of $\mathbb{Z}$ by $n\mathbb{Z}$, i.e., the set of all cosets of $n\mathbb{Z}$ in $\mathbb{Z}$. Since $\mathbb{Z}/n\mathbb{Z}$ is isomorphic to $\mathbb{Z}_n$, both notations are commonly used to refer to the integers modulo n.

■ **Example 4.19** — $\mathbb{R}/\mathbb{Z}$. For the quotient group $\mathbb{R}/\mathbb{Z}$,

- $\mathbb{R}$ is the group of real numbers under addition.

- $\mathbb{Z}$ is the subgroup of integers (also under addition).

- Because $\mathbb{Z}$ is normal in $\mathbb{R}$ (in fact, all subgroups of an Abelian group are normal), we can form the quotient group $\mathbb{R}/\mathbb{Z}$.

- Each coset of $\mathbb{Z}$ in $\mathbb{R}$ is a set of real numbers that differ from one another by an integer.

With $\mathbb{R}/\mathbb{Z}$, we are forming the set of equivalence classes of real numbers "modulo integers," i.e., identifying any two reals that differ by an integer.

Consequently, $\mathbb{R}/\mathbb{Z}$ can be visualized as a circle of circumference 1, where every point corresponds to one coset—i.e., one "fractional part" class. ■

Definition 4.24 — **Cyclic Group.** A group G is cyclic if there exists an element $g \in G$ (called a *generator* of G) such that every element of G can be expressed as powers (i.e., repeated applications) of g. In other words, G can be entirely generated by g, denoted as $G = \langle g \rangle$.

Every subgroup of a cyclic group is also cyclic. A cyclic group can be finite or infinite, depending on whether the powers of g eventually repeat or not.

■ **Example 4.20** The number 2 is a generator for the cyclic group of powers of 2, which are all positive real numbers of the form 2^n where n is an integer.

The group $\mathbb{Z}/4\mathbb{Z} = \{0, 1, 2, 3\}$ under addition modulo 4 has a subgroup $\{0, 2\}$, which is also a cyclic subgroup generated by 2. ■

4.2.4 ✶ Rings and Fields

Definition 4.25 — **Ring** A ring is a set R equipped with two operations $+$ and $\times$ satisfying the following properties:

1. $(R, +)$ forms an abelian group.

2. Associativity of $\times$: For every $a, b, c \in R$,
$$a \times (b \times c) = (a \times b) \times c.$$

3. Distributive properties: For every $a, b, c \in R$,
$$a \times (b + c) = (a \times b) + (a \times c),$$

$$(b + c) \times a = (b \times a) + (c \times a).$$

Note: A ring does not necessarily have a multiplicative identity.

■ **Example 4.21** Both the examples $\mathbb{Z}/n\mathbb{Z}$ and $\mathbb{Z}$ from before are also *rings*. Note that we don't require multiplicative inverses.

$\mathbb{Z}[x]$, all polynomials with integer coefficients, is a ring. Multiplication and addition is the usual multiplication and addition of polynomials.   ■

Definition 4.26 — Field. A field is a set F with two operations $+$ and $\times$ where:

1. $(F, +)$ is an abelian group,

2. $(F - \{0\}, \times)$ is an abelian group (every non-zero element has a multiplicative inverse),

3. The distributive laws hold as in rings.

■ **Example 4.22** $\mathbb{Q}$, the set of rational numbers, $\mathbb{R}$, the set of real numbers, and $\mathbb{C}$, the set of complex numbers are all infinite fields.   ■

■ **Example 4.23** $\mathbb{Z}/p\mathbb{Z}$ (where p is a prime number) is a field, since $\mathbb{Z}/p\mathbb{Z}$ is an additive group and $(\mathbb{Z}/p\mathbb{Z}) - \{0\}$ is a group under multiplication.

Why does p have to be a prime? Consider a counterexample, $\mathbb{Z}/6\mathbb{Z}$. The element 2 does not have an inverse x such that $2x = 1 \pmod 6$.   ■

Exercise 4.8 Explain why $\mathbb{Z}$ is not a field.

4.3 Functions

Functions are one of the core concepts in mathematics, serving as mappings from one set to another that assign each element in the domain to exactly one element in the codomain. They are essential for describing relationships between quantities and for formulating mathematical models. Functions can be used to express a wide variety of phenomena in mathematics and science, making them a critical tool in both theoretical and applied contexts.

4.3.1 Image, Range, and Domain

Definition 4.27 — Function. Let f be a function from set A to set B. A *function from A to B*, denoted $f : A \to B$, is an assignment of exactly one element of B to each element of A. We write $f(a) = b$ to denote the assignment of b to an element a of A by the function f.

■ **Example 4.24** $f(x) = x^2$ is a function $f : \mathbb{R} \to \mathbb{R}$. $r = \sqrt{x^2 + y^2}$ is a function $f : \mathbb{R} \times \mathbb{R} \to \mathbb{R}$.   ■

Definition 4.28 — Image, Range, and Domain. Let f be a function from A to B.

• We say that A is the *domain* of f and B is the *codomain* of f.

- If $f(a) = b$, b is the *image* of a and a is a *pre-image* of b.

- The *range* of f (a subset of B) is the set of all images of elements of A.

- Let S be a subset of A. The *image of S* is a subset of B that consists of the images of the elements of S. We denote the image of S by $f(S)$, so that

$$f(S) = \{f(s) \mid s \in S\}.$$

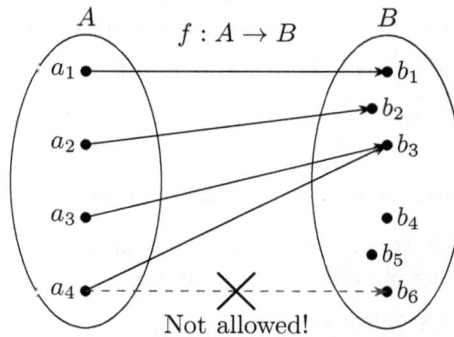

Figure 4.2: Illustration of a Function

■ **Example 4.25** In Fig. 4.2, the range of f is $\{b_1, b_2, b_3\}$, which is a subset of B. The image of $\{a_1, a_2\}$ under f is $\{b_1, b_2\}$. The set of pre-images of b_3 is $\{a_3, a_4\}$. The mapping from a_4 to b_6 is not allowed as part of the function f because a_4 is already mapped to b_3. ■

4.3.2 Function Types and Inverse

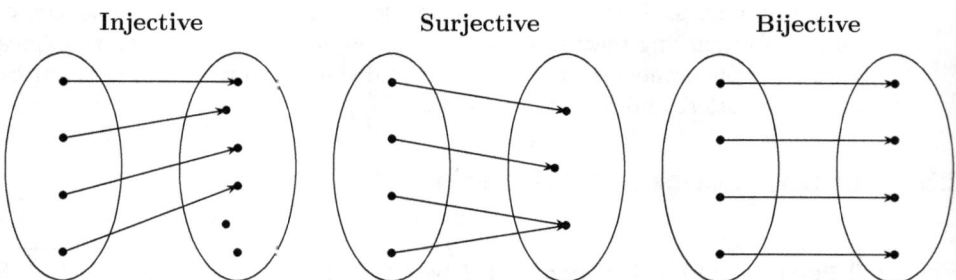

Figure 4.3: Injective, Surjective, and Bijective Functions

Injective, surjective, and bijective are terms used to describe certain properties of functions, as illustrated in Fig. 4.3:

Definition 4.29 — Injective Function. A function $f : A \to B$ is injective if for every $a_1, a_2 \in A$, whenever $f(a_1) = f(a_2)$, it follows that $a_1 = a_2$.

An injective function, also known as a one-to-one function, assigns distinct elements of the domain to distinct elements of the codomain.

> **Definition 4.30 — Surjective Function.** A surjective function, also called an onto function, covers the entire codomain. That is, for every element $b \in B$, there exists at least one element $a \in A$ such that $f(a) = b$.

The range of a surjective function is equal to its codomain. Surjective functions can be many-to-one functions.

> **Definition 4.31 — Bijective Function.** A bijective function is both injective and surjective.

A bijective function pairs each element of the domain with a distinct and unique element of the codomain. It establishes a one-to-one correspondence between the elements of the domain and the codomain, meaning it is both one-to-one and onto.

> **Definition 4.32 — Inverse Function.** Let $f : A \to B$ be a bijective function with domain A and codomain B. The inverse function of f, denoted by $f^{-1} : B \to A$, is defined such that for every $b \in B$, $f^{-1}(b) = a$ if and only if $f(a) = b$ for some $a \in A$. In other words, $f^{-1}(f(a)) = a$ for all $a \in A$, and $f(f^{-1}(b)) = b$ for all $b \in B$.

■ **Example 4.26** The function $f(x) = x^2$ is not injective over the set of all real numbers $\mathbb{R}$ because it maps both a positive and a negative number to the same value (for example, $f(2) = 4$ and $f(-2) = 4$).

However, it is injective over the set of non-negative real numbers $\mathbb{R}^+$ since every non-negative real number has a unique non-negative square root. In fact, $f : \mathbb{R}^+ \to \mathbb{R}^+$ with $f(x) = x^2$ is bijective, and its inverse function is $f^{-1}(x) = \sqrt{x}$. ■

Exercise 4.9 Define $f : \mathbb{Z} \to \mathbb{Z}$ by $f(n) = 3n - 1$, where $\mathbb{Z}$ is the set of integers.

(a) Is f one-to-one?

(b) Is f onto?

4.4 Common Functions and Asymptotic Behavior

4.4.1 Commonly Used Functions

Here is a quick review of real function (meaning $\mathbb{R} \to \mathbb{R}$) commonly used in quantum mechanics and quantum computing:

1 **Power Functions**

Power functions are defined as $f(x) = x^p$, where $x \geq 0$ and p is a real number. The behavior of the function varies significantly with the exponent p:

- For $p > 0$, $f(x)$ increases as x increases. $f(x)$ exhibits a more rapid growth with a larger p.

- For $p < 0$, $f(x)$ decreases as x increases.

- When $p = 0$, $f(x) = 1$, regardless of x (excluding $x = 0$), which is a constant function.

- For $p = 1$, $f(x) = x$, representing a linear relationship.

Key properties of power functions include the rules for exponentiation:

- Multiplying powers with the same base: $x^a \cdot x^b = x^{a+b}$.
- Dividing powers with the same base: $x^a / x^b = x^{a-b}$.

2   Polynomial Functions

A polynomial function is a sum of terms $a_i x^i$, where i is a non-negative integer:

$$f(x) = a_0 + a_1 x + a_2 x^2 + \cdots + a_n x^n.$$

Its behavior for large x values is predominantly determined by its highest power term, x^n, where n is the degree of the polynomial.

An nth-degree polynomial has n complex roots (counting multiplicities), and according to Vieta's formulas, the sum of these roots is equal to $-a_{n-1}/a_n$ and their product $(-1)^n a_0/a_n$.

■ **Example 4.27** Consider the polynomial function $f(x) = x^3 - 7x^2 + 14x - 8$. It can be factored as:

$$f(x) = (x - 1)(x - 2)(x - 4).$$

The roots of this polynomial are $x = 1$, $x = 2$, and $x = 4$, which can be found by solving the equations $(x - 1) = 0$, $(x - 2) = 0$, and $(x - 4) = 0$. According to Vieta's formulas, the sum of the roots is:

$$1 + 2 + 4 = 7 = -\frac{-7}{1},$$

and the product of the roots is:

$$1 \cdot 2 \cdot 4 = 8 = (-1)^3 \cdot \frac{-8}{1}.$$

■

Exercise 4.10 How is the coefficient of x, 14, in the above example related to the roots of the polynomial?

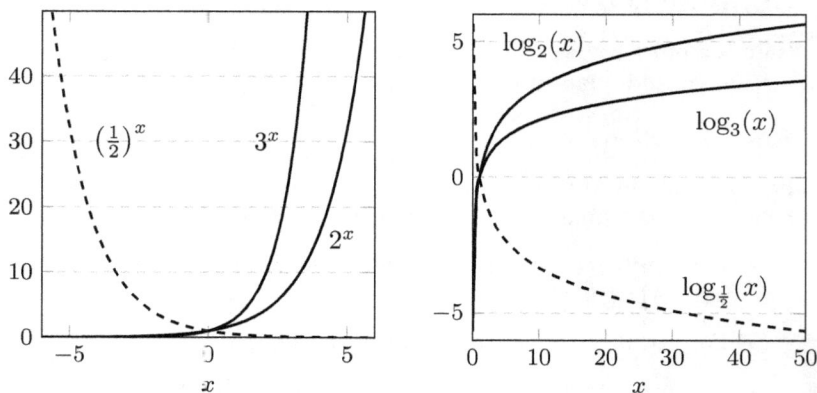

Figure 4.4: Exponential and Logarithmic Functions

3 Exponential Functions

Exponential functions are defined as $f(x) = b^x$, where b is a positive constant (called the base) and $b \neq 1$. The variable x is the exponent. The key characteristic is that the variable is in the exponent. Some important notes:

- Growth and Decay:

    - If $b > 1$, $f(x)$ exhibits exponential growth – increasing rapidly as x increases. Larger bases lead to faster growth.

    - If $0 < b < 1$, $f(x)$ shows exponential decay – decreasing towards zero as x increases.

- Always Positive: Exponential functions are always positive for any real-valued input x.

- Horizontal Asymptote: They approach zero for one direction of x (negative infinity for growth, positive infinity for decay).

- Base e: The natural exponential function with $b = e$ (Euler's number, ≈ 2.718), i.e, e^x, also denoted as $\exp(x)$, has special significance across mathematics.

4 Logarithmic Functions

Logarithmic functions are the inverses of exponential functions. They are defined as $f(x) = \log_b(x)$, where b is a positive constant $(b \neq 1)$ and $x > 0$. Some key points:

- Reversing Exponentiation: If $b^y = x$ then $\log_b(x) = y$.

- Growth and Behavior

    - For $b > 1$, $\log_b(x)$ increases as x increases, but very slowly.

    - For $0 < b < 1$, $\log_b(x)$ decreases as x increases.

- Vertical Asymptote: Logarithmic functions have a vertical asymptote at $x = 0$.

- Logarithms of 1 and the Base: $\log_b(1) = 0$ and $\log_b(b) = 1$.

- The natural logarithm, written as $\ln(x)$ has the base e.

Key Properties:

- The Product Rule: $\log_b(xy) = \log_b(x) + \log_b(y)$

- Logarithms "Break" Exponents: $\log_b(x^y) = y \cdot \log_b(x)$

- Changing Bases: $\log_b(x) = \log_a(x) / \log_a(b)$

Exercise 4.11 Solve the following equations:

(a) $17 = 10x^{-2} + 12$

(b) $2^{3-8x} - 5 = 11$

(c) $2\ln(x) - \ln(x+6) = 0$

(d) $\log_2(x) + \log_2(x-1) = 1$

5 Trigonometric Functions

Trigonometric functions are detailed separately in Chapter 2.

4.4.2 ∗ Asymptotic Behavior and the Big O Notation

Asymptotic, or scaling, behavior examines the evolution of function curves and their response to large inputs. We often use the *Big O notation* to describe the asymptotic behavior of a function $\tilde{f}(x)$, typically in terms of a simpler function $g(x)$. When we say that $f(x)$ *is* $O(g(x))$ as x approaches infinity, it means that there exists a positive constant c such that $f(x)$ does not grow faster than $c \cdot g(x)$ for sufficiently large x. This is written as

$$f(x) = O(g(x)),$$

indicating that $c \cdot g(x)$ serves as an "upper bound" on $f(x)$ in the limit.

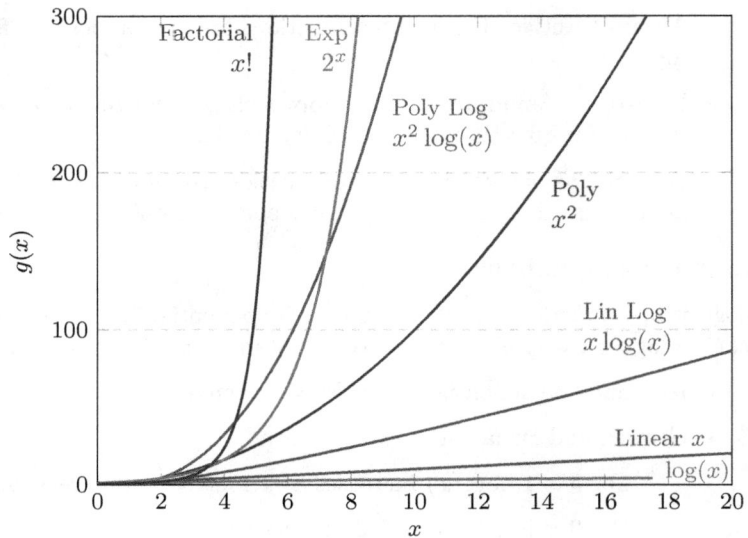

Figure 4.5: Scaling Behavior of Common Functions

For example, consider $f(x) = 6x^3 + 3x^2 + 2$. We can say that $f(x)$ is $O(x^3)$ as $x \to \infty$. Here, the term x^3 is the dominant term, significantly influencing the growth rate of $f(x)$ as x becomes very large, while the coefficients and the lower degree terms become less significant. This illustrates how Big O notation focuses on the most significant term affecting the growth rate of the function as the variable grows large. This simplification is particularly useful in numerical analysis, complexity theory, and other fields of mathematics and computer science.

The following is a summary of limiting functions $(g(x))$ commonly used to analyze algorithmic efficiency. These functions provide a theoretical framework for understanding how the time complexity of an algorithm scales with the size of the input. By classifying algorithms according to these functions, we can predict their performance and identify the most efficient algorithms for large datasets.

1. Log-log: $g(x) = \log\log(x)$

   - Exhibits extremely slow growth. Algorithms within this complexity class increase their running time at a negligible rate with input size escalation.

   - Applications include specialized computational geometry problems.

2. Log: $g(x) = \log(x)$

- Denotes high efficiency. The execution time grows much slower than the input size.

- Examples include binary search in sorted arrays and operations on certain balanced tree data structures.

3. Sublinear: $g(x) = x^p$, $0 < p < 1$

- Exhibits growth slower than linear but faster than logarithmic.

- Common examples include the Grover's search algorithm in quantum computing, which has a complexity of approximately $O(\sqrt{x})$, and some algorithms that utilize probabilistic methods to achieve faster-than-linear performance on average.

4. Linear: $g(x) = x$

- Indicates direct proportionality. Doubling the input size doubles the running time.

- Common examples are searching in unsorted lists and identifying max/min elements in a list.

5. Polynomial: $g(x) = x^p$, $p > 1$

- The growth rate is influenced by the exponent p. Higher values lead to rapid increases in running time with input size.

- Examples: Bubble sort and insertion sort (quadratic complexity), matrix multiplication algorithms (cubic complexity or better).

6. Poly-log: $g(x) = x^p \log(x)$, $p \geq 1$

- Less efficient than the corresponding poly (or linear for $p = 1$) but still considered scalable.

- Fast Fourier Transform (FFT) algorithms are a prime example of algorithms with linear-log complexity. Some fast sorting algorithms also approach this performance.

7. Exponential: $g(x) = b^x$, $b > 1$

- Characterized by rapid growth. Algorithms in this class quickly become impractical for moderate input sizes.

- Examples: Brute-force approaches to the Traveling Salesman Problem. Currently known classical algorithms for integer factorization.

8. Factorial: $g(x) = x!$

- Exhibits extremely rapid growth, surpassing even exponential functions in rate. Practical for only very small input sizes.

- Example: Generating all permutations of a set.

9. Hyper-exponential: $g(x) = x^x$, $g(x) = b^{a^x}$, $g(x) = b^{x!}$, and $g(x) = b^{x^x}$, etc., where $a, b > 1$

- Exhibits growth that is even more rapid than factorial functions.

- Example: Modeling scenarios with extremely high growth rates, beyond combinatorial complexity.

Exercise 4.12 For each of the following functions, determine the dominant term as x grows large and express the growth rate using Big O notation.

(a) $4x + x^2$

(b) $0.3x + 5x^{1.5} + 5x^{1.8}$

(c) $10x + x \ln x + 0.1x \ln(\ln x)$

(d) $x \log_3 x + x \log_2 x$

Problem Set 4

4.1 For each of the following cases, draw a Venn diagram.

(a) $A \subseteq B$, $C \subseteq B$, $A \cap C = \emptyset$

(b) $A \supseteq C$, $B \cap C = \emptyset$.

4.2 Investigate how many partitions there are of a set A with n elements, specifically for $n \leq 4$.

4.3 Let $A = \{1, 2, 3\}$, $B = \{x, y\}$, and $C = \{m, n\}$. List the elements in $A \times B \times C$.

4.4 Let A, B, and C be subsets of U. Prove or disprove the following:

(a) $A - B$ and $B - C$ are disjoint (meaning empty intersection).

(b) $B - A$ and $B - C$ are disjoint.

(c) $A - (B \cup C)$ and $B - (A \cup C)$ are disjoint.

(d) $A - (B \cap C)$ and $B - (A \cap C)$ are disjoint.

4.5 Explain by checking all the group axioms:

(a) The set of all even integers $2\mathbb{Z} = \{2n \mid n \in \mathbb{Z}\}$ forms a group under addition.

(b) The set of all odd integers $1 + 2\mathbb{Z} = \{2n + 1 \mid n \in \mathbb{Z}\}$ is not a group under addition.

(c) Dihedral groups, representing the symmetries (rotations and reflections) of a regular polygon (a triangle, square, etc.), are non-abelian groups. (Hint: Rotations don't commute with reflections in general. For example, rotating a square 90 degrees and then flipping it yields a different result than flipping first and then rotating.)

(d) The special orthogonal group in 3 dimensions, SO(3), consisting of all rotations around the origin of a Cartesian coordinate system, is a non-abelian group. (Hint: Generally, rotating around the x-axis does not commute with rotating around the y-axis.)

4.6 In each of the following cases, is f one-to-one?

(a) $f(x) = \frac{x}{x^2+1}$ where $f : \mathbb{R} \rightarrow \mathbb{R}$

(b) $f(x) = \frac{2x+1}{x-2}$ where $f : (\mathbb{R} \setminus \{2\}) \rightarrow \mathbb{R}$

4.7 Let $f : X \to Y$ and $g : Y \to Z$. The composition of f and g is denoted as $g \circ f : X \to Z$, or $z = g(f(x))$.

(a) If $g \circ f$ is onto, must f and g be onto?

(b) If $g \circ f$ is one-to-one, must f and g be one-to-one?

4.8 Solve the following equations:

(a) $2t - t2^{6t-1} = 0$

(b) $15e^{6-x} + e^{12x-7} = 0$

(c) $1 - 8 \ln \left(\frac{2x-1}{8} \right) = 13$

(d) $\ln(y - 1) = 1 + \ln(3y + 2)$

(e) $\left(\frac{1}{6} \right)^{-3x-2} = 36^{x+1}$

(f) $e^x e^2 = \frac{e^4}{e^x + 1}$

(g) $4x + 1 = (12x + 3)e^{x^2 - 2}$

4.9 ✳ For each of the following functions, determine the dominant term as x grows large and express the growth rate using Big O notation.

(a) $5 + 0.001x^3 + 0.025x$

(b) $1 + 0.1x^{0.3} + 0.05\sqrt{x}$

(c) $500x + 100x^{1.5} + 50x \log_{10} x$

(d) $x^2 \log_2 x + x(\log_2 x)^2$

(e) $3 \log_2 x + \log_2 \log_2 \log_2 x$

(f) $100x + 0.01x^2$

(g) $2x + x^{0.5} \log_2 x + 0.5x^{1.25}$

(h) $0.01x \log_2 x + x(\log_2 x)^2$

(i) $100x \log_3 x + x^3 + 100x$

(j) $0.003 \log_4 x + \log_2 \log_2 x$

4.10 ✳ Is each of the following statements true or false? Explain. If false, provide a counter example.

(a) $O(f + g) = O(f) + O(g)$

(b) $O(f \cdot g) = O(f)O(g)$

(c) $O(f/g) = O(f)/O(g)$

(d) $O(f(g)) = O(f(O(g)))$

II

Vectors, Matrices and Linear Spaces

The study of linear algebra forms the backbone of the theory of vectors and matrices. In quantum computing, vectors and matrices, along with their associated computational rules, are indispensable for describing the physical states, operations, and measurements of qubits. A solid grasp of linear algebra is therefore essential for mastering the language of quantum computing; without it, understanding the field is impossible. Part II of this book provides the necessary foundation in linear algebra to prepare readers for more advanced topics.

Quantum computing stands apart from other disciplines in its distinctive application of linear algebra. It operates over complex numbers rather than real numbers, employs Dirac notation (bra-ket notation) in place of conventional matrix notation, and makes extensive use of tensor products of vector spaces—topics that are rarely covered in standard linear algebra textbooks.

To familiarize readers with the specialized linear algebra used in quantum computing, this Part adopts a structured approach: each concept is first introduced using real vectors to build intuition and is then generalized to complex vectors, which are critical for practical applications. Examples involving both real and complex vectors are included to ensure a smooth and comprehensive learning experience.

The prerequisite knowledge for Part II includes familiarity with complex numbers, basic analytic geometry, and fundamental concepts from set theory, group theory, and functions—most of which are covered in Part I of this book. Readers encountering unfamiliar concepts or terminology are encouraged to refer to the earlier chapters for review.

5. Vectors and Vector Spaces

Contents

The concepts of vectors and vector spaces are fundamental to the study of quantum computing, as they provide the mathematical framework for representing and manipulating quantum states. In quantum mechanics, state vectors in complex vector spaces describe phenomena such as superposition and entanglement, which are essential to the power of quantum computation. By understanding vector spaces, we can better comprehend how quantum algorithms function and how to harness the unique capabilities of quantum systems. This chapter systematically introduces vectors, vector spaces, and their properties, laying the foundation for more advanced concepts built upon them.

5.1 Real Vectors and Complex Vectors

Numbers frequently encountered in everyday life, such as three apples, minus ten degrees Celsius, or 12.3 centimeters, often suffice to describe quantities. However, certain situations require more than one number for a complete description. For instance, to precisely pinpoint New York City's location on Earth, at least two numbers are needed: 40.71°N and 74.01°W, representing latitude and longitude, respectively. This concept leads us to vectors, which combine multiple ordered numbers into a single mathematical entity.

5.1.1 Algebraic Representations of Vectors

A vector is an ordered sequence of numbers, illustrated by the following examples:

$$(1,3), \quad \left(\frac{\sqrt{2}}{2}, \frac{\sqrt{2}}{2} \right), \quad (0,0,0), \quad (-1.1, 2.3, 0.8, -1.5), \quad (1+i, 1-i).$$

> *i*   Although vectors can, in principle, contain elements other than numeric values—such as symbols, letters, or even other vectors or matrices—our discussion will focus on vectors with numeric elements for now.

Definition 5.1 — Ordered n-Tuple. Given a positive integer n, an ordered sequence of n real (or complex) numbers $(v_1, v_2, \ldots, v_n)$ is called an ordered n-tuple.

Definition 5.2 — Euclidean Space. The set comprising all n-tuples of real numbers is called n-space or Euclidean space, and is denoted as $\mathbb{R}^n$.

The set comprising all n-tuples of complex numbers is called complex Euclidean space and is denoted as $\mathbb{C}^n$.

Generally, a vector in $\mathbb{R}^n$ or $\mathbb{C}^n$ can be represented as an n-tuple:

$$\boldsymbol{v} = (v_1, v_2, \ldots, v_n), \tag{5.1}$$

where $v_i \in \mathbb{R}$ if $\boldsymbol{v} \in \mathbb{R}^n$, or $v_i \in \mathbb{C}$ if $\boldsymbol{v} \in \mathbb{C}^n$ for $i = 1, 2, \ldots, n$. This is known as the **algebraic representation** of a vector.

In this representation, each number in the sequence $v_1, v_2, \ldots, v_n$ is called a **component** or a **coordinate** (refer to § 5.4 for more details). In the context of quantum computing, vectors are typically finite-dimensional, meaning n is finite.

Definition 5.3 — Zero Vector. The zero vector in $\mathbb{R}^n$ or $\mathbb{C}^n$, denoted as $\boldsymbol{0}$, is defined as the vector where all components equal zero. Algebraically, it is expressed as:

$$\boldsymbol{0} = (0, 0, \ldots, 0).$$

A vector in $\mathbb{R}^n$ is called a **real vector**, and a vector in $\mathbb{C}^n$ is called a **complex vector**. Since $\mathbb{R} \subset \mathbb{C}$, a vector in $\mathbb{R}^n$ is also in $\mathbb{C}^n$.

> ⓘ **Notation**: In this book, vectors are denoted by boldface types such as $\boldsymbol{u}, \boldsymbol{v}, \boldsymbol{w}$.
> Scalars (single numbers) are denoted by lowercase italic types such as a, b, α, β.

5.1.2 Geometric Representations of Real Vectors

For real vectors in $\mathbb{R}^2$ or $\mathbb{R}^3$, they can also be represented geometrically. These are known as the geometric representations of vectors, or more briefly, **geometric vectors**. They are commonly used in engineering and physics, where physical quantities often possess both **magnitude** and **direction**, such as force, velocity, and heat flow. These quantities are sometimes referred to as **directed quantities** to emphasize their dual nature.

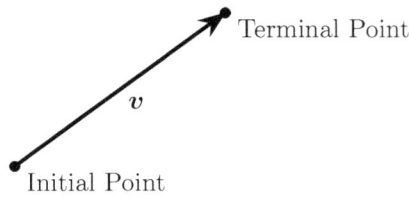

Figure 5.1: Geometric Representation of a Vector

Directed quantities can be visually represented by arrows, as shown in Fig. 5.1. The direction of the arrowhead indicates the **direction** of the vector, while the length of the arrow represents its **magnitude**. The tail and tip of the arrow, often highlighted with dots, are referred to as the **initial point** and **terminal point** of the vector, respectively.

In linear algebra, vectors with identical length and direction are considered equivalent, irrespective of their initial and terminal points (Fig. 5.2). Geometric vectors are therefore not fixed in position and can be freely translated. (In physics, however, this principle may not always apply, as with force vectors.)

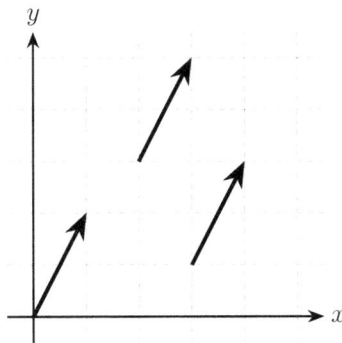

Figure 5.2: Equivalent Geometric Vectors

Since geometric vectors allow free translation, being *collinear* and *parallel* are equivalent statements, as illustrated in Fig. 5.3.

When the initial point of a geometric vector is fixed at the origin of a Cartesian coordinate system, the vector is completely determined (and represented) by the

Figure 5.3: Collinear and Parallel Geometric Vectors

coordinate of its terminal point. Generally, we express $v = (v_1, v_2)$ in 2D (two-dimensional space) and $v = (v_1, v_2, v_3)$ in 3D (three-dimensional space), as illustrated in Fig. 5.4. These coordinates correspond exactly to the algebraic representation of the given vector.

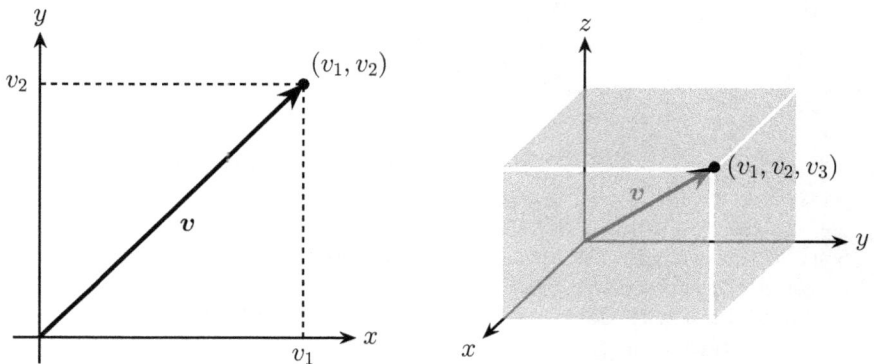

Figure 5.4: Representation of 2D and 3D Geometric Vectors

> A zero geometric vector has coincidental initial and terminal points. It has a length of zero and is assigned an arbitrary direction due to the absence of a natural direction.

The introduction of geometric vectors serves to visualize the fundamental properties and rules of vectors in $\mathbb{R}^2$ and $\mathbb{R}^3$. Such visualization helps establish an intuitive understanding of key concepts in linear algebra. This intuition can often be extended to $\mathbb{R}^n$ ($n > 3$) or $\mathbb{C}^n$, for which geometric interpretations are no longer directly applicable.

Exercise 5.1 Draw the following vectors in a two-dimensional coordinate system, then use principles of analytic geometry to calculate both the length and direction of each vector:

(a) $(2, 0)$,

(b) $(0, 2)$,

(c) $(1, 1)$,

(d) $(-1, \sqrt{3})$.

Determine the direction of each vector by calculating the angle from the positive x-axis to align with the vector.

5.2 Basic Vector Algebra

In this section, we investigate the fundamental operations on vectors: vector addition, subtraction, and scalar multiplication. These operations are studied algebraically in $\mathbb{R}^n$ or $\mathbb{C}^n$, and geometric interpretations in 2D or 3D are provided wherever they offer helpful insights.

5.2.1 Equivalent Vectors

Definition 5.4 — Vector Equality. Given two vectors $\boldsymbol{u} = (u_1, u_2, \ldots, u_n)$ and $\boldsymbol{v} = (v_1, v_2, \ldots, v_n)$ in $\mathbb{R}^n$ or $\mathbb{C}^n$, the two vectors are equal if and only if $u_1 = v_1, u_2 = v_2, \ldots, u_n = v_n$. This is denoted as $\boldsymbol{u} = \boldsymbol{v}$.

▪ **Example 5.1** Find real numbers α and β such that $\left(\frac{\alpha+\beta}{\sqrt{2}}, \frac{\alpha-\beta}{\sqrt{2}}\right)$ is equal to $(1, 0)$.

Solution. For $\left(\frac{\alpha+\beta}{\sqrt{2}}, \frac{\alpha-\beta}{\sqrt{2}}\right) = (1, 0)$, α and β must satisfy:

$$\begin{cases} \frac{\alpha+\beta}{\sqrt{2}} = 1 \\ \frac{\alpha-\beta}{\sqrt{2}} = 0 \end{cases}.$$

Solving this linear system, we find:

$$\begin{cases} \alpha = \frac{\sqrt{2}}{2} \\ \beta = \frac{\sqrt{2}}{2} \end{cases}.$$

▪

Exercise 5.2 Find complex numbers α and β such that $\left(\frac{\alpha+\beta}{\sqrt{2}}, \frac{\alpha-\beta}{\sqrt{2}}\right)$ is equal to $(1, i)$.

5.2.2 Complex Conjugate

For each complex vector, there exists a unique complex conjugate defined as follows.

Definition 5.5 — Complex Conjugate of a Vector. The complex conjugate of a complex vector $\boldsymbol{v} = (v_1, v_2, \ldots, v_n)$ is denoted as $\boldsymbol{v}^*$, and is given by:

$$\boldsymbol{v}^* = (v_1^*, v_2^*, \ldots, v_n^*),$$

where v_i^* is the complex conjugate of v_i for $i = 1, 2, \ldots, n$.

▪ **Example 5.2** Find the complex conjugate of $\boldsymbol{v} = (1, i, 1 + i)$.

Solution. Taking the complex conjugate of each component of $\boldsymbol{v}$ gives $\boldsymbol{v}^* = (1, -i, 1 - i)$.

▪

Exercise 5.3 Judge whether the following statement is true or false: if $v = v^*$, then v must be a real vector.

5.2.3 Vector Addition

Addition of two vectors is defined component-wise.

> **Definition 5.6 — Vector Sum.** Given two vectors $u = (u_1, u_2, \ldots, u_n)$ and $v = (v_1, v_2, \ldots, v_n)$ in $\mathbb{R}^n$ or $\mathbb{C}^n$, their vector sum is:
>
> $$u + v = (u_1 + v_1, u_2 + v_2, \ldots, u_n + v_n).$$

> **Theorem 5.1** Vector addition satisfies the following properties:
>
> $$u + v = v + u \qquad \text{(Commutativity)}$$
> $$u + (v + w) = (u + v) + w \qquad \text{(Associativity)}$$
> $$v + 0 = v \qquad \text{(Identity)}.$$

The proof of these properties is straightforward using Defs. 5.3 and 5.6, and is omitted here for brevity.

In two- or three-dimensional spaces, vector addition can be interpreted geometrically.

> **Theorem 5.2** Given two geometric vectors u and v in 2D or 3D, the vector sum $u + v$ is represented by the diagonal of a parallelogram formed with u and v as adjacent sides (Fig. 5.5).

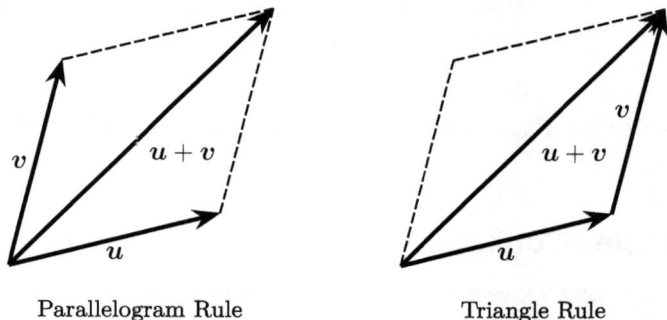

Parallelogram Rule   Triangle Rule

Figure 5.5: Parallelogram and Triangle Rules for Vector Addition

These methods, known as the **Parallelogram Rule** and **Triangle Rule**, are consistent with the algebraic definition of vector addition. As demonstrated by Fig. 5.6, the addition of two geometric vectors in 2D can be effectively carried out through component-wise summation, consistent with Def. 5.6 for algebraic representation.

Exercise 5.4 Given vectors $u = (1, 0)$ and $v = (0, 1)$, calculate $u + v$. Illustrate your result using geometric vectors.

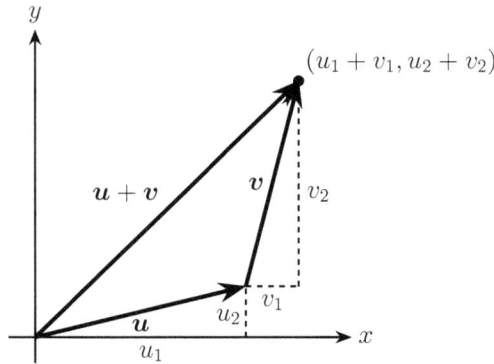

Figure 5.6: Vector Addition by Components on the x-y Plane

Exercise 5.5 Given vectors $u = (1, i)$ and $v = (-i, 1)$, calculate $u + v$. Can you visualize your result geometrically? Explain why.

5.2.4 Vector Subtraction

The negative of a geometric vector v, denoted as $-v$, is the vector with the same magnitude but opposite direction to v (Fig. 5.7). Generally, vector negation is defined as follows.

Definition 5.7 — Vector Negation. Given $v = (v_1, v_2, \ldots, v_n)$ in $\mathbb{R}^n$ or $\mathbb{C}^n$, the negative of v, denoted $-v$, is defined as:

$$-v = (-v_1, -v_2, \ldots, -v_n).$$

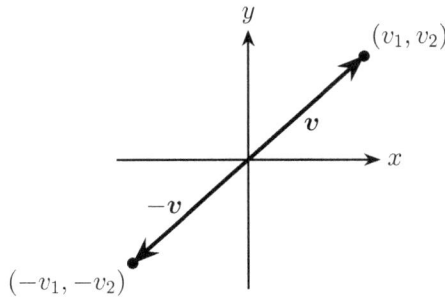

Figure 5.7: Vector Negation for Geometric Vectors

With vector negation defined, we can introduce vector subtraction as follows:

Definition 5.8 — Vector Subtraction. Given two vectors $u = (u_1, u_2, \ldots, u_n)$ and $v = (v_1, v_2, \ldots, v_n)$ in $\mathbb{R}^n$ or $\mathbb{C}^n$, the vector subtraction $u - v$ is the vector sum of u and $-v$:

$$u - v = u + (-v) = (u_1 - v_1, u_2 - v_2, \ldots, u_n - v_n). \tag{5.2}$$

For geometric vectors, $u - v$ can be calculated more directly by drawing a vector from the terminal point of v to the terminal point of u, as illustrated in Fig. 5.8.

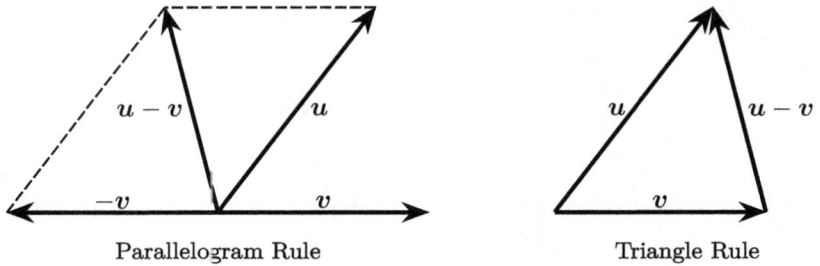

Parallelogram Rule   Triangle Rule

Figure 5.8: Parallelogram and Triangle Rules for Vector Subtraction

5.2.5 Scalar Multiplication of Vectors

A fundamental operation on vectors is to multiply them by scalars, which are real or complex numbers. This process, called **scalar multiplication**, scales the vector's magnitude and may reverse its direction depending on the scalar's sign. For example, multiplying a vector by 2 doubles its magnitude without changing its orientation, whereas multiplying by -2 doubles the magnitude and reverses its direction. Geometrically, the scalar product kv for a 2D vector $v = (v_1, v_2)$ results in a vector (kv_1, kv_2), as depicted in Fig. 5.9.

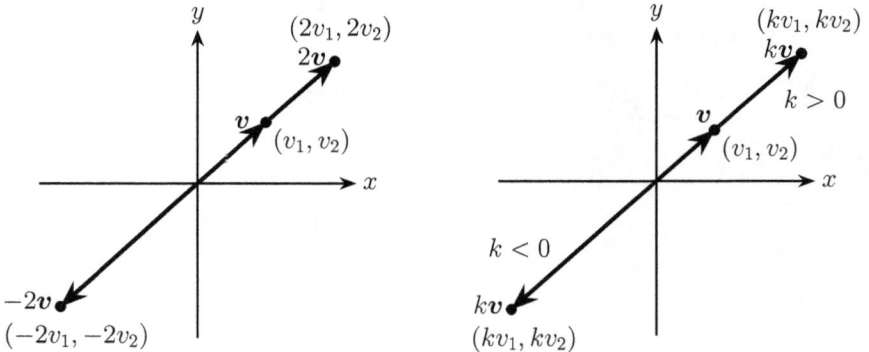

Figure 5.9: Scalar Multiplication of a Geometric Vector

Algebraically, scalar multiplication is defined as follows:

Definition 5.9 — Scalar-Vector Product. Given $v = (v_1, v_2, \ldots, v_n)$ in $\mathbb{R}^n$ or $\mathbb{C}^n$ and a scalar k, the scalar product of v by k, denoted as kv, is defined as:

$$kv = (kv_1, kv_2, \ldots, kv_n).$$

Exercise 5.6 Prove that if $kv = 0$ for a scalar k and a vector v in $\mathbb{R}^n$ or $\mathbb{C}^n$, then either $k = 0$ or $v = 0$.

ⓘ In Def. 5.9, scalars k are typically chosen from $\mathbb{R}$ when $v \in \mathbb{R}^n$, and from $\mathbb{C}$ when $v \in \mathbb{C}^n$. This choice aligns with the vector space under study, which will be explored further in the next section.

5.2.6 Linear Combinations

In linear algebra, the significance of addition and scalar multiplication lies in their role in forming linear combinations.

Definition 5.10 — Linear Combination. Consider a vector w and a set of r vectors $\{v_1, v_2, \ldots, v_r\}$ in $\mathbb{R}^n$ or $\mathbb{C}^n$. We say that w is a linear combination of $\{v_1, v_2, \ldots, v_r\}$ if w can be expressed as:

$$w = k_1 v_1 + k_2 v_2 + \cdots + k_r v_r = \sum_{i=1}^{r} k_i v_i, \tag{5.3}$$

where each k_i (for $i = 1, 2, \ldots, r$) is a scalar, referred to as the **coefficients** of the linear combination.

■ **Example 5.3** Given $v_1 = (1, 1)$ and $v_2 = (-1, 1)$ in $\mathbb{R}^2$, express $w = (-1, 5)$ as a linear combination of $\{v_1, v_2\}$ with real scalars.

Solution. To find the coefficients of the linear combination, we solve the vector equation $w = k_1 v_1 + k_2 v_2$, which gives:

$$k_1 v_1 + k_2 v_2 = k_1(1, 1) + k_2(-1, 1) = (k_1 - k_2, k_1 + k_2) = (-1, 5).$$

This results in two scalar equations:

$$\begin{cases} k_1 - k_2 &= -1, \\ k_1 + k_2 &= 5. \end{cases}$$

Solving this linear system gives $k_1 = 2$ and $k_2 = 3$. Thus,

$$w = 2v_1 + 3v_2.$$

■

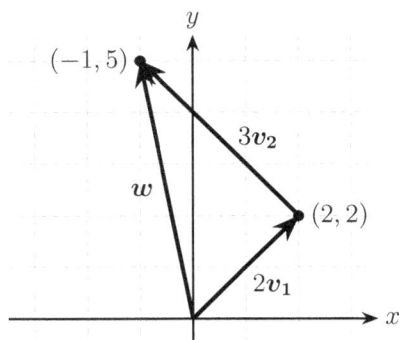

Figure 5.10: Linear Combination of Geometric Vectors

A geometric interpretation of Example 5.3 is shown in Fig. 5.10. Moving from the origin to $w = (-1, 5)$ can be visualized as a two-step translation: first move along v_1 by 2 units (with each unit defined as the length of v_1) to reach $(2, 2)$, then move along v_2 by 3 units to reach $(-1, 5)$.

■ **Example 5.4** Given $v_1 = (1, i)$ and $v_2 = (i, 1)$ in $\mathbb{C}^2$, express $w = (1, -1)$ as a linear combination of $\{v_1, v_2\}$ with complex scalars.

Solution. Using the same approach as in Example 5.3, we solve:

$$w = k_1 v_1 + k_2 v_2, \quad \text{or} \quad \begin{cases} k_1 + ik_2 &= 1, \\ ik_1 + k_2 &= -1. \end{cases}$$

Solving this linear system gives $k_1 = \frac{1+i}{2}$ and $k_2 = \frac{-1-i}{2}$. Therefore,

$$w = \frac{1+i}{2} v_1 + \frac{-1-i}{2} v_2.$$

■

■ **Example 5.5** Express an arbitrary vector $v = (v_1, v_2)$ in $\mathbb{C}^2$ as a linear combination of $(1, i)$ and $(i, 1)$:

$$v = \alpha(1, i) + \beta(i, 1).$$

Find the complex scalars α and β.

Solution. By matching components, we obtain:

$$\begin{cases} v_1 &= \alpha + i\beta, \\ v_2 &= i\alpha + \beta. \end{cases}$$

Solving for β, we find:

$$v_1 + iv_2 = (\alpha + i\beta) + i^2\alpha + i\beta = 2i\beta \quad \Rightarrow \quad \beta = \frac{v_1 + iv_2}{2i} = \frac{v_2 - iv_1}{2}.$$

Similarly, for α:

$$v_1 - iv_2 = (\alpha + i\beta) - i^2\alpha - i\beta = 2\alpha \quad \Rightarrow \quad \alpha = \frac{v_1 - iv_2}{2}.$$

■

ⓘ In Examples 5.3 to 5.5, there is a unique solution for the linear combination. However, in general, a linear combination may have no solution or infinitely many solutions. This will be discussed further in § 5.4.

Exercise 5.7 Given vectors $v_1 = (1, 1)$ and $v_2 = (2, 3)$, demonstrate that any vector (a, b) in $\mathbb{R}^2$ can be expressed as a linear combination of $\{v_1, v_2\}$ in exactly one way. Can you also use geometric vectors to analyze this?

Exercise 5.8 Express $\left(\frac{1}{\sqrt{2}}, \frac{i}{\sqrt{2}}\right)$ as a linear combination of $\left(\frac{1}{\sqrt{2}}, \frac{1}{\sqrt{2}}\right)$ and $\left(\frac{1}{\sqrt{2}}, \frac{-1}{\sqrt{2}}\right)$ using complex scalars.

In linear algebra and quantum computing, the concept of linear combinations is a cornerstone. It is deeply intertwined with key topics such as vector spaces, linear independence, basis, coordinates, dimensionality, and matrix transformations. We will revisit and explore these ideas in greater depth in subsequent sections and chapters.

5.3 Vector Spaces, Subspaces, and Span

Having laid the groundwork with basic vector operations, we now focus on vector spaces. A vector space is a collection of vectors—objects that can be added together and multiplied by scalars—satisfying specific axioms. Vector spaces form a foundational concept in linear algebra, providing a structured way to handle and manipulate multiple dimensions.

5.3.1 Real and Complex Vector Spaces

A vector space is characterized by four key components: (i) vectors, (ii) scalars, (iii) vector addition, and (iv) scalar multiplication. The following definition abstracts these components, providing a general framework for vector spaces.

Definition 5.11 — Vector Space. A vector space (or linear space) is a set V together with a scalar field F (either $\mathbb{R}$ or $\mathbb{C}$) such that the following two operations are defined:

(a) Vector addition: For any $u, v \in V$, the sum $u + v \in V$.

(b) Scalar multiplication: For any $c \in F$ and $v \in V$, the product $cv \in V$.

These operations must satisfy the following axioms for all $u, v, w \in V$ and scalars $a, b \in F$:

(c) Associativity of addition: $(u + v) + w = u + (v + w)$.

(d) Commutativity of addition: $u + v = v + u$.

(e) Additive identity: There exists an element $0 \in V$ such that $v + 0 = v$ for all $v \in V$.

(f) Additive inverse: For each $v \in V$, there exists $-v \in V$ such that $v + (-v) = 0$.

(g) Distributivity of scalar multiplication over vector addition: $a(u + v) = au + av$.

(h) Distributivity of scalar multiplication over scalar addition: $(a + b)v = av + bv$.

(i) Compatibility of scalar multiplication: $(ab)v = a(bv)$.

(j) Multiplicative identity: There exists $1 \in F$ such that $1v = v$ for all $v \in V$.

---

i

A more succinct, albeit less elementary, definition of a vector space replaces properties (a) and (c)-(f) with the assertion that $(V, +)$ forms an Abelian group (Def. 4.19). Scalar multiplication can alternatively be conceptualized as a binary operation $F \times V \to V$ that inherently satisfies property (b), making its explicit assertion redundant.

■ **Example 5.6** Prove that $\mathbb{R}^n$ satisfies property (d) in Def. 5.11.

Proof. Let $u = (u_1, u_2, \ldots, u_n)$ and $v = (v_1, v_2, \ldots, v_n)$ be two vectors in $\mathbb{R}^n$. Using the definition of vector addition:

$$u + v = (u_1 + v_1, u_2 + v_2, \ldots, u_n + v_n),$$
$$v + u = (v_1 + u_1, v_2 + u_2, \ldots, v_n + u_n).$$

By the commutativity of real numbers, $u_i + v_i = v_i + u_i$ for each i. Thus, $\boldsymbol{u} + \boldsymbol{v} = \boldsymbol{v} + \boldsymbol{u}$, proving commutativity of addition. ∎

■ **Example 5.7** Prove that $\mathbb{R}^n$ satisfies properties (a) and (b) in Def. 5.11 over the scalar field $\mathbb{R}$.

Proof. For vector addition, $\boldsymbol{u} + \boldsymbol{v} = (u_1 + v_1, u_2 + v_2, \ldots, u_n + v_n)$ results in another vector in $\mathbb{R}^n$. Similarly, for scalar multiplication, $k\boldsymbol{v} = (kv_1, kv_2, \ldots, kv_n)$ also belongs to $\mathbb{R}^n$ for $k \in \mathbb{R}$. Hence, closure under both operations is satisfied. ∎

The axioms of Def. 5.11 are central to the structure of vector spaces. Properties (a) and (b) are referred to as **closure under addition** and **closure under scalar multiplication**, respectively. Together, they guarantee that any linear combination of vectors in a vector space remains within the vector space.

Exercise 5.9 Prove that $\mathbb{R}^n$ satisfies properties (c)-(j) in Def. 5.11 over the field $\mathbb{R}$.

Exercise 5.10 Show that $\mathbb{R}^n$ does not satisfy all properties of Def. 5.11 over the field $\mathbb{C}$. Identify which property is violated.

For $\mathbb{R}^n$ under the standard operations of vector addition (Def. 5.6) and scalar multiplication (Def. 5.9), all axioms in Def. 5.11 are satisfied. Therefore, $\mathbb{R}^n$ constitutes a vector space over $\mathbb{R}$. In this text, unless otherwise stated, discussions about $\mathbb{R}^n$ assume the scalar field $\mathbb{R}$.

Exercise 5.11 Over the scalar field $\mathbb{C}$, prove that $\mathbb{C}^n$ satisfies all axioms of Def. 5.11, making it a vector space.

> **Definition 5.12 — Real and Complex Vector Spaces.** A vector space is called a real vector space if its scalar field is $\mathbb{R}$, and a complex vector space if its scalar field is $\mathbb{C}$.

ⓘ Note that the classification of a vector space as real or complex is determined by the scalar field F, rather than the specific properties of the vector set V.

In principle, F can represent any field as defined in Def. 4.26, giving rise to the concept of F-vector spaces. However, in this text, we restrict our discussion to vector spaces over the fields $\mathbb{R}$ or $\mathbb{C}$.

As demonstrated in Exercise 5.11, $\mathbb{C}^n$ is a vector space over $\mathbb{C}$. Since $\mathbb{R} \subset \mathbb{C}$, $\mathbb{C}^n$ is also a vector space over $\mathbb{R}$. According to Def. 5.12, $\mathbb{C}^n$ can function as either a complex vector space or a real vector space, depending on the choice of the scalar field.

In the context of quantum computing, $\mathbb{C}^n$ is consistently treated as a complex vector space. We will adopt this assumption throughout, unless otherwise specified.

5.3.2 ✳ Further Discussion on Vector Space Definition

The properties (c)-(j) delineated in Def. 5.11 may initially appear trivial, especially since their validity seems self-evident for numbers and ordered pairs. This resemblance may lead to questions about their necessity. However, these properties are critical—they provide the logical foundation needed to deduce all other "desirable" properties of numbers and vectors that we often take for granted, such as $0v = 0$ or $(-1)v = -v$. The following example highlights this importance.

■ **Example 5.8** Let V be a vector space over the field F. Given a vector v in V and a scalar $k \in F$, use the axioms in Def. 5.11 to prove that $k0 = 0$.

Proof. From property (g), we have:

$$k(0 + 0) = k0 + k0.$$

From property (e), we know $0 + 0 = 0$, so:

$$k0 = k0 + k0.$$

By property (f), there exists an additive inverse $-k0 \in V$. Adding $-k0$ to both sides gives:

$$k0 + (-k0) = (k0 + k0) + (-k0).$$

The left-hand side simplifies to 0, and using properties (d) and (f) on the right-hand side:

$$0 = k0 + [k0 + (-k0)] = k0 + 0 = k0.$$

This completes the proof. Note that properties (c) and (e) are also used.   ■

To deepen understanding, readers are encouraged to engage with exercises (Problem 5.8) exploring these properties without predefined rules for vector addition and scalar multiplication. Such exploration provides valuable insights into the essential nature of these axioms.

5.3.3   Subspaces

A subspace is a subset of a vector space that forms a vector space under the same operations. Not every subset of a vector space is a subspace. For example, the set of vectors in $\mathbb{R}^2$ with a positive x-component satisfies closure under addition but violates closure under scalar multiplication (why?). Subspaces are rigorously defined as follows.

Definition 5.13 — Subspace. A non-empty subset W of a vector space V is called a subspace of V if W is a vector space under the same scalar field F and the same operations of vector addition and scalar multiplication as in V.

In Def. 5.11, properties (c)-(d) and (g)-(j) relate directly to the scalar field F and the defined operations. These properties are inherently preserved in subsets of V with the same operations, so they need not be re-verified. Additionally, properties (e)-(f) can be derived from the remaining properties (Problem 5.9). To verify whether a subset is a subspace, it suffices to check properties (a) and (b), as summarized below.

Theorem 5.3 If W is a non-empty subset of a vector space V, then W is a subspace of V if and only if W satisfies closure under addition and closure under scalar multiplication.

■ **Example 5.9** Prove that the set containing only the zero vector, $\{0\}$, is a subspace of $\mathbb{R}^n$ (or $\mathbb{C}^n$).

Proof. The given set contains a single element, 0. For any scalar k, we have:

$$0 + 0 = 0, \quad k0 = 0.$$

Thus, closure under addition and scalar multiplication are satisfied. By Theorem 5.3, $\{0\}$ is a subspace. ∎

Definition 5.14 — Zero Subspace. The subset $W = \{0\}$ is called the zero subspace of a vector space V, where 0 is the zero vector in V.

■ **Example 5.10** Show that $S = \{x \mid x = (a, a, a), a \in \mathbb{R}\}$ is a subspace of $\mathbb{R}^3$.

Proof. Let $v = (a, a, a)$ and $w = (b, b, b)$ be vectors in S, and let $k \in \mathbb{R}$. Then:

$$v + w = (a + b, a + b, a + b),$$
$$kv = (ka, ka, ka),$$

which are still in S. Therefore, S satisfies closure under addition and scalar multiplication, making it a subspace by Theorem 5.3. ∎

Exercise 5.12 Show that $S = \{x \mid x = (a, b, a + b), a, b \in \mathbb{R}\}$ is a subspace of $\mathbb{R}^3$.

Exercise 5.13 Show that $S = \{x \mid x = (a, a, 1), a \in \mathbb{R}\}$ is not a subspace of $\mathbb{R}^3$.

To determine whether a subset is a subspace, the key criteria are closure under addition and scalar multiplication. These properties ensure that any linear combination of vectors in the subset also belongs to the subset. For example, a line through the origin in $\mathbb{R}^2$ is a subspace because the linear combination of any two vectors on this line remains on the line. In contrast, a line that does not pass through the origin is not a subspace, as illustrated in Fig. 5.11.

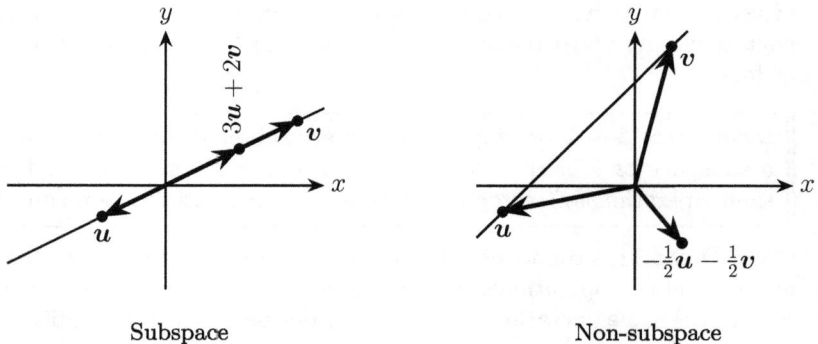

Subspace   Non-subspace

Figure 5.11: Subspace and Non-subspace in $\mathbb{R}^2$

Recall that the identity of a vector is uniquely defined by the coordinates of its terminal point, assuming its initial point is at the origin. In this framework, when we refer to a line or a plane as representing a set of vectors, we are specifically discussing vectors that originate from the origin and terminate at points located on that line or plane.

| **Exercise 5.14** Offer a geometric interpretation of Exercises 5.12 and 5.13.

■ **Example 5.11** Show that $S = \{x \mid x = (z, z^*), z \in \mathbb{C}\}$ is not a subspace of $\mathbb{C}^2$.

Proof. Let $v = (z, z^*)$ be a vector in S and $k \in \mathbb{C}$. Then:

$$kv = (kz, kz^*).$$

When k has a non-zero imaginary part, we find:

$$k \neq k^* \quad \Rightarrow \quad (kz)^* = k^* z^* \neq kz^* \quad \Rightarrow \quad kv \notin S.$$

Therefore, S violates closure under scalar multiplication and is not a subspace of $\mathbb{C}^2$.
■

| **Exercise 5.15** In Example 5.11, we treat $\mathbb{C}^2$ as a complex vector space. Show that if we treat $\mathbb{C}^2$ as a real vector space, then S is a subspace of $\mathbb{C}^2$.

| **Exercise 5.16** Given a subset of $\mathbb{C}^2$, $S = \{x \mid x = (a + ib, b + ia), a, b \in \mathbb{R}\}$. Is S a subspace of $\mathbb{C}^2$?

5.3.4 Span

Vector spaces and their subspaces share the crucial property of closure under arbitrary linear combinations. This property is fundamental to the development of many key concepts in linear algebra, such as basis, dimension, and linear transformations (represented by matrices). As observed, not every subset of a vector space is a subspace. In this subsection, we will explore how subspaces can be constructed by considering the span of vectors from an existing vector space.

Theorem 5.4 If $S = \{v_1, v_2, \ldots, v_r\}$ is a set of vectors from a vector space V, then all possible linear combinations of the vectors in S form a subspace of V.

Proof. Let W be the set of all possible linear combinations of the vectors in S, and let $a, b \in W$. Then, a and b can be expressed as:

$$a = a_1 v_1 + a_2 v_2 + \cdots + a_r v_r,$$
$$b = b_1 v_1 + b_2 v_2 + \cdots + b_r v_r,$$

where $a_i, b_i \in F$ for all i. Then, for any scalar $k \in F$, we compute:

$$a + b = (a_1 + b_1)v_1 + (a_2 + b_2)v_2 + \cdots + (a_r + b_r)v_r,$$
$$ka = (ka_1)v_1 + (ka_2)v_2 + \cdots + (ka_r)v_r.$$

Both results are linear combinations of the vectors in S, which implies closure under addition and scalar multiplication. Therefore, W is a subspace of V by Theorem 5.3.
□

Definition 5.15 — Span. Given a set $S = \{v_1, v_2, \ldots, v_r\}$ of vectors from a vector space V, and let W be the subspace of V that contains all possible linear combinations of the vectors in S. Then W is termed the span of the vectors in S, or simply the span of S. This relationship is mathematically expressed as:

$$W = \text{span}(S) = \text{span}\{v_1, v_2, \ldots, v_r\}.$$

Accordingly, when W is the span of S, we say that S **spans** W.

■ **Example 5.12** Prove that $\hat{\imath} = (1,0)$ and $\hat{\jmath} = (0,1)$ span $\mathbb{R}^2$.

Proof. Consider a general vector $(a,b) \in \mathbb{R}^2$, where $a, b \in \mathbb{R}$. We can express this vector as:
$$(a,b) = a(1,0) + b(0,1) = a\hat{\imath} + b\hat{\jmath}.$$

Since every vector in $\mathbb{R}^2$ can be written as a linear combination of $\hat{\imath}$ and $\hat{\jmath}$, we conclude that $\{\hat{\imath}, \hat{\jmath}\}$ spans $\mathbb{R}^2$.   ■

Exercise 5.17 Prove that $\hat{\imath} = (1,0)$ and $\hat{\jmath} = (0,1)$ span $\mathbb{C}^2$.

i | $\hat{\imath} = (1,0)$ and $\hat{\jmath} = (0,1)$ are referred to as the **standard basis vectors** for $\mathbb{R}^2$ or $\mathbb{C}^2$. A detailed discussion of basis vectors will follow in § 5.4.

■ **Example 5.13** Prove that $v_1 = (1,1)$ and $v_2 = (1,-1)$ span $\mathbb{R}^2$ or $\mathbb{C}^2$.

Proof. The standard basis vectors $\hat{\imath} = (1,0)$ and $\hat{\jmath} = (0,1)$ can be written as linear combinations of v_1 and v_2:

$$(1,0) = \frac{1}{2}(1,1) + \frac{1}{2}(1,-1),$$

$$(0,1) = \frac{1}{2}(1,1) - \frac{1}{2}(1,-1).$$

Therefore, any vector $(a,b) \in \mathbb{R}^2$ or $\mathbb{C}^2$ can be expressed as:

$$(a,b) = a(1,0) + b(0,1)$$
$$= a\left[\frac{1}{2}(1,1) + \frac{1}{2}(1,-1)\right] + b\left[\frac{1}{2}(1,1) - \frac{1}{2}(1,-1)\right]$$
$$= \frac{a+b}{2}(1,1) + \frac{a-b}{2}(1,-1)$$
$$= \frac{a+b}{2}v_1 + \frac{a-b}{2}v_2.$$

This proves that $\{v_1, v_2\}$ spans $\mathbb{R}^2$ or $\mathbb{C}^2$.   ■

Exercise 5.18 Prove that $\{(1,0,0), (1,1,0), (1,1,1)\}$ spans $\mathbb{R}^3$ or $\mathbb{C}^3$.

5.4 Linear Independence, Basis and Dimension

Basis is a fundamental concept in linear algebra associated with vector spaces. To describe all vectors and their operations in a vector space, one needs to choose a basis. For example, $\hat{\imath} = (1,0)$ and $\hat{\jmath} = (0,1)$ form a basis for $\mathbb{R}^2$. Any vector in this space can be expressed *uniquely* as a linear combination of these basis vectors:

$$(a,b) = a(1,0) + b(0,1).$$

This linear combination introduces the concepts of coordinates and dimension. In this section, we will rigorously define the basis and related concepts in preparation for studying orthonormal bases in the forthcoming chapters, which are among the most foundational mathematical tools in quantum computing.

5.4.1 Linear Independence

A prerequisite for defining basis is establishing the concept of linear independence. The question to be raised here is whether any "redundant" vector in a set can be removed without affecting the span. For instance, the span of $\{(1,0), (0,1)\}$ over a complex field is the whole $\mathbb{C}^2$ since any complex vector (α, β) can be expressed as:

$$(\alpha, \beta) = \alpha(1,0) + \beta(0,1).$$

What happens if we add another vector, such as $(1,1)$, to the set? The span of $\{(1,0), (0,1), (1,1)\}$ remains $\mathbb{C}^2$ because adding this vector does not extend the span. Conversely, $(1,1)$ can be removed without affecting the span:

$$\text{span}(\{(1,0), (0,1), (1,1)\}) = \text{span}(\{(1,0), (0,1)\}).$$

This is because $(1,1)$ can be expressed as a linear combination of $(1,0)$ and $(0,1)$:

$$(1,1) = (1,0) + (0,1).$$

Thus, any linear combination of $\{(1,0), (0,1), (1,1)\}$ can also be written as a linear combination of $\{(1,0), (0,1)\}$:

$$\alpha(1,0) + \beta(0,1) + \gamma(1,1) = (\alpha + \gamma)(1,0) + (\beta + \gamma)(0,1).$$

Exercise 5.19 Show that we can remove $(1,0)$ or $(0,1)$ from $\{(1,0), (0,1), (1,1)\}$ without affecting the span.

Given a set of vectors from a vector space, if removing any vector from the set would result in a change in its span, the set is linearly independent; otherwise, it is linearly dependent, as defined below.

Definition 5.16 — Linear Dependence. Let $S = \{v_1, v_2, \ldots, v_n\}$ be a set of vectors from a vector space V. The set S is said to be linearly dependent if there exist scalars $c_1, c_2, \ldots, c_n$, not all zero, such that:

$$c_1 v_1 + c_2 v_2 + \cdots + c_n v_n = 0. \tag{5.4}$$

If no such scalars exist, the set is linearly independent.

Definition 5.16 states that if any nonzero coefficients c_i exist, the corresponding vector v_i can be expressed as a linear combination of the other vectors in S:

$$v_i = \sum_{\substack{j=1 \\ j \neq i}}^{n} -\frac{c_j}{c_i} v_j.$$

Thus, removing v_i does not change the span, indicating that S is linearly dependent.

As a special case, a set containing only one vector $\{v_1\}$ is defined to be linearly independent, except when $v_1 = 0$.

■ **Example 5.14** Show that $S = \{(1,2,3), (4,5,6), (7,8,9)\}$ is linearly dependent in $\mathbb{R}^3$.

Proof. To use Def. 5.16, solve the vector equation:

$$k_1(1, 2, 3) + k_2(4, 5, 6) + k_3(7, 8, 9) = (0, 0, 0). \tag{5.5}$$

Rearranging the equation in components results in a linear system:

$$\begin{cases} k_1 + 4k_2 + 7k_3 & = 0, \\ 2k_1 + 5k_2 + 8k_3 & = 0, \\ 3k_1 + 6k_2 + 9k_3 & = 0. \end{cases}$$

Solving this system yields a parametric solution:

$$(k_1, k_2, k_3) = (t, -2t, t) \quad \text{for any } t \in \mathbb{R}.$$

This shows that nonzero solutions, such as $k_1 = 1, k_2 = -2, k_3 = 1$, exist for Eq. 5.5. Thus, S is linearly dependent according to Def. 5.16. ∎

Exercise 5.20 Express $(7, 8, 9)$ as a linear combination of $\{(1, 2, 3), (4, 5, 6)\}$ based on Example 5.14.

Exercise 5.21 Given $\{(1, 0, 1), (1, 1, 2), (1, -1, 0)\}$ in $\mathbb{R}^3$, determine whether it is linearly independent or dependent.

5.4.2 Basis

A basis is a minimal set of vectors that spans a vector space. Its rigorous definition incorporates linear independence.

> **Definition 5.17 — Basis.** A set of vectors $S = \{v_1, v_2, \ldots, v_n\}$ from a vector space V is a basis of V if S spans V and is linearly independent.

Definition 5.17 requires two properties for a basis: (i) spanning the vector space V, and (ii) linear independence. Together, these ensure the following:

> **Theorem 5.5 — Uniqueness of Basis Representation.** If S is a basis of a vector space V, then every vector $w \in V$ has a unique representation as a linear combination of S.

Proof. Let $w \in V$ and $S = \{v_1, v_2, \ldots, v_n\}$. Since S spans V, we can express w as:

$$w = k_1 v_1 + k_2 v_2 + \cdots + k_n v_n.$$

Assume another representation exists:

$$w = m_1 v_1 + m_2 v_2 + \cdots + m_n v_n.$$

Subtracting gives:

$$0 = (k_1 - m_1)v_1 + \cdots + (k_n - m_n)v_n.$$

Since S is linearly independent (Def. 5.16), we must have:

$$k_1 - m_1 = k_2 - m_2 = \cdots = k_n - m_n = 0 \quad \Rightarrow \quad k_i = m_i \text{ for all } i.$$

Thus, the representation is unique. ☐

▪ **Example 5.15** Show that $S = \{(1,0,0),(1,1,0),(1,1,1)\}$ is a basis of $\mathbb{R}^3$.

Proof. For a general vector $v = (a,b,c) \in \mathbb{R}^3$, write:

$$(a,b,c) = k_1(1,0,0) + k_2(1,1,0) + k_3(1,1,1),$$

where $k_1, k_2, k_3 \in \mathbb{R}$. This leads to the linear system:

$$\begin{cases} k_1 + k_2 + k_3 = a \\ \qquad k_2 + k_3 = b \;. \\ \qquad\qquad k_3 = c \end{cases}$$

Solving, we find:

$$k_1 = a - b, \quad k_2 = b - c, \quad k_3 = c.$$

Since the system has a unique solution for all $(a,b,c) \in \mathbb{R}^3$, S spans $\mathbb{R}^3$. Further, the only solution to:

$$k_1(1,0,0) + k_2(1,1,0) + k_3(1,1,1) = (0,0,0)$$

is $k_1 = k_2 = k_3 = 0$, proving linear independence. Hence, S is a basis.   ▪

The example above also demonstrates the converse of Theorem 5.5. In fact, if we can establish that every vector in a vector space V can be uniquely expressed as a linear combination of the vectors in S, then S satisfies both the spanning property and linear independence, making it a basis of V.

▎**Exercise 5.22** Show that $S = \{(i,0,0),(1,i,0),(1,1,i)\}$ is a basis of $\mathbb{C}^3$.

5.4.3 Coordinates and Standard Basis

1 Coordinates

Following Theorem 5.5, there exists a unique set of coefficients when expressing any vector in terms of a basis for a given vector space. These coefficients form a vector that is defined as a coordinate.

Definition 5.18 — Coordinate. Express w in terms of a basis $S = \{v_1, v_2, \ldots, v_n\}$ from a vector space V over a field $\mathbb{R}$ (or $\mathbb{C}$):

$$w = k_1 v_1 + k_2 v_2 + \cdots + k_n v_n.$$

The unique vector $(k_1, k_2, \ldots, k_n)$ formed from the scalar coefficients in $\mathbb{R}^n$ (or $\mathbb{C}^n$) is said to be the coordinate vector, or simply coordinate of w relative to S, denoted as:

$$(w)_S = (k_1, k_2, \ldots, k_n). \tag{5.6}$$

To understand basis and coordinates in a more intuitive way, consider $\mathbb{R}^2$ and its standard basis $\hat{i} = (1,0)$ and $\hat{j} = (0,1)$. Express any vector (a,b) in $\mathbb{R}^2$ in terms of the standard basis gives:

$$(a,b) = a(1,0) + b(0,1), \tag{5.7}$$

hence its coordinate relative to the standard basis is also (a,b):

$$(a,b)_{\{\hat{i},\hat{j}\}} = (a,b).$$

This is because when we define coordinates in a Cartesian coordinate system, we actually use the unit vectors along the x-axis ($\hat{i}$) and the y-axis ($\hat{j}$) as our measurement basis. Therefore, the Cartesian coordinates used in analytic geometry are precisely the coordinates relative to the standard basis in $\mathbb{R}^2$ (or similarly in $\mathbb{R}^3$).

Figure 5.12 illustrates Eq. 5.7 using 2D geometric vectors. To reach (a, b) from the origin, one can move along the x-axis by a units and then along the y-axis by b units (Fig. 5.12a). Since a, b are arbitrary, any point on the x-y plane can be reached by moving along the x-axis and y-axis alone, reflecting the fact that $\{\hat{i}, \hat{j}\}$ spans $\mathbb{R}^2$. These axes, representing two *principal directions*, define the standard basis.

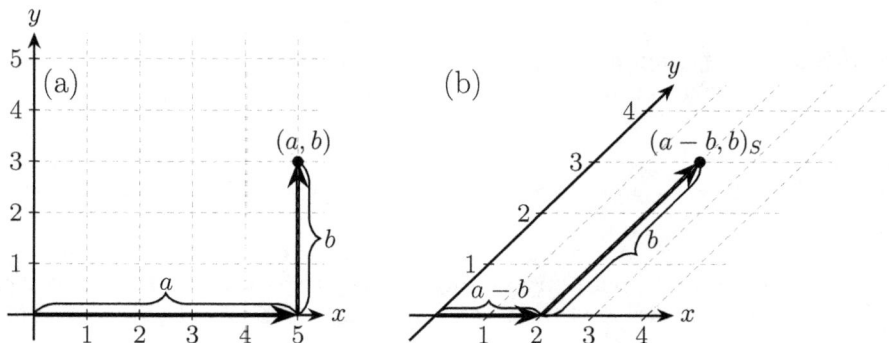

Figure 5.12: Orthonormal and Non-orthonormal Basis in $\mathbb{R}^2$

Alternatively, different principal directions can be chosen to span the x-y plane. For example, the basis $S = \{(1, 0), (1, 1)\}$ allows any vector (a, b) to be expressed as:

$$(a, b) = (a - b)(1, 0) + b(1, 1),$$

hence the coordinate of (a, b) relative to S is:

$$(a, b)_S = (a - b, b). \tag{5.8}$$

Geometrically, to reach (a, b) from the origin, one moves along the x-axis by $(a - b)$ units and then along $(1, 1)$ by b "units" (where one "unit" is the length of $(1, 1)$), as illustrated in Fig. 5.12b. This basis is *non-orthonormal* since the basis vectors are neither orthogonal nor unit vectors.

> **Exercise 5.23** Calculate the coordinate of a general vector (a, b) in $\mathbb{R}^2$ relative to the basis $S = \{(1, -1), (1, 1)\}$. Explain your result with geometric vectors.

2   The Standard Basis

In general, the standard basis for $\mathbb{R}^n$ (or $\mathbb{C}^n$) is defined as:

$$e_1 = (1, 0, \ldots, 0), \ e_2 = (0, 1, \ldots, 0), \ \ldots, \ e_n = (0, 0, \ldots, 1),$$

where each e_i (for $i = 1, 2, \ldots, n$) has the ith component as one and all others as zero. A general vector $v = (v_1, v_2, \ldots, v_n)$ can be *uniquely* expressed as:

$$v = v_1 e_1 + v_2 e_2 + \cdots + v_n e_n = \sum_{i=1}^{n} v_i e_i.$$

Thus, $\{e_1, e_2, \ldots, e_n\}$ is the **standard basis** of $\mathbb{R}^n$ (or $\mathbb{C}^n$). By default, the coordinate of a vector in $\mathbb{R}^n$ or $\mathbb{C}^n$ is relative to the standard basis:

$$(v_1, v_2, \ldots, v_n)_{\{e_1, e_2, \ldots, e_n\}} = (v_1, v_2, \ldots, v_n). \tag{5.9}$$

The standard basis is fundamental in quantum mechanics and quantum computing. Dirac notation, introduced later, will further simplify working with standard basis vectors.

5.4.4 Dimension

1 Dimension of a Finite-dimensional Vector Space

Intuitively, we recognize a line as a one-dimensional entity, a plane as two-dimensional, and the space that we inhabit as three-dimensional. In the field of linear algebra, the concept of dimension is rigorously defined for finite-dimensional vector spaces through the following theorem.

> **Theorem 5.6** For a finite-dimensional vector space, all bases possess the same number of vectors.

We omit a formal proof of Theorem 5.6 as it requires a background in linear systems and matrix algebra. To develop an intuitive understanding, consider any basis $\{v_1, v_2, \ldots, v_n\}$ of a finite-dimensional vector space V. Adding an extra vector to this set would render it linearly dependent, as the new vector could be expressed as a linear combination of the original vectors. Conversely, removing any vector from the set would reduce the span, thereby failing to cover the entire vector space V. Consequently, if different bases of V contained different numbers of vectors, a contradiction would arise regarding their linear independence.

Theorem 5.6 confirms that the dimension of a vector space can be consistently defined as follows.

> **Definition 5.19 — Dimension.** The dimension of a finite-dimensional vector space V is the number of vectors in a basis, denoted as $\dim(V)$.

> (i) Since $\{0\}$ is considered a linearly dependent set, there is no non-empty linearly independent set of vectors for the zero subspace. Therefore, the zero subspace of a vector space is defined to be **zero-dimensional**. In $\mathbb{R}^n$, the zero subspace corresponds to the single point of origin in Euclidean space.

It is evident that $\dim(\mathbb{R}^n) = n$ since there are n standard basis vectors. However, for $\mathbb{C}^n$, its dimension depends on the chosen scalar field, as shown in the following example.

▪ **Example 5.16** Show that $\dim(\mathbb{C}^n) = n$ over $\mathbb{C}$ and $\dim(\mathbb{C}^n) = 2n$ over $\mathbb{R}$.

Proof. Consider $\mathbb{C}^n$ as a complex vector space. A basis for $\mathbb{C}^n$ is the standard basis vectors $\{e_1, e_2, \ldots, e_n\}$ introduced in § 5.4.3.2, because any complex vector $v = (v_1, v_2, \ldots, v_n)$ $(v_1, v_2, \ldots, v_n \in \mathbb{C})$ can be uniquely expressed as a linear combination of $\{e_1, e_2, \ldots, e_n\}$:

$$v = v_1 e_1 + v_2 e_2 + \cdots + v_n e_n. \tag{5.10}$$

This verifies that $\dim(\mathbb{C}^n) = n$ over $\mathbb{C}$.

On the other hand, if we consider $\mathbb{C}^n$ as a real vector space, $\{e_1, e_2, \ldots, e_n\}$ is insufficient as a basis since Eq. 5.10 requires complex coefficients. To span $\mathbb{C}^n$ with real coefficients, we need an additional n basis vectors:

$$i_1 = (i, 0, \ldots, 0), \quad i_2 = (0, i, \ldots, 0), \quad \ldots, \quad i_n = (0, 0, \ldots, i).$$

Given an arbitrary vector $v = (a_1 + ib_1, a_2 + ib_2, \ldots, a_n + ib_n)$ in $\mathbb{C}^n$ $(a_i, b_i \in \mathbb{R}$ for $i = 1, 2, \ldots, n)$, it can be uniquely expressed as:

$$v = a_1 e_1 + a_2 e_2 + \cdots + a_n e_n + b_1 i_1 + b_2 i_2 + \cdots + b_n i_n.$$

Therefore, $\{e_1, e_2, \ldots, e_n, i_1, i_2, \ldots, i_n\}$ is a basis for $\mathbb{C}^n$ over $\mathbb{R}$. There are a total of $2n$ basis vectors, thus $\dim(\mathbb{C}^n) = 2n$ over $\mathbb{R}$.   ∎

In quantum computing, $\mathbb{C}^n$ is treated as a complex vector space, so it has a dimension of n and is equipped with the standard basis.

2   Dimension of Subspaces

Subspaces can be understood as lower-dimensional constructs of a vector space. For example, there are 1D lines and 2D planes in a three-dimensional space. Algebraically, we can apply Def. 5.19 to find the dimension of subspaces.

■ **Example 5.17** Show that in $\mathbb{R}^3$, the set $S = \{(x, y, z) \mid x + y - z = 0\}$ constitutes a two-dimensional plane.

Proof. It can be shown that S is a subspace of $\mathbb{R}^3$ (Problem 5.10c). For any vector (x, y, z) in S, $z = x + y$ must hold. Parameterizing x and y as $x = t$ and $y = s$, we can express (x, y, z) as:

$$(x, y, z) = (t, s, t + s) = t(1, 0, 1) + s(0, 1, 1).$$

This shows that any vector in S is a linear combination of $\{(1, 0, 1), (0, 1, 1)\}$, which serves as a basis of S. Therefore, $\dim(S) = 2$.   ∎

In Example 5.17, t and s can be understood as two "free parameters" to describe a vector in S. Following Def. 5.18, they serve as the "coordinates" (or principal directions) of S, respectively associated with the basis vectors $(1, 0, 1)$ and $(0, 1, 1)$. A geometric interpretation of Example 5.17 is given in Fig. 5.13.

In Fig. 5.13, the plane $x + y - z = 0$ is spanned by the two basis vectors $(1, 0, 1)$ and $(0, 1, 1)$. These vectors are linearly independent, confirming that they form a basis for the plane. By linear combinations of these basis vectors, all vectors residing on the plane can be accessed. The illustration uses a dashed representation of the partial z-axis to indicate that the plane passes through the origin and intersects in front of the z-axis, relative to the perspective adopted in the figure.

Exercise 5.24 Find the dimension and a basis of $S = \{(x, y, z, w) \mid ax + by + cz - w = 0\}$ as a subspace of $\mathbb{R}^4$, where a, b, c are given real numbers.

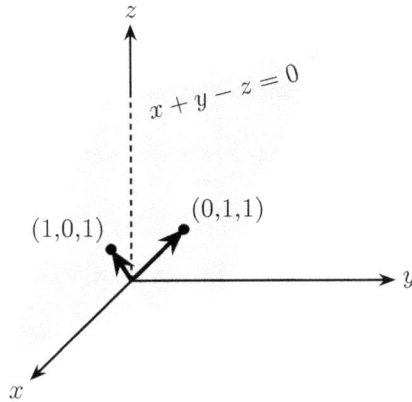

Figure 5.13: 2D Plane as a Subspace of $\mathbb{R}^3$

Problem Set 5

5.1 Draw the following vectors in a two-dimensional coordinate system: $(2, 1)$, $(-4, -2)$, and $(1, -2)$. Analyze and describe the geometric relationships that exist between each pair of these three vectors.

5.2 Considering two vectors u and v of equal length in $\mathbb{R}^2$, determine the angle between u and v when $u + v$ also has the same length as u and v. Demonstrate the result using an appropriate geometric representation.

5.3 $*$ Given five vectors originating from the center of a regular pentagon to each vertex, prove that their vector sum is zero using a geometric approach.

5.4 Given $v = (1 + i, 1 - i)$, express v as a linear combination of:

(a) $(1, 1)$ and $(1, -1)$ using complex scalars;

(b) $(1, 1)$ and $(-i, i)$ using real scalars.

This exercise highlights the distinction between complex vectors and the use of complex versus real scalars as vector components.

5.5 Given vectors $v_1 = (1, 1)$ and $v_2 = (2, 2)$, demonstrate that $w = (3, 4)$ cannot be expressed as a linear combination of $\{v_1, v_2\}$ using real scalars. For a general vector (a, b) in $\mathbb{R}^2$, what conditions must a and b satisfy to be a linear combination of $\{v_1, v_2\}$? Perform your analysis both algebraically and geometrically.

5.6 Given $z = (1, i)$:

(a) Find z^*.

(b) Calculate $\frac{1}{2}(z + z^*)$ and $\frac{1}{2}(z - z^*)$.

(c) Can z^* be expressed as a linear combination of z?

5.7 Consider a non-zero complex vector $z = (z_1, z_2)$ in $\mathbb{C}^2$. If z^* is a linear combination of z, then

$$z^* = \alpha z.$$

(a) What condition must z satisfy for a complex α to exist?

(b) What condition must z satisfy for a real α to exist?

5.8 $*$ Let V be a vector space over the field F. Given a vector v in V and a scalar k in F, use the axioms in Def. 5.11 to prove the following identities:

(a) $0v = 0$.

(b) $(-1)v = -v$.

(c) If $kv = 0$, then $k = 0$ or $v = 0$.

5.9 $*$ Consider W, a subset of a vector space V, adhering to the prescribed definitions of vector addition and scalar multiplication. Prove that properties (c) and (d) in Def. 5.11 are satisfied for W if all other properties hold true. (Hint: Use the properties proved in Problem 5.8.)

5.10 Determine whether each of the following sets is a subspace of $\mathbb{R}^2$ or $\mathbb{R}^3$. Assume all geometric vectors originate from the origin.

(a) The line $y = 2x$ in the x-y plane.

(b) The line $y = 2x + 1$ in the x-y plane.

(c) The plane $x + y - z = 0$ in three-dimensional space.

(d) The plane $x + y - z = 1$ in three-dimensional space.

(e) $S = \{x \mid x = (a, b), a \geq 0, b \geq 0\}$.

(f) $S = \{x \mid x = (a, b), ab \geq 0\}$.

(g) $S = \{x \mid x = (a, b), a^2 + b^2 = 1\}$.

5.11 Geometrically interpret the span of the following vector sets as a subspace of $\mathbb{R}^2$ or $\mathbb{R}^3$:

(a) $\{(1, 1)\}$.

(b) $\{(1, 1), (-1, -1)\}$.

(c) $\{(1, 1), (1, 0)\}$.

(d) $\{(1, 1, 1)\}$.

(e) $\{(1, 0, 0), (0, 1, 0)\}$.

(f) $\{(1, 0, 0), (0, 1, 0), (0, 0, 1)\}$.

5.12 Find a set of vectors that span the following subspaces:

(a) The line $y = -2x$ in the x-y plane.

(b) The plane $z = 2x + y$ in three-dimensional space.

(c) All vectors of the form $(a, b, a + b)$ where a, b are scalars.

(d) The zero subspace of $\mathbb{R}^n$.

5.13 Show that $\{(1, 0, 0, 1), (0, 1, 1, 0), (0, -i, i, 0), (1, 0, 0, -1)\}$ spans $\mathbb{C}^4$.

5.14 Calculate the coordinate of $(1, 0)$ relative to the basis $\{(\frac{1}{\sqrt{2}}, \frac{i}{\sqrt{2}}), (\frac{1}{\sqrt{2}}, -\frac{i}{\sqrt{2}})\}$ in $\mathbb{C}^2$.

5.15 Find the dimension and a basis of the following subspaces of the given vector space:

(a) $S = \text{span}((1,1))$ in $\mathbb{R}^2$.

(b) $S = \text{span}((1,i),(i,-1))$ in $\mathbb{C}^2$.

(c) $S = \text{span}((1,1,-2),(1,-2,1),(-2,1,1))$ in $\mathbb{R}^3$.

(d) $S = \text{span}((1,1,i),(1,i,1),(i,1,1))$ in $\mathbb{C}^3$.

(e) $S = \text{span}(e_1,e_2,\ldots,e_r)$ in $\mathbb{C}^n$ for $r < n$.

(f) $S = \{(x,y,z) \mid y = 2x = 3z\}$ in $\mathbb{C}^3$.

5.16 Given a non-orthogonal basis S for $\mathbb{R}^2$, if $u = (a,b)_S$ and $v = (c,d)_S$, geometrically show that $u + v = (a+c, b+d)_S$ still follows the parallelogram rule of vector addition under S.

6. Inner Product Spaces

Contents

In quantum computing, the state vectors of an n-qubit system reside in the vector space $\mathbb{C}^{2^n}$ [1], a complex vector space as discussed in the previous chapter. According to the first postulate of quantum mechanics, these state vectors must be unit vectors—vectors with a length of one. To describe, transform, and measure these states within the framework of quantum mechanics, orthonormal bases are essential. These bases consist of mutually orthogonal unit vectors.

To rigorously define orthonormal bases, we extend our study from complex vector spaces to complex inner product spaces, which serve as the foundation of Hilbert spaces. These inner product spaces are vector spaces equipped with an inner product, a mathematical structure that generalizes geometric notions of length (norm) and orthogonality to higher-dimensional spaces. These concepts are fundamental to understanding and constructing orthonormal bases. Table 6.1 provides a concise comparison of vector spaces, inner product spaces, and Hilbert spaces.

Additionally, this chapter introduces Dirac notation (also known as bra-ket notation), a compact and elegant framework developed by Paul Dirac in 1939. Dirac notation is integral to modern quantum mechanics and quantum computing, offering an intuitive way to represent quantum states and operations. While this notation is central to the subject, we will continue to use matrix notation for detailed calculations and derivations. To ensure a smooth learning experience, Dirac notation will be gradually incorporated as the chapter progresses.

Property	Vector Space	Inner Product Space	Hilbert Space
Structure	Set with vector operations	Vector space with inner product	Complete inner product space
Operations	Addition, scalar multiplication	Addition, scalar multiplication, inner product	Same as inner product space
Completeness	Not required	Not required	Complete under the norm
Norm Definition	Not required	Induced by the inner product	Same as inner product space

Table 6.1: Comparison of Vector Space, Inner Product Space, and Hilbert Space

6.1   Basics of Dirac Notation

In 1939, Paul Dirac introduced the bra-ket notation in his work *A New Notation for Quantum Mechanics* [7]. This innovative notation system, now bearing Dirac's name, has proven exceptionally advantageous and insightful for performing calculations in quantum mechanics and quantum computing. In this section, we explore the fundamentals of Dirac notation, laying the groundwork for its application in the manipulation and computation of vectors.

6.1.1   Column Vectors and Row Vectors

Until now, our discussions have primarily utilized ordered tuples to represent vectors. In linear algebra practice, however, vectors are often expressed as column vectors or row vectors using matrix notation. This subsection introduces these forms as a precursor to discussing Dirac notation. A more comprehensive exploration of matrices will follow in the subsequent chapter.

> **Definition 6.1 — Matrix.** A matrix is a rectangular array of numbers, which can be either real or complex. A matrix with m rows and n columns is represented as:
>
> $$A = \begin{bmatrix} a_{11} & a_{12} & \cdots & a_{1n} \\ a_{21} & a_{22} & \cdots & a_{2n} \\ \vdots & \vdots & \ddots & \vdots \\ a_{m1} & a_{m2} & \cdots & a_{mn} \end{bmatrix}_{m \times n}, \tag{6.1}$$
>
> where the entries $a_{11}, a_{12}, \ldots, a_{mn}$ are called **elements**. The matrix is referred to as an $m \times n$ matrix, indicating its size.

When a matrix contains only one column (or one row), it is called a column vector (or a row vector):

> **Definition 6.2 — Column and Row Vectors.** A matrix with only one column ($n \times 1$ in size) is called a column vector, denoted as:
>
> $$\begin{bmatrix} v_1 \\ v_2 \\ \vdots \\ v_n \end{bmatrix}.$$
>
> Similarly, a matrix with only one row ($1 \times m$ in size) is called a row vector, denoted as:
>
> $$\begin{bmatrix} v_1 & v_2 & \cdots & v_m \end{bmatrix}.$$

The process of matrix transposition, which interchanges the rows and columns of a matrix, is a fundamental operation in matrix algebra, defined as follows.

> **Definition 6.3 — Transpose.** Given an $m \times n$ matrix A as given in Eq. 6.1, the transpose of A, denoted as A^T, is an $n \times m$ matrix defined as:
>
> $$A^T = \begin{bmatrix} a_{11} & a_{21} & \cdots & a_{m1} \\ a_{12} & a_{22} & \cdots & a_{m2} \\ \vdots & \vdots & \ddots & \vdots \\ a_{1n} & a_{2n} & \cdots & a_{mn} \end{bmatrix}_{n \times m}, \tag{6.2}$$
>
> resulting from interchanging the rows and columns of A.

From Def. 6.3, the first row of A corresponds to the first column of A^T, the second row of A corresponds to the second column of A^T, and so forth.

■ **Example 6.1** Consider $A = \begin{bmatrix} 0 & i \\ -i & 0 \end{bmatrix}$. According to Def. 6.3,

$$A^T = \begin{bmatrix} 0 & -i \\ i & 0 \end{bmatrix}.$$

■

Exercise 6.1 Determine the size and transpose of the following matrices:

$$\text{(a) } \begin{bmatrix} 1 & 0 & 0 \end{bmatrix}, \text{ (b) } \begin{bmatrix} 1 & 0 \\ 0 & -1 \end{bmatrix}, \text{ (c) } \begin{bmatrix} 0 \\ 0 \\ 1 \end{bmatrix}.$$

It is straightforward to see that the transpose of a row vector is a column vector, and vice versa.

Notation: To avoid ambiguity, we will generally use v for column vectors and v^T for row vectors in matrix notation, unless otherwise specified. For example, given $v = (1, i)$, its matrix representations are:

$$v = \begin{bmatrix} 1 \\ i \end{bmatrix}, \ v^T = \begin{bmatrix} 1 & i \end{bmatrix}.$$

We may sometimes transpose a row vector to conveniently depict a column vector, as in

$$v = \begin{bmatrix} 1 & i \end{bmatrix}^T,$$

which represents $(1, i)$ as a column vector.

6.1.2   Kets for Column Vectors

Conventionally, state vectors in quantum computing are represented as column vectors [1]. In the framework of Dirac notation, a specific symbol known as a "ket" is employed to represent column vectors in $\mathbb{C}^n$.

Definition 6.4 — Ket. Given a general vector $v = (v_1, v_2, \ldots, v_n)$ in $\mathbb{C}^n$, we use $|v\rangle$ to denote its column vector form:

$$|v\rangle \equiv \begin{bmatrix} v_1 \\ v_2 \\ \vdots \\ v_n \end{bmatrix}, \tag{6.3}$$

where $|v\rangle$ is referred to as a ket.

Notation: Any ket used to represent a vector will specifically denote a complex vector in $\mathbb{C}^n$ unless otherwise specified.

In Def. 6.4, the letter "v" in $|v\rangle$ is a chosen label for the vector it represents. The choice of labels can be arbitrary, but it usually follows some practical guidelines, similar to how functions, variables, or constants are named:

1. Symbolic Representation: Labels often reflect the vector's role or properties within the quantum system, akin to common function labels such as f or g in algebra. Examples include:

   - $|\psi\rangle$ for a general state vector;

   - $|\lambda_i\rangle$ for eigenvectors corresponding to the eigenvalue λ_i;

- $|j\rangle, |k\rangle$ $(j, k \in \{0, 1, 2, \dots, n-1\})$ to denote computational basis vectors in $\mathbb{C}^n$.

2. Reserved Nomenclature: Certain labels are reserved for specific vectors of computational or physical significance, akin to the roles of constants like π and e. These include:

   - $|0\rangle, |1\rangle$ for the computational basis vectors in single-qubit systems;

   - $|V\rangle, |H\rangle$ for rectilinear polarization states of a photon;

   - $|\Phi^+\rangle, |\Phi^-\rangle, |\Psi^+\rangle, |\Psi^-\rangle$ for Bell states of two qubits.

The standard basis vectors of $\mathbb{C}^n$ (particularly in $\mathbb{C}^2$) are indispensable in quantum computing. Therefore, it is necessary to explicitly define their notations for clarity and uniformity.

Definition 6.5 — Computational Basis. For a single-qubit system in $\mathbb{C}^2$, the standard basis vectors, commonly referred to as the **computational basis**, are defined as the following ket vectors:

$$|0\rangle \equiv \begin{bmatrix} 1 \\ 0 \end{bmatrix}, \ |1\rangle \equiv \begin{bmatrix} 0 \\ 1 \end{bmatrix}. \tag{6.4}$$

For a one-qudit system (with d distinct levels) in $\mathbb{C}^d$, the computational basis includes vectors:

$$|0\rangle \equiv \begin{bmatrix} 1 \\ 0 \\ \vdots \\ 0 \end{bmatrix}, \ |1\rangle \equiv \begin{bmatrix} 0 \\ 1 \\ \vdots \\ 0 \end{bmatrix}, \ \cdots, \ |d-1\rangle \equiv \begin{bmatrix} 0 \\ 0 \\ \vdots \\ 1 \end{bmatrix}. \tag{6.5}$$

$\boxed{i}$ **Notation:** Throughout this book, $|0\rangle$ and $|1\rangle$ will denote the computational basis for $\mathbb{C}^2$, as defined in Eq. 6.4, unless explicitly stated otherwise.

Using Dirac notation, vector addition and scalar multiplication retain their familiar forms. For example:

$$\frac{1}{\sqrt{2}}|0\rangle + \frac{1}{\sqrt{2}}|1\rangle = \frac{1}{\sqrt{2}}\begin{bmatrix} 1 \\ 0 \end{bmatrix} + \frac{1}{\sqrt{2}}\begin{bmatrix} 0 \\ 1 \end{bmatrix} = \begin{bmatrix} \frac{1}{\sqrt{2}} \\ \frac{1}{\sqrt{2}} \end{bmatrix}.$$

A general vector in $\mathbb{C}^2$ can therefore be expressed as:

$$\begin{bmatrix} \alpha \\ \beta \end{bmatrix} = \alpha|0\rangle + \beta|1\rangle, \ \text{for } \alpha, \beta \in \mathbb{C}. \tag{6.6}$$

Exercise 6.2 Express $|\psi\rangle = \begin{bmatrix} 1 \\ i \end{bmatrix}$ as a linear combination of the computational basis using Dirac notation.

6.1.3 Bra Vectors and Hermitian Adjoint

For each column vector $|\psi\rangle$ in $\mathbb{C}^n$, there is an associated vector obtained through an operation known as the conjugate transpose. For instance, the conjugate transpose of

$$\begin{bmatrix} 1+i \\ 1-i \end{bmatrix}$$

is computed as:

$$\left(\begin{bmatrix} 1+i \\ 1-i \end{bmatrix}^* \right)^T = \begin{bmatrix} (1+i)^* & (1-i)^* \end{bmatrix} = \begin{bmatrix} 1-i & 1+i \end{bmatrix}.$$

This operation, critical in the context of quantum computing, is referred to as the Hermitian adjoint. It plays a fundamental role in the formulation of quantum mechanics:

> **Definition 6.6 — Hermitian Adjoint.** Consider a column vector in $\mathbb{C}^n$:
>
> $$v = \begin{bmatrix} v_1 \\ v_2 \\ \vdots \\ v_n \end{bmatrix},$$
>
> its Hermitian adjoint, or simply adjoint, is defined as its conjugate transpose, denoted as:
>
> $$v^\dagger \equiv (v^*)^T = (v^T)^* = \begin{bmatrix} v_1^* & v_2^* & \cdots & v_n^* \end{bmatrix}. \tag{6.7}$$

> **Exercise 6.3** Compute $v^\dagger$ for $v = \begin{bmatrix} 1 \\ i \\ 2+3i \end{bmatrix}$.

Dirac notation encapsulates the concept of adjoint vectors by transforming a ket into a bra. This transformation is essential for expressing quantum states and operations succinctly.

> **Definition 6.7 — Bra.** The bra, denoted as $\langle v|$, represents the Hermitian adjoint of $|v\rangle$, formally defined as:
>
> $$\langle v| \equiv |v\rangle^\dagger. \tag{6.8}$$

The reverse relationship also holds:

$$|v\rangle \equiv \langle v|^\dagger. \tag{6.9}$$

■ **Example 6.2** Find the matrix forms of $\langle 0|$ and $\langle 1|$.

Solution. Using the definition of adjoint:

$$\langle 0| = |0\rangle^\dagger = \begin{bmatrix} 1 \\ 0 \end{bmatrix}^\dagger = \begin{bmatrix} 1 & 0 \end{bmatrix},$$

$$\langle 1| = |1\rangle^\dagger = \begin{bmatrix} 0 \\ 1 \end{bmatrix}^\dagger = \begin{bmatrix} 0 & 1 \end{bmatrix}.$$

■

The adjoint operation, which transforms kets into bras (or vice versa), exhibits conjugate-linearity. These properties are encapsulated in the following theorem:

> **Theorem 6.1** Given vectors $|u\rangle$, $|v\rangle$ in $\mathbb{C}^n$ and scalars $\alpha, \beta \in \mathbb{C}$, bras exhibit the following conjugate-linear properties:
>
> $$\langle u + v| = \langle u| + \langle v|, \tag{6.10a}$$
> $$\langle \alpha v| = \alpha^* \langle v|, \tag{6.10b}$$
> $$\langle \alpha u + \beta v| = \alpha^* \langle u| + \beta^* \langle v|. \tag{6.10c}$$

The proof of Theorem 6.1 is straightforward using the properties of complex numbers: $(\alpha + \beta)^* = \alpha^* + \beta^*$ and $(\alpha\beta)^* = \alpha^*\beta^*$. This proof is left to the reader as an exercise (Problem 6.1).

■ **Example 6.3** Use the computational basis to calculate $\langle\psi|$ for:

$$|\psi\rangle = \frac{1+i}{2}|0\rangle + \frac{1-i}{2}|1\rangle.$$

Solution. Applying Eq. 6.10c to $|\psi\rangle$:

$$\langle\psi| = \left(\frac{1+i}{2}\right)^* \langle 0| + \left(\frac{1-i}{2}\right)^* \langle 1| = \frac{1-i}{2}\langle 0| + \frac{1+i}{2}\langle 1|.$$

■

Through Example 6.3, the given vector is expressed as a linear combination of the computational basis, either in kets or bras. This methodology, commonly employed in Dirac notation, offers significant advantages for efficiently handling inner products, outer products, and tensor products, as will be seen in subsequent sections.

ⓘ Matrix algebra restricts addition to matrices of the same size, rendering the operation $\langle z| + |w\rangle$ invalid. A comprehensive introduction to matrix algebra will be presented in the next chapter.

Exercise 6.4 Given $|\psi\rangle = i|1\rangle - i|0\rangle$, compute $\langle\psi|$ using Dirac notation and matrix notation.

6.2 Norm and Unit Vectors

Our everyday experience in a three-dimensional world makes us familiar with geometric notions such as "length" or "distance". This familiarity can be traced back over two millennia to Euclid's theories in two- and three-dimensional geometry. In this section, we study how to calculate length and distance more generally for vectors in $\mathbb{R}^n$ or $\mathbb{C}^n$, drawing an analogy with Euclidean geometry. This exploration leads to the notion of "unit vectors", which are fundamental to quantum computing.

6.2.1 Vector Norm

1 Norm of Real Vectors

A geometric vector is fundamentally characterized by two properties: magnitude and direction. The magnitude of a geometric vector is defined by its length in Euclidean geometry. Consider a vector $v = (v_1, v_2)$ in $\mathbb{R}^2$. Its length can be calculated using the Pythagorean Theorem:

$$\|\boldsymbol{v}\| = \sqrt{v_1^2 + v_2^2}. \tag{6.11}$$

The length of $\boldsymbol{v}$ is a real scalar and denoted as $\|\boldsymbol{v}\|$, known as "the norm of $\boldsymbol{v}$". The term "norm" is synonymous with "length" or "magnitude", but it extends to generalized vectors such as matrices and functions.

In the three-dimensional space, the formula for calculating the norm of $\boldsymbol{v} = (v_1, v_2, v_3)$ in $\mathbb{R}^3$ is given by:

$$\|\boldsymbol{v}\| = \sqrt{v_1^2 + v_2^2 + v_3^2}. \tag{6.12}$$

An illustration of Eqs. 6.11 and 6.12 is given in Fig. 6.1.

Observing a clear pattern from Eqs. 6.11 and 6.12, we can generalize the norm for a vector in $\mathbb{R}^n$ as follows:

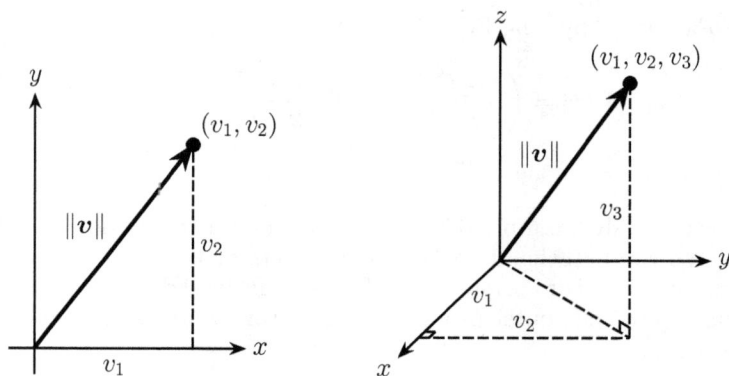

Figure 6.1: Norms of Geometric Vectors in 2D and 3D

Definition 6.8 — Norm of a Real Vector. For a vector $\boldsymbol{v} = (v_1, v_2, \ldots, v_n)$ in $\mathbb{R}^n$, its norm (or length), denoted as $\|\boldsymbol{v}\|$, is defined by the formula:

$$\|\boldsymbol{v}\| = \sqrt{v_1^2 + v_2^2 + v_3^2 + \cdots + v_n^2}. \tag{6.13}$$

■ **Example 6.4** Calculate the norm of the following vectors:

(a) $\left(-\frac{\sqrt{2}}{2}, \frac{\sqrt{2}}{2}\right)$

(b) $\left(\frac{1}{3}, \frac{1}{3}, \frac{1}{3}\right)$

Solution.

(a) $\left\|\left(-\frac{\sqrt{2}}{2}, \frac{\sqrt{2}}{2}\right)\right\| = \sqrt{\left(-\frac{\sqrt{2}}{2}\right)^2 + \left(\frac{\sqrt{2}}{2}\right)^2} = \sqrt{\frac{1}{2} + \frac{1}{2}} = 1.$

(b) $\left\|\left(\frac{1}{3}, \frac{1}{3}, \frac{1}{3}\right)\right\| = \sqrt{\left(\frac{1}{3}\right)^2 + \left(\frac{1}{3}\right)^2 + \left(\frac{1}{3}\right)^2} = \frac{\sqrt{3}}{3}.$

Exercise 6.5 Calculate the norm of $\left(\frac{1}{3}, \frac{2}{3}, -\frac{2}{3}\right)$ and $\left(\frac{2}{7}, \frac{3}{7}, \frac{6}{7}\right)$.

Theorem 6.2 The norm of a vector in $\mathbb{R}^n$ (or $\mathbb{C}^n$) exhibits the following properties:

(a) $\|\boldsymbol{v}\| \geq 0$

(b) $\|\boldsymbol{v}\| = 0$ if and only if $\boldsymbol{v} = \boldsymbol{0}$

(c) $\|k\boldsymbol{v}\| = |k| \|\boldsymbol{v}\|$

For real vectors, the first two properties of Theorem 6.2 can be easily proved using Eq. 6.13. Property (a) asserts that the length of a vector is non-negative, and property (b) asserts that the only vector with zero length is the zero vector itself.

Property (c) of Theorem 6.2 illustrates that multiplying a vector by a scalar k effectively scales its length by a factor of $|k|$, irrespective of the sign of k.

Exercise 6.6 Prove Theorem 6.2c for real vectors using Eq. 6.13.

(i) The term "Euclidean geometry" encompasses the studies of two-dimensional plane geometry and three-dimensional solid geometry. Named after the ancient Greek mathematician Euclid, Euclidean geometry aligns closely with human intuition, as it mirrors the perceptions of the "flat" physical space that we live in.

Euclidean geometry is founded upon a comprehensive axiomatic system, which facilitates the logical derivation of theorems. These theorems can be algebraically generalized to $\mathbb{R}^n$, facilitating the extension of geometric arguments to higher dimensions. For this reason, $\mathbb{R}^n$ is often referred to as "Euclidean space" in the context of linear algebra.

2 Norm of Complex Vectors

For complex vectors, we still expect norm to be a non-negative real-valued function to be interpreted as a sensible "length". Utilizing complex conjugates, norm of a complex vector is defined as follows.

Definition 6.9 — Norm of a Complex Vector. For a vector $\boldsymbol{v} = (v_1, v_2, \ldots, v_n)$ in $\mathbb{C}^n$, its norm is defined by the formula:

$$\|\boldsymbol{v}\| = \sqrt{v_1^* v_1 + v_2^* v_2 + \cdots + v_n^* v_n}. \tag{6.14}$$

Note that Eq. 6.14 reduces to Eq. 6.13 when $\boldsymbol{v}$ is real.

Theorem 6.3 For a complex vector $\boldsymbol{v} = (v_1, v_2, \ldots, v_n)$ in $\mathbb{C}^n$, the following identity holds:

$$\|\boldsymbol{v}\|^2 = |v_1|^2 + |v_2|^2 + \cdots + |v_n|^2. \tag{6.15}$$

Proof. Squaring Eq. 6.14 and using the definition of the absolute value of a complex number: $|z| = \sqrt{z^* z}$ or $|z|^2 = z^* z$ gives

$$\|\boldsymbol{v}\|^2 = v_1^* v_1 + v_2^* v_2 + \cdots + v_n^* v_n = |v_1|^2 + |v_2|^2 + \cdots + |v_n|^2.$$

$\square$

Theorem 6.2 applies not only to real vectors but also to complex ones. Properties (a) and (b) are straightforward to prove following Def. 6.9. We offer a proof of property (c) below.

Proof. Given $v = (v_1, v_2, \ldots, v_n)$ in $\mathbb{C}^n$, we want to prove that $\|kv\| = |k|\|v\|$ for $k \in \mathbb{C}$. Starting from the left-hand side, $kv = (kv_1, kv_2, \ldots, kv_n)$ and using Eq. 6.15 gives:

$$\|kv\|^2 = |kv_1|^2 + |kv_2|^2 + \cdots + |kv_n|^2$$
$$= |k|^2|v_1|^2 + |k|^2|v_2|^2 + \cdots + |k|^2|v_n|^2$$
$$= |k|^2(|v_1|^2 + |v_2|^2 + \cdots + |v_n|^2).$$

Note that in the derivation above we apply the properties of complex numbers $|z_1 z_2|^2 = (z_1 z_2)^*(z_1 z_2) = z_1^* z_1 z_2^* z_2 = |z_1|^2|z_2|^2$. Taking the square root of the equation above completes the proof:

$$\|kv\| = |k|\sqrt{(|v_1|^2 + |v_2|^2 + \cdots + |v_n|^2)} = |k|\|v\|.$$

$\square$

■ **Example 6.5** Calculate the norm of $v = (1 + 2i, i)$.

Solution. Using Eq. 6.15:

$$\|v\|^2 = |1 + 2i|^2 + |i|^2 = (1^2 + 2^2) + 1^2 = 6,$$

Therefore, $v = \sqrt{6}$. Note that $|a + ib|^2 = a^2 + b^2$ for $a, b \in \mathbb{R}$.   ■

Exercise 6.7 Calculate the norm of $|v\rangle = \begin{bmatrix} \frac{1+i}{2} \\ -\frac{1}{2} \\ -\frac{i}{2} \end{bmatrix}$.

3   Dirac Notation of Vector Norm

In Dirac notation, we use $\|v\|$ to denote the norm of a vector $|v\rangle$. Note that the symbol v is not rendered in boldface. In addition, $\langle v|$ and $|v\rangle$ have the same length. Thus,

$$\|v\| \equiv \|\,|v\rangle\,\| = \|\,\langle v|\,\|.$$

6.2.2   Unit Vectors

Vectors of norm one are referred to as unit vectors. Unit vectors are of prominent importance in quantum computing, as the postulate of quantum mechanics demands that all quantum states be described by unit complex vectors.

1   Unit Real Vectors

For any non-zero geometric vector v in $\mathbb{R}^2$, it is evident that there is a unique unit vector that aligns with the direction of v. This principle extends to any non-zero vectors in $\mathbb{R}^n$. The method to derive this unique unit vector from a given vector is known as normalizing a vector.

> **Theorem 6.4** For any non-zero vector v in $\mathbb{R}^n$, the following formula yields the unit (or normalized) vector that aligns with the direction of v, denoted as $\hat{v}$:
>
> $$\hat{v} = \frac{1}{\|v\|} v. \tag{6.16}$$

Proof. The following derivation demonstrates that $\hat{v}$, as given in Eq. 6.16, is a unit vector.

$$\|\hat{v}\| = \left\| \frac{1}{\|v\|} v \right\| = \left| \frac{1}{\|v\|} \right| \|v\| = \frac{1}{\|v\|} \|v\| = 1.$$

In this derivation, the second step utilizes property (c) from Theorem 6.2. Furthermore, since $1/\|v\|$ only scales v by a positive scalar, $\hat{v}$ aligns with the direction of v. This completes the proof. $\qquad\square$

■ **Example 6.6** Normalize the following vectors in $\mathbb{R}^2$ or $\mathbb{R}^3$:

(a) $\left(\frac{1}{2}, 1\right)$

(b) $(1, 1, -2)$

Solution.

(a) The norm of the vector is $\sqrt{\left(\frac{1}{2}\right)^2 + 1^2} = \frac{\sqrt{5}}{2}$, therefore, the normalized vector is

$$\frac{2}{\sqrt{5}} \left(\frac{1}{2}, 1\right) = \left(\frac{1}{\sqrt{5}}, \frac{2}{\sqrt{5}}\right).$$

(b) The norm of $(1, 1, -2)$ is $\sqrt{1^2 + 1^2 + (-2)^2} = \sqrt{6}$, therefore, the normalized vector is

$$\frac{1}{\sqrt{6}} (1, 1, -2) = \left(\frac{1}{\sqrt{6}}, \frac{1}{\sqrt{6}}, -\frac{2}{\sqrt{6}}\right).$$

∎

> **Exercise 6.8** Normalize the following vectors:
>
> (a) $(\sqrt{2}, 1)$
>
> (b) $(\sqrt{2}, -\sqrt{3}, \sqrt{5})$

2 Unit Complex Vectors

Analogous to unit real vectors, complex vectors of norm one are called unit complex vectors. The ensuing samples and exercises are intended to familiarize readers with typical unit vector configurations in $\mathbb{C}^2$ (a single-qubit system).

■ **Example 6.7** Which of the following complex vectors are unit vectors in $\mathbb{C}^2$?

$$\text{(a) } |v_1\rangle = \begin{bmatrix} 1 \\ i \end{bmatrix}, \quad \text{(b) } |v_2\rangle = \begin{bmatrix} \frac{1}{\sqrt{2}} \\ \frac{-i}{\sqrt{2}} \end{bmatrix}, \quad \text{(c) } |v_3\rangle = \begin{bmatrix} \frac{1+i}{2} \\ \frac{1-i}{2} \end{bmatrix}.$$

Solution. To ascertain which vectors are unit vectors, we will evaluate whether $|v|^2 = 1$ for each vector using Eq. 6.15:

(a) $|v_1|^2 = |1|^2 + |i|^2 = 1 + 1 = 2,$

(b) $|v_2|^2 = \left|\frac{1}{\sqrt{2}}\right|^2 + \left|\frac{-i}{\sqrt{2}}\right|^2 = \frac{1}{2} + \frac{1}{2} = 1,$

(c) $|v_3|^2 = \left|\frac{1+i}{2}\right|^2 + \left|\frac{1-i}{2}\right|^2 = \left[\left(\frac{1}{2}\right)^2 + \left(\frac{1}{2}\right)^2\right] + \left[\left(\frac{1}{2}\right)^2 + \left(-\frac{1}{2}\right)^2\right] = 1.$

Therefore, $|v_2\rangle$ and $|v_3\rangle$ are unit vector while $|v_1\rangle$ is not. ∎

■ Example 6.8

(a) Given $|v\rangle = \begin{bmatrix} \alpha \\ \beta \end{bmatrix}$ where $\alpha, \beta \in \mathbb{C}$, what conditions must α, β satisfy for $|v\rangle$ to be a unit vector?

(b) Given $|v\rangle = \begin{bmatrix} a + ib \\ c + id \end{bmatrix}$ where $a, b, c, d \in \mathbb{R}$, what conditions must a, b, c, d satisfy for $|v\rangle$ to be a unit vector?

Solution.

(a) For $|v\rangle = \begin{bmatrix} \alpha \\ \beta \end{bmatrix}$ to be a unit complex vector, we must have $|v|^2 = 1$. Using Eq. 6.15, this gives:

$$|v|^2 = 1 \quad \Rightarrow \quad |\alpha|^2 + |\beta|^2 = 1. \tag{6.17}$$

(b) Following the result in (a), $|v\rangle = \begin{bmatrix} a + ib \\ c + id \end{bmatrix}$ is a unit vector if and only if

$$|a + ib|^2 + |c + id|^2 = 1 \quad \Rightarrow \quad a^2 + b^2 + c^2 + d^2 = 1.$$

∎

Exercise 6.9 Identify the condition that complex numbers α and β must satisfy to ensure that $|\psi\rangle = \alpha |0\rangle + \beta |1\rangle$ represents a unit vector.

Exercise 6.10 Show that given any unit vector $|v\rangle \in \mathbb{C}^n$, $e^{i\phi} |v\rangle$ is still a unit vector for $\phi \in \mathbb{R}$.

3 Transition Between Unit Vectors

For a vector (a, b) in $\mathbb{R}^2$ to be a unit vector, it must satisfy the condition $a^2 + b^2 = 1$. This equation represents a unit circle in the two-dimensional coordinate system, as illustrated in Fig. 6.2. Similarly, in $\mathbb{R}^3$, all unit vectors reside on the surface of a unit sphere centered at the origin.

Generally, all unit vectors in $\mathbb{R}^n$ lies on the surface of a unit hypersphere. Based on this understanding, transitioning from one unit vector to another one is equivalent to performing a specific rotational transformation on the surface of a circle, sphere or hypersphere.

In quantum computing and quantum information, transitioning between quantum states is a fundamental aspect of quantum circuits, typically realized through

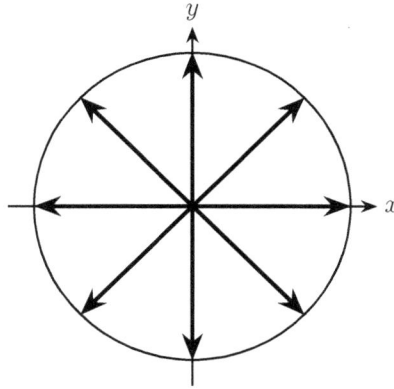

Figure 6.2: Unit Vectors in 2D

quantum gates. Analogous to real vectors, such transitions can be understood as rotational transformations in $\mathbb{C}^n$, represented by unitary matrices. Further discussion on this topic will be provided in § 8.3.

6.2.3 Distance

In the two-dimensional Cartesian coordinate system, the distance between two points $P(w_1, w_2)$ and $Q(v_1, v_2)$ is:

$$|PQ| = \sqrt{(v_1 - w_1)^2 + (v_2 - w_2)^2}. \tag{6.18}$$

Considering two vectors in $\mathbb{R}^2$, $\boldsymbol{v} = (v_1, v_2)$ and $\boldsymbol{w} = (w_1, w_2)$, we define their distance as the Euclidean distance between their terminal points, P and Q, each originating from the origin. Utilizing Eq. 6.18 allows us to calculate this distance as follows:

$$\sqrt{(v_1 - w_1)^2 + (v_2 - w_2)^2} = \|(v_1 - w_1, v_2 - w_2)\| = \|\boldsymbol{v} - \boldsymbol{w}\|. \tag{6.19}$$

A geometric illustration of Eq. 6.19 is given in Fig. 6.3.

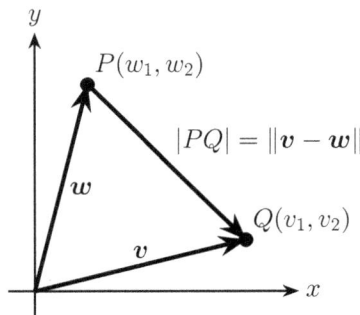

Figure 6.3: Distance Between Vectors in 2D

Extending Eq. 6.19 to $\mathbb{R}^n$ or $\mathbb{C}^n$, we arrive at a generalized definition for vector distance for real or complex Euclidean spaces:

Definition 6.10 — Distance. Given two vectors $u = (u_1, u_2, \ldots, u_n)$ and $v = (v_1, v_2, \ldots, v_n)$ in $\mathbb{R}^n$ or $\mathbb{C}^n$, the distance between u and v is defined as:

$$d(u, v) = \|u - v\| \qquad (6.20)$$

■ **Example 6.9** Calculate the distance between:

(a) $u = (1, 0, 0, 0)$ and $v = (0, 1, 1, 1)$;

(b) $|u\rangle = \begin{bmatrix} 0 \\ i \end{bmatrix}$ and $|v\rangle = \begin{bmatrix} \frac{\sqrt{3}}{2} \\ \frac{1}{2} \end{bmatrix}$.

Solution.

(a) Following Eq. 6.20, the distance between u and v is:

$$\|u - v\| = \sqrt{(1-0)^2 + (0-1)^2 + (0-1)^2 + (0-1)^2} = 2.$$

(b) $|u\rangle - |v\rangle = \begin{bmatrix} -\frac{\sqrt{3}}{2} \\ \frac{-1+2i}{2} \end{bmatrix}$. Therefore, the distance between $|u\rangle$ and $|v\rangle$ is:

$$\sqrt{\left|-\frac{\sqrt{3}}{2}\right|^2 + \left|\frac{-1+2i}{2}\right|^2} = \sqrt{\frac{3}{4} + \left(\frac{1}{4} + 1\right)} = \sqrt{2}.$$

■

Exercise 6.11 Determine the distance between $u = (1, 0)$ and $v = (\frac{1}{2}, \frac{\sqrt{3}}{2})$. Illustrate u, v, and $u - v$ within a two-dimensional coordinate system, and explain the result geometrically.

Establishing the concept of distance is crucial for the definition of "projection" in $\mathbb{R}^n$. Specifically, projecting one vector v orthogonally onto another vector a is equivalent to finding the smallest distance between v and any vector in the span of a. This projection problem will be discussed in § 6.4.

6.3 Complex Inner Product Spaces

A vector space becomes an inner product space when equipped with an operation called the "inner product" between two vectors. The inner product serves several purposes: it defines the norm of vectors, provides a way to measure angles and lengths, and introduces the concept of orthogonality. In this section, we begin by exploring the inner product in the context of $\mathbb{R}^n$, commonly known as the dot product, and then extend this concept to $\mathbb{C}^n$. This extension leads to the definition of a complex inner product space, which, in the context of quantum computing, is a Hilbert space.

6.3.1 Dot Products

The dot product is the most familiar form of inner product. It is a fundamental operation in quantum computing, playing a crucial role in numerous calculations and derivations.

1 Real Dot Products

The dot product between two vectors in $\mathbb{R}^n$ is calculated by multiplying corresponding components of the vectors and subsequently summing these products. The formal definition is presented as follows:

Definition 6.11 — Real Dot Product. For vectors $u = (u_1, u_2, \ldots, u_n)$ and $v = (v_1, v_2, \ldots, v_n)$ in $\mathbb{R}^n$, their dot product is defined as:

$$u \cdot v = u_1 v_1 + u_2 v_2 + \cdots + u_n v_n. \tag{6.21}$$

This is also known as the Euclidean inner product.

■ **Example 6.10** Given $u = (1, 1, -1, -1)$, $v = (1, -1, 1, -1)$, compute $u \cdot v$.

Solution. Following Eq. 6.21,

$$u \cdot v = 1 \times 1 + 1 \times (-1) + (-1) \times 1 + (-1) \times (-1) = 1 - 1 - 1 + 1 = 0.$$

■

The norm of a real vector and the distance between two real vectors can both be expressed as specific dot products, as shown in the following formulas.

Theorem 6.5 For vectors u and v in $\mathbb{R}^n$:

$$\|v\|^2 = v \cdot v \tag{6.22a}$$

$$\|u - v\|^2 = (u - v) \cdot (u - v) \tag{6.22b}$$

Proof. For Eq. 6.22a, let $v = (v_1, v_2, \ldots, v_n)$, then

$$v \cdot v = v_1 \cdot v_1 + v_2 \cdot v_2 + \cdots + v_n \cdot v_n = v_1^2 + v_2^2 + \cdots + v_n^2 = \|v\|^2.$$

By replacing v with $u - v$ in Eq. 6.22a, we obtain

$$(u - v) \cdot (u - v) = \|u - v\|^2,$$

which proves Eq. 6.22b.   □

Dot product exhibits several properties analogous to those of multiplication between scalars. These properties are encapsulated in the following theorem.

Theorem 6.6 Given vectors u, v and w in $\mathbb{R}^n$ and a scalar k in $\mathbb{R}$, the following algebraic identities are true:

$u \cdot v = v \cdot u$	(Commutative Law)	(6.23a)
$u \cdot (v + w) = u \cdot v + u \cdot w$	(Left Distributive Law)	(6.23b)
$(u + v) \cdot w = u \cdot w + v \cdot w$	(Right Distributive Law)	(6.23c)
$(ku) \cdot v = k(u \cdot v)$	(Left Homogeneity Law)	(6.23d)
$u \cdot (kv) = k(u \cdot v)$	(Right Homogeneity Law)	(6.23e)

Here we provide a proof of Eq. 6.23a. Proofs of other properties are left as exercises to readers.

Proof. Given $u = (u_1, u_2, \ldots, u_n)$ and $v = (v_1, v_2, \ldots, v_n)$,

$$u \cdot v = u_1 v_1 + u_2 v_2 + \cdots + u_n v_n = v_1 u_1 + v_2 u_2 + \cdots + v_n u_n = v \cdot u.$$

$\square$

Exercise 6.12 Prove the left- and the right-homogeneity rules in Theorem 6.6.

Theorem 6.6 enables algebraic manipulations of dot products in a manner akin to scalar multiplication, as demonstrated below:

$$(a + b) \cdot (u + v) = (a + b) \cdot u + (a + b) \cdot v$$
$$= a \cdot u + b \cdot u + a \cdot v + b \cdot v. \tag{6.24}$$

$$\|u + v\|^2 = (u + v) \cdot (u + v)$$
$$= u \cdot u + u \cdot v + v \cdot u + v \cdot v \tag{6.25}$$
$$= \|u\|^2 + 2u \cdot v + \|v\|^2.$$

Exercise 6.13 Prove the Parallelogram Law in $\mathbb{R}^n$:

$$\|u + v\|^2 + \|u - v\|^2 = 2(\|u\|^2 + \|v\|^2). \tag{6.26}$$

2   Complex Dot Products

In $\mathbb{C}^n$, the definition of the complex dot product differs slightly from its real counterpart, as it involves the use of the complex conjugate in the calculation.

Definition 6.12 — Complex Dot Product. Given two vectors $u = (u_1, u_2, \ldots, u_n)$, $v = (v_1, v_2, \ldots, v_n)$ in $\mathbb{C}^n$, the complex dot product of u and v is defined as:

$$u \cdot v = u_1^* v_1 + u_2^* v_2 + \cdots + u_n^* v_n = \sum_{i=1}^n u_i^* v_i. \tag{6.27}$$

■ **Example 6.11** Compute $u \cdot v$ for $u = (1, i)$ and $v = (i, -i)$.

Solution. $u^* = (1, -i)$; hence by Eq. 6.27

$$u \cdot v = 1 \times i + (-i) \times (-i) = -1 + i.$$

■

(i) | Although the dot product between two real vectors always results in a real scalar, the complex dot product generally results in a complex scalar.

Exercise 6.14 Compute $u \cdot v$ for $u = (1 + i, i)$ and $v = (i, 1 - i)$.

(i) | In some other contexts, the complex dot product adopts an alternative definition: $u \cdot v = u_1 v_1^* + u_2 v_2^* + \cdots + u_n v_n^*$. Definition 6.12 aligns with conventions predominantly used in quantum mechanics and quantum computing.

One may wonder why complex conjugates are used in Def. 6.12. It is included to ensure that the norm defined in Def. 6.9 is still the dot product of a complex vector with itself.

Theorem 6.7 Given a complex vector $v = (v_1, v_2, \ldots, v_n)$, its norm can be expressed as:
$$\|v\| = \sqrt{v \cdot v}, \tag{6.28}$$
or equivalently,
$$\|v\|^2 = v \cdot v. \tag{6.29}$$

Proof. Following Eqs. 6.15 and 6.27,
$$v \cdot v = v_1^* v_1 + v_2^* v_2 + \cdots + v_n^* v_n = |v_1|^2 + |v_2|^2 + \cdots + |v_n|^2 = \|v\|^2.$$

$\qquad\qquad\qquad\qquad\qquad\qquad\qquad\qquad\qquad\qquad\qquad\qquad\qquad\qquad\qquad\quad$ □

6.3.2 Complex Inner Product Spaces

The real and complex dot products discussed earlier are specific examples of inner products. In general, an inner product is a scalar-valued operation associated with a vector space that satisfies certain properties. When a vector space is equipped with an inner product, it is referred to as an inner product space. Common examples include $\mathbb{R}^n$ (a real inner product space) and $\mathbb{C}^n$ (a complex inner product space). In the context of quantum computing, complex inner product spaces are particularly important due to their role in representing quantum states and operators.

1 Complex Inner Product Spaces

A precise definition of complex inner product space is as below:

Definition 6.13 — Complex Inner Product Space. A complex inner product space is a complex vector space V equipped with an inner product, which is a binary function resulting in a scalar $\langle \cdot, \cdot \rangle : V \times V \to \mathbb{C}$.

The inner product must satisfy the following properties for all vectors u, v, w in V, and any scalar k in $\mathbb{C}$:

(a) $\langle u, v \rangle = \langle v, u \rangle^*$

(b) $\langle u, v + w \rangle = \langle u, v \rangle + \langle u, w \rangle$

(c) $\langle u, kv \rangle = k \langle u, v \rangle$

(d) $\langle v, v \rangle$ is a real, non-negative number, and $\langle v, v \rangle = 0$ if and only if $v = 0$.

i We use $\langle \cdot, \cdot \rangle$ to denote a general inner product (for example, $\langle A, B \rangle$, which represents the inner product between two matrices, as discussed later). For the (real or complex) dot product in $\mathbb{R}^n$ or $\mathbb{C}^n$ in matrix notation, we use $u \cdot v$.

In § 5.3, we established that $\mathbb{C}^n$ is a complex vector space. Furthermore, when equipped with the complex dot product defined in Def. 6.12 as its inner product, $\mathbb{C}^n$ evidently qualifies as a complex inner product space, as proved below.

Theorem 6.8 $\mathbb{C}^n$ is a complex inner product space when the complex dot product is used as the inner product.

Proof. Take V to be $\mathbb{C}^n$. We need to prove that the complex dot product qualifies as an inner product by demonstrating that it satisfies all four properties in Def. 6.13.

Let $\boldsymbol{u} = (u_1, u_2, \ldots, u_n)$, $\boldsymbol{v} = (v_1, v_2, \ldots, v_n)$, $\boldsymbol{w} = (w_1, w_2, \ldots, w_n)$. We will use Sigma Notation to simplify the computation process.

(a)

$$\boldsymbol{u} \cdot \boldsymbol{v} = \sum_{i=1}^{n} u_i^* v_i = \sum_{i=1}^{n} (u_i v_i^*)^* = \left(\sum_{i=1}^{n} u_i v_i^* \right)^* = (\boldsymbol{v} \cdot \boldsymbol{u})^*.$$

(b)

$$\boldsymbol{u} \cdot (\boldsymbol{v} + \boldsymbol{w}) = \sum_{i=1}^{r} u_i^* (v_i + w_i) = \sum_{i=1}^{n} u_i^* v_i + \sum_{i=1}^{n} u_i^* w_i = \boldsymbol{u} \cdot \boldsymbol{v} + \boldsymbol{u} \cdot \boldsymbol{w}.$$

(c)

$$\boldsymbol{u} \cdot (k\boldsymbol{v}) = \sum_{i=1}^{n} u_i^* (k v_i) = k \sum_{i=1}^{n} u_i^* v_i = k(\boldsymbol{u} \cdot \boldsymbol{v}).$$

(d)

$$\boldsymbol{v} \cdot \boldsymbol{v} = v_1^* v_1 + v_2^* v_2 + \cdots + v_n^* v_n = |v_1|^2 + |v_2|^2 + \cdots + |v_n|^2 \geq 0 \text{ for any } \boldsymbol{v}.$$

Moreover, if $\boldsymbol{v} \cdot \boldsymbol{v} = 0$, we must have

$$|v_1| = |v_2| = \cdots = |v_n| = 0 \quad \Rightarrow \quad v_1 = v_2 = \cdots = v_n = 0 \quad \Rightarrow \quad \boldsymbol{v} = \boldsymbol{0}.$$

□

Apart from the properties specified in its definition, a complex inner product also follows these additional identities.

Theorem 6.9 Given a complex inner product space V equipped with an inner product $\langle \cdot, \cdot \rangle$, for any vectors $\boldsymbol{u}, \boldsymbol{v}, \boldsymbol{w} \in V$ and any scalar $k \in \mathbb{C}$, the following identities hold true:

$$\langle \boldsymbol{0}, \boldsymbol{v} \rangle = \langle \boldsymbol{v}, \boldsymbol{0} \rangle = 0 \tag{6.30a}$$
$$\langle \boldsymbol{u} + \boldsymbol{v}, \boldsymbol{w} \rangle = \langle \boldsymbol{u}, \boldsymbol{w} \rangle + \langle \boldsymbol{v}, \boldsymbol{w} \rangle \tag{6.30b}$$
$$\langle k\boldsymbol{u}, \boldsymbol{v} \rangle = k^* \langle \boldsymbol{u}, \boldsymbol{v} \rangle \tag{6.30c}$$

These properties can be easily verified using the formula for the complex dot product (Problem 6.10). In the next, we provide a general proof of Eq. 6.30c which is valid for all complex inner products. The proofs for the remaining two properties are left as exercises for the reader (Problem 6.11).

Proof. Consider the left-hand side of the identity:

$$\langle k\boldsymbol{u}, \boldsymbol{v} \rangle = \langle \boldsymbol{v}, k\boldsymbol{u} \rangle^*$$
$$= (k \langle \boldsymbol{v}, \boldsymbol{u} \rangle)^*$$
$$= k^* (\langle \boldsymbol{v}, \boldsymbol{u} \rangle)^*$$
$$= k^* \langle \boldsymbol{u}, \boldsymbol{v} \rangle.$$

In this derivation properties (a) and (c) in Def. 6.13 are utilized.   □

Exercise 6.15 Prove the following identities for a complex inner product:

(a) Linearity in the second argument:

$$\langle u, kv + mw \rangle = k\langle u, v \rangle + m\langle u, w \rangle \tag{6.31}$$

(b) Sesquilineararity:

$$\langle ku + mv, w \rangle = k^*\langle u, w \rangle + m^*\langle v, w \rangle \tag{6.32}$$

> ⓘ Definition 6.13 preserves linearity in the second argument, aligning with the definition of the complex dot product in Def. 6.12. If the complex dot product is alternatively defined as $z \cdot w = z_1 w_1^* + z_2 w_2^* + \cdots + z_n w_n^*$, the definition of the complex inner product space must be modified accordingly to preserve linearity in the first argument.

2 Real Inner Product Spaces

Real inner product spaces are specific cases of complex inner product spaces where all vectors and scalars are restricted to be real.

> **Definition 6.14 — Real Inner Product Space.** A real inner product space is a real vector space V equipped with an inner product, which is a binary function resulting in a scalar $\langle \cdot, \cdot \rangle : V \times V \to \mathbb{R}$. The inner product must satisfy the following properties for all vectors u, v, w in V, and any scalar k in $\mathbb{R}$:
>
> (a) $\langle u, v \rangle = \langle v, u \rangle$
>
> (b) $\langle u, v + w \rangle = \langle u, v \rangle + \langle u, w \rangle$
>
> (c) $\langle u, kv \rangle = k\langle u, v \rangle$
>
> (d) $\langle v, v \rangle$ is a real, non-negative number, and $\langle v, v \rangle = 0$ if and only if $v = 0$.

Comparing Def. 6.14 with Def. 6.13, the key difference lies in the first property: in a complex inner product space, this property is known as conjugate symmetry. In contrast, for a real inner product space, it simplifies to the symmetry property because $\langle u, v \rangle$ and $\langle v, u \rangle$ are always real. As a result, a real inner product is linear in both arguments.

> **Theorem 6.10** For a real inner product, the following identities are true:
> $$\langle u, kv + mw \rangle = k\langle u, v \rangle + m\langle u, w \rangle \tag{6.33a}$$
> $$\langle ku + mv, w \rangle = k\langle u, w \rangle + m\langle v, w \rangle \tag{6.33b}$$

The proof of Theorem 6.10 is straightforward by taking the scalars to be real in Eqs. 6.31 and 6.32.

3 Norm and Distance

In general, the norm and distance can be defined for any inner product space, including $\mathbb{R}^n$ and $\mathbb{C}^n$.

> **Definition 6.15 — Norm and Distance.** Given a vector v in an inner product space V, the norm (or length) of v, denoted by $\|v\|$, is defined by
>
> $$\|v\| = \sqrt{\langle v, v \rangle}. \tag{6.34}$$
>
> The distance between two vectors v and w in V, denoted by $d(v, w)$, is defined by
>
> $$d(v, w) = \|v - w\| = \sqrt{\langle v - w, v - w \rangle}. \tag{6.35}$$

6.3.3 Dirac Notation of Inner Products

Dirac notation (or bra-ket notation), introduced in § 6.1, can represent complex inner products and greatly simplify their computations. To understand this notation, we first need to examine the dot product as a matrix product between a row vector and a column vector.

> **Definition 6.16 — Product of Row and Column Vectors.** Given a $1 \times n$ row vector r and an $n \times 1$ column vector c in $\mathbb{C}^n$:
>
> $$r = \begin{bmatrix} r_1 & r_2 & \cdots & r_n \end{bmatrix}, \quad c = \begin{bmatrix} c_1 \\ c_2 \\ \vdots \\ c_n \end{bmatrix},$$
>
> their matrix product, denoted as rc, is defined as:
>
> $$rc = r_1 c_1 + r_2 c_2 + \cdots + r_n c_n = \sum_{i=1}^{n} r_i c_i. \tag{6.36}$$

i In Definition 6.16, the row vector and the column vector are treated as matrices. While matrix multiplication will be discussed in more detail in the next chapter, this early introduction aims to facilitate a solid understanding of the bra-ket inner product notation. Important points to note include:

(a) Matrix multiplication is typically represented without an explicit multiplication symbol, a convention that underscores the operation's streamlined nature.

(b) To correctly apply Def. 6.16, the row vector must be placed to the left of the column vector.

Exercise 6.16 Demonstrate that for any two column vectors $u, v \in \mathbb{R}^n$, the dot product $u \cdot v = u^T v = v^T u$.

The complex dot product finds a new, simplified expression through the bra-ket notation. Given two vectors

$$|u\rangle = u = \begin{bmatrix} u_1 & u_2 & \cdots & u_n \end{bmatrix}^T,$$
$$|v\rangle = v = \begin{bmatrix} v_1 & v_2 & \cdots & v_n \end{bmatrix}^T,$$

the following derivation is valid using Defs. 6.7, 6.12 and 6.16:

$$\boldsymbol{u} \cdot \boldsymbol{v} = u_1^* v_1 + u_2^* v_2 + \cdots + u_n^* v_n = \begin{bmatrix} u_1^* & u_2^* & \cdots & u_n^* \end{bmatrix} \begin{bmatrix} v_1 \\ v_2 \\ \vdots \\ v_n \end{bmatrix} = \langle u | \, | v \rangle \equiv \langle u | v \rangle .$$

Here, for simplicity, we denote $\langle u | \, | v \rangle$ as $\langle u | v \rangle$. This notation, further defined below, streamlines complex quantum calculations and enhances readability.

Definition 6.17 — Dirac Notation of Inner Product. Given two vectors $|u\rangle = \boldsymbol{u}$ and $|v\rangle = \boldsymbol{v}$ in $\mathbb{C}^n$, their inner product is denoted as:

$$\langle u | v \rangle \equiv \langle \boldsymbol{u}, \boldsymbol{v} \rangle = \boldsymbol{u} \cdot \boldsymbol{v}. \tag{6.37}$$

Exercise 6.17 Explore the fundamental properties of the computational basis by verifying that:
  (a) $\langle 0 | 0 \rangle = 1$,   (c) $\langle 1 | 0 \rangle = 0$,
  (b) $\langle 0 | 1 \rangle = 0$,   (d) $\langle 1 | 1 \rangle = 1$.

The basic properties of complex inner products can be concisely expressed in Dirac notation through the following theorem:

Theorem 6.11 Given vectors $|u\rangle$, $|v\rangle$, $|w\rangle$ in $\mathbb{C}^n$ and a scalar $\alpha \in \mathbb{C}$, the following identities are true:

$$\langle u | v \rangle = \langle v | u \rangle^* \tag{6.38a}$$
$$\langle u + v | w \rangle = \langle u | w \rangle + \langle v | w \rangle \tag{6.38b}$$
$$\langle u | v + w \rangle = \langle u | v \rangle + \langle u | w \rangle \tag{6.38c}$$
$$\langle u | \alpha v \rangle = \alpha \langle u | v \rangle \tag{6.38d}$$
$$\langle \alpha u | v \rangle = \alpha^* \langle u | v \rangle \tag{6.38e}$$
$$\| v \|^2 = \langle v | v \rangle \tag{6.38f}$$

Equation 6.38f rewrites Eq. 6.29 in the bra-ket notation. This leads to the following theorem for a unit vector.

Theorem 6.12 Any unit vector $|v\rangle$ satisfies:

$$\langle v | v \rangle = 1. \tag{6.39}$$

Proof. Since $|v\rangle$ is a unit vector, its length must be one:

$$\| v \| = 1.$$

Then according to Eq. 6.38f,

$$\langle v | v \rangle = \| v \|^2 = 1.$$

$\square$

■ **Example 6.12** Let $|+\rangle = \frac{1}{\sqrt{2}}(|0\rangle + |1\rangle)$ and $|-\rangle = \frac{1}{\sqrt{2}}(|0\rangle - |1\rangle)$. Use Dirac notation to compute

(a) $\langle +|0\rangle$,   (b) $\langle +|+\rangle$.

Solution. We apply Theorems 6.1 and 6.11 and the properties of computational basis from Exercise 6.17:

(a)
$$\langle +|0\rangle = \langle +|\,|0\rangle = \frac{1}{\sqrt{2}}\left((\langle 0| + \langle 1|)\,|0\rangle\right)$$
$$= \frac{1}{\sqrt{2}}\left(\langle 0|0\rangle + \langle 1|0\rangle\right)$$
$$= \frac{1}{\sqrt{2}}(1+0)$$
$$= \frac{1}{\sqrt{2}}.$$

(b)
$$\langle +|+\rangle = \frac{1}{\sqrt{2}}\left((\langle 0| + \langle 1|)\frac{1}{\sqrt{2}}(|0\rangle + |1\rangle)\right)$$
$$= \frac{1}{2}\left(\langle 0|0\rangle + \langle 0|1\rangle + \langle 1|0\rangle + \langle 1|1\rangle\right)$$
$$= \frac{1}{2}(1+0+0+1)$$
$$= 1.$$

The results above can be verified using matrix notation:

$$|+\rangle = \frac{1}{\sqrt{2}}(|0\rangle + |1\rangle) = \begin{bmatrix}\frac{1}{\sqrt{2}}\\\frac{1}{\sqrt{2}}\end{bmatrix}, \quad |0\rangle = \begin{bmatrix}1\\0\end{bmatrix}.$$

Using the complex dot product formula (Def. 6.12), we find:

$$\langle +|0\rangle = \frac{1}{\sqrt{2}}\cdot 1 + \frac{1}{\sqrt{2}}\cdot 0 = \frac{1}{\sqrt{2}},$$
$$\langle +|+\rangle = \frac{1}{\sqrt{2}}\cdot\frac{1}{\sqrt{2}} + \frac{1}{\sqrt{2}}\cdot\frac{1}{\sqrt{2}} = \frac{1}{2} + \frac{1}{2} = 1.$$

∎

Exercise 6.18 Define $|-\rangle = \frac{1}{\sqrt{2}}(|0\rangle - |1\rangle)$. Use Dirac notation to compute

(a) $\langle -|1\rangle$,   (b) $\langle +|-\rangle$,   (c) $\langle -|-\rangle$.

Verify your answers using matrix notation.

When working with complex components, it's important to carefully apply the formulas from Eqs. 6.1Cc, 6.38a and 6.38e, especially where complex conjugates are involved. An illustrative example is given below.

■ **Example 6.13** Let $|+_i\rangle = \frac{1}{\sqrt{2}}(|0\rangle + i\,|1\rangle)$, $|-_i\rangle = \frac{1}{\sqrt{2}}(|0\rangle - i\,|1\rangle)$. Use Dirac notation to compute

(a) $\langle +_i|+_i\rangle$,   (b) $\langle +_i|-_i\rangle$.

Solution. Applying Eq. 6.10c, the bra of $|+_i\rangle$ is $\langle+_i| = \frac{1}{\sqrt{2}}(\langle 0| - i\langle 1|)$. Hence, the computations unfold as follows:

(a)
$$\langle+_i|+_i\rangle = \frac{1}{\sqrt{2}}(\langle 0| - i\langle 1|)\frac{1}{\sqrt{2}}(|0\rangle + i|1\rangle)$$
$$= \frac{1}{2}(\langle 0|0\rangle + i\langle 0|1\rangle - i\langle 1|0\rangle - i^2\langle 1|1\rangle)$$
$$= \frac{1}{2}(1 + 0 + 0 + 1)$$
$$= 1.$$

(b)
$$\langle+_i|-_i\rangle = \frac{1}{\sqrt{2}}(\langle 0| - i\langle 1|)\frac{1}{\sqrt{2}}(|0\rangle - i|1\rangle)$$
$$= \frac{1}{2}(\langle 0|0\rangle - i\langle 0|1\rangle - i\langle 1|0\rangle + i^2\langle 1|1\rangle)$$
$$= \frac{1}{2}(1 + 0 + 0 - 1)$$
$$= 0.$$

∎

Exercise 6.19 Verify the results in Example 6.13 using matrix notation.

Exercise 6.20 Compute $\langle+_i|+\rangle$ and $\langle-|-_i\rangle$ using both Dirac notation and matrix notation.

(i) The notation $|+\rangle, |-\rangle, |+_i\rangle, |-_i\rangle$ is often employed to denote specific spin states for a spin-$\frac{1}{2}$ particle. Together with $|0\rangle$ and $|1\rangle$, these states constitute a fundamental framework to study the basics of quantum computing. Mastering the properties of these states is essential for delving into more advanced topics within the field.

6.3.4 * Hilbert Spaces

In quantum mechanics and quantum computing, the term "Hilbert space" refers to the mathematical framework where quantum states reside. While Hilbert spaces are often infinite-dimensional, the complex inner product spaces $\mathbb{C}^n$ discussed earlier are special cases of finite-dimensional Hilbert spaces.

1 Infinite-Dimensional Complex Vector Spaces $\mathbb{C}^\infty$

The vector space $\mathbb{C}^\infty$ extends the finite-dimensional complex vector spaces $\mathbb{C}^n$. Whereas $\mathbb{C}^n$ consists of column vectors with n complex entries, $\mathbb{C}^\infty$ consists of infinite sequences of complex numbers:

$$\boldsymbol{v} = (v_1, v_2, v_3, \ldots), \quad v_i \in \mathbb{C}.$$

These spaces form the foundation for infinite-dimensional Hilbert spaces, generalizing finite-dimensional linear algebra concepts to handle infinite sequences or functions.

The inner product in $\mathbb{C}^\infty$ is defined analogously to $\mathbb{C}^n$. For two vectors $\boldsymbol{u}, \boldsymbol{v} \in \mathbb{C}^\infty$, the inner product is:

$$\langle \boldsymbol{u}, \boldsymbol{v}\rangle = \sum_{i=1}^{\infty} u_i^* v_i,$$

where u_i^* denotes the complex conjugate of u_i. For the inner product to be well-defined, the series must converge, leading to the concept of square-summable sequences (ℓ^2).

2   Functional Inner Spaces

Functional spaces, also known as spaces of functions, consist of functions with specific properties. When equipped with an inner product, these spaces become functional inner product spaces. For example, the inner product for functions $f(x)$ and $g(x)$ over an interval $[a, b]$ is:

$$\langle f, g \rangle = \int_a^b f^*(x) g(x)\, dx. \tag{6.40}$$

One of the most important examples of functional inner spaces is $L^2([a, b])$, the space of all square-integrable functions on $[a, b]$. Formally:

$$L^2([a, b]) = \left\{ f : [a, b] \to \mathbb{C} \mid \int_a^b |f(x)|^2\, dx < \infty \right\}.$$

The corresponding norm, called the L^2-norm, is defined as:

$$\|f\| = \sqrt{\int_a^b |f(x)|^2\, dx}. \tag{6.41}$$

Spaces of square-integrable functions like $L^2([a, b])$ play a central role in functional analysis and are examples of infinite-dimensional Hilbert spaces. These generalizations allow concepts like orthogonality and projection to extend naturally to functions.

Example: Fourier Series

A Fourier series exemplifies how infinite-dimensional spaces work. On the interval $[0, 2\pi]$, the basis functions are the complex exponentials:

$$\phi_k(x) = \frac{1}{\sqrt{2\pi}} e^{ikx}, \quad k \in \mathbb{Z}. \tag{6.42}$$

These basis functions satisfy the orthonormality condition:

$$\langle \phi_k, \phi_j \rangle = \int_0^{2\pi} \phi_k^*(x) \phi_j(x)\, dx = \delta_{kj}, \tag{6.43}$$

where δ_{kj} is the Kronecker delta.

Any square-integrable function $f(x) \in L^2([0, 2\pi])$ can be expressed as:

$$f(x) = \sum_{k=-\infty}^{\infty} c_k \phi_k(x), \tag{6.44}$$

where the Fourier coefficients c_k are computed using the projection formula:

$$c_k = \langle \phi_k, f \rangle = \int_0^{2\pi} \frac{1}{\sqrt{2\pi}} e^{-ikx} f(x)\, dx. \tag{6.45}$$

Note that the basis functions and their orthonormality condition depend on the interval. For instance, $\sin x$ and $\cos x$ are orthogonal on $[0, \pi]$, but not on $[\pi/4, 5\pi/4]$, even though the two intervals have the same range.

> **Exercise 6.21** Establish Eqs. 6.42 and 6.43 for the Fourier series on an arbitrary interval $[a, b]$.

3 Hilbert Spaces

Building on the concepts of infinite-dimensional complex vector spaces and functional inner product spaces, we are now ready to explore Hilbert spaces. A Hilbert space is a generalization of inner product spaces that includes the crucial property of *completeness*.

> **Definition 6.18 — Hilbert Space.** A Hilbert space is an inner product space that is complete with respect to the norm induced by its inner product. Completeness means that every Cauchy sequence in the space converges to a point within the space.

In the context of quantum computing and information, where the state space for an n-qubit system is a finite-dimensional complex inner product space of dimension 2^n, the terms "complex inner product space" and "Hilbert space" are interchangeable. This equivalence is formalized by the following theorem:

> **Theorem 6.13** Every finite-dimensional real or complex inner product space is a Hilbert space.

In quantum mechanics, infinite-dimensional Hilbert spaces are also common. Completeness is particularly significant in such infinite-dimensional spaces. It ensures that any linear combination or superposition of quantum states remains a valid quantum state. This property is essential for the consistency of quantum mechanics, enabling superposition and linear evolution within the space.

Space Completeness vs. Basis Completeness

In the Fourier series example discussed earlier, completeness of the Hilbert space $L^2([0, 2\pi])$ ensures that the Fourier series (Eq. 6.44) converges within the space. This property underpins the concept of a complete basis.

The basis $\{\phi_k(x)\}$ is said to be complete when the basis functions fully span $L^2([0, 2\pi])$. This means that any square-integrable function $f(x)$ can be decomposed uniquely into an infinite series of these basis functions. Without basis completeness, the basis might fail to span the entire space, leaving some functions unrepresentable.

The basis completeness condition in this case can also be expressed as:

$$\sum_{k=-\infty}^{\infty} \phi_k^*(x')\phi_k(x) = \delta(x - x'), \tag{6.46}$$

where $\delta(x - x')$ is the Dirac delta function. This serves as an analogue of the basis completeness condition in finite dimensions (see Eq. 8.30):

$$\sum_{k=1}^{n} |\phi_k\rangle\langle\phi_k| = I. \tag{6.47}$$

For further details on completeness and Cauchy sequences, as well as proofs of Theorem 6.13, readers are encouraged to consult advanced texts such as *Introduction to Hilbert Spaces with Applications* by Lokenath Debnath and Piotr Mikusinski [2].

6.4 Orthogonality and Projection

Orthogonality, defined by the inner product, is a key concept alongside norm in establishing orthonormal bases. It is closely related to the projection problem, which plays a crucial role in quantum computing for determining the measurement probabilities of different outcome states. This section explores orthogonality and the projection problem for complex vectors. To provide an intuitive understanding, we begin with an introduction using real geometric vectors.

6.4.1 Orthogonality for Real Vectors

1 Angle Between Geometric Vectors

From a geometric standpoint, the angle between two non-zero vectors in $\mathbb{R}^2$ (and similarly in $\mathbb{R}^3$) is defined as the angle formed by their arrow representations when their initial points coincide. This angle must satisfy the condition that $0 \leq \theta \leq \pi$, as depicted in Fig. 6.4. This constraint ensures that the angle is uniquely determined by its cosine value.

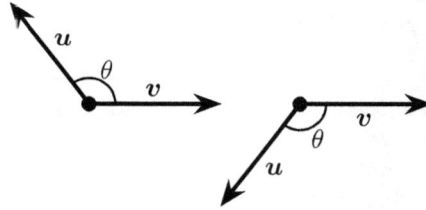

Figure 6.4: Angles Between Geometric Vectors

> i Since the direction of a zero vector is arbitrary, the angle between a zero vector and any vector is also arbitrary.

2 Dot Product and Angle

The dot product is closely related to the angle for real vectors. Let's look at an example in $\mathbb{R}^2$. Consider two non-zero vectors $\boldsymbol{u} = (u_1, u_2)$ and $\boldsymbol{v} = (v_1, v_2)$. The angle θ between them can be determined by examining the triangle formed by $\boldsymbol{u}$, $\boldsymbol{v}$, and $\boldsymbol{u} - \boldsymbol{v}$, as depicted in Fig. 6.5. By applying the law of cosines (see § 2.6), we obtain:

$$\|\boldsymbol{u} - \boldsymbol{v}\|^2 = \|\boldsymbol{u}\|^2 + \|\boldsymbol{v}\|^2 - 2\|\boldsymbol{u}\|\|\boldsymbol{v}\| \cos \theta. \tag{6.48}$$

Rearranging this equation yields:

$$\|\boldsymbol{u}\|\|\boldsymbol{v}\| \cos \theta = \frac{1}{2} \left(\|\boldsymbol{u}\|^2 + \|\boldsymbol{v}\|^2 - \|\boldsymbol{u} - \boldsymbol{v}\|^2 \right). \tag{6.49}$$

It can be shown that the right-hand side of Eq. 6.49 equals the dot product of $\boldsymbol{u}$ and $\boldsymbol{v}$ (Problem 6.15a), resulting in the following theorem.

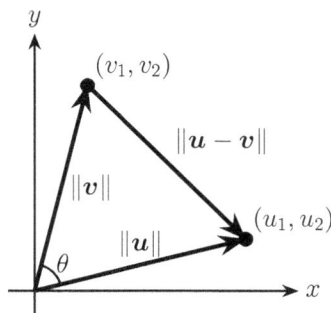

Figure 6.5: A Triangle Formed by u, v and $u - v$

Theorem 6.14 For vectors u and v in $\mathbb{R}^2$ or $\mathbb{R}^3$, the following identity is true:

$$u \cdot v = \|u\|\|v\| \cos \theta, \tag{6.50}$$

where θ is the geometric angle between u and v.

For non-zero vectors, rearranging Theorem 6.14 gives the formula for $\cos \theta$:

$$\cos \theta = \frac{u \cdot v}{\|u\|\|v\|}. \tag{6.51}$$

3   Generalized Angle and Orthogonality in $\mathbb{R}^n$

Equation 6.51 is derived based on Euclidean geometry in two or three dimensions. In higher dimensions, this equation serves as the definition of the generalized angle between two real vectors.

Definition 6.19 — Angle Between Two Real Vectors. For any non-zero real vectors u and v in $\mathbb{R}^n$, the angle θ between them is defined to lie within the interval $[0, \pi]$, and the cosine of θ is given by the formula:

$$\cos \theta = \frac{u \cdot v}{\|u\|\|v\|}.$$

■ **Example 6.14** Determine the angle between $(1, 1, 0, 0)$ and $(1, 0, 1, 0)$.

Solution. Let θ be the angle between the given vectors. Using Eq. 6.51, we obtain:

$$\cos \theta = \frac{1 \times 1 + 1 \times 0 + 0 \times 1 + 0 \times 0}{\sqrt{1^2 + 1^2 + 0^2 + 0^2}\sqrt{1^2 + 0^2 + 1^2 + 0^2}} = \frac{1}{2}.$$

Therefore, $\theta = \cos^{-1} \frac{1}{2} = \frac{\pi}{3}$.   ■

ⓘ For Def. 6.19 to hold, the computed cosine value must lie within the interval $[-1, 1]$, as required by the range of the cosine function. This condition is always satisfied due to the **Cauchy-Schwarz Inequality**, which states:

$$-\|u\|\|v\| \leq u \cdot v \leq \|u\|\|v\|. \tag{6.52}$$

A more general discussion of this important inequality will be provided in § 11.1.3.

It is straightforward to imply from Definition 6.19 that $\theta = \frac{\pi}{2}$ is equivalent to $u \cdot v = 0$. In this scenario, u and v are said to be **orthogonal**.

■ **Example 6.15** Find the unit vectors that are orthogonal to $(1, -2)$.

Solution. Let (a, b) be a unit vector orthogonal to $(1, -2)$. It is evident that a, b have to satisfy two conditions:

(a) Orthogonality: $(a, b) \cdot (1, -2) = a - 2b = 0$

(b) Normalization: $a^2 + b^2 = 1$

The first condition demands $a = 2b$. Substituting this into the second condition gives $5b^2 = 1 \quad \Rightarrow \quad b = \pm\frac{1}{\sqrt{5}}$. Therefore, there are two unit vectors as the answer:

$$\left(\frac{2}{\sqrt{5}}, \frac{1}{\sqrt{5}}\right), \ \left(-\frac{2}{\sqrt{5}}, -\frac{1}{\sqrt{5}}\right).$$

■

| **Exercise 6.22** Find the unit vectors that are orthogonal to $(3, 4)$.

6.4.2 Orthogonality in Complex Inner Product Spaces

For complex vectors, $u \cdot v$ is generally a complex number. As a result, Eq. 6.51 no longer serves as a valid definition for the generalized angle, since the right-hand side can be complex. On the other hand, the orthogonality condition $u \cdot v = 0$ for real vectors remains valid and provides useful insights for complex vectors, as we will see in the projection problem discussed later in this section.

Definition 6.20 — Orthogonality in $\mathbb{C}^n$. Given two vectors $u, v \in \mathbb{C}^n$, we say u and v are orthogonal when $u \cdot v = v \cdot u = 0$.

Definition 6.21 — Orthogonality in Inner Product Spaces. Generally, for u and v in a complex inner product space V, we say u and v are orthogonal when $\langle u, v \rangle = \langle v, u \rangle = 0$.

Theorem 6.15 — Generalized Pythagorean Theorem. For $u, v \in \mathbb{C}^n$, if u and v are orthogonal, then

$$\|u\|^2 + \|v\|^2 = \|u + v\|^2. \tag{6.53}$$

Proof. Starting from the right-hand side of Eq. 6.53 and using the orthogonality condition $u \cdot v = v \cdot u = 0$ gives:

$$\|u + v\|^2 = (u + v) \cdot (u + v) = \|u\|^2 + u \cdot v + v \cdot u + \|v\|^2 = \|u\|^2 + \|v\|^2.$$

□

■ **Example 6.16** Given $u = (1, 0)$, find a unit vector $v \in \mathbb{C}^2$ that is orthogonal to u.

Solution. Let $v = (v_1, v_2)$. The given orthogonality condition gives

$$u \cdot v = 0 \quad \Rightarrow \quad 1 \cdot v_1 + 0 \cdot v_2 = 0 \quad \Rightarrow \quad v_1 = 0.$$

Furthermore, to ensure v to be a unit vector, it must hold that:

$$\|v\|^2 = |v_1|^2 + |v_2|^2 = 0 + |v_2|^2 = 1 \quad \Rightarrow \quad |v_2| = 1,$$

This leads us to a general form of $w = (0, v_2)$, where $|v_2| = 1$. Therefore, we can explicitly express v as:

$$v = (0, e^{i\phi}) \text{ for } \phi \in \mathbb{R}.$$

∎

The result of Example 6.16 uncovers an intriguing aspect about complex vectors—there exist an infinite number of unit vectors orthogonal to $(1, 0)$ in $\mathbb{C}^2$, in stark contrast to $\mathbb{R}^2$, where only two such vectors can be found. This distinction arises from the fact that multiplication by a phase factor $e^{i\phi}$ maintains the magnitude of a complex number z: $|ze^{i\phi}| = |z|$. This concept is further expanded upon in the subsequent theorem, which provides a broader application.

Theorem 6.16 Given two unit vectors $z, w \in \mathbb{C}^n$ that are orthogonal, there are infinitely many unit vectors in the span of w that are orthogonal to z in the form of $e^{i\phi}w$ for $\phi \in \mathbb{R}$.

Proof. Consider a vector w' within the span of w, represented by $w' = \lambda w$ for $\lambda \in \mathbb{C}$. The orthogonality of w' to z can be demonstrated as follows:

$$z \cdot w' = z \cdot (\lambda w) = \lambda(z \cdot w) = \lambda \cdot 0 = 0.$$

Furthermore, to calculate the norm of w', we have:

$$\|w'\| = \|\lambda w\| = |\lambda| \cdot \|w\| = |\lambda|.$$

Here we utilize the property of complex vectors and the given condition $\|w\| = 1$. For w' to be a unit vector, we must have $|\lambda| = 1$, implying that λ is in the form of $e^{i\phi}$ ($\phi \in \mathbb{R}$).

□

ⓘ In quantum computing, quantum states are represented by unit vectors in $\mathbb{C}^n$. According to Theorem 6.16, the act of multiplying any unit vector by a factor of $e^{i\phi}$ leaves its modulus unchanged, preserving its status as a unit vector, as well as maintaining existing orthogonality relationships. These multipliers are identified as **global phase factors**, underscoring their uniform effect across the quantum state.

The following theorem expresses Def. 6.20 in the bra-ket notation, based on Def. 6.17:

Theorem 6.17 Given two orthogonal vectors $|u\rangle$ and $|v\rangle$ in $\mathbb{C}^n$, they satisfy:

$$\langle u|v \rangle = \langle v|u \rangle = 0. \tag{6.54}$$

■ **Example 6.17** Show that $|+\rangle = \frac{1}{\sqrt{2}}(|0\rangle + |1\rangle)$ and $|-\rangle = \frac{1}{\sqrt{2}}(|0\rangle - |1\rangle)$ are orthogonal.

Proof. Using the properties of the computational basis: $\langle 0|0\rangle = \langle 1|1\rangle = 1$, $\langle 0|1\rangle = \langle 1|0\rangle = 0$:

$$\langle +|-\rangle = \frac{1}{\sqrt{2}}(\langle 0| + \langle 1|)\frac{1}{\sqrt{2}}(|0\rangle - |1\rangle)$$

$$= \frac{1}{2}(\langle 0| + \langle 1|)(|0\rangle - |1\rangle)$$

$$= \frac{1}{2}(\langle 0|0\rangle + \langle 1|0\rangle - \langle 0|1\rangle - \langle 1|1\rangle)$$

$$= \frac{1}{2}(1 + 0 - 0 - 1)$$

$$= 0.$$

Therefore, $|+\rangle$ and $|-\rangle$ are orthogonal according to Theorem 6.17.   ∎

Exercise 6.23 Show that $|u\rangle = \frac{1}{\sqrt{2}}(i\,|0\rangle + |1\rangle)$ and $|v\rangle = \frac{1}{\sqrt{2}}(i\,|0\rangle - |1\rangle)$ are orthogonal.

6.4.3   Orthogonal Projections

1   The Projection Theorem

Orthogonality is closely related to projection. Typically, when the term "projection" is used, it refers to "orthogonal projection". An example in $\mathbb{R}^2$ is given in Fig. 6.6.

Given a vector $v = (v_1, v_2)$, we hope to find its orthogonal projection w onto a subspace of $\mathbb{R}^2$ as established by the span of a given non-zero vector a (the dashed line), as illustrated in Fig. 6.6a.

In Euclidean geometry, this is equivalent to finding the coordinate of the perpendicular foot (w_1, w_2) (Fig. 6.6b) from (v_1, v_2) to the line where a resides.

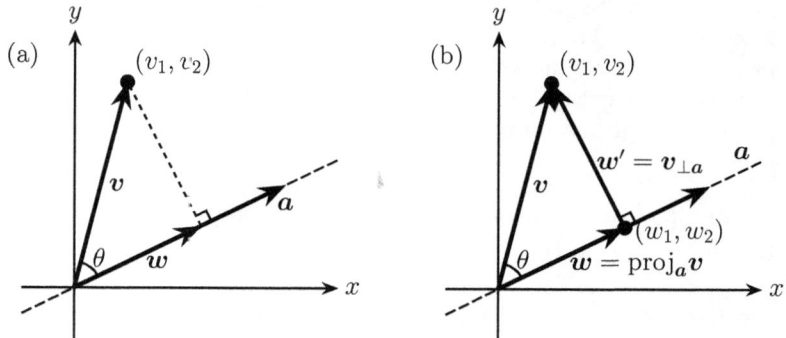

Figure 6.6: Vector Projection in 2D

Knowledge in Euclidean geometry tells us that there is a unique solution to this problem. The following theorem provides a general solution that applies not only to $\mathbb{R}^2$, but also to $\mathbb{R}^n$ and $\mathbb{C}^n$.

Theorem 6.18 — Projection Theorem. Given v and a non-zero vector a in $\mathbb{R}^n$ or $\mathbb{C}^n$, there is a unique way to decompose v into two vector components $v = w + w'$, where w is in the span of a and w' is orthogonal to a.

w is known as the **orthogonal projection of v onto a**, denoted as $\mathrm{proj}_a v$ or $v_{\|a}$, and w' is known as the **component of v orthogonal to a**, denoted as $v_{\perp a}$. They are given by the following formulas:

$$\mathrm{proj}_a v = v_{\|a} = \frac{a \cdot v}{\|a\|^2} a, \tag{6.55a}$$

$$v_{\perp a} = v - \frac{a \cdot v}{\|a\|^2} a. \tag{6.55b}$$

Proof. Since w is in the span of a, it can be written as a scalar multiple of a: $w = \lambda a$, where $\lambda \in \mathbb{C}$. To determine the value of λ, we take the dot product of both sides of the equation $v = w + w'$ with a:

$$a \cdot v = a \cdot (w + w') = a \cdot w + a \cdot w' = a \cdot w = a \cdot (\lambda a) = \lambda (a \cdot a). \tag{6.56}$$

Note that $a \cdot w' = 0$ since a and w' are orthogonal by construction. Solving for λ, we get:

$$\lambda = \frac{a \cdot v}{a \cdot a} = \frac{a \cdot v}{\|a\|^2}. \tag{6.57}$$

Therefore, we have:

$$w = \lambda a = \frac{a \cdot v}{\|a\|^2} a, \quad w' = v - w = v - \frac{a \cdot v}{\|a\|^2} a.$$

The decomposition is unique, as Eq. 6.57 gives a unique solution for λ in terms of the dot product. $\square$

Exercise 6.24 If $v, a \in \mathbb{R}^n$, can we replace the term $a \cdot v$ with $v \cdot a$ in Eqs. 6.55a and 6.55b? What if $v, a \in \mathbb{C}^n$?

■ **Example 6.18** Find $\mathrm{proj}_a v$ and $v_{\perp a}$ for $v = (4, 1)$ and $a = (1, 2)$.

Solution. Plugging v and a into Eqs. 6.55a and 6.55b gives:

$$\mathrm{proj}_a v = \frac{a \cdot v}{\|a\|^2} a = \frac{1 \times 4 + 2 \times 1}{1^2 + 2^2} (1, 2) = \left(\frac{6}{5}, \frac{12}{5} \right),$$

$$v_{\perp a} = v - \mathrm{proj}_a v = (4, 1) - (\frac{6}{5}, \frac{12}{5}) = \left(\frac{14}{5}, -\frac{7}{5} \right).$$

■

■ **Example 6.19** Find $\mathrm{proj}_a v$ and $v_{\perp a}$ for $v = (0, 1)$ and $a = (1, i)$.

Solution. Plugging v and a into Eqs. 6.55a and 6.55b gives:

$$\mathrm{proj}_a v = \frac{a \cdot v}{\|a\|^2} a = \frac{1 \times 0 + (-i) \times 1}{1^2 + |i|^2} (1, i) = \left(-\frac{i}{2}, \frac{1}{2} \right),$$

$$v_{\perp a} = v - \mathrm{proj}_a v = (0, 1) - (-\frac{i}{2}, \frac{1}{2}) = \left(\frac{i}{2}, \frac{1}{2} \right).$$

■

Exercise 6.25 Find $\mathrm{proj}_a v$ and $v_{\perp a}$ for $v = (3, 1)$ and $a = (1, 1)$. Draw a graph to geometrically interpret your result.

| **Exercise 6.26** Find $\text{proj}_a v$ and $v_{\perp a}$ for $v = \frac{1}{2}(1+i, 1-1)$ and $a = (0, i)$.

2 Projection Between Unit Complex Vectors

In quantum computing, since quantum state vectors are unit vectors, the projection between unit complex vectors is crucial. The following theorem establishes the fundamental formula for this operation.

Theorem 6.19 — Projection Theorem Between Unit Complex Vectors. Consider $|u\rangle$ and $|v\rangle$ to be two unit vectors in $\mathbb{C}^n$. The projection of $|v\rangle$ onto $|u\rangle$ is given by the formula:

$$\text{proj}_{|u\rangle} |v\rangle = \langle u|v\rangle |u\rangle, \tag{6.58}$$

where $|\langle u|v\rangle|^2 \leq 1$.

Proof.   Rewriting Eq. 6.58 in Dirac notation gives:

$$\text{proj}_{|u\rangle} |v\rangle = \frac{\langle u|v\rangle}{\langle u|u\rangle} |u\rangle.$$

Since $|u\rangle$ is a unit vector, $\langle u|u\rangle = \|u\|^2 = 1$, which simplifies to Eq. 6.58.

To prove $|\langle u|v\rangle| \leq 1$, we apply the Cauchy-Schwarz inequality:

$$|\langle u|v\rangle|^2 \leq \langle u|u\rangle \langle v|v\rangle. \tag{6.59}$$

Since $|u\rangle$ and $|v\rangle$ are unit vectors, $\langle u|u\rangle = \langle v|v\rangle = 1$. Thus, we have:

$$|\langle u|v\rangle|^2 \leq 1.$$

□

The Born Rule in quantum mechanics states that upon measuring an observable M on a quantum state $|\psi\rangle$, the probability of obtaining each eigenvalue λ_i is given by $|\langle \lambda_i|\psi\rangle|^2$, where $|\lambda_i\rangle$ is the corresponding eigenvector [1]. (Refer to Chapter 9 for more details about eigenvalues and eigenvectors.) Theorem 6.19 ensures that $|\langle \lambda_i|\psi\rangle|^2 \leq 1$, thus guaranteeing that it represents a valid probability.

3 The Cauchy-Schwarz Inequality

As discussed above, the cosine value between two vectors in $\mathbb{R}^n$, given by Def. 6.19, must lie within $[-1, 1]$, and the projected length of a unit vector onto another, given in Eq. 6.58, must lie within $[0, 1]$ for the Born rule to hold. These conditions are guaranteed by the Cauchy-Schwarz inequality.

The Cauchy-Schwarz inequality is a fundamental theorem in linear algebra and plays a crucial role in validating the mathematical framework underlying quantum mechanics and quantum computing. Its general form is given as follows.

Theorem 6.20 — Cauchy-Schwarz Inequality. Consider any inner product space V and two vectors $u, v \in V$. The following inequality holds:

$$|\langle u, v\rangle|^2 \leq \langle u, u\rangle\langle v, v\rangle. \tag{6.60}$$

The Cauchy-Schwarz inequality appears in several specific forms. When V is taken to be $\mathbb{C}^n$ and Dirac notation is used, Eq. 6.60 transforms into Eq. 6.59.

Similarly, when V is taken to be $\mathbb{R}^n$, it becomes Eq. 6.52, which applies to real inner product spaces.

We will further explore the Cauchy-Schwarz inequality, along with its proof and related inequalities, in § 11.1.3.

6.5 Orthonormal Bases

Recall from § 5.4.2 that a basis is a minimal set of vectors that spans a given vector space. A set of vectors $S = \{v_1, v_2, \ldots, v_n\}$ is said to be a basis of a vector space V if S spans V and is linearly independent.

In this section, we extend the concept of bases to inner product spaces, which possess additional properties such as norm and orthogonality. In these spaces, orthonormal bases—sets of vectors that are both orthogonal and normalized—are fundamental to the theoretical framework of quantum mechanics and quantum computing. They simplify the representation of quantum states and key concepts such as superposition, entanglement, state evolution, and measurement. In these fields, Dirac notation is the standard tool for working with orthonormal bases.

6.5.1 Definition and Basic Properties

An orthonormal basis is defined as a set of basis vectors that are both normalized and mutually orthogonal. A notable example is the computational basis, $\{|0\rangle, |1\rangle\}$, in $\mathbb{C}^2$. The complete definition of an orthonormal basis follows.

Definition 6.22 — Orthonormal Basis. A basis S of an n-dimensional complex inner product space V is said to be orthonormal when all vectors $\{|b_1\rangle, |b_2\rangle, \ldots, |b_n\rangle\}$ in S are normalized (each having a norm of one) and orthogonal to each other:

$$\||b_1\|| = \||b_2\|| = \cdots = \||b_n\|| = 1, \tag{6.61a}$$

$$\langle b_i|b_j\rangle = 0 \quad \text{when } i \neq j. \tag{6.61b}$$

Since $\||b_i\||^2 = \langle b_i|b_i\rangle$, we can integrate Eqs. 6.61a and 6.61b into a fundamental property for any orthonormal basis:

Theorem 6.21 — Fundamental Property of Orthonormal Bases. Given a basis $S = \{|b_1\rangle, |b_2\rangle, \ldots, |b_n\rangle\}$ of an n-dimensional complex inner product space, S is orthonormal if and only if all inner products between pairs of basis vectors satisfy:

$$\langle b_i|b_j\rangle = \delta_{ij}, \quad \text{where } \delta_{ij} = \begin{cases} 0, & \text{when } i \neq j, \\ 1, & \text{when } i = j. \end{cases} \tag{6.62}$$

■ **Example 6.20** Show that $|+\rangle = \frac{1}{\sqrt{2}}(|0\rangle + |1\rangle)$ and $|-\rangle = \frac{1}{\sqrt{2}}(|0\rangle - |1\rangle)$ form an orthonormal basis of $\mathbb{C}^2$, referred to as the **Hadamard Basis**.

Proof. The results from Example 6.12 and Exercise 6.18 show that

$$\langle +|+\rangle = 1, \ \langle -|-\rangle = 1, \ \langle +|-\rangle = 0.$$

Therefore, based on Theorem 6.21, $|+\rangle, |-\rangle$ form an orthonormal basis of $\mathbb{C}^2$. ■

Exercise 6.27 Show that $|+_i\rangle = \frac{1}{\sqrt{2}}(|0\rangle + i|1\rangle)$, $|-_i\rangle = \frac{1}{\sqrt{2}}(|0\rangle - i|1\rangle)$ form an orthonormal basis of $\mathbb{C}^2$.

When expressing any unit vector in terms of orthonormal bases, the following key property holds true.

Theorem 6.22 Let $|\psi\rangle$ be a **unit vector** in $\mathbb{C}^n$, and let $\{|b_i\rangle\}$, $i = 1, 2, \ldots, n$, be an orthonormal basis for this space. Suppose $|\psi\rangle$ is expressed as a linear combination of the basis vectors:

$$|\psi\rangle = c_1|b_1\rangle + c_2|b_2\rangle + \cdots + c_n|b_n\rangle,$$

where $c_1, c_2, \ldots, c_n \in \mathbb{C}$. Then, the squared magnitudes of the coefficients satisfy the following constraint:

$$\sum_{i=1}^{n} |c_i|^2 = 1. \tag{6.63}$$

Proof. Using rules of the complex dot product and the properties of an orthonormal basis gives:

$$\langle\psi|\psi\rangle = \left(c_1^*\langle b_1| + c_2^*\langle b_2| + \cdots + c_n^*\langle b_n|\right)\left(c_1|b_1\rangle + c_2|b_2\rangle + \cdots + c_n|b_n\rangle\right)$$

$$= \sum_{i=1}^{n} c_i^* c_i \langle b_i|b_i\rangle + \sum_{i \neq j} c_i^* c_j \langle b_i|b_j\rangle$$

$$= \sum_{i=1}^{n} c_i^* c_i \cdot 1 + \sum_{i \neq j} c_i^* c_j \cdot 0$$

$$= \sum_{i=1}^{n} |c_i|^2.$$

Since $|\psi\rangle$ is a unit vector, $\langle\psi|\psi\rangle = \||\psi\rangle\|^2 = 1$. Therefore, $\sum_{i=1}^{n} |c_i|^2 = 1$. □

Superposition and Probability in Quantum Mechanics

In quantum mechanics, the expression $|\psi\rangle = c_1|b_1\rangle + c_2|b_2\rangle + \cdots + c_n|b_n\rangle$ represents the vector $|\psi\rangle$ as a **superposition state** within the basis $\{|b_i\rangle\}$. When $|\psi\rangle$ is measured in the same basis, the probability associated with each basis state $|b_i\rangle$ is $|c_i|^2$. Theorem 6.22 asserts that the sum of these probabilities equals one, ensuring consistency with the principles of probability.

$\boxed{i}$ Starting this section, we adopt the convention that all complex vectors expressed in Dirac notation are normalized (unit vectors), unless stated otherwise. This is a standard practice in quantum computing literature.

6.5.2   Decomposition of Vectors in Orthonormal Bases

In quantum computing, state vectors $|\psi\rangle$ are typically expressed as superpositions within an orthonormal basis. For example, the state $|+\rangle$ is written as $|+\rangle = \frac{1}{\sqrt{2}}(|0\rangle + |1\rangle)$, rather than in matrix form $|+\rangle = \begin{bmatrix} \frac{1}{\sqrt{2}} & \frac{1}{\sqrt{2}} \end{bmatrix}^T$. This notation simplifies operations such as outer products and tensor products, which will be covered in later chapters. In this subsection, we focus on decomposing vectors into orthonormal bases.

Theorem 6.23 Given a unit vector $|\psi\rangle$ and an orthonormal basis $S = \{|b_i\rangle\}$, $i = 1, 2, \ldots, n$, in an n-dimensional complex inner product space, $|\psi\rangle$ can be decomposed as a superposition of the basis vectors $\{|b_i\rangle\}$:

$$|\psi\rangle = c_1 |b_1\rangle + c_2 |b_2\rangle + \cdots + c_n |b_n\rangle = \sum_{i=1}^{n} c_i |b_i\rangle, \tag{6.64}$$

where each coefficient c_i is given by:

$$c_i = \langle b_i|\psi\rangle. \tag{6.65}$$

Proof. Since S is an orthonormal basis, the vector $|\psi\rangle$ can be expressed as a linear combination of the basis vectors:

$$|\psi\rangle = c_1 |b_1\rangle + c_2 |b_2\rangle + \cdots + c_n |b_n\rangle.$$

To find c_i, we perform the bra-ket inner product between each $|b_i\rangle$ and $|\psi\rangle$:

$$\begin{aligned}
\langle b_i|\psi\rangle &= \langle b_i| \left(c_1 |b_1\rangle + c_2 |b_2\rangle + \cdots + c_n |b_n\rangle\right) \\
&= c_1 \langle b_i|b_1\rangle + c_2 \langle b_i|b_2\rangle + \cdots + c_n \langle b_i|b_n\rangle \\
&= c_i \langle b_i|b_i\rangle + \sum_{j \neq i} c_j \langle b_i|b_j\rangle \\
&= c_i \cdot 1 + \sum_{j \neq i} c_j \cdot 0 \\
&= c_i.
\end{aligned}$$

Therefore, $c_i = \langle b_i|\psi\rangle$ for $i = 1, 2, \ldots, n$. ◻

Exercise 6.28 In Theorem 6.23, explain why $c_i = \langle b_i|\psi\rangle$ and not $\langle\psi|b_i\rangle$.

Theorem 6.23 can also be rewritten as:

$$|\psi\rangle = \langle b_1|\psi\rangle |b_1\rangle + \langle b_2|\psi\rangle |b_2\rangle + \cdots + \langle b_n|\psi\rangle |b_n\rangle = \sum_{i=1}^{n} \langle b_i|\psi\rangle |b_i\rangle. \tag{6.66}$$

By comparing Eq. 6.66 with Eq. 6.58, we observe that each component $\langle b_i|\psi\rangle |b_i\rangle$ corresponds to $\text{proj}_{|b_i\rangle} |\psi\rangle$, the projection of $|\psi\rangle$ onto $|b_i\rangle$.

In $\mathbb{R}^2$, a general vector $\boldsymbol{v} = (v_1, v_2)$ can be decomposed as:

$$\boldsymbol{v} = v_1\hat{\boldsymbol{i}} + v_2\hat{\boldsymbol{j}}, \tag{6.67}$$

where $\{\hat{\boldsymbol{i}}, \hat{\boldsymbol{j}}\}$ represents the standard basis, defined as $\hat{\boldsymbol{i}} = (1, 0)$ and $\hat{\boldsymbol{j}} = (0, 1)$. Using the dot product, we find that:

$$v_1 = \hat{\boldsymbol{i}} \cdot \boldsymbol{v}, \quad v_2 = \hat{\boldsymbol{j}} \cdot \boldsymbol{v}.$$

Substituting these into the decomposition gives:

$$\boldsymbol{v} = (\hat{\boldsymbol{i}} \cdot \boldsymbol{v})\hat{\boldsymbol{i}} + (\hat{\boldsymbol{j}} \cdot \boldsymbol{v})\hat{\boldsymbol{j}}. \tag{6.68}$$

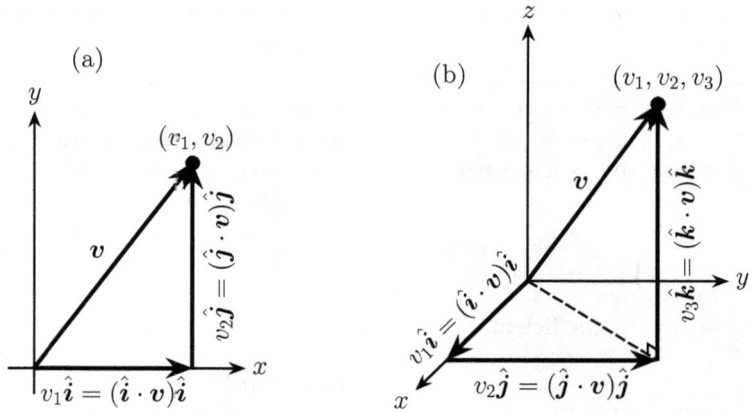

Figure 6.7: Decomposition of Vectors in Orthonormal Bases

Similarly, in $\mathbb{R}^3$, a general vector $v = (v_1, v_2, v_3)$ can be decomposed as:

$$v = v_1\hat{i} + v_2\hat{j} + v_3\hat{k}, \tag{6.69}$$

where the standard basis vectors are $\hat{i} = (1,0,0)$, $\hat{j} = (0,1,0)$, and $\hat{k} = (0,0,1)$. Using the dot product:

$$v_1 = \hat{i} \cdot v, \quad v_2 = \hat{j} \cdot v, \quad v_3 = \hat{k} \cdot v.$$

The decomposition then becomes:

$$v = (\hat{i} \cdot v)\hat{i} + (\hat{j} \cdot v)\hat{j} + (\hat{k} \cdot v)\hat{k}. \tag{6.70}$$

The same principle applies to complex inner product spaces, as stated in Theorem 6.23. In this context, the coefficient $c_i = \langle b_i | \psi \rangle$ represents the ith coordinate of $|\psi\rangle$ relative to the orthonormal basis $S = \{|b_i\rangle\}$. Each coordinate is obtained by orthogonally projecting $|\psi\rangle$ onto the corresponding basis vector $|b_i\rangle$.

This demonstrates a direct analogy between the decomposition of vectors in $\mathbb{R}^2$ or $\mathbb{R}^3$ and the superposition representation of vectors in $\mathbb{C}^n$.

■ **Example 6.21** Decompose $|0\rangle$ in the orthonormal basis $\{|+\rangle, |-\rangle\}$.

Solution. The decomposition can be obtained using Theorem 6.23:

$$|0\rangle = \langle +|0\rangle\, |+\rangle + \langle -|0\rangle\, |-\rangle$$
$$= \frac{1}{\sqrt{2}}|+\rangle + \frac{1}{\sqrt{2}}|-\rangle.$$

Here, $\langle +|0\rangle = \langle -|0\rangle = \frac{1}{\sqrt{2}}$, calculated based on the definitions $|+\rangle = \frac{1}{\sqrt{2}}(|0\rangle + |1\rangle)$ and $|-\rangle = \frac{1}{\sqrt{2}}(|0\rangle - |1\rangle)$. ■

❘ **Exercise 6.29** Decompose $|1\rangle$ in the orthonormal basis $\{|+_i\rangle, |-_i\rangle\}$.

6.5.3 ✳Gram-Schmidt Process

Every finite-dimensional inner product space, except for the trivial zero space, can be shown to possess an orthonormal basis. In some situations, it is crucial to create an orthonormal basis from a non-orthonormal one. The Gram-Schmidt Process provides a methodical algorithm for this task. This process is especially beneficial when working with degenerate eigenvalues (see § 9.1.3 for further details). This subsection delves into the Gram-Schmidt Process, explaining its steps with illustrative examples and exercises.

1 Orthogonal Basis

To fully understand the Gram-Schmidt Process, we must first introduce a few theorems that lay the theoretical foundation required for a detailed exploration of the process.

> **Theorem 6.24 — Basis from Linearly Independent Vectors.** Given a set S with exactly n vectors from a vector space V of dimension n, S is a basis of V if S is linearly independent.

Proof. Assume S is linearly dependent. Then, at least one vector in S can be expressed as a linear combination of the others, implying that it is possible to remove at least one vector from S without changing its span.

This reduction results in a set of fewer than n vectors that purportedly spans V. This contradicts Theorem 5.6, which states that the minimum number of vectors required to span V is precisely n. Therefore, S must be linearly independent. ◻

> **Theorem 6.25 — Orthogonal Basis.** If $S = \{v_1, v_2, \ldots, v_n\}$ is a set of nonzero vectors in an inner product space V of dimension n, and these vectors are orthogonal to each other such that $\langle v_i | v_j \rangle = 0$ for all $i \neq j$, then S constitutes a basis of V. Such a set is termed an orthogonal basis.

Proof. To establish that S forms a basis of V, we need to demonstrate that S is linearly independent, based on Theorem 6.24. Consider the equation for scalars k_i, $i = 1, 2, \ldots, n$:

$$k_1 v_1 + k_2 v_2 + \cdots + k_n v_n = 0.$$

Taking the left dot product of this equation with each v_i, $i = 1, 2, \ldots, n$, the right-hand side becomes $\langle v_i | 0 \rangle = 0$. For the left-hand side:

$$\langle v_i | k_1 v_1 + k_2 v_2 + \cdots + k_n v_n \rangle = k_1 \langle v_i | v_1 \rangle + k_2 \langle v_i | v_2 \rangle + \cdots + k_n \langle v_i | v_n \rangle$$
$$= k_i \langle v_i | v_i \rangle,$$

since $\langle v_i | v_j \rangle = 0$ for all $i \neq j$. Therefore:

$$k_i \langle v_i | v_i \rangle = 0.$$

Since $\langle v_i | v_i \rangle = \|v_i\|^2 \neq 0$ for nonzero vectors, it follows that $k_i = 0$. Hence, S is linearly independent. ◻

Normalizing each basis vector in an orthogonal basis naturally leads to an orthonormal basis. This transformation is formalized in the following theorem:

Theorem 6.26 Given an orthogonal basis $S = \{v_1, v_2, \ldots, v_n\}$ for an inner product space V, an orthonormal basis $\{\hat{v}_1, \hat{v}_2, \ldots, \hat{v}_n\}$ can be obtained by normalizing each basis vector in S:

$$\hat{v}_1 = \frac{v_1}{\|v_1\|}, \ \hat{v}_2 = \frac{v_2}{\|v_2\|}, \ \ldots, \ \hat{v}_n = \frac{v_n}{\|v_n\|}.$$

▪ **Example 6.22** Transform the orthogonal basis $\{(i, -1), (1, -i)\}$ for $\mathbb{C}^2$ into an orthonormal basis.

Solution. Normalizing each basis vector gives:

$$\frac{(i, -1)}{\|(i, -1)\|} = \frac{(i, -1)}{\sqrt{|i|^2 + |1|^2}} = (\frac{i}{\sqrt{2}}, -\frac{1}{\sqrt{2}}),$$

$$\frac{(1, -i)}{\|(1, -i)\|} = \frac{(1, -i)}{\sqrt{|1|^2 + |-i|^2}} = (\frac{1}{\sqrt{2}}, -\frac{i}{\sqrt{2}}).$$

The orthonormal basis is $\{(\frac{i}{\sqrt{2}}, -\frac{1}{\sqrt{2}}), (\frac{1}{\sqrt{2}}, -\frac{i}{\sqrt{2}})\}$.   ▪

Exercise 6.30 Given $S = \{(1, 1, i), (0, 1, -i), (2i, -i, 1)\}$ in $\mathbb{C}^3$,

(a) Show that S is an orthogonal basis for $\mathbb{C}^3$;

(b) Transform S into an orthonormal basis.

2   Gram-Schmidt Process

The Gram-Schmidt process systematically transforms any non-orthogonal basis into an orthogonal basis. By applying the principles outlined in Theorem 6.26, this orthogonal basis can then be normalized to form an orthonormal basis. This procedure is applicable to any finite-dimensional inner product space. In the context of quantum computing, we will specifically detail the application of this process within $\mathbb{C}^n$.

Theorem 6.27 — The Gram-Schmidt Process. The following computational steps, known as the Gram-Schmidt process, transform any basis $\{u_1, u_2, \ldots, u_n\}$ of an inner product space V into an orthogonal basis $\{v_1, v_2, \ldots, v_n\}$:

1. $v_1 = u_1$

2. $v_2 = u_2 - \text{proj}_{v_1} u_2$

3. $v_3 = u_3 - \text{proj}_{v_1} u_3 - \text{proj}_{v_2} u_3$

4. $v_4 = u_4 - \text{proj}_{v_1} u_4 - \text{proj}_{v_2} u_4 - \text{proj}_{v_3} u_4$

   $\vdots$

n. $v_n = u_n - \sum_{i=1}^{n-1} \text{proj}_{v_i} u_n$

Proof. In this proof, we utilize the complex dot product as the inner product. Given that $\{u_1, u_2, \ldots, u_n\}$ is a basis for V, and hence $\dim(V) = n$, we aim to demonstrate that $\{v_1, v_2, \ldots, v_n\}$ forms an orthogonal basis. This follows from Theorem 6.25 by showing that $v_i \cdot v_j = 0$ for all $i \neq j$.

Using Eq. 6.55a, we establish the following relationships:

$$\boldsymbol{v}_i \cdot \mathrm{proj}_{\boldsymbol{v}_i} \boldsymbol{u}_j = \boldsymbol{v}_i \cdot \left(\frac{\boldsymbol{v}_i \cdot \boldsymbol{u}_j}{\boldsymbol{v}_i \cdot \boldsymbol{v}_i} \boldsymbol{v}_i \right) = \frac{\boldsymbol{v}_i \cdot \boldsymbol{u}_j}{\boldsymbol{v}_i \cdot \boldsymbol{v}_i} \boldsymbol{v}_i \cdot \boldsymbol{v}_i = \boldsymbol{v}_i \cdot \boldsymbol{u}_j, \qquad (6.71\mathrm{a})$$

$$\boldsymbol{v}_i \cdot \mathrm{proj}_{\boldsymbol{v}_k} \boldsymbol{u}_j = \boldsymbol{v}_i \cdot \left(\frac{\boldsymbol{v}_k \cdot \boldsymbol{u}_j}{\boldsymbol{v}_k \cdot \boldsymbol{v}_k} \boldsymbol{v}_k \right) = \frac{\boldsymbol{v}_k \cdot \boldsymbol{u}_j}{\boldsymbol{v}_k \cdot \boldsymbol{v}_k} \boldsymbol{v}_i \cdot \boldsymbol{v}_k = 0 \quad \text{if } \boldsymbol{v}_i \cdot \boldsymbol{v}_k = 0. \qquad (6.71\mathrm{b})$$

With these formulas, we proceed with the proof using mathematical induction.

Base Case:

For $n = 2$, the vectors $\boldsymbol{v}_1$ and $\boldsymbol{v}_2$ are given by:

$$\boldsymbol{v}_1 = \boldsymbol{u}_1, \quad \boldsymbol{v}_2 = \boldsymbol{u}_2 - \mathrm{proj}_{\boldsymbol{v}_1} \boldsymbol{u}_2.$$

Compute the inner product $\boldsymbol{v}_1 \cdot \boldsymbol{v}_2$:

$$\begin{aligned}
\boldsymbol{v}_1 \cdot \boldsymbol{v}_2 &= \boldsymbol{v}_1 \cdot (\boldsymbol{u}_2 - \mathrm{proj}_{\boldsymbol{v}_1} \boldsymbol{u}_2) \\
&= \boldsymbol{v}_1 \cdot \boldsymbol{u}_2 - \boldsymbol{v}_1 \cdot \mathrm{proj}_{\boldsymbol{v}_1} \boldsymbol{u}_2 \\
&= \boldsymbol{v}_1 \cdot \boldsymbol{u}_2 - \boldsymbol{v}_1 \cdot \boldsymbol{u}_2 \\
&= 0.
\end{aligned}$$

Thus, $\boldsymbol{v}_1$ and $\boldsymbol{v}_2$ are orthogonal.

Inductive Step:

Assume that $\boldsymbol{v}_1, \boldsymbol{v}_2, \ldots, \boldsymbol{v}_k$ are mutually orthogonal. We prove that $\boldsymbol{v}_{k+1}$ is orthogonal to all previous $\boldsymbol{v}_i$ for $i = 1, 2, \ldots, k$. By definition:

$$\boldsymbol{v}_{k+1} = \boldsymbol{u}_{k+1} - \sum_{i=1}^{k} \mathrm{proj}_{\boldsymbol{v}_i} \boldsymbol{u}_{k+1}.$$

Compute the inner product $\boldsymbol{v}_j \cdot \boldsymbol{v}_{k+1}$ for $j = 1, 2, \ldots, k$:

$$\begin{aligned}
\boldsymbol{v}_j \cdot \boldsymbol{v}_{k+1} &= \boldsymbol{v}_j \cdot \left(\boldsymbol{u}_{k+1} - \sum_{i=1}^{k} \mathrm{proj}_{\boldsymbol{v}_i} \boldsymbol{u}_{k+1} \right) \\
&= \boldsymbol{v}_j \cdot \boldsymbol{u}_{k+1} - \sum_{i=1}^{k} \boldsymbol{v}_j \cdot \mathrm{proj}_{\boldsymbol{v}_i} \boldsymbol{u}_{k+1}.
\end{aligned}$$

For $i \neq j$, $\boldsymbol{v}_j \cdot \mathrm{proj}_{\boldsymbol{v}_i} \boldsymbol{u}_{k+1} = 0$ because $\boldsymbol{v}_j \cdot \boldsymbol{v}_i = 0$ by the induction hypothesis. For $i = j$:

$$\boldsymbol{v}_j \cdot \mathrm{proj}_{\boldsymbol{v}_j} \boldsymbol{u}_{k+1} = \boldsymbol{v}_j \cdot \boldsymbol{u}_{k+1}.$$

Thus:

$$\boldsymbol{v}_j \cdot \boldsymbol{v}_{k+1} = \boldsymbol{v}_j \cdot \boldsymbol{u}_{k+1} - \boldsymbol{v}_j \cdot \boldsymbol{u}_{k+1} = 0.$$

By induction, $\boldsymbol{v}_1, \boldsymbol{v}_2, \ldots, \boldsymbol{v}_{k+1}$ are mutually orthogonal.

Conclusion:

By mathematical induction, the Gram-Schmidt process produces an orthogonal basis $\{\boldsymbol{v}_1, \boldsymbol{v}_2, \ldots, \boldsymbol{v}_n\}$ from the given basis $\{\boldsymbol{u}_1, \boldsymbol{u}_2, \ldots, \boldsymbol{u}_n\}$. $\qquad \square$

Exercise 6.31 Explain why the Gram-Schmidt Process, as outlined in Theorem 6.27, would never result in a zero vector for any v_i in the sequence of vectors produced.

■ **Example 6.23** Transform the basis $\{(1,1,0),(1,0,1),(0,1,i)\}$ for $\mathbb{C}^3$ into an orthogonal one using the Gram-Schmidt Process.

Solution. Let $u_1 = (1,1,0)$, $u_2 = (1,0,1)$, $u_3 = (0,1,i)$. We apply the Gram-Schmidt Process as outlined in Theorem 6.27 to the given basis:

$$v_1 = u_1 = (1,1,0),$$

$$v_2 = u_2 - \text{proj}_{v_1} u_2 = u_2 - \frac{v_1 \cdot u_2}{v_1 \cdot v_1} v_1,$$

$$= (1,0,1) - \frac{(1,1,0) \cdot (1,0,1)}{(1,1,0) \cdot (1,1,0)}(1,1,0),$$

$$= (1,0,1) - \frac{1}{2}(1,1,0) = \left(\frac{1}{2}, -\frac{1}{2}, 1\right),$$

$$v_3 = u_3 - \text{proj}_{v_1} u_3 - \text{proj}_{v_2} u_3,$$

$$= (0,1,i) - \frac{v_1 \cdot u_3}{v_1 \cdot v_1} v_1 - \frac{v_2 \cdot u_3}{v_2 \cdot v_2} v_2,$$

$$= \left(-\frac{1+i}{3}, \frac{1+i}{3}, \frac{1+i}{3}\right).$$

Hence, $\{v_1, v_2, v_3\}$ forms an orthogonal basis of $\mathbb{C}^3$.   ■

Exercise 6.32 Transform the basis $\{(1,0,i),(0,1,0),(0,0,1)\}$ for $\mathbb{C}^3$ into an orthogonal one using the Gram-Schmidt Process, and then normalize each vector to form an orthonormal basis.

3   Orthogonal Projection on a Subspace

An illustration of the Gram-Schmidt Process in $\mathbb{R}^2$ and $\mathbb{R}^3$ is depicted in Fig. 6.8. In $\mathbb{R}^2$, v_2 represents the component of u_2 that is orthogonal to v_1 (Fig. 6.8a). Similarly, in $\mathbb{R}^3$, v_3 is the component of u_3 orthogonal to the subspace W spanned by v_1 and v_2 (Fig. 6.8b).

(i)   For a vector in a vector space V to be considered orthogonal to a subspace W, it must be orthogonal to every vector contained within W. This condition is met if and only if the inner product of the vector with each basis vector in W is zero.

Generally, we can define the projection of a vector on a subspace as:

Definition 6.23 — Projection. Let u be a vector in an inner product space V, and let W be a subspace of V. If $\{v_1, v_2, \ldots, v_r\}$ is an orthogonal basis for W, then the projection of u onto W, denoted as $\text{proj}_W u$, is defined as:

$$\text{proj}_W u = \sum_{i=1}^{r} \text{proj}_{v_i} u = \sum_{i=1}^{r} \frac{\langle v_i, u \rangle}{\langle v_i, v_i \rangle} v_i. \tag{6.72}$$

The component of u orthogonal to W, denoted $\text{proj}_{W \perp} u$, is:

$$\text{proj}_{W \perp} u = u - \text{proj}_W u. \tag{6.73}$$

(a)

$$u_2$$

$$\text{proj}_{v_1} u_2$$

$$v_1 = u_1$$

$$v_2 = u_2 - \text{proj}_{v_1} u_2$$

(b)

$$v_3 = u_3 - \text{proj}_{v_1} u_3 - \text{proj}_{v_2} u_3$$

$$u_3$$

$$v_2 \qquad \text{proj}_{v_2} u_3$$

$$v_1 \qquad\qquad\qquad W$$

$$\text{proj}_{v_1} u_3$$

$$\text{proj}_{v_1} u_3 + \text{proj}_{v_2} u_3$$

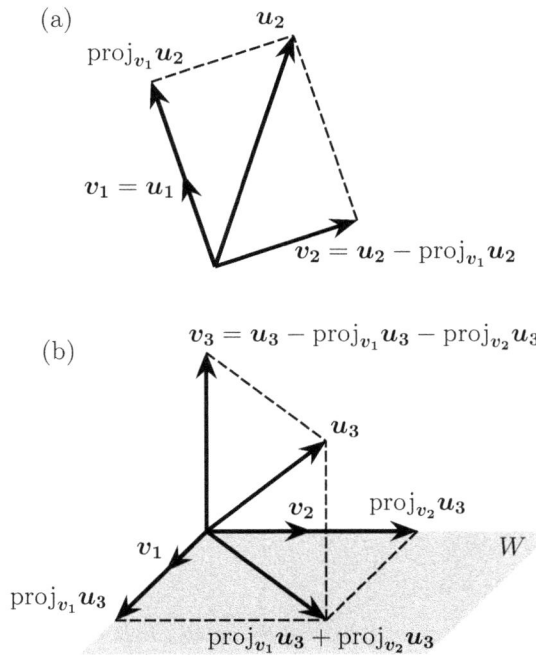

Figure 6.8: Illustration of the Gram-Schmidt Process

In Def. 6.23, $W^\perp$ is called the **orthogonal complement** of W. The vectors in $W^\perp$ are precisely those orthogonal to every vector in W. This concept will be explored further in § 11.2.4.2.

Following Def. 6.23, the Gram-Schmidt Process, as presented in Theorem 6.27, can be succinctly summarized:

1. $v_1 = u_1$,

2. $v_2 = \text{proj}_{W_1^\perp} u_2$, where $W_1 = \text{span}\{v_1\}$,

3. $v_3 = \text{proj}_{W_2^\perp} u_3$, where $W_2 = \text{span}\{v_1, v_2\}$,

4. $v_4 = \text{proj}_{W_3^\perp} u_4$, where $W_3 = \text{span}\{v_1, v_2, v_3\}$,

$\vdots$

n. $v_n = \text{proj}_{W_{n-1}^\perp} u_n$, where $W_{n-1} = \text{span}\{v_1, v_2, \ldots, v_{n-1}\}$,

where W_i represents the subspace spanned by $v_1, v_2, \ldots, v_i$.

Problem Set 6

6.1 Prove Theorem 6.1.

6.2 Given two vectors $u = (a, b)$ and $v = (-b, a)$ in $\mathbb{R}^2$, prove that $\|u\|^2 + \|v\|^2 = \|u - v\|^2$. Draw the three vectors u, v, and $u - v$ on a two-dimensional coordinate system and explain the result geometrically.

6.3 Given a unit vector (a, b), prove that $(\frac{a+b}{\sqrt{2}}, \frac{a-b}{\sqrt{2}})$ is also a unit vector for

    (a) $a, b \in \mathbb{R}$;

    (b) $a, b \in \mathbb{C}$.

6.4 Given two unit vectors (α, β) and $(\gamma, \delta) \in \mathbb{C}^2$, show that $(\alpha\gamma, \alpha\delta, \beta\gamma, \beta\delta)$ (we will learn later that this is called the tensor product of these two vectors) is a unit vector in $\mathbb{C}^4$.

6.5 Show that $|\psi\rangle = \cos\frac{\theta}{2}|0\rangle + \sin\frac{\theta}{2}e^{i\phi}|1\rangle$ is a unit vector for $\theta, \phi \in \mathbb{R}$.

6.6 Given the quantum state $|\psi\rangle = \alpha|0\rangle + \beta|1\rangle$ on the Bloch Sphere, determine the positions of the following states relative to $|\psi\rangle$:

    (a) $\alpha|0\rangle - \beta|1\rangle$;

    (b) $-\alpha|0\rangle - \beta|1\rangle$;

    (c) $\beta|0\rangle + \alpha|1\rangle$;

6.7 Given two vectors $u = (\cos\theta_1, \sin\theta_1)$ and $v = (\cos\theta_2, \sin\theta_2)$, prove that

    (a) u and v are unit vectors.

    (b) $u \cdot v = \cos(\theta_1 - \theta_2)$.

6.8 Prove that if u and v are unit vectors in $\mathbb{R}^n$, then

$$(u + v) \cdot (u - v) = 0.$$

Explain this result geometrically in $\mathbb{R}^2$. Does this result hold true when $u, v \in \mathbb{C}^n$?

6.9 Prove the left- and the right-distributive rules in Theorem 6.6.

6.10 Verify Theorem 6.9 using formulas of the complex dot product.

6.11 Prove Eqs. 6.30a and 6.30b in Theorem 6.9.

6.12 $*$ Which of the following formulas is still true for $u = (a, b)_S$ and $v = (c, d)_S$, where S is a non-orthogonal basis of $\mathbb{R}^2$?

    (a) $u \cdot v = ac + bd$

    (b) $\|u\|^2 = a^2 + b^2$

    (c) $u \cdot v = \|u\|\|v\|\cos\theta$, where θ is the angle between u and v.

6.13 $*$ Verify that $C[a, b]$, the set of all continuous real-valued functions defined on a closed interval $[a, b]$, is a real inner product space with the inner product defined by

$$\langle f, g \rangle = \int_a^b f(x)g(x)dx,$$

where $f, g \in C[a, b]$.

6.14 Given two general vectors in $\mathbb{C}^2$: $|\psi\rangle = \alpha|0\rangle + \beta|1\rangle$, $|\phi\rangle = \gamma|0\rangle + \delta|1\rangle$, compute

    (a) $\langle\psi|\phi\rangle$;

(b) $\langle\phi|\psi\rangle$;

(c) $\langle\psi|\psi\rangle$;

(d) $\langle\phi|\phi\rangle$.

Verify your answers using matrix notation.

6.15 (a) Given two real vectors $u = (u_1, u_2)$ and $v = (v_1, v_2)$, prove the following identity:

$$\frac{1}{2}(\|u\|^2 + \|v\|^2 - \|u - v\|^2) = u_1 v_1 + u_2 v_2 = u \cdot v.$$

(b) Generally prove the following identity for $u, v \in \mathbb{R}^n$ or $\mathbb{C}^n$:

$$\frac{1}{2}(\|u\|^2 + \|v\|^2 - \|u - v\|^2) = u \cdot v.$$

6.16 ✳ Prove that the sum of the angles in any "hyper-triangle" formed by u, v, and $u - v$ in $\mathbb{R}^n$ is exactly 180 degrees.

6.17 Use the projection theorem to prove the following equation:

$$\tan^{-1} 1 + \tan^{-1} \frac{1}{2} + \tan^{-1} \frac{1}{3} = \frac{\pi}{2}.$$

Hint: refer to the graph for Exercise 6.25.

6.18 For $v, a \in \mathbb{R}^n$, let θ be the angle between them. Prove that:

$$\|v\| \cos\theta = \|\text{proj}_a v\|.$$

Hint: This equation is geometrically clear in $\mathbb{R}^2$ (Fig. 6.6). Here, you are required to prove it generally for $\mathbb{R}^n$ using the definition of the generalized angle in Eq. 6.51.

6.19 ✳ Given $v, a \in \mathbb{C}^n, v, a \neq 0$, prove that among all scalars $\lambda \in \mathbb{C}$, the value given by the projection theorem $\lambda = \frac{a \cdot v}{\|a\|^2}$ minimizes $\|v - \lambda a\|$. In other words, the norm of $v_{\perp a}$ is the shortest distance among all distances between v and the span of a.

6.20 Given a unit vector $|u\rangle = \alpha |0\rangle + \beta |1\rangle$ in $\mathbb{C}^2$, find the general form of $|v\rangle$ such that $\{|u\rangle, |v\rangle\}$ is an orthonormal basis.

6.21 Given $|\psi\rangle = \cos\frac{\theta_1}{2} |0\rangle + \sin\frac{\theta_1}{2} |1\rangle$, $|\phi\rangle = \cos\frac{\theta_2}{2} |0\rangle + \sin\frac{\theta_2}{2} |1\rangle$ for $\theta_1, \theta_2 \in \mathbb{R}$, compute $\langle\psi|\psi\rangle$, $\langle\psi|\phi\rangle$, and $\langle\phi|\phi\rangle$. What conditions must θ_1, θ_2 satisfy for $\{|\psi\rangle, |\phi\rangle\}$ to be an orthonormal basis?

6.22 ✳ Given $|\psi\rangle = \cos\frac{\theta_1}{2} |0\rangle + \sin\frac{\theta_1}{2} e^{i\phi_1} |1\rangle$, $|\phi\rangle = \cos\frac{\theta_2}{2} |0\rangle + \sin\frac{\theta_2}{2} e^{i\phi_2} |1\rangle$ for $\theta_1, \theta_2, \phi_1, \phi_2 \in \mathbb{R}$, compute $\langle\psi|\psi\rangle$, $\langle\psi|\phi\rangle$, and $\langle\phi|\phi\rangle$. What conditions must $\theta_1, \theta_2, \phi_1, \phi_2$ satisfy for $\{|\psi\rangle, |\phi\rangle\}$ to be an orthonormal basis?

6.23 Decompose a general vector $|\psi\rangle = \alpha |0\rangle + \beta |1\rangle$ in the orthonormal basis of

(a) $\{|+\rangle, |-\rangle\}$

(b) $\{|+_i\rangle, |-_i\rangle\}$

defined in Examples 6.12 and 6.13.

6.24 ∗ Use the Gram-Schmidt Process to find an orthogonal basis of $\mathbb{C}^4$ that includes:

(a) $(1, 1, 1, 1)$;

(b) $(1, i, 1, i)$.

7. Fundamentals of Matrix Algebra

Contents

Matrices are a crucial mathematical structure with applications across various scientific disciplines, including physics, engineering, and data science. In quantum computing, matrices play an even more fundamental role: they represent quantum evolution, measurements, and mixed quantum states. A solid understanding of matrix algebra and its applications is essential for both foundational and advanced topics in quantum computing.

The relationship between matrices and vectors is profound. Matrices can be viewed as linear operators acting on vectors, while vectors serve as the building blocks for constructing matrices. With a solid foundation in vectors and vector spaces from the previous chapters, we now turn to a detailed study of matrix algebra.

This chapter introduces the fundamental principles of matrix algebra. Subsequent chapters will explore matrices as linear operators, as well as the concepts of eigenvalues and diagonalization, which are essential in quantum computing. By the end of these chapters, readers will have a deeper understanding of how matrices and linear algebra provide the foundational mathematical framework for articulating quantum mechanics and quantum computing.

7.1 Matrix Basics

We introduced matrices as well as row and column vectors in § 6.1. In this section, we will first revisit these definitions and subsequently provide a thorough exploration of all fundamental concepts essential for understanding the subsequent sections and chapters.

7.1.1 Definitions

Definition 7.1 — Matrix. A matrix is defined as a rectangular array comprising numbers, either real or complex. A matrix consisting of m rows and n columns is represented as follows:

$$A = \begin{bmatrix} a_{11} & a_{12} & \cdots & a_{1n} \\ a_{21} & a_{22} & \cdots & a_{2n} \\ \vdots & \vdots & \ddots & \vdots \\ a_{m1} & a_{m2} & \cdots & a_{mn} \end{bmatrix}_{m \times n}, \tag{7.1}$$

where the numbers in the matrix $a_{11}, a_{12}, \ldots$ are called **entries** or **elements**. The matrix is said to be an $m \times n$ matrix, or $m \times n$ in size.

Notation: In linear algebra, matrices are conventionally denoted by capitalized Latin or Greek letters, such as A, B, U, V, Σ, and Λ. This convention largely holds in quantum computing as well, with a notable exception: density matrices are typically denoted by ρ.

The individual entries within a matrix, which are scalars, are commonly represented by corresponding lowercase letters with subscripted row and column indices. For instance, the entry in the ith row and jth column of a matrix $A_{m \times n}$ is denoted as a_{ij}, where $i = 1, 2, \ldots, m$ and $j = 1, 2, \ldots, n$.

Additionally, the notation $[a_{ij}]$ is sometimes employed to denote the matrix A, and $(A)_{ij}$ is used to refer specifically to the entries of A. Here, i and j are used solely as indices and should not be confused with the imaginary unit i for complex numbers.

A matrix composed entirely of real numbers is termed a real matrix. Correspondingly, a matrix containing complex numbers is designated as a complex matrix. We will adopt Eq. 7.1 as the general representation for an arbitrary matrix. Typically, the dimensions $m \times n$ of a matrix are omitted unless it is particularly insightful to explicitly specify them.

Definition 7.2 — Column Vector and Row Vector. A matrix with only one column ($n \times 1$ in size) is called a column vector, denoted as:

$$\begin{bmatrix} v_1 \\ v_2 \\ \vdots \\ v_n \end{bmatrix}.$$

Similarly, a matrix with only one row ($1 \times m$ in size) is called a row vector, denoted as:

$$\begin{bmatrix} v_1 & v_2 & \cdots & v_m \end{bmatrix}.$$

ⓘ In matrix algebra, a scalar can be considered as a 1×1 matrix when this interpretation fits the context of the operations. This allows scalars to be seamlessly incorporated into matrix operations, especially when dealing with matrices of varying dimensions, as we will see in the next section.

Definition 7.3 — Transpose. Given an $m \times n$ matrix A as given in Eq. 7.1, the transpose of A, denoted as A^T, is an $n \times m$ matrix defined as:

$$A^T = \begin{bmatrix} a_{11} & a_{21} & \cdots & a_{m1} \\ a_{12} & a_{22} & \cdots & a_{m2} \\ \vdots & \vdots & \ddots & \vdots \\ a_{1n} & a_{2n} & \cdots & a_{mn} \end{bmatrix}_{n \times m}, \tag{7.2}$$

resulting from the rows and columns of A being interchanged.

Definition 7.4 — Conjugate. For a complex matrix A given in Eq. 7.1, its conjugate is defined as follows:

$$A^* = \begin{bmatrix} a_{11}^* & a_{12}^* & \cdots & a_{1n}^* \\ a_{21}^* & a_{22}^* & \cdots & a_{2n}^* \\ \vdots & \vdots & \ddots & \vdots \\ a_{m1}^* & a_{m2}^* & \cdots & a_{mn}^* \end{bmatrix}. \tag{7.3}$$

Definition 7.5 — Adjoint. For a complex matrix A, as specified in Eq. 7.1, its Hermitian adjoint, or adjoint, is defined as its conjugate transpose, denoted by

$A^\dagger$:

$$A^\dagger = (A^*)^T = (A^T)^* = \begin{bmatrix} a_{11}^* & a_{21}^* & \cdots & a_{m1}^* \\ a_{12}^* & a_{22}^* & \cdots & a_{m2}^* \\ \vdots & \vdots & \ddots & \vdots \\ a_{1n}^* & a_{2n}^* & \cdots & a_{mn}^* \end{bmatrix}_{n \times m}. \tag{7.4}$$

For a real matrix, its conjugate is the matrix itself, and its adjoint is identical to its transpose.

The transpose or adjoint of a row vector results in a column vector, and vice versa. Taking the transpose or adjoint of a matrix twice returns it to its original form:

Theorem 7.1 For any matrix A, the following property holds:

$$(A^T)^T = A, \quad (A^\dagger)^\dagger = A. \tag{7.5}$$

i **Notation:** Readers should note that in different mathematical conventions, $\overline{A}$ may be used to denote the conjugate of a matrix, while A^* may instead refer to the conjugate transpose. It is essential to distinguish between these notations when interpreting matrix operations across various texts.

■ **Example 7.1** Given a matrix A, compute its transpose A^T, conjugate A^*, and Hermitian adjoint $A^\dagger$, along with the dimensions of these matrices, for

$$A = \begin{bmatrix} 1 & i & 1-i \\ -2i & 2 & 0 \end{bmatrix}.$$

Solution. The matrix A is of size 2×3, which implies that A^* will also be 2×3, while A^T and $A^\dagger$ will have dimensions 3×2. These matrices are given by:

$$A^* = \begin{bmatrix} 1 & -i & 1+i \\ 2i & 2 & 0 \end{bmatrix}, \quad A^T = \begin{bmatrix} 1 & -2i \\ i & 2 \\ 1-i & 0 \end{bmatrix}, \quad A^\dagger = \begin{bmatrix} 1 & 2i \\ -i & 2 \\ 1+i & 0 \end{bmatrix}.$$

■

Exercise 7.1 Determine A^T, A^*, $A^\dagger$, and the sizes of these matrices for

$$\text{(a) } A = \begin{bmatrix} 1 \\ i \\ -1+3i \end{bmatrix}, \quad \text{(b) } A = \begin{bmatrix} a+ib & c+id & e+if \\ g & h & k \end{bmatrix},$$

where $a, b, c, \ldots, h, k \in \mathbb{R}$.

Definition 7.6 — Square Matrix. A matrix that has an equal number of rows and columns is termed a square matrix, or an $n \times n$ matrix.

Definition 7.7 — Main Diagonal. The main diagonal of a square matrix consists of the entries where the row and column indices are the same. In an $n \times n$ matrix

A, the entry a_{ij} is on the main diagonal if $i = j$.

Definition 7.8 — Matrix Equality. Two matrices of the same size, $A_{m \times n}$ and $B_{m \times n}$, are defined to be equal if all corresponding entries are identical:

$$a_{ij} = b_{ij}, \quad \text{for all } i = 1, 2, \ldots, m, \text{ and } j = 1, 2, \ldots, n. \tag{7.6}$$

If two matrices A and B are identical, this relationship is denoted simply by

$$A = B.$$

Definition 7.8 allows for the validation of matrix identities by demonstrating that the matrices in question are of the same size and that Eq. 7.6 holds true for all corresponding pairs of i and j, as will be demonstrated in various examples throughout this chapter.

An $n \times n$ square matrix has a transpose that is also $n \times n$ in size, which allows it to potentially be identical to the original matrix. When this occurs, the matrix is referred to as a symmetric matrix.

Definition 7.9 — Symmetric Matrix. A square matrix A is termed a symmetric matrix if $A = A^T$.

Symmetric matrices are particularly significant in the analysis of real matrices. For complex matrices, however, it is more pertinent to define a Hermitian matrix, which remains invariant when taking the Hermitian adjoint.

Definition 7.10 — Hermitian Matrix. A square complex matrix A is said to be Hermitian or a Hermitian matrix if $A = A^\dagger$.

Hermitian matrices are of crucial importance in quantum mechanics and quantum computing, where they represent operators for quantum measurements. A comprehensive study of Hermitian matrices will be provided in § 9.4.

■ **Example 7.2** Determine whether each of the following matrices is symmetric, Hermitian, or neither:

(a) $A = \begin{bmatrix} 1 & 2 \\ 2 & 0 \end{bmatrix}$
 (b) $B = \begin{bmatrix} 1 & 2 & 0 \\ 2 & 1 & 0 \end{bmatrix}$

(c) $C = \begin{bmatrix} 1 & i \\ -i & 0 \end{bmatrix}$
 (d) $D = \begin{bmatrix} 1 + 2i \end{bmatrix}$

Solution. A matrix is symmetric if it is square and its entries are symmetric about the main diagonal. Here, matrices A and D are symmetric, with D considered symmetric because any 1×1 matrix is trivially symmetric.

A matrix is Hermitian if it is square, with entries that are conjugate symmetric about the main diagonal. This means the entries on the main diagonal must be real, and entries off the main diagonal must be the complex conjugate of their corresponding mirror entry. Both A and C satisfy these conditions, making them Hermitian. Note that any real symmetric matrix is also Hermitian. Matrix D is not Hermitian because its entry contains a nonzero imaginary part.

Matrix B is neither symmetric nor Hermitian, as it is not square.   ■

Exercise 7.2 Determine which of the following matrices are symmetric, Hermitian, or neither:

(a) $A = \begin{bmatrix} i & 1+2i \\ 1-2i & -1 \end{bmatrix}$ 　　　　(b) $B = \begin{bmatrix} 0 & 0 & 0 \\ 0 & 0 & 0 \end{bmatrix}$

(c) $C = \begin{bmatrix} i & i \\ i & i \end{bmatrix}$ 　　　　(d) $D = \begin{bmatrix} 3 \end{bmatrix}$

Exercise 7.3 Determine the condition that a scalar k must meet so that the matrix kA remains Hermitian, given that A is a Hermitian matrix.

7.1.2　Basic Algebraic Operations

1　Addition, Subtraction, and Scalar Multiplication

Similar to vectors, the fundamental algebraic operations for matrices include addition and scalar multiplication, which are defined as follows.

Definition 7.11 — Matrix Addition. Addition is defined only for two matrices of the same size, $A_{m \times n}$ and $B_{m \times n}$. Their sum, denoted by $A + B$, retains the same size, $m \times n$. The sum is computed as follows:

$$A + B = \begin{bmatrix} a_{11}+b_{11} & a_{12}+b_{12} & \cdots & a_{1n}+b_{1n} \\ a_{21}+b_{21} & a_{22}+b_{22} & \cdots & a_{2n}+b_{2n} \\ \vdots & \vdots & \ddots & \vdots \\ a_{m1}+b_{m1} & a_{m2}+b_{m2} & \cdots & a_{mn}+b_{mn} \end{bmatrix}. \tag{7.7}$$

Definition 7.12 — Scalar Multiplication of Matrices. Multiplying a matrix $A_{m \times n}$ by a scalar k results in a matrix of the same size, denoted as kA, computed as follows:

$$kA = \begin{bmatrix} ka_{11} & ka_{12} & \cdots & ka_{1n} \\ ka_{21} & ka_{22} & \cdots & ka_{2n} \\ \vdots & \vdots & \ddots & \vdots \\ ka_{m1} & ka_{m2} & \cdots & ka_{mn} \end{bmatrix} \tag{7.8}$$

Definitions 7.11 and 7.12 lead to the following definitions of the negative of a matrix and matrix subtraction:

Definition 7.13 — Negative of a Matrix. The negative of a matrix $A_{m \times n}$, denoted as $-A$, is defined as the product of -1 and A, given by:

$$-A = \begin{bmatrix} -a_{11} & -a_{12} & \cdots & -a_{1n} \\ -a_{21} & -a_{22} & \cdots & -a_{2n} \\ \vdots & \vdots & \ddots & \vdots \\ -a_{m1} & -a_{m2} & \cdots & -a_{mn} \end{bmatrix} \tag{7.9}$$

Definition 7.14 — Matrix Subtraction. Matrix subtraction $A - B$ is defined as the addition of A and the negative of B, computed by:

$$A - B = \begin{bmatrix} a_{11} - b_{11} & a_{12} - b_{12} & \cdots & a_{1n} - b_{1n} \\ a_{21} - b_{21} & a_{22} - b_{22} & \cdots & a_{2n} - b_{2n} \\ \vdots & \vdots & \ddots & \vdots \\ a_{m1} - b_{m1} & a_{m2} - b_{m2} & \cdots & a_{mn} - b_{mn} \end{bmatrix} \qquad (7.10)$$

Exercise 7.4 Given $A = \begin{bmatrix} 1 & i \\ -i & 0 \end{bmatrix}$, compute the following:

(a) $A + A^T$ (b) $A - A^T$ (c) $A + A^*$

(d) $A - A^*$ (e) $A + A^\dagger$ (f) $A - A^\dagger$

Exercise 7.5 Given $A = \begin{bmatrix} 1 - i & 1 + i \\ 0 & 1 \end{bmatrix}$ and $B = \begin{bmatrix} 1 & -1 \\ i & 0 \end{bmatrix}$, compute the following:

(a) $A + iB$ (b) $iA - B$ (c) $2A^\dagger - (1+i)B^T$

2   Element-wise Notation

Equations 7.7 to 7.10 are visually comprehensive but can be cumbersome. Typically, the same operations are more concisely expressed in an element-wise notation:

Theorem 7.2 Given two matrices $A_{m \times n} = [a_{ij}]$ and $B_{m \times n} = [b_{ij}]$ of the same size, and a scalar k, the basic algebraic operations can be computed element-wise as follows:

$$(A + B)_{ij} = a_{ij} + b_{ij}, \qquad (7.11a)$$
$$(kA)_{ij} = ka_{ij}, \qquad (7.11b)$$
$$(-A)_{ij} = -a_{ij}, \qquad (7.11c)$$
$$(A - B)_{ij} = a_{ij} - b_{ij}, \qquad (7.11d)$$

for all $i = 1, 2, \ldots, m$, and $j = 1, 2, \ldots, n$.

For transpose and Hermitian of matrices, the element-wise formula is as follows:

Theorem 7.3 Given $A_{m \times n} = [a_{ij}]$, the entries of A^T and $A^\dagger$ are given element-wise by:

$$(A^T)_{ji} = a_{ij}, \qquad (7.12a)$$
$$(A^\dagger)_{ji} = a_{ij}^*, \qquad (7.12b)$$

for all $i = 1, 2, \ldots, m$, and $j = 1, 2, \ldots, n$.

Exercise 7.6 Verify Theorem 7.3 with the matrix $A = \begin{bmatrix} 1 & -1 & 0 \\ i & -i & 1+i \end{bmatrix}$.

■ **Example 7.3** Use element-wise notation to prove the identity:

$$(A + B)^T = A^T + B^T. \qquad (7.13)$$

Proof. Let A, B both be $m \times n$ in size, and let $C = (A + B)^T$. Using Eqs. 7.11a and 7.12a, for all $i = 1, 2, \ldots, m$, and $j = 1, 2, \ldots, n$, we have

$$c_{ji} = (A + B)_{ji}^T = (A + B)_{ij} = a_{ij} + b_{ij} = (A^T)_{ji} + (B^T)_{ji} = (A^T + B^T)_{ji}.$$

By Def. 7.8, we obtain that $C = A^T + B^T$.

Note that in c_{ji}, j should be the row index and i should be the column index to match the size of matrix C $(n \times m)$.   ∎

Element-wise notation is particularly valuable for proofs involving matrix multiplication, which will be discussed in the next section.

Exercise 7.7 Use element-wise notation to prove the identity:

$$(A + B)^\dagger = A^\dagger + B^\dagger. \tag{7.14}$$

Exercise 7.8 Prove that for any square matrix A:

(a) $A + A^T$ is a symmetric matrix,

(b) $A + A^\dagger$ is a Hermitian matrix.

7.1.3 The Vector Spaces of Matrices

Since matrices adhere to the same fundamental operations as vectors, they naturally form vector spaces according to Def. 5.11. To complete this definition, we first need to define the zero matrix.

Definition 7.15 — Zero Matrix. Among all matrices of a specific size $m \times n$, the unique zero matrix, denoted by 0, is defined as the matrix whose entries are all zero:

$$0_{m \times n} = \begin{bmatrix} 0 & 0 & \cdots & 0 \\ 0 & 0 & \cdots & 0 \\ \vdots & \vdots & \ddots & \vdots \\ 0 & 0 & \cdots & 0 \end{bmatrix}_{m \times n}, \tag{7.15}$$

or in element-wise notation:

$$(0)_{ij} = 0, \quad \text{for all } i = 1, 2, \ldots, m, \text{ and } j = 1, 2, \ldots, n. \tag{7.16}$$

i **Notation:** Given that $0_{m \times n}$ is unique for any specified size, we generally use 0 to denote the zero matrix throughout this book, as this typically presents no ambiguity. However, when differentiation between the zero matrix and the zero scalar is necessary, the italic symbol 0 will be employed.

Theorem 7.4 The following properties hold true for any matrices A, B, C, 0 of the same size and scalars k, m:

$$A + B = B + A, \tag{7.17a}$$
$$A + 0 = 0 + A = A, \tag{7.17b}$$
$$(A + B) + C = A + (B + C), \tag{7.17c}$$
$$k(A + B) = kA + kB, \tag{7.17d}$$
$$(k + m)A = kA + mA, \tag{7.17e}$$
$$(km)A = k(mA), \tag{7.17f}$$
$$0A = 0, \tag{7.17g}$$
$$1A = A. \tag{7.17h}$$

All identities in Theorem 7.4 remain valid when any instance of addition is replaced by subtraction. The proofs of Theorem 7.4 are omitted since they are straightforward. It is readily verifiable that matrices of a specific size $m \times n$ satisfy all the axioms required for a vector space as given in Def. 5.11. Consequently, we state the following theorem.

Theorem 7.5 All matrices of size $m \times n$ constitute a vector space. For real matrices over $\mathbb{R}$, this vector space is denoted as $\mathbb{R}^{m \times n}$, and for complex matrices over $\mathbb{C}$, it is denoted as $\mathbb{C}^{m \times n}$.

■ **Example 7.4** Find a basis and determine the dimension for both $\mathbb{R}^{m \times n}$ and $\mathbb{C}^{m \times n}$.

Solution. A natural basis for $\mathbb{R}^{m \times n}$ and $\mathbb{C}^{m \times n}$ is the set $\{B_{ij}\}$ for all $i = 1, 2, \ldots, m$ and $j = 1, 2, \ldots, n$, where each B_{ij} is an $m \times n$ matrix with a 1 at the (i, j)-th entry and 0 elsewhere. For example, the basis for $\mathbb{R}^{2 \times 2}$ or $\mathbb{C}^{2 \times 2}$ is:

$$B_{11} = \begin{bmatrix} 1 & 0 \\ 0 & 0 \end{bmatrix}, \quad B_{12} = \begin{bmatrix} 0 & 1 \\ 0 & 0 \end{bmatrix}, \quad B_{21} = \begin{bmatrix} 0 & 0 \\ 1 & 0 \end{bmatrix}, \quad B_{22} = \begin{bmatrix} 0 & 0 \\ 0 & 1 \end{bmatrix}.$$

Therefore, any 2×2 matrix can be expressed as:

$$\begin{bmatrix} c_{11} & c_{12} \\ c_{21} & c_{22} \end{bmatrix} = c_{11}B_{11} + c_{12}B_{12} + c_{21}B_{21} + c_{22}B_{22},$$

where $c_{ij} \in \mathbb{R}$ for real 2×2 matrices and $c_{ij} \in \mathbb{C}$ for complex 2×2 matrices.

In general, there are mn basis matrices for the vector spaces $\mathbb{R}^{m \times n}$ or $\mathbb{C}^{m \times n}$, so the dimension is mn. ■

Tensors
Vectors can be conceptualized as a one-dimensional array of numbers, residing in the vector space of $\mathbb{R}^n$ or $\mathbb{C}^n$. Similarly, square matrices can be understood as a two-dimensional array of numbers, residing in $\mathbb{R}^{n \times n}$ or $\mathbb{C}^{n \times n}$. Extending this concept, a three-dimensional array of numbers would reside in the vector space of $\mathbb{R}^{n \times n \times n}$ or $\mathbb{C}^{n \times n \times n}$, and higher-dimensional arrays of numbers are generally termed **tensors** in mathematics. By this definition, vectors are 1D tensors, matrices are 2D tensors, and there are generally k-dimensional tensors which reside in $\mathbb{R}^{n \times n \times \cdots \times n}$ (k times) or $\mathbb{C}^{n \times n \times \cdots \times n}$ (k times). An example of 3D tensors with multipartite systems is given in § 13.3.

7.2 Matrix Multiplication

The previous section established the foundational principles of matrix algebra; however, up to this point, the behavior of matrices has seemed largely analogous to that of vectors. Operations such as addition, subtraction, and scalar multiplication apply to both vectors and matrices following the same element-wise rules.

Nevertheless, matrix multiplication—which involves the product of two matrices under specific conditions—significantly differentiates matrices from vectors. This operation not only parallels the dot product between two vectors but also exhibits more complex and significant properties. This section is dedicated to a systematic exploration of matrix multiplication, aiming to cultivate a robust understanding of this pivotal operation.

In the context of quantum computing, square matrices and vectors are of particular importance. Therefore, the examples and exercises in this section are specifically designed to emphasize square matrices and vectors, ensuring that readers are well-prepared for more advanced topics in subsequent sections and chapters.

7.2.1 Matrix Partition

Before we delve into the definition of matrix multiplication, it is crucial to understand how to partition a matrix into its row vectors or column vectors. It is evident that a matrix of $m \times n$ dimensions consists of m rows and n columns, with each row or column forming a vector. Thus, a matrix can be partitioned in two fundamental ways:

Definition 7.16 — Partition into Row and Column Vectors. Given a matrix $A_{m \times n}$, it can be partitioned into its row vectors as:

$$A = \begin{bmatrix} a_{11} & a_{12} & \cdots & a_{1n} \\ a_{21} & a_{22} & \cdots & a_{2n} \\ \vdots & \vdots & \ddots & \vdots \\ a_{m1} & a_{m2} & \cdots & a_{mn} \end{bmatrix} = \begin{bmatrix} r_1 \\ r_2 \\ \vdots \\ r_m \end{bmatrix}, \tag{7.18}$$

where $r_1, r_2, \ldots, r_m$ represent the row vectors of A. The matrix can also be partitioned into its column vectors as:

$$A = \begin{bmatrix} a_{11} & a_{21} & \cdots & a_{m1} \\ a_{12} & a_{22} & \cdots & a_{m2} \\ \vdots & \vdots & \ddots & \vdots \\ a_{1n} & a_{2n} & \cdots & a_{mn} \end{bmatrix} = \begin{bmatrix} c_1 & c_2 & \cdots & c_n \end{bmatrix}. \tag{7.19}$$

Here $c_1, c_2, \ldots, c_n$ are the column vectors of A.

Matrix partitioning will be instrumental in understanding many concepts and properties to be introduced later. We will adopt the convention that r_i and c_j represent the row vectors and column vectors of a given matrix, respectively, unless otherwise noted.

Exercise 7.9 Given a matrix of $m \times n$ in size, answer the following questions:

(a) How many row vectors are there in this matrix? How many column vectors?

(b) How many components are there in each row vector? In each column vector?

7.2.2 Definition

In § 6.1.3, we introduced the matrix multiplication between a row vector and a column vector, each consisting of the same number of components. Recall that for a row vector $r = \begin{bmatrix} r_1 & r_2 & \cdots & r_n \end{bmatrix}$ and a column vector $c = \begin{bmatrix} c_1 & c_2 & \cdots & c_n \end{bmatrix}^T$, the matrix product rc is defined as the dot product between these vectors:

$$rc = \begin{bmatrix} r_1 & r_2 & \cdots & r_n \end{bmatrix} \begin{bmatrix} c_1 \\ c_2 \\ \vdots \\ c_n \end{bmatrix} = r_1 c_1 + r_2 c_2 + \cdots + r_n c_n = r \cdot c.$$

We now extend this concept to general matrix multiplication. Consider two matrices, $A_{m \times p}$ and $B_{p \times n}$. For the multiplication AB to be possible, the number of columns in A must match the number of rows in B. This ensures that each row vector of A and each column vector of B contain p components, enabling the computation of their dot products.

Partition matrix A into its row vectors $r_1, r_2, \ldots, r_m$, and matrix B into its column vectors $c_1, c_2, \ldots, c_n$. Given that all r_i and c_j vectors have p components, the dot product $r_i \cdot c_j$ is valid for each pair $i = 1, 2, \ldots, m$ and $j = 1, 2, \ldots, n$. This enables us to formally define the matrix product AB:

Definition 7.17 — Matrix Product. The matrix product AB is defined when the number of columns in $A_{m \times p}$ matches the number of rows in $B_{p \times n}$. Partition A into its row vectors and B into its column vectors:

$$A = \begin{bmatrix} r_1 \\ r_2 \\ \vdots \\ r_m \end{bmatrix}, \quad B = \begin{bmatrix} c_1 & c_2 & \cdots & c_n \end{bmatrix}.$$

The product AB is consequently defined as an $m \times n$ matrix, computed as follows:

$$(AB)_{m \times n} = \begin{bmatrix} r_1 \cdot c_1 & r_1 \cdot c_2 & \cdots & r_1 \cdot c_n \\ r_2 \cdot c_1 & r_2 \cdot c_2 & \cdots & r_2 \cdot c_n \\ \vdots & \vdots & \ddots & \vdots \\ r_m \cdot c_1 & r_m \cdot c_2 & \cdots & r_m \cdot c_n \end{bmatrix}. \tag{7.20}$$

From Def. 7.17, we observe that the product AB comprises dot products between each row vector of A and each column vector of B. More precisely, the entry $(AB)_{ij}$, located at the ith row and the jth column of AB, is calculated by the dot product of the ith row vector of A and the jth column vector of B (Fig. 7.1). This relationship can be succinctly expressed using element-wise notation:

Theorem 7.6 Given two matrices $A_{m \times p}$ and $B_{p \times n}$, and $C_{m \times n} = AB$, the entries of C are given by the **row-column rule** for matrix multiplication:

$$c_{ij} = a_{i1}b_{1j} + a_{i2}b_{2j} + \cdots + a_{ip}b_{pj} = \sum_{k=1}^{p} a_{ik}b_{kj}, \qquad (7.21)$$

where $i = 1, 2, \ldots, m$ and j$= 1, 2, \ldots, n$.

$$AB = \begin{bmatrix} a_{11} & a_{12} & \cdots & a_{1p} \\ a_{21} & a_{22} & \cdots & a_{2p} \\ \vdots & \vdots & \ddots & \vdots \\ a_{i1} & a_{i2} & \cdots & a_{ip} \\ \vdots & \vdots & \ddots & \vdots \\ a_{m1} & a_{m2} & \cdots & a_{mp} \end{bmatrix} \begin{bmatrix} b_{11} & b_{12} & \cdots & b_{1j} & \cdots & b_{1n} \\ b_{21} & b_{22} & \cdots & b_{2j} & \cdots & b_{2n} \\ \vdots & \vdots & \ddots & \vdots & \ddots & \vdots \\ b_{p1} & b_{p2} & \cdots & b_{pj} & \cdots & b_{pn} \end{bmatrix}$$

Figure 7.1: The Row-Column Rule for Matrix Multiplication

When verifying the validity of matrix multiplication, one should consider the rule of matching the "inner sizes." Specifically, the product $A_{m \times p}B_{p \times n}$ is valid because the inner sizes (p from $A_{m \times p}$ and p from $B_{p \times n}$) of A and B match each other.

This rule can be readily applied to a chain of multiplications. For instance, the product $A_{m \times p}B_{p \times q}C_{q \times n}$ results in a valid $m \times n$ matrix since all inner sizes between two adjacent matrices align, and the outermost sizes m and n determine the final size of the resulting matrix.

Exercise 7.10 Given matrices $A_{4 \times 3}$ and $B_{3 \times 4}$, determine whether the following matrix products are valid. If valid, determine the size of the resulting product:

(a) AB (b) BA (c) AB^T

(d) $A^T B$ (e) $A^T B^T$ (f) $B^T A^T$

(g) ABA (h) $BB^T A^T A$

Exercise 7.11 Given $A_{n \times n}$ and $v_{n \times 1}$, determine whether the following matrix products are valid. If valid, determine the size of the resulting product:

(a) Av (b) $Av^\dagger$ (c) vA

(d) $v^\dagger A$ (e) $v^\dagger Av$ (f) $v^\dagger A^\dagger Av$

Exercise 7.12 Given two column vectors $u_{3 \times 1}$ and $v_{3 \times 1}$, determine whether the following matrix products are valid. If valid, specify the size of the resulting product.

(a) uv (b) $uv^\dagger$ (c) $u^\dagger v$

(d) $v^\dagger u$ (e) $vu^\dagger$ (f) $u^\dagger v^\dagger$

■ **Example 7.5** Given $A = \begin{bmatrix} 1 & 2 \\ 1 & 1 \end{bmatrix}$ and $B = \begin{bmatrix} 0 & 1 \\ 2 & 1 \end{bmatrix}$, compute AB and BA.

Solution. Both AB and BA are 2×2 matrices. Using Eq. 7.20 to compute each entry of the results:

$$AB = \begin{bmatrix} 1 \cdot 0 + 2 \cdot 2 & 1 \cdot 1 + 2 \cdot 1 \\ 1 \cdot 0 + 1 \cdot 2 & 1 \cdot 1 + 1 \cdot 1 \end{bmatrix} = \begin{bmatrix} 4 & 3 \\ 2 & 2 \end{bmatrix},$$

$$BA = \begin{bmatrix} 0 \cdot 1 + 1 \cdot 1 & 0 \cdot 2 + 1 \cdot 1 \\ 2 \cdot 1 + 1 \cdot 1 & 2 \cdot 2 + 1 \cdot 1 \end{bmatrix} = \begin{bmatrix} 1 & 1 \\ 3 & 5 \end{bmatrix}.$$

■

Exercise 7.13 Given $A = \begin{bmatrix} 1 & i \\ i & 1 \end{bmatrix}$ and $B = \begin{bmatrix} -i & 1 \\ -1 & i \end{bmatrix}$, compute AB and BA.

Exercise 7.14 Given $A = \begin{bmatrix} 5 & 0 \\ 0 & 3 \\ -2 & 2 \end{bmatrix}$, $B = \begin{bmatrix} 3 & -2 \\ 0 & 4 \end{bmatrix}$, $C = \begin{bmatrix} 2 & 5 & 0 \\ 3 & 1 & 1 \end{bmatrix}$, compute

the following matrix products if they are valid:

(a) $AA^T - B$ (b) $2B^T + CC^T$ (c) $(AB)C$ (d) $A(BC)$

7.2.3   Commutator and Anti-Commutator

The examples from Exercises 7.10 and 7.12 and Example 7.5 illustrate the non-commutativity of matrix multiplication. Generally, $AB \neq BA$, and this can occur in three distinct ways:

(i) AB is defined while BA is not, or vice versa.

(ii) Both AB and BA are defined, but they differ in size, as demonstrated in Exercise 7.10.

(iii) Both AB and BA are defined, are of the same size, but are not equal, as shown in Example 7.5.

The non-commutativity of matrix multiplication contrasts sharply with the multiplication of scalars and the dot products of vectors, which are commutative or conjugate commutative. Understanding this key difference is crucial for comprehending matrix operations and their properties. This topic will be explored further in the next chapter, where matrices are discussed as linear operators on vectors.

In quantum computing, the special case when two matrices commute, meaning $AB = BA$, holds significant importance. To formally express this relationship, we use the commutator, a specialized operator defined as follows:

Definition 7.18 — Commutator. Given two square matrices of the same size, their commutator is denoted as $[A, B]$ and defined by:

$$[A, B] = AB - BA. \tag{7.22}$$

When $[A, B] = 0$, or equivalently $AB = BA$, the matrices A and B are said to commute.

Building on the concept of commutation, where matrix multiplication is symmetric, we now explore its counterpart: the anti-commutator. This operator provides an alternative perspective on matrix interactions.

Definition 7.19 — Anti-Commutator. Given two square matrices of the same size, their anti-commutator is denoted as $\{A, B\}$ and defined by:

$$\{A, B\} = AB + BA. \tag{7.23}$$

When $\{A, B\} = 0$, or equivalently $AB = -BA$, the matrices A and B are said to anticommute.

Commutation and anticommutation relations are particularly relevant when working with Pauli matrices, which will be thoroughly covered in Chapter 12.

■ **Example 7.6** Verify that $A = \begin{bmatrix} 0 & -1 \\ 1 & 0 \end{bmatrix}$ and $B = \begin{bmatrix} -1 & 0 \\ 0 & -1 \end{bmatrix}$ commute.

Solution. Compute AB and BA:

$$AB = \begin{bmatrix} 0 & -1 \\ 1 & 0 \end{bmatrix} \begin{bmatrix} -1 & 0 \\ 0 & -1 \end{bmatrix} = \begin{bmatrix} 0 & 1 \\ -1 & 0 \end{bmatrix},$$

$$BA = \begin{bmatrix} -1 & 0 \\ 0 & -1 \end{bmatrix} \begin{bmatrix} 0 & -1 \\ 1 & 0 \end{bmatrix} = \begin{bmatrix} 0 & 1 \\ -1 & 0 \end{bmatrix}.$$

Since $AB = BA$, the matrices A and B commute.   ■

Exercise 7.15 Given $A = \begin{bmatrix} 0 & 1 \\ 1 & 0 \end{bmatrix}$ and $B = \begin{bmatrix} 0 & -i \\ i & 0 \end{bmatrix}$, verify that A and B anticommute.

7.2.4 Properties of Matrix Multiplication

Despite the non-commutativity of matrix multiplication, it shares many properties with scalar multiplication. The following theorem lists several key properties of matrix multiplication.

Theorem 7.7 Assuming all matrix multiplications are valid, the following identities hold true:

$$A(B + C) = AB + AC \quad \text{(Left Distributive Law)} \tag{7.24a}$$
$$(A + B)C = AC + BC \quad \text{(Right Distributive Law)} \tag{7.24b}$$
$$(AB)C = A(BC) \quad \text{(Associative Law)} \tag{7.24c}$$

All identities in Theorem 7.7 remain valid when addition is replaced by subtraction. Here we provide a proof of Eq. 7.24a using element-wise notation. The other two proofs are left to readers as exercises (Problems 7.5 and 7.6).

Proof.   For the operation $A(B + C)$ to be valid, consider matrices $A_{m \times p} = [a_{ik}]$, $B_{p \times n} = [b_{kj}]$, and $C_{p \times n} = [c_{kj}]$ for $i = 1, 2, \ldots, m$, $k = 1, 2, \ldots, p$, and $j = 1, 2, \ldots, n$.

Define $D = A(B + C) = [d_{ij}]$ and $F = AB + AC = [f_{ij}]$ (both are $m \times n$ matrices). Compute the entries of D and F using element-wise notation for matrix multiplication:

$$d_{ij} = \sum_{k=1}^{p} a_{ik}(b_{kj} + c_{kj}),$$

$$f_{ij} = \sum_{k=1}^{p} a_{ik}b_{kj} + \sum_{k=1}^{p} a_{ik}c_{kj}.$$

By applying the distributive laws of scalars within the summation:

$$d_{ij} = \sum_{k=1}^{p} a_{ik}(b_{kj} + c_{kj}) = \sum_{k=1}^{p}(a_{ik}b_{kj} + a_{ik}c_{kj}) = \sum_{k=1}^{p} a_{ik}b_{kj} + \sum_{k=1}^{p} a_{ik}c_{kj} = f_{ij},$$

for all $i = 1, 2, \ldots, m$ and $j = 1, 2, \ldots, n$. Hence, by Def. 7.8, $D = F$. □

Matrix multiplication often involves compositions with transpose or adjoint operations, where the following identities prove to be very useful:

Theorem 7.8 Assuming all matrix multiplications are valid, the following identities hold true:

$$(AB)^* = A^*B^* \tag{7.25a}$$
$$(AB)^T = B^T A^T \tag{7.25b}$$
$$(AB)^\dagger = B^\dagger A^\dagger \tag{7.25c}$$

We leave the proof of Eq. 7.25a to readers (Problem 7.7). Here we provide the proofs of Eqs. 7.25b and 7.25c.

Proof. To prove Eq. 7.25b, consider $A_{m \times p} = [a_{ik}]$ and $B_{p \times n} = [b_{kj}]$ for $i = 1, 2, \ldots, m$, $k = 1, 2, \ldots, p$, and $j = 1, 2, \ldots, n$.

Let $C_{n \times m} = (AB)^T = [c_{ji}]$, $D_{n \times m} = B^T A^T = [d_{ji}]$. Using Eq. 7.21, compute the entries of C and D as follows:

$$c_{ji} = \left((AB)^T\right)_{ji} = (AB)_{ij} = \sum_{k=1}^{p} a_{ik}b_{kj},$$

$$d_{ji} = (B^T A^T)_{ji} = \sum_{k=1}^{p}(B^T)_{jk}(A^T)_{ki} = \sum_{k=1}^{p} b_{kj}a_{ik}.$$

Since multiplication of scalar entries is commutative, $a_{ik}b_{kj} = b_{kj}a_{ik}$ for all i, j, k, it follows that $c_{ji} = d_{ji}$. Therefore, $C = D$, proving that $(AB)^T = B^T A^T$.

To prove Eq. 7.25c, take the complex conjugate of Eq. 7.25b and use Eq. 7.25a:

$$(AB)^\dagger = ((AB)^*)^T = (A^*B^*)^T = (B^*)^T(A^*)^T = B^\dagger A^\dagger.$$

□

Exercise 7.16 Assuming all matrix products are valid, prove the general property of transposes (or adjoints) of matrix products by demonstrating that:

$$(A_1 A_2 \cdots A_{n-1} A_n)^T = A_n^T A_{n-1}^T \cdots A_2^T A_1^T, \tag{7.26}$$
$$(A_1 A_2 \cdots A_{n-1} A_n)^\dagger = A_n^\dagger A_{n-1}^\dagger \cdots A_2^\dagger A_1^\dagger. \tag{7.27}$$

■ **Example 7.7** Given $n \times n$ matrices A, B, prove that $(A + B)(A - B) = A^2 - B^2$ if and only if A and B commute.

Proof. Using Theorem 7.7, expand the product $(A + B)(A - B)$ as follows:

$$(A + B)(A - B) = (A + B)A - (A + B)B = AA + BA - AB - BB$$
$$= A^2 - B^2 + (BA - AB) = A^2 - B^2.$$

Notice that in the last step, $BA - AB = 0$ since A and B commute. Here, the products AA and BB are respectively written as A^2 and B^2 for simplicity. A formal definition of matrix powers will be provided in the following section.   ■

Example 7.7 demonstrates that the scalar identity $(a + b)(a - b) = a^2 - b^2$ does not generally hold for matrices due to non-commutativity. This emphasizes the importance of carefully verifying algebraic formulas before applying them in matrix operations. Problem 7.8 offers exercises to explore properties that hold for numbers but may not apply to matrices.

7.2.5 Column-Row Expansion

Matrix multiplication can also be performed using a technique known as the **column-row expansion**. This method provides an alternative way to compute the product of matrices by expanding it into a sum of individual matrix products.

Theorem 7.9 — Column-Row Expansion. Consider matrices $A_{m \times p}$ and $B_{p \times n}$. Partition A into its column vectors and B into its row vectors:

$$A = \begin{bmatrix} c_1 & c_2 & \cdots & c_p \end{bmatrix}, \quad B = \begin{bmatrix} r_1 \\ r_2 \\ \vdots \\ r_p \end{bmatrix}.$$

Then, the product AB can be calculated by summing the products of these vectors:

$$AB = c_1 r_1 + c_2 r_2 + \cdots + c_p r_p = \sum_{i=1}^{p} c_i r_i. \tag{7.28}$$

In Eq. 7.28, each term $c_{i\,m \times 1} r_{i\,1 \times n}$ is an $m \times n$ matrix, and thus the entire product AB is expressed as a sum of p such matrices, each of the same dimension. This method of matrix multiplication is aptly named the "column-row" expansion because it involves taking columns from matrix A and rows from matrix B to form the product. This is in contrast to the method described in Eq. 7.21, referred to as the "row-column" rule, where the rows of A and the columns of B are used to compute each individual entry of the final matrix product.

The proof of Theorem 7.9 is left to readers as an exercise (Problem 7.9).

■ **Example 7.8** Redo Example 7.5 using column-row expansion.

Solution. Given $A = \begin{bmatrix} 1 & 2 \\ 1 & 1 \end{bmatrix}$ and $B = \begin{bmatrix} 0 & 1 \\ 2 & 1 \end{bmatrix}$, using Eq. 7.28 gives:

$$AB = \begin{bmatrix} 1 \\ 1 \end{bmatrix} \begin{bmatrix} 0 & 1 \end{bmatrix} + \begin{bmatrix} 2 \\ 1 \end{bmatrix} \begin{bmatrix} 2 & 1 \end{bmatrix} = \begin{bmatrix} 0 & 1 \\ 0 & 1 \end{bmatrix} + \begin{bmatrix} 4 & 2 \\ 2 & 1 \end{bmatrix} = \begin{bmatrix} 4 & 3 \\ 2 & 2 \end{bmatrix}.$$

■

Exercise 7.17 Redo Exercise 7.13 using column-row expansion.

7.2.6 Matrix Multiplication in Dirac Notation

In the field of quantum computing, the matrices and vectors we often discuss are primarily square matrices and column vectors. In this subsection, we will delve into some of the most frequently encountered matrix multiplication scenarios in quantum computing, utilizing the bra-ket notation.

Throughout this section, we will consistently work with $n \times n$ square matrices, denoted by symbols such as A, B, C, etc. Recalling what we have learned from § 6.1, $n \times 1$ column vectors, referred to as kets, will be denoted by $|u\rangle, |v\rangle$, etc., and $1 \times n$ row vectors, known as bras, will be denoted by $\langle u|, \langle v|$, etc. These bras represent the adjoints of their corresponding kets.

1 Inner Product

As we learned from § 6.3.3, inner products are a special form of matrix multiplication between a row vector on the left and a column vector on the right. The result is a 1×1 matrix, effectively a scalar:

$$\langle u|v \rangle = \langle u| \, |v \rangle = \begin{bmatrix} u_1^* & u_2^* & \cdots & u_n^* \end{bmatrix} \begin{bmatrix} v_1 \\ v_2 \\ \vdots \\ v_n \end{bmatrix} = u_1^* v_1 + u_2^* v_2 + \cdots + u_n^* v_n.$$

■ **Example 7.9** Given $|u\rangle = |0\rangle + i\,|1\rangle$ and $|v\rangle = |0\rangle - |1\rangle$, compute $\langle u|v\rangle + \langle v|u\rangle$.

Solution. Note that $\langle u| = (|0\rangle + i\,|1\rangle)^\dagger = \langle 0| - i\,\langle 1|$. Consequently,

$$\begin{aligned} \langle u|v \rangle &= ((\langle 0| - i\,\langle 1|)(|0\rangle - |1\rangle) \\ &= \langle 0|0\rangle - \langle 0|1\rangle - i\,\langle 1|0\rangle + i\,\langle 1|1\rangle \\ &= 1 + i, \end{aligned}$$

where $\langle 0|0\rangle = \langle 1|1\rangle = 1$ and $\langle 0|1\rangle = \langle 1|0\rangle = 0$, in accordance with the properties of orthonormal bases (Theorem 6.21).

Note that $\langle v|u\rangle = \langle u|v\rangle^* = 1 - i$, adhering to the conjugate symmetry of the complex inner product. Thus,

$$\langle u|v \rangle + \langle v|u \rangle = (1 + i) + (1 - i) = 2.$$

■

Exercise 7.18 Compute $\langle u|v\rangle - \langle v|u\rangle$ for $|u\rangle = i\,|0\rangle - |1\rangle$ and $|v\rangle = |0\rangle + i\,|1\rangle$.

2 Matrix-Vector Multiplication

Matrix-vector multiplication involves multiplying an $n \times n$ matrix by an $n \times 1$ column vector, yielding another $n \times 1$ column vector. In Dirac notation, this operation is typically expressed as:

$$A\,|u\rangle = |v\rangle. \tag{7.29}$$

In this context, A is referred to as a **transformation matrix** because it transforms the vector $|u\rangle$ into $|v\rangle$ within $\mathbb{C}^n$. The matrix representation of this transformation is shown as:

$$
\begin{bmatrix}
a_{11} & a_{12} & \cdots & a_{1n} \\
a_{21} & a_{22} & \cdots & a_{2n} \\
\vdots & \vdots & \ddots & \vdots \\
a_{n1} & a_{n2} & \cdots & a_{nn}
\end{bmatrix}
\begin{bmatrix}
u_1 \\ u_2 \\ \vdots \\ u_n
\end{bmatrix}
=
\begin{bmatrix}
v_1 \\ v_2 \\ \vdots \\ v_n
\end{bmatrix}. \tag{7.30}
$$

This operation is fundamental in quantum computing, particularly in the context of quantum gates. The vector $|u\rangle$ represents an input quantum state, A embodies the matrix representation of a quantum gate, and $|v\rangle$ is the transformed output quantum state. The implications of this transformation are further detailed in § 8.1.

3   Outer Product of Two Vectors

When we swap the positions of the row vector and the column vector in an inner product, we obtain an outer product of vectors. Outer products result from matrix multiplication with a column vector on the left and a row vector on the right, producing an $n \times n$ matrix.

> **Definition 7.20 — Outer Product of Two Vectors.** Given two vectors $|u\rangle$ and $|v\rangle$ in $\mathbb{C}^n$, their outer product is an $n \times n$ matrix defined as:
>
> $$
> |u\rangle\langle v| =
> \begin{bmatrix}
> u_1 \\ u_2 \\ \vdots \\ u_n
> \end{bmatrix}
> \begin{bmatrix}
> v_1^* & v_2^* & \cdots & v_n^*
> \end{bmatrix}
> =
> \begin{bmatrix}
> u_1 v_1^* & u_1 v_2^* & \cdots & u_1 v_n^* \\
> u_2 v_1^* & u_2 v_2^* & \cdots & u_2 v_n^* \\
> \vdots & \vdots & \ddots & \vdots \\
> u_n v_1^* & u_n v_2^* & \cdots & u_n v_n^*
> \end{bmatrix}. \tag{7.31}
> $$

■ **Example 7.10** Compute $|0\rangle\langle 0|$, $|0\rangle\langle 1|$, $|1\rangle\langle 0|$, and $|1\rangle\langle 1|$ in matrix representation.

Solution. Using Def. 7.20, we find:

$$
|0\rangle\langle 0| = \begin{bmatrix} 1 \\ 0 \end{bmatrix} \begin{bmatrix} 1 & 0 \end{bmatrix} = \begin{bmatrix} 1 & 0 \\ 0 & 0 \end{bmatrix}, \quad
|0\rangle\langle 1| = \begin{bmatrix} 1 \\ 0 \end{bmatrix} \begin{bmatrix} 0 & 1 \end{bmatrix} = \begin{bmatrix} 0 & 1 \\ 0 & 0 \end{bmatrix},
$$

$$
|1\rangle\langle 0| = \begin{bmatrix} 0 \\ 1 \end{bmatrix} \begin{bmatrix} 1 & 0 \end{bmatrix} = \begin{bmatrix} 0 & 0 \\ 1 & 0 \end{bmatrix}, \quad
|1\rangle\langle 1| = \begin{bmatrix} 0 \\ 1 \end{bmatrix} \begin{bmatrix} 0 & 1 \end{bmatrix} = \begin{bmatrix} 0 & 0 \\ 0 & 1 \end{bmatrix}.
$$

■

> **Exercise 7.19** Given the vectors $|u\rangle = |0\rangle + i\,|1\rangle$ and $|v\rangle = i\,|0\rangle - |1\rangle$, compute the matrix representations for $|v\rangle\langle u|$ and $|u\rangle\langle v|$.

A useful formula involving outer products is:

$$
|v\rangle\langle u| = (|u\rangle\,\langle v|)^\dagger. \tag{7.32}
$$

This can be demonstrated by applying the conjugate transpose property Eq. 7.25c:

$$
(|u\rangle\,\langle v|)^\dagger = ((\langle v|)^\dagger(|u\rangle))^\dagger = |v\rangle\,\langle u|.
$$

> **Exercise 7.20** Express the outer product $|v\rangle\,\langle u|$ in matrix form for the vectors $|u\rangle$ and $|v\rangle$ as specified in Eq. 7.31. Confirm that this matrix is equal to the adjoint (conjugate transpose) of the matrix representation of $|u\rangle\,\langle v|$.

4   Decomposing Matrices Using Outer Products

Example 7.10 suggests that a general 2×2 matrix $A = \begin{bmatrix} a_{00} & a_{01} \\ a_{10} & a_{11} \end{bmatrix}$ can be expressed in terms of outer products as:

$$A = a_{00} \, |0\rangle\langle 0| + a_{01} \, |0\rangle\langle 1| + a_{10} \, |1\rangle\langle 0| + a_{11} \, |1\rangle\langle 1| . \tag{7.33}$$

This approach can be generalized to any matrix, as outlined in the following theorem.

Theorem 7.10 — Matrix Basis Decomposition. Let $\{|0\rangle, |1\rangle, \ldots, |d-1\rangle\}$ be the computational basis of $\mathbb{C}^d$. For a general matrix $A \in \mathbb{C}^{d \times d}$ given by:

$$A = \begin{bmatrix} a_{00} & a_{01} & \cdots & a_{0(d-1)} \\ a_{10} & a_{11} & \cdots & a_{1(d-1)} \\ \vdots & \vdots & \ddots & \vdots \\ a_{(d-1)0} & a_{(d-1)1} & \cdots & a_{(d-1)(d-1)} \end{bmatrix},$$

it can be decomposed into a sum of outer products:

$$A = \sum_{i,j} a_{ij} \, |i\rangle\langle j| , \quad \text{for } i, j = 0, 1, \ldots, d-1. \tag{7.34}$$

(i) In Theorem 7.10, the indices i and j start from zero, aligning with the convention for computational basis labels. While this theorem is presented for square matrices, it also applies to non-square matrices, though the focus here remains on square matrices, reflecting contexts frequently encountered in quantum computing.

■ **Example 7.11** Given the matrix A as described in Eq. 7.34, demonstrate that

$$A \, |k\rangle = \sum_i a_{ik} \, |i\rangle . \tag{7.35}$$

Proof.

$$A \, |k\rangle = \left(\sum_{i,j} a_{ij} \, |i\rangle\langle j| \right) |k\rangle$$

$$= \sum_{i,j} a_{ij} \, |i\rangle \, \langle j|k\rangle$$

$$= \sum_{i,j} a_{ij} \, |i\rangle \, \delta_{jk}$$

$$= \sum_i a_{ik} \, |i\rangle ,$$

where the equality $\langle j|k\rangle = \delta_{jk}$ (the Kronecker delta) simplifies the expression by collapsing the sum over j to only the term where $j = k$. ■

Exercise 7.21 Using Theorem 7.10, express the matrix

$$A = \begin{bmatrix} 1 & 0 & 0 & 0 \\ 0 & 0 & 0 & 1 \\ 0 & 0 & 1 & 0 \\ 0 & 1 & 0 & 0 \end{bmatrix}$$

in terms of a sum of outer products, utilizing the computational basis of $\mathbb{C}^4$.

5 * Basis Change Matrices

When considering basis changes, matrices in quantum computing are often expressed as sums of outer products between two different sets of basis vectors. As an example, consider the basis transition matrix from the computational basis $\{|0\rangle, |1\rangle\}$ to the basis $\{|+\rangle, |-\rangle\}$:

$$U = |+\rangle \langle 0| + |-\rangle \langle 1| . \tag{7.36}$$

This matrix U can be verified directly using bra-ket notation to ensure it performs the intended basis transformation:

$$U |0\rangle = (|+\rangle \langle 0| + |-\rangle \langle 1|) |0\rangle = |+\rangle \langle 0|0\rangle + |-\rangle \langle 1|0\rangle = |+\rangle ,$$
$$U |1\rangle = (|+\rangle \langle 0| + |-\rangle \langle 1|) |1\rangle = |+\rangle \langle 0|1\rangle + |-\rangle \langle 1|1\rangle = |-\rangle .$$

This computation illustrates that U maps $|0\rangle$ to $|+\rangle$ and $|1\rangle$ to $|-\rangle$ as expected.

Exercise 7.22 Express the basis change matrix that maps $|0\rangle$ to $|-\rangle$ and $|1\rangle$ to $|+\rangle$ in terms of outer products and as a matrix representation.

The derivation above utilizes the associative property of matrix multiplication, as illustrated in the calculation $(|+\rangle \langle 0|) |0\rangle = |+\rangle \langle 0|0\rangle$. This result can be generalized to the following formula:

Theorem 7.11 Given a matrix U as an outer product of $U = |u\rangle \langle v|$ and a vector $|w\rangle$, the product $U |w\rangle$ is computed as:

$$U |w\rangle = \langle v|w\rangle |u\rangle , \tag{7.37}$$

resulting in a scalar multiple of $|u\rangle$.

Proof. $U |w\rangle = (|u\rangle \langle v|) |w\rangle = |u\rangle (\langle v| |w\rangle) = |u\rangle \langle v|w\rangle = \langle v|w\rangle |u\rangle$. Since $\langle v|w\rangle$ is a scalar, the result is a scalar multiple of $|u\rangle$. □

A comprehensive discussion on change of basis in quantum computing will be provided later in § 8.4.

Exercise 7.23 Given the operator matrix $U = |0\rangle\langle+| + |1\rangle\langle-|$ and the quantum state $|\psi\rangle = \frac{1}{\sqrt{2}}(|+\rangle + i|-\rangle)$, compute $U |\psi\rangle$ without using matrix representation.

6 Inner Products Involving Matrix Transformations

Sometimes, we need to compute the inner product between two vectors, where one or both vectors are the result of a matrix-vector product. Below are a few examples illustrating this, presented in both bra-ket and matrix notations:

$$\langle Ax|y\rangle = (Ax) \cdot y,$$
$$\langle x|Ay\rangle = x \cdot (Ay),$$
$$\langle Ax|Ay\rangle = (Ax) \cdot (Ay).$$

Here, A is an $n \times n$ matrix, and $|x\rangle$, $|y\rangle$ are $n \times 1$ vectors. For brevity, the notations $|Ax\rangle \equiv A|x\rangle$ and $\langle Ax| \equiv (A|x\rangle)^\dagger$ are adopted.

These forms naturally arise in linear algebra and quantum computing (see § 11.1.5 for an example). The following identities are often employed when dealing with these forms:

Theorem 7.12 The inner products on the left-hand side can be expressed as matrix products on the right-hand side:

$$\langle x|Ay\rangle = \langle x|A|y\rangle, \tag{7.38a}$$
$$\langle Ax|y\rangle = \langle x|A^\dagger|y\rangle, \tag{7.38b}$$
$$\langle Ax|Ay\rangle = \langle x|A^\dagger A|y\rangle. \tag{7.38c}$$

Proof. These relationships follow from Eq. 7.25c and the rules of bra-ket notation:

$$\langle x|Ay\rangle = \langle x|\,|Ay\rangle = \langle x|A|y\rangle,$$
$$\langle Ax|y\rangle = \langle Ax|\,|y\rangle = (|Ax\rangle)^\dagger\,|y\rangle = |x\rangle^\dagger\,A^\dagger\,|y\rangle = \langle x|A^\dagger|y\rangle,$$
$$\langle Ax|Ay\rangle = \langle Ax|\,|Ay\rangle = (|Ax\rangle)^\dagger A\,|y\rangle = \langle x|A^\dagger A|y\rangle.$$

□

For a given matrix A, the expression $\langle x|A|y\rangle$ defines a mapping from $\mathbb{C}^n \times \mathbb{C}^n$ to $\mathbb{C}$. This mapping is linear in its second argument $|y\rangle$ and conjugate linear in its first argument $|x\rangle$. Hence, $\langle x|A|y\rangle$ is often referred to as a sesquilinear form.

■ **Example 7.12** Given the matrix $A = |0\rangle\langle 1| + |1\rangle\langle 0|$, compute the expression $\langle 0|A|1\rangle$ using both matrix representation and Dirac notation.

Solution. Using matrix representation:

$$\langle 0|A|1\rangle = \begin{bmatrix} 1 & 0 \end{bmatrix} \begin{bmatrix} 0 & 1 \\ 1 & 0 \end{bmatrix} \begin{bmatrix} 0 \\ 1 \end{bmatrix}$$
$$= \begin{bmatrix} 0 & 1 \end{bmatrix} \begin{bmatrix} 0 \\ 1 \end{bmatrix} = 0 \cdot 0 + 1 \cdot 1 = 1.$$

Using Dirac notation:

$$\langle 0|A|1\rangle = \langle 0|(|0\rangle\langle 1| + |1\rangle\langle 0|)|1\rangle = \langle 0|0\rangle\,\langle 1|1\rangle + \langle 0|1\rangle\,\langle 0|1\rangle = 1 \cdot 1 + 0 \cdot 0 = 1.$$

■

Exercise 7.24 Consider a matrix A expressed as $A = \sum_{i,j} a_{ij}\,|i\rangle\langle j|$. Prove that the identity $\langle i|A|j\rangle = a_{ij}$ holds.

Exercise 7.25 Prove that

$$\langle x|A^\dagger|y\rangle = (\langle y|A|x\rangle)^*, \tag{7.39}$$

$$\langle x|Ay\rangle = \langle A^\dagger x|y\rangle. \tag{7.40}$$

When A is Hermitian and vectors $|x\rangle, |y\rangle$ are equal, the forms in Theorem 7.12 simplify to a **Hermitian quadratic form**:

$$\langle \psi|M|\psi\rangle, \quad \text{where } M \text{ is Hermitian.}$$

This form is critical in quantum computing as it represents the statistical average of an observable M on the state $|\psi\rangle$, always yielding a real scalar. This is explored further in the proofs and examples below.

Theorem 7.13 Given a Hermitian matrix $M \in \mathbb{C}^{n\times n}$ and a complex vector $|\psi\rangle \in \mathbb{C}^n$, $\langle \psi|M|\psi\rangle$ is always real.

Proof. Taking $|x\rangle = |y\rangle = |\psi\rangle$ in Eq. 7.39, we have:

$$\langle \psi|M^\dagger|\psi\rangle = (\langle \psi|M|\psi\rangle)^*.$$

Since M is Hermitian, it holds that $M = M^\dagger$, and thus:

$$\langle \psi|M|\psi\rangle = \langle \psi|M^\dagger|\psi\rangle.$$

This equality implies:

$$\langle \psi|M|\psi\rangle = (\langle \psi|M|\psi\rangle)^* \quad \Rightarrow \quad \langle \psi|M|\psi\rangle \text{ is real.}$$

$\square$

■ **Example 7.13** Given $A = \begin{bmatrix} 0 & i \\ -i & 0 \end{bmatrix}$ and $|v\rangle = \begin{bmatrix} \frac{1}{\sqrt{2}} \\ \frac{i}{\sqrt{2}} \end{bmatrix}$. Compute $\langle v|A|v\rangle$.

Solution. In matrix representation:

$$\langle v|A|v\rangle = v^\dagger A v = \begin{bmatrix} \frac{1}{\sqrt{2}} & -\frac{i}{\sqrt{2}} \end{bmatrix} \begin{bmatrix} 0 & i \\ -i & 0 \end{bmatrix} \begin{bmatrix} \frac{1}{\sqrt{2}} \\ \frac{i}{\sqrt{2}} \end{bmatrix}$$

$$= \begin{bmatrix} -\frac{1}{\sqrt{2}} & \frac{i}{\sqrt{2}} \end{bmatrix} \begin{bmatrix} \frac{1}{\sqrt{2}} \\ \frac{i}{\sqrt{2}} \end{bmatrix}$$

$$= -\frac{1}{\sqrt{2}} \cdot \frac{1}{\sqrt{2}} + \frac{i}{\sqrt{2}} \cdot \frac{i}{\sqrt{2}} = -1.$$

■

Exercise 7.26 Given $A = \begin{bmatrix} 0 & 1 \\ -1 & 0 \end{bmatrix}$ and $|v\rangle = \begin{bmatrix} \frac{1}{\sqrt{2}} \\ \frac{i}{\sqrt{2}} \end{bmatrix}$. Compute $\langle v|A|v\rangle$. Explain why the result is not a real number.

7.3 Matrix Inverses

Similar to how every non-zero number z possesses a unique multiplicative inverse $1/z$, such that their product is unity, square matrices under certain conditions also exhibit unique inverses. These inverses, when multiplied by their corresponding matrices, yield the identity matrix, colloquially referred to as "the one" of matrix algebra.

Matrix inverses are highly useful in matrix algebra and analysis, often facilitating the simplification of computations. This section explores the theory of matrix inverses and addresses related concepts, including diagonal matrices, matrix powers, and matrix polynomials.

7.3.1 Identity Matrices

The number "1" acts as the multiplicative identity within the fields of $\mathbb{R}$ or $\mathbb{C}$ (refer to § 4.2.4), ensuring that any scalar multiplied by one remains unchanged. Analogously, in the context of matrix operations, identity matrices serve a similar fundamental role.

Definition 7.21 — Identity Matrices. The identity matrix, denoted by I_n, is an $n \times n$ square matrix characterized by ones on its main diagonal and zeros in all off-diagonal positions:

$$I_n = \begin{bmatrix} 1 & 0 & 0 & \cdots & 0 \\ 0 & 1 & 0 & \cdots & 0 \\ 0 & 0 & 1 & \cdots & 0 \\ \vdots & \vdots & \vdots & \ddots & \vdots \\ 0 & 0 & 0 & \cdots & 1 \end{bmatrix}. \tag{7.41}$$

The identity matrix I_n can be concisely represented using the Kronecker delta function:

$$I_n = [\delta_{ij}], \text{ where } \delta_{ij} = \begin{cases} 0 & \text{if } i \neq j, \\ 1 & \text{if } i = j, \end{cases} \text{ for } i, j = 1, 2, \ldots, n. \tag{7.42}$$

Theorem 7.14 For any matrices $A_{m \times n}$ and $B_{n \times p}$, the identity matrix I_n satisfies the properties $AI_n = A$ and $I_n B = B$.

Proof. Let $A = [a_{ij}]$ with $i = 1, 2, \ldots, m$ and $j = 1, 2, \ldots, n$. Consider the product $C = AI_n = [c_{ij}]$ calculated using Theorem 7.6:

$$c_{ij} = \sum_{k=1}^{n} a_{ik} \delta_{kj}.$$

The only non-zero term in the summation occurs when $k = j$, yielding:

$$c_{ij} = a_{ij} \delta_{jj} = a_{ij},$$

for all i, j, affirming $AI_n = A$. A similar method can be applied to demonstrate $I_n B = B$. □

Theorem 7.14 demonstrates that any matrix multiplied by an identity matrix results in the original matrix, whether on the left or right. Consequently, I_n functions as the multiplicative identity within the context of matrix multiplication.

Exercise 7.27 Consider matrices $A = \begin{bmatrix} 1 & i \\ 0 & 1 \end{bmatrix}$ and $B = \begin{bmatrix} 1 & -i \\ 0 & 1 \end{bmatrix}$. Verify that $AB = BA = I_2$.

(i) **Notation:** When the context clearly identifies the size of the identity matrix I_n, we may omit the subscript and denote it simply as I.

7.3.2 Inverse of a Matrix

Following the introduction of the identity matrix, we proceed to define the inverse of a matrix, which is conceptually analogous to the inverse of a scalar. If two square matrices multiply to yield the identity matrix, they are inverses of each other.

Definition 7.22 — Matrix Inverse. Consider a square matrix $A_{n \times n}$. If there exists a square matrix $B_{n \times n}$ such that $AB = BA = I_n$, then A is deemed **invertible**, and B is identified as the **inverse** of A, denoted as A^{-1}. If no such matrix B can be found, A is described as **singular**.

As stipulated in Def. 7.22, the inverse of a matrix, if it exists, is unique. This property can be rigorously demonstrated as follows.

Theorem 7.15 The inverse of a matrix is unique if it exists.

Proof. Assume B and C are both inverses of a matrix A. By definition, $BA = AB = CA = AC = I$. Considering these conditions, we can perform the following calculations:

$$BAC = (BA)C = IC = C.$$

Simultaneously, applying the associative law of matrix multiplication, we obtain:

$$BAC = B(AC) = BI = B.$$

Consequently, since both expressions for BAC must equal each other, we conclude that $B = C$. This demonstrates the uniqueness of the inverse matrix. □

Exercise 7.27 illustrates an instance where two matrices are inverse to each other. Specifically, when B is the inverse of A, A is also the inverse of B. In practice, it is sufficient to verify either $AB = I$ or $BA = I$ to establish their inverse relationship, as detailed in the following theorem:

Theorem 7.16 For two square matrices A and B, if $AB = I$, then it necessarily follows that $BA = I$, establishing that A and B are inverses of each other.

The proof of this theorem is omitted because it relies on an understanding of linear systems, which has not yet been introduced in this textbook.

Exercise 7.28 Verify that $A = \begin{bmatrix} i & 1 \\ 1 & i \end{bmatrix}$ and $B = \dfrac{1}{2} \begin{bmatrix} -i & 1 \\ 1 & -i \end{bmatrix}$ are inverses of each other.

For 2×2 matrices, a general formula exists for finding A^{-1} (Problem 7.14). However, deriving A^{-1} for a general $n \times n$ matrix is significantly more complex. Many algorithms have been developed to efficiently compute the inverse, A^{-1}, of large matrices, although a detailed exploration of these algorithms is beyond the focus of this text.

7.3.3 ✳ Singular Matrices

Not all square matrices are invertible. The zero matrix is an obvious example of a singular matrix, since any product involving the zero matrix results in the zero matrix, making it impossible to obtain the identity matrix. However, the set of singular matrices extends beyond just the zero matrix. Unlike numbers, where zero is the only element without a multiplicative inverse, there are infinitely many non-zero matrices that are singular. The example below illustrates this point clearly.

▪ **Example 7.14** Show that $A = \begin{bmatrix} 1 & 2 \\ 2 & 4 \end{bmatrix}$ is singular.

Proof. Notice that the row (and column) vectors of A are linearly dependent. We will demonstrate that it is not possible for $AB = I$ for any matrix $B = \begin{bmatrix} a & b \\ c & d \end{bmatrix}$.

Compute AB by matrix multiplication:

$$AB = \begin{bmatrix} 1 & 2 \\ 2 & 4 \end{bmatrix} \begin{bmatrix} a & b \\ c & d \end{bmatrix} = \begin{bmatrix} a + 2c & b + 2d \\ 2a + 4c & 2b + 4d \end{bmatrix}.$$

It can be observed that the two row (and column) vectors of AB are also linearly dependent since $\begin{bmatrix} 2a + 4c & 2b + 4d \end{bmatrix} = 2 \begin{bmatrix} a + 2c & b + 2d \end{bmatrix}$. This property is inconsistent with I, where the row (and column) vectors are linearly independent. This completes the proof by Def. 7.22 that A is singular.   ▪

> **Theorem 7.17** A square matrix A is singular if its row vectors or column vectors are linearly dependent.

Proof. Consider a square matrix $A_{n \times n}$ with n linearly dependent column vectors. Let S be the span of these vectors. If these vectors are linearly dependent, then $\dim(S) < n$, otherwise these vectors would form a basis for S and must be linearly independent.

Next, consider any square matrix $B_{n \times n}$ partitioned into column vectors $B = \begin{bmatrix} b_1 & b_2 & \cdots & b_n \end{bmatrix}$. Define $C = AB$ with column vectors expressed as (Problem 7.10):

$$\begin{bmatrix} Ab_1 & Ab_2 & \cdots & Ab_n \end{bmatrix}.$$

Each product Ab_i is a linear combination of the column vectors of A (Problem 7.11), thus each Ab_i lies in S for $i = 1, 2, \ldots, n$. Because $\dim(S) < n$, the column vectors of C are necessarily linearly dependent, making it not possible for C to be the identity matrix I_n, which requires n linearly independent column vectors. Since this argument holds for any matrix B, matrix A must be singular.

A similar argument applies when A has linearly dependent row vectors. By analyzing the row vectors of $C = BA$, it can be shown that these vectors are also linearly dependent, implying $C \neq I$ for any B. Hence, A is singular.   □

> **Invertibility**
>
> Invertibility is a fundamental property of square matrices. Determining whether a matrix is invertible or singular is a key task in linear algebra. Various theorems, such as Theorem 7.17, provide methods for assessing this attribute.
>
> In quantum computing, however, the need to examine matrix invertibility arises less frequently compared to other disciplines. This is primarily because unitary matrices, which represent quantum gates, are inherently invertible by definition (see § 8.3 for more details).
>
> As a result, this text does not focus as extensively on invertibility as a typical linear algebra textbook might. Instead, the emphasis lies on unitary matrices and their properties, which are central to quantum computing.

7.3.4   Matrix Equations

In practice, it is often desirable to represent the solution of a problem as a vector. In the realm of linear algebra, such problems are often formulated using matrix equations. The following theorem explores one of the most fundamental and essential types of matrix equations, particularly emphasizing the role of matrix inverses.

> **Theorem 7.18** Given an invertible matrix $A \in \mathbb{C}^{n \times n}$ and a vector $|v\rangle \in \mathbb{C}^n$, the matrix equation
> $$A |x\rangle = |v\rangle , \tag{7.43}$$
> admits a unique solution:
> $$|x\rangle = A^{-1} |v\rangle . \tag{7.44}$$

Proof.   By multiplying Eq. 7.43 by A^{-1} on the left, we obtain:
$$A^{-1}A |x\rangle = A^{-1} |v\rangle \quad \Rightarrow \quad I |x\rangle = A^{-1} |v\rangle \quad \Rightarrow \quad |x\rangle = A^{-1} |v\rangle .$$

Thus, $|x\rangle = A^{-1} |v\rangle$ is the unique solution, as stipulated by the equation.   □

■ **Example 7.15** Solve the matrix equation
$$ABC |x\rangle = |v\rangle ,$$

where A, B, C are invertible matrices.

Proof.   Multiplying $C^{-1}B^{-1}A^{-1}$ on the left side of both sides of the equation yields:
$$C^{-1}B^{-1}A^{-1}ABC |x\rangle = C^{-1}B^{-1}A^{-1} |v\rangle .$$

Noting that the product $C^{-1}B^{-1}A^{-1}ABC$ simplifies as follows:
$$C^{-1}B^{-1}A^{-1}ABC = C^{-1}B^{-1}IBC = C^{-1}B^{-1}BC = C^{-1}IC = C^{-1}C = I,$$

we obtain
$$I |x\rangle = C^{-1}B^{-1}A^{-1} |v\rangle \quad \Rightarrow \quad |x\rangle = C^{-1}B^{-1}A^{-1} |v\rangle .$$

■

> **Theorem 7.19** Given a series of invertible matrices $A_1, A_2, \ldots, A_{n-1}, A_n$ of the

same dimensions, the following identity holds true:

$$(A_1 A_2 \cdots A_{n-1} A_n)^{-1} = A_n^{-1} A_{n-1}^{-1} \cdots A_2^{-1} A_1^{-1}. \tag{7.45}$$

Proof. We verify the identity by showing that the product of the sequence of matrices and their inverses results in the identity matrix, I:

$$
\begin{aligned}
(A_n^{-1} A_{n-1}^{-1} \cdots A_2^{-1} A_1^{-1})(A_1 A_2 \cdots A_{n-1} A_n) &= A_n^{-1} A_{n-1}^{-1} \cdots A_2^{-1} I A_2 \cdots A_{n-1} A_n \\
&= A_n^{-1} A_{n-1}^{-1} \cdots A_3^{-1} I A_3 \cdots A_{n-1} A_n \\
&= \cdots \\
&= A_n^{-1} A_{n-1}^{-1} A_{n-1} A_n \\
&= A_n^{-1} I A_n = A_n^{-1} A_n = I,
\end{aligned}
$$

Therefore, by Theorem 7.16, we conclude that

$$(A_1 A_2 \cdots A_{n-1} A_n)^{-1} = A_n^{-1} A_{n-1}^{-1} \cdots A_2^{-1} A_1^{-1}.$$

□

Theorem 7.20 If A is an invertible matrix, then its adjoint, $A^\dagger$, is also invertible, and the inverse of $A^\dagger$ is given by:

$$(A^\dagger)^{-1} = (A^{-1})^\dagger. \tag{7.46}$$

Exercise 7.29 Use Theorem 7.16 to prove Theorem 7.20.

Exercise 7.30 Simplify the following matrix expressions using matrix algebra and properties of matrix inverses, assuming all involved matrices are invertible:

(a) $(AC)^{-1}(DC^{-1})^{-1}DA$

(b) $A(B^{-1} + A^{-1})B(B + A)^{-1}$

7.3.5 Diagonal Matrices

Diagonal matrices, defined as square matrices with non-zero entries only on the main diagonal, play a significant role in linear algebra theory. Exploring their properties is necessary in preparation for many future topics.

1 Definition

Definition 7.23 — Diagonal Matrix. A diagonal matrix is a square matrix where all off-diagonal entries are zero. A general $n \times n$ diagonal matrix is represented as

$$\Lambda = \begin{bmatrix} \lambda_1 & 0 & \cdots & 0 \\ 0 & \lambda_2 & \cdots & 0 \\ 0 & 0 & \ddots & 0 \\ 0 & 0 & \cdots & \lambda_n \end{bmatrix}. \tag{7.47}$$

(i) **Notation:** The symbols Λ or D are commonly used to denote diagonal matrices. Given that an $n \times n$ diagonal matrix is defined by its entries on the main diagonal, the notation $\text{diag}(\lambda_1, \lambda_2, \ldots, \lambda_n)$ is often employed to succinctly represent the diagonal matrix described in Eq. 7.47.

2 Matrix Multiplication

A key advantage of using diagonal matrices is that matrix algebra operations involving them are exceptionally straightforward. The following theorem delineates the general rules for matrix multiplication involving a diagonal matrix.

Theorem 7.21 Given an $n \times n$ diagonal matrix $\Lambda = \text{diag}(\lambda_1, \lambda_2, \ldots, \lambda_n)$ and an $m \times n$ matrix A partitioned into column vectors $A = \begin{bmatrix} c_1 & c_2 & \cdots & c_n \end{bmatrix}$, the matrix product $A\Lambda$ is obtained by multiplying each column of A by the corresponding λ_i in Λ:

$$A\Lambda = \begin{bmatrix} \lambda_1 c_1 & \lambda_2 c_2 & \cdots & \lambda_n c_n \end{bmatrix}. \tag{7.48}$$

Similarly, for an $n \times p$ matrix B partitioned into row vectors $B = \begin{bmatrix} r_1 \\ r_2 \\ \vdots \\ r_n \end{bmatrix}$, the matrix product ΛB is obtained by multiplying each row vector of B by the corresponding λ_i in Λ:

$$\Lambda B = \begin{bmatrix} \lambda_1 r_1 \\ \lambda_2 r_2 \\ \vdots \\ \lambda_n r_n \end{bmatrix}. \tag{7.49}$$

The proof of Theorem 7.21 can be effectively demonstrated using element-wise notation and is left as an exercise for the reader. For two diagonal matrices, they still multiply to another diagonal matrix given by:

$$\begin{bmatrix} \lambda_1 & 0 & \cdots & 0 \\ 0 & \lambda_2 & \cdots & 0 \\ 0 & 0 & \ddots & 0 \\ 0 & 0 & \cdots & \lambda_n \end{bmatrix} \begin{bmatrix} \mu_1 & 0 & \cdots & 0 \\ 0 & \mu_2 & \cdots & 0 \\ 0 & 0 & \ddots & 0 \\ 0 & 0 & \cdots & \mu_n \end{bmatrix} = \begin{bmatrix} \lambda_1\mu_1 & 0 & \cdots & 0 \\ 0 & \lambda_2\mu_2 & \cdots & 0 \\ 0 & 0 & \ddots & 0 \\ 0 & 0 & \cdots & \lambda_n\mu_n \end{bmatrix}. \tag{7.50}$$

3 Inverse

Checking the invertibility or computing the inverse of a diagonal matrix is straightforward, as illustrated in the following theorem.

Theorem 7.22 A diagonal matrix $\Lambda = \text{diag}(\lambda_1, \lambda_2, \ldots, \lambda_n)$ is invertible if and only if $\lambda_i \neq 0$ for all $i = 1, 2, \ldots, n$. The inverse of such a matrix is given by:

$$\Lambda^{-1} = \text{diag}\left(\frac{1}{\lambda_1}, \frac{1}{\lambda_2}, \ldots, \frac{1}{\lambda_n}\right) = \begin{bmatrix} \frac{1}{\lambda_1} & 0 & \cdots & 0 \\ 0 & \frac{1}{\lambda_2} & \cdots & 0 \\ 0 & 0 & \ddots & 0 \\ 0 & 0 & \cdots & \frac{1}{\lambda_n} \end{bmatrix}. \tag{7.51}$$

Theorem 7.22 can be easily proved by utilizing Eq. 7.50 to compute the product of Λ and Λ^{-1} as specified in Eq. 7.51, which results in the identity matrix I.

4 Dirac Notation

In Dirac notation, diagonal matrices are often expressed as a sum of outer products, leveraging the clarity and convenience this representation offers.

Theorem 7.23 Let $\{|0\rangle, |1\rangle, \ldots, |d-1\rangle\}$ be the computational basis of $\mathbb{C}^d$. A general diagonal matrix $\Lambda = \mathrm{diag}(k_0, k_1, \ldots, k_{d-1})$ can be expressed as:

$$\begin{bmatrix} k_0 & 0 & \cdots & 0 \\ 0 & k_1 & \cdots & 0 \\ 0 & 0 & \ddots & 0 \\ 0 & 0 & \cdots & k_{d-1} \end{bmatrix} = k_0 |0\rangle\langle 0| + k_1 |1\rangle\langle 1| + \cdots + k_{d-1} |d-1\rangle\langle d-1|. \quad (7.52)$$

Theorem 7.23 represents a special case of Theorem 7.10, wherein only the diagonal terms $a_{ii} |i\rangle\langle i|$ are retained in the summation.

Exercise 7.31 Compute the matrix representation of $3 |0\rangle\langle 0| - 2 |1\rangle\langle 1| + 4 |2\rangle\langle 2|$, where $|0\rangle \equiv \begin{bmatrix} 1 \\ 0 \\ 0 \end{bmatrix}$, $|1\rangle \equiv \begin{bmatrix} 0 \\ 1 \\ 0 \end{bmatrix}$, and $|2\rangle \equiv \begin{bmatrix} 0 \\ 0 \\ 1 \end{bmatrix}$ form the computational basis for $\mathbb{C}^3$.

Diagonal matrices possess many other properties that greatly facilitate computation and derivation. Those properties will be introduced later in the appropriate context.

7.3.6 Matrix Powers and Polynomials

1 Matrix Power

For square matrices $A_{n \times n}$, repeated multiplication by itself is common, as in expressions like AA (denoted A^2) or AAA (denoted A^3). By convention, the zeroth power of any matrix is defined as the identity matrix I, analogous to $a^0 = 1$ for numbers.

Definition 7.24 — Matrix Power. For a square matrix A, non-negative integer powers are defined as follows:

$$A^0 = I, \quad A^n = AA \cdots A \quad \text{(with } n \text{ factors of } A\text{)}.$$

For invertible matrices, negative integer powers can also be defined:

Definition 7.25 — Negative Power. For an invertible matrix A, negative integer powers are defined as:

$$A^{-n} = A^{-1}A^{-1} \cdots A^{-1} \quad \text{(with } n \text{ factors of } A^{-1}\text{)}.$$

The definitions in Definitions 7.24 and 7.25 naturally extend the laws for numbers to matrices, yielding the following properties:

> **Theorem 7.24** For any matrix A, the following properties hold:
>
> $$A^p A^q = A^{p+q}, \quad (A^p)^q = A^{pq},$$
>
> provided all powers involved are defined.

(i) Square matrices can also take square roots or other non-integer powers under certain conditions. Detailed explanations can be found in § 11.3.1.

2 Idempotent Matrices

A special type of matrix related to matrix powers is the idempotent matrix, defined as follows:

> **Definition 7.26 — Idempotent Matrix.** A square matrix A is called idempotent if $A^2 = A$.

The word "idempotent" comes from Latin, where "idem" means "same" and "potent" means "power." In this context, it signifies that repeatedly applying the matrix (multiplying it by itself) yields the same result as the original matrix.

> **Exercise 7.32** Verify that any 2×2 matrix of the form
>
> $$\begin{bmatrix} a & a \\ 1-a & 1-a \end{bmatrix}$$
>
> is idempotent, where c is any scalar.

Idempotent matrices frequently arise in the context of projection matrices, defined as follows:

> **Definition 7.27 — Projection Matrix.** A projection matrix onto an $n \times 1$ vector $|u\rangle$ is defined as:
>
> $$P_u = |u\rangle\langle u|, \tag{7.53}$$
>
> such that $P_u |v\rangle$ projects any $n \times 1$ vector $|v\rangle$ onto $|u\rangle$. Here, $|u\rangle$ and $|v\rangle$ are normalized vectors.

Proof. Consider the matrix multiplication $P_u |v\rangle$ and use the outer product formula (Eq. 7.37):

$$P_u |v\rangle = |u\rangle\langle u| |v\rangle = \langle u|v\rangle |u\rangle.$$

The result, $\langle u|v\rangle |u\rangle$, is precisely the projection of $|v\rangle$ onto $|u\rangle$, consistent with Theorem 6.19.   □

> **Theorem 7.25** The projection matrix $P_u = |u\rangle\langle u|$ is idempotent.

Proof. To verify idempotency, compute P_u^2:

$$P_u^2 = |u\rangle\langle u| |u\rangle\langle u| = |u\rangle \langle u|u\rangle \langle u| = |u\rangle\langle u| = P_u,$$

where the normalization $\langle u|u\rangle = 1$ was used.   □

The idempotency of P_u follows because applying P_u repeatedly results in the same projection. This occurs since $P_u |v\rangle$ is already within the span of $|u\rangle$ after the first application. Further details on projection matrices will be discussed in § 11.2.4.

3   Power of Diagonal Matrices

The computation of powers for diagonal matrices is straightforward and follows a consistent formula:

> **Theorem 7.26** Given a diagonal matrix $\Lambda = \text{diag}(\lambda_1, \lambda_2, \ldots, \lambda_n)$, its dth power is given by:
> $$\Lambda^d = \begin{bmatrix} \lambda_1^d & 0 & \cdots & 0 \\ 0 & \lambda_2^d & \cdots & 0 \\ 0 & 0 & \ddots & 0 \\ 0 & 0 & \cdots & \lambda_n^d \end{bmatrix}, \tag{7.54}$$
> where $d \in \mathbb{Z}$. If d is negative, Λ must be invertible.

■ **Example 7.16** Compute A^{99} for the matrix $A = \begin{bmatrix} 0 & -1 \\ 1 & 0 \end{bmatrix}$.

Solution. Although A is not diagonal, it satisfies $A^2 = \begin{bmatrix} -1 & 0 \\ 0 & -1 \end{bmatrix}$, which is diagonal. Using the properties of matrix powers:

$$A^4 = (A^2)^2 = I,$$
$$A^{99} = A^{96}A^3 = (A^4)^{24}A^3 = I^{24}A^3 = A^3.$$

Finally, compute A^3:

$$A^3 = A^2 A = \begin{bmatrix} -1 & 0 \\ 0 & -1 \end{bmatrix} \begin{bmatrix} 0 & -1 \\ 1 & 0 \end{bmatrix} = \begin{bmatrix} 0 & 1 \\ -1 & 0 \end{bmatrix}.$$

■

The simplicity of computing powers of diagonal matrices is particularly beneficial when combined with matrix decomposition techniques. Matrix decomposition techniques allow the efficient computation of powers for matrices that can be diagonalized:

■ **Example 7.17** Compute A^n (where $n \in \mathbb{N}$) for a square matrix A that can be diagonalized as $A = P\Lambda P^{-1}$. Here, P is an invertible matrix and Λ is diagonal.

Solution. Using the decomposition, expand A^n:

$$A^n = (P\Lambda P^{-1})^n = P\Lambda P^{-1} P\Lambda P^{-1} \cdots P\Lambda P^{-1}$$
$$= P\Lambda^n P^{-1},$$

where Λ^n is computed by raising its diagonal entries to the nth power. ■

> **Exercise 7.33** Given $A = \begin{bmatrix} 3 & 4 \\ -2 & -3 \end{bmatrix} = P\Lambda P^{-1}$, where $P = \begin{bmatrix} 2 & -1 \\ -1 & 1 \end{bmatrix}$ and $\Lambda = \text{diag}(1, -1)$, compute A^{99} following Example 7.17.

4   Matrix Polynomials

Analogous to polynomials of real or complex variables, polynomials can also be constructed from square matrices. This extension broadens the scope of algebraic methods available for matrix analysis.

> **Definition 7.28 — Matrix Polynomial.** Given a polynomial function
>
> $$p(x) = c_0 + c_1 x + c_2 x^2 + \cdots + c_n x^n,$$
>
> the polynomial function of a square matrix A is defined as:
>
> $$p(A) = c_0 I + c_1 A + c_2 A^2 + \cdots + c_n A^n, \qquad (7.55)$$
>
> where I is the identity matrix of the same size as A, and $p(A)$ is a square matrix of the same size as A.

Since matrices $\{I, A, A^2, \dots\}$ are associative under multiplication, matrix polynomials follow the same algebraic rules as polynomials of scalar variables. For instance, the identity:

$$1 - x^n = (1 - x)(1 + x + x^2 + \cdots + x^{n-1}), \qquad (7.56)$$

has the corresponding matrix formula:

$$I - A^n = (I - A)(I + A + A^2 + \cdots + A^{n-1}), \qquad (7.57)$$

which holds for any square matrix A.

Similarly, matrix polynomials can be factored as polynomial functions, provided constants are replaced by scalar multiples of I. For example:

$$x^2 + 3x + 2 = (x + 1)(x + 2) \quad \Rightarrow \quad A^2 + 3A + 2I = (A + I)(A + 2I).$$

■ **Example 7.18** Show that if a matrix A satisfies $A^2 - 2A + I = 0$, then A is invertible. Subsequently, find the inverse of A.

Proof. Starting with the equation $A^2 - 2A + I = 0$, rearrange it to express the identity matrix:

$$A^2 - 2A + I = 0 \quad \Rightarrow \quad I = 2A - A^2.$$

Factorizing the right-hand side yields:

$$I = (2I - A)A.$$

This demonstrates that A is invertible and that $2I - A$ is its inverse:

$$A^{-1} = 2I - A.$$

∎

> **Exercise 7.34** Show that a matrix A is invertible if it satisfies $A^2 - 3A + 2I = 0$, and find its inverse.

Matrix polynomials are a powerful tool for analyzing square matrices, particularly in spectral decomposition, diagonalization, and the Cayley-Hamilton theorem, which are covered in § 11.3.1.5.

7.4 Trace and Determinant

Square matrices are characterized by two important scalar quantities: the trace and the determinant, both foundational in linear algebra and quantum computing. This section introduces their basic properties, laying the groundwork for understanding their fundamental roles and implications in both classical and quantum contexts. In Chapter 10, we will revisit traces and determinants to explore advanced properties involving tensor products.

7.4.1 Matrix Trace

The trace of a square matrix is defined as the sum of all entries on its main diagonal:

Definition 7.29 — Trace. The trace of an $n \times n$ square matrix $A = [a_{ij}]$, denoted as $\mathrm{tr}(A)$, is defined as:

$$\mathrm{tr}(A) = a_{11} + a_{22} + \cdots + a_{nn} = \sum_{i=1}^{n} a_{ii}. \tag{7.58}$$

Theorem 7.27 The following identities demonstrate the basic properties of the trace function for any square matrices A, B and scalar k:

$$\mathrm{tr}(A^T) = \mathrm{tr}(A), \tag{7.59a}$$
$$\mathrm{tr}(A^\dagger) = \mathrm{tr}(A^*) = \mathrm{tr}(A)^*, \tag{7.59b}$$
$$\mathrm{tr}(kA) = k\,\mathrm{tr}(A), \tag{7.59c}$$
$$\mathrm{tr}(A + B) = \mathrm{tr}(A) + \mathrm{tr}(B), \tag{7.59d}$$
$$\mathrm{tr}(AB) = \mathrm{tr}(BA). \tag{7.59e}$$

The proofs of Eqs. 7.59a to 7.59d are straightforward applications of Def. 7.29 and are left as exercises for the reader (Problem 7.20). Here, we provide a proof of Eq. 7.59e.

Proof.   Let $A = [a_{ij}]$ and $B = [b_{ij}]$, where $i, j = 1, 2, \ldots, n$. Using Eq. 7.21, compute the entries of AB and BA as follows:

$$(AB)_{ij} = \sum_{k=1}^{n} a_{ik} b_{kj},$$

$$(BA)_{ij} = \sum_{k=1}^{n} b_{ik} a_{kj}.$$

The traces of AB and BA are then:

$$\mathrm{tr}(AB) = \sum_{i=1}^{n} (AB)_{ii} = \sum_{i=1}^{n} \sum_{k=1}^{n} a_{ik} b_{ki},$$

$$\mathrm{tr}(BA) = \sum_{i=1}^{n} (BA)_{ii} = \sum_{i=1}^{n} \sum_{k=1}^{n} b_{ik} a_{ki}.$$

Since summations are symmetric with respect to i and k, swapping i and k in the summation for $\mathrm{tr}(BA)$ yields:

$$\mathrm{tr}(BA) = \sum_{i=1}^{n} \sum_{k=1}^{n} b_{ik} a_{ki} = \sum_{i=1}^{n} \sum_{k=1}^{n} b_{ki} a_{ik} = \mathrm{tr}(AB).$$

$\square$

Exercise 7.35 Use general 2×2 matrices $A = \begin{bmatrix} a_{11} & a_{12} \\ a_{21} & a_{22} \end{bmatrix}$ and $B = \begin{bmatrix} b_{11} & b_{12} \\ b_{21} & b_{22} \end{bmatrix}$ to verify Eq. 7.59e by directly computing $\mathrm{tr}(AB)$ and $\mathrm{tr}(BA)$.

Equation 7.59e leads to the following important properties of the trace function:

Theorem 7.28 The following identities hold true for square matrices A, B, C, and an invertible matrix P:

$$\text{tr}(ABC) = \text{tr}(BCA) = \text{tr}(CAB), \qquad \text{(Cyclic invariance)} \qquad (7.60a)$$
$$\text{tr}(P^{-1}AP) = \text{tr}(A), \qquad \text{(Similarity invariance)}. \qquad (7.60b)$$

Proof. Using the cyclic property of the trace, we can rearrange the product inside the trace function:

$$\text{tr}(ABC) = \text{tr}(A(BC)) = \text{tr}((BC)A) = \text{tr}(BCA),$$
$$\text{tr}(ABC) = \text{tr}((AB)C) = \text{tr}(C(AB)) = \text{tr}(CAB),$$
$$\text{tr}(P^{-1}AP) = \text{tr}(APP^{-1}) = \text{tr}(AI) = \text{tr}(A).$$

□

The cyclic invariance property holds for the trace of a product of square matrices of any length. For example:

$$\text{tr}(ABCD) = \text{tr}(BCDA) = \text{tr}(CDAB) = \text{tr}(DABC).$$

However, in general, $\text{tr}(ABC) \neq \text{tr}(ACB)$, as this does not follow the cyclic rule (Problem 7.23).

Equation 7.59e applies not only to square matrices but also to cases where A and B have compatible dimensions such that AB and BA are both valid.

Theorem 7.29 For matrices $A \in \mathbb{R}^{m \times n}$ and $B \in \mathbb{C}^{n \times m}$, it holds that:

$$\text{tr}(AB) = \text{tr}(BA). \qquad (7.61)$$

The proof of Theorem 7.29 is similar to that of Eq. 7.59e and is left as an exercise for the reader (Problem 7.24).

Theorem 7.29 leads to several results useful in quantum computing.

Theorem 7.30 The following identities hold for a matrix $A \in \mathbb{C}^{n \times n}$ and vectors $|u\rangle, |v\rangle \in \mathbb{C}^n$:

$$\text{tr}(|u\rangle\langle v|) = \langle v|u\rangle, \qquad (7.62)$$
$$\text{tr}(A|u\rangle\langle v|) = \langle v|A|u\rangle. \qquad (7.63)$$

The proof of Theorem 7.30 follows directly from Theorem 7.29, noting that the trace of a scalar is equal to the scalar itself.

Exercise 7.36 Verify Theorem 7.30 using the vectors $|u\rangle = \begin{bmatrix} i \\ 1 \end{bmatrix}$, $|v\rangle = \begin{bmatrix} 1 \\ 1+i \end{bmatrix}$, and the matrix $A = \begin{bmatrix} 1 & i \\ i & 1 \end{bmatrix}$.

7.4.2 　*Applications of Trace

1　Introduction to Density Matrices

Theorem 7.30 is fundamental in describing density matrices (also known as density operators) in quantum computing [1].

In quantum mechanics, a **pure state** is represented by a unit vector in a Hilbert space (or in $\mathbb{C}^n$ for finite dimensions). A **mixed state** represents a statistical ensemble of such pure states.

A mixed state is described by a density matrix defined as:

$$\rho = \sum_i p_i |\psi_i\rangle\langle\psi_i|, \tag{7.64}$$

where $p_i \geq 0$ and $\sum_i p_i = 1$. Here, $|\psi_i\rangle$ represents a pure state, and when the sum includes multiple terms, the state is mixed, with p_i representing the probability of observing the pure state $|\psi_i\rangle$.

A pure state is a special case of a mixed state, expressible as a single outer product: $\rho = |\psi\rangle\langle\psi|$. Density matrices for pure states exhibit specific properties: $\text{tr}(\rho) = 1$ and $\text{tr}(\rho^2) = 1$.

The unit trace of ρ is readily proven using Eq. 7.62:

$$\text{tr}(|\psi\rangle\langle\psi|) = \langle\psi|\psi\rangle = 1, \tag{7.65}$$

since $|\psi\rangle$ is normalized, as required for a quantum state.

Additionally, for pure states, it follows that $\rho^2 = \rho$, which implies $\text{tr}(\rho^2) = \text{tr}(\rho) = 1$.

For mixed states, $\text{tr}(\rho) = 1$, and $0 < \text{tr}(\rho^2) \leq 1$. The proof of the first property is as follows:

$$\text{tr}\left(\sum_i p_i |\psi_i\rangle\langle\psi_i|\right) = \sum_i \text{tr}(p_i |\psi_i\rangle\langle\psi_i|) \qquad \text{(by Eq. 7.59d)}$$

$$= \sum_i p_i \,\text{tr}(|\psi_i\rangle\langle\psi_i|) \qquad \text{(by Eq. 7.59c)}$$

$$= \sum_i p_i \qquad \text{(using Eq. 7.65)}$$

$$= 1.$$

To demonstrate that $0 < \text{tr}(\rho^2) \leq 1$, consider the case where $\{|\psi_i\rangle\}$ are orthonormal pure states with $\langle\psi_i|\psi_j\rangle = \delta_{ij}$. Since ρ is Hermitian (Exercise 7.37), it can always be decomposed as in Eq. 7.64 with orthonormal pure states, according to the spectral decomposition theorem (to be discussed in § 9.3). We derive:

$$\text{tr}(\rho^2) = \text{tr}\left(\sum_i p_i |\psi_i\rangle\langle\psi_i| \sum_j p_j |\psi_j\rangle\langle\psi_j|\right)$$

$$= \text{tr}\left(\sum_i \sum_j p_i p_j |\psi_i\rangle\langle\psi_i| \,|\psi_j\rangle\langle\psi_j|\right)$$

$$= \operatorname{tr}\left(\sum_i \sum_j p_i p_j \left|\psi_i\right\rangle \left\langle\psi_i\middle|\psi_j\right\rangle \left\langle\psi_j\right|\right)$$

$$= \operatorname{tr}\left(\sum_i \sum_j p_i p_j \delta_{ij} \left|\psi_i\right\rangle\left\langle\psi_j\right|\right)$$

$$= \operatorname{tr}\left(\sum_i p_i^2 \left|\psi_i\right\rangle\left\langle\psi_i\right|\right) \quad (\because \delta_{ij} = 0 \text{ when } i \neq j)$$

$$= \sum_i p_i^2 \operatorname{tr}(\left|\psi_i\right\rangle\left\langle\psi_i\right|)$$

$$= \sum_i p_i^2.$$

Since $p_i \geq 0$ and $\sum_i p_i = 1$, it follows that $0 < \sum_i p_i^2 \leq \sum_i p_i = 1$.

Exercise 7.37 Show that the density matrix ρ, as defined in Eq. 7.64, is always Hermitian.

2 Quadratic Forms via Density Matrices

The trace function is crucial for expressing the expected value of an observable A in a quantum state represented by a density matrix ρ. For a pure state $|\psi\rangle$, the expected value of A is traditionally calculated as a Hermitian quadratic form:

$$\langle A \rangle = \langle\psi|A|\psi\rangle. \tag{7.66}$$

Using Eq. 7.63, this can be reformulated as:

$$\langle A \rangle = \langle\psi|A|\psi\rangle = \operatorname{tr}(A\,|\psi\rangle\langle\psi|) = \operatorname{tr}(A\rho), \tag{7.67}$$

where $\rho = |\psi\rangle\langle\psi|$ is the density matrix corresponding to the pure state $|\psi\rangle$.

Equation 7.67 also applies to mixed states. For a mixed state represented by $\rho = \sum p_i |\psi_i\rangle\langle\psi_i|$, the expected value of A is a weighted average of the expected values for its constituent pure states:

$$\langle A \rangle = \sum_i p_i \langle\psi_i|A|\psi_i\rangle = \sum_i p_i \operatorname{tr}(A\,|\psi_i\rangle\langle\psi_i|) = \operatorname{tr}(A\sum_i p_i |\psi_i\rangle\langle\psi_i|) = \operatorname{tr}(A\rho). \tag{7.68}$$

This derivation illustrates how the trace function bridges pure and mixed state formulations, facilitating the calculation of expected values in quantum measurements.

3 Inner Product of Two Matrices

The trace function is instrumental in defining the inner product between matrices, as outlined below.

> **Definition 7.30 — Matrix Inner Product.** Given two complex matrices $A_{m\times n}$ and $B_{m\times n}$ of the same size, their inner product is defined as:
>
> $$\langle A, B \rangle \equiv \operatorname{tr}(A^\dagger B). \tag{7.69}$$
>
> This inner product, known as the Hilbert-Schmidt inner product or Frobenius inner product, is sometimes denoted as $\langle A, B \rangle_F$ to distinguish it from other types

of inner products.

This matrix inner product generalizes the vector dot product by aggregating the products of corresponding elements $a_{ij}^* b_{ij}$ across all indices i and j. For example, for matrices $A_{2\times 2} = [a_{ij}]$ and $B_{2\times 2} = [b_{ij}]$, the inner product is:

$$\langle A, B \rangle = \text{tr}(A^\dagger B) = a_{11}^* b_{11} + a_{12}^* b_{12} + a_{21}^* b_{21} + a_{22}^* b_{22}.$$

This concept is formalized in the following theorem:

Theorem 7.31 For matrices $A_{m\times n} = [a_{ij}]$ and $B_{m\times n} = [b_{ij}]$, the Hilbert-Schmidt inner product is given by:

$$\text{tr}(A^\dagger B) = \sum_{i=1}^{m}\sum_{j=1}^{n} a_{ij}^* b_{ij}. \tag{7.70}$$

Proof. The adjoint $A^\dagger$ is an $n \times m$ matrix with elements $(A^\dagger)_{ji} = a_{ij}^*$. Let the product $C = A^\dagger B$ be an $n \times n$ matrix with elements c_{jk} given by:

$$c_{jk} = \sum_{i=1}^{m} (A^\dagger)_{ji} b_{ik} = \sum_{i=1}^{m} a_{ij}^* b_{ik}.$$

The trace of C is the sum of its diagonal entries:

$$\text{tr}(A^\dagger B) = \text{tr}(C) = \sum_{j=1}^{n} c_{jj} = \sum_{j=1}^{n}\left(\sum_{i=1}^{m} a_{ij}^* b_{ij}\right) = \sum_{i=1}^{m}\sum_{j=1}^{n} a_{ij}^* b_{ij}.$$

This completes the proof. $\square$

The Hilbert-Schmidt inner product also induces a natural norm for matrices, analogous to the Euclidean norm for vectors.

Definition 7.31 — Frobenius Norm. The Frobenius norm of a matrix $A_{m\times n}$, denoted as $\|A\|_F$, is defined as:

$$\|A\|_F = \sqrt{\langle A, A \rangle_F} = \sqrt{\text{tr}(A^\dagger A)} = \sqrt{\sum_{i=1}^{m}\sum_{j=1}^{n} |a_{ij}|^2}. \tag{7.71}$$

4 Inner Product Space of Matrices

The Hilbert-Schmidt inner product defines an inner product space structure on the set of all $m \times n$ complex matrices, $\mathbb{C}^{m\times n}$. This endows $\mathbb{C}^{m\times n}$ with the structure of a vector space and facilitates expressing any matrix as a linear combination of basis matrices.

In quantum computing, a significant example is the decomposition of a single-qubit density matrix ρ as a linear combination of the matrices $\{I, X, Y, Z\}$, where I is the identity matrix and X, Y, Z are the Pauli matrices (see § 12.1.1). These matrices form an orthogonal basis. We will delve into this topic in § 12.1.2 and § 12.4.

Exercise 7.38 Given matrices $I = \begin{bmatrix} 1 & 0 \\ 0 & 1 \end{bmatrix}$, $X = \begin{bmatrix} 0 & 1 \\ 1 & 0 \end{bmatrix}$, $Y = \begin{bmatrix} 0 & -i \\ i & 0 \end{bmatrix}$, and $Z = \begin{bmatrix} 1 & 0 \\ 0 & -1 \end{bmatrix}$, verify that $\left\{ \frac{1}{\sqrt{2}}I, \frac{1}{\sqrt{2}}X, \frac{1}{\sqrt{2}}Y, \frac{1}{\sqrt{2}}Z \right\}$ forms an orthonormal basis for $\mathbb{C}^{2 \times 2}$.

7.4.3   Determinants

The determinant is a scalar-valued function defined exclusively for square matrices. It encodes important properties of matrices and the linear mappings they represent. Unlike the trace, which is computed as a simple sum, the determinant involves a more intricate formulation.

In quantum computing, the focus is typically on the applications and implications of determinants rather than the derivation of their formulas. Therefore, this section emphasizes key properties and examples of determinants without delving into detailed proofs. For a thorough exploration, readers are encouraged to consult standard linear algebra texts, such as *Introduction to Linear Algebra* by Gilbert Strang [3].

1   Definition

> **Definition 7.32 — Determinant.** The determinant of an $n \times n$ square matrix $A = [a_{ij}]$, denoted as det A, det(A), or $|A|$, is a scalar defined by:
>
> $$\det(A) = \sum_{i_1, i_2, \ldots, i_n} \epsilon_{i_1 i_2 \cdots i_n} a_{1 i_1} a_{2 i_2} \cdots a_{n i_n}, \tag{7.72}$$
>
> where ϵ is the Levi-Civita symbol, defined as:
>
> $$\epsilon_{i_1 i_2 \cdots i_n} = \begin{cases} +1 & \text{if } (i_1, i_2, \cdots, i_n) \text{ is an even permutation of } (1, 2, \ldots, n), \\ -1 & \text{if } (i_1, i_2, \cdots, i_n) \text{ is an odd permutation of } (1, 2, \ldots, n), \\ 0 & \text{otherwise.} \end{cases} \tag{7.73}$$

In Def. 7.32, each term in the summation $a_{1 i_1} a_{2 i_2} \cdots a_{n i_n}$ is a product of n entries from the matrix A. Each entry is selected from a different row, and the indices $i_1, i_2, \ldots, i_n$ correspond to distinct columns, ensuring that no row or column is repeated within a single product.

The Levi-Civita symbol $\epsilon_{i_1 i_2 \cdots i_n}$ determines the sign of each term based on the parity of the permutation of indices $(i_1, i_2, \ldots, i_n)$. Specifically:

- If the permutation is even (i.e., obtained from $(1, 2, \ldots, n)$ by an even number of swaps), $\epsilon = +1$.

- If the permutation is odd (i.e., obtained by an odd number of swaps), $\epsilon = -1$.

- If any index repeats, $\epsilon = 0$, causing the term to vanish.

Exercise 7.39 Determine whether the permutation $(4, 3, 2, 1)$ of $(1, 2, 3, 4)$ is even or odd.

To illustrate Def. 7.32, we compute determinants for smaller matrices as examples.

For a 2×2 matrix:

$$\begin{vmatrix} a_{11} & a_{12} \\ a_{21} & a_{22} \end{vmatrix} = a_{11}a_{22} - a_{12}a_{21}, \tag{7.74}$$

derived from the permutations $(i_1, i_2) \in \{(1,2), (2,1)\}$ of $(1,2)$.

For a 3×3 matrix, there are $3! = 6$ terms:

$$\begin{vmatrix} a_{11} & a_{12} & a_{13} \\ a_{21} & a_{22} & a_{23} \\ a_{31} & a_{32} & a_{33} \end{vmatrix} = a_{11}a_{22}a_{33} + a_{12}a_{23}a_{31} + a_{13}a_{21}a_{32}$$

$$- a_{13}a_{22}a_{31} - a_{12}a_{21}a_{33} - a_{11}a_{23}a_{32}. \tag{7.75}$$

The column indices of each term correspond to all possible permutations of $(1,2,3)$:

$$(1,2,3), \ (2,3,1), \ (3,1,2), \ (3,2,1), \ (2,1,3), \ (1,3,2),$$

where the first three are even and contribute positively, while the last three are odd and contribute negatively, as specified in Def. 7.32.

Exercise 7.40 Evaluate the following determinants using Def. 7.32:

$$\text{(a)} \ \begin{vmatrix} 1 & -1 \\ 1 & 1 \end{vmatrix}, \quad \text{(b)} \ \begin{vmatrix} 1 & 2 & 3 \\ 4 & 5 & 6 \\ 7 & 8 & 9 \end{vmatrix}.$$

For a general $n \times n$ matrix, Eq. 7.72 involves $n!$ terms. Each term consists of a product of n entries from the matrix, with signs determined by the parity of the corresponding permutation. While this formulation may seem complex, the determinant can often be computed more efficiently using **cofactor expansion**, an inductive approach that simplifies analysis.

Although the details of cofactor expansion are omitted here, it is an essential tool in practical computations and can be studied in regular linear algebra texts.

(i) **Notation:** For a complex matrix A, the determinant can be a complex scalar. When referring to the absolute value of this scalar, we use $|\det(A)|$. The notation $|A|$ alone should not be used to denote determinants, as it may lead to confusion with absolute values.

2 Properties of Determinants

One of the fundamental properties of the determinant is its multiplicative nature:

Theorem 7.32 Given square matrices A and B of the same size,

$$\det(AB) = \det(A)\det(B). \tag{7.76}$$

A full proof of Theorem 7.32 typically involves cofactor expansion and is omitted here. However, a verification for 2×2 matrices is provided as an exercise (Problem 7.27).

Exercise 7.41 Consider the matrix $A = \begin{bmatrix} 1 & 1 \\ i & 1 \end{bmatrix}$. Compute the determinant of A and A^2. Verify that $\det(A^2) = \det(A)^2$.

The determinant of the identity matrix holds particular significance, consistently evaluating to one:

Theorem 7.33 For identity matrices of any size I_n, the determinant is:

$$\det(I_n) = 1. \tag{7.77}$$

The proof of Theorem 7.33 follows directly from Def. 7.32: the only nonzero term in the determinant's expansion is the product of the diagonal entries, which are all 1.

Similarly, determinants of diagonal matrices are easy to compute:

Theorem 7.34 The determinant of a diagonal matrix $\Lambda = \text{diag}(\lambda_1, \lambda_2, \ldots, \lambda_n)$ is the product of its diagonal entries:

$$\begin{vmatrix} \lambda_1 & 0 & \cdots & 0 \\ 0 & \lambda_2 & \cdots & 0 \\ \vdots & \vdots & \ddots & \vdots \\ 0 & 0 & \cdots & \lambda_n \end{vmatrix} = \lambda_1 \lambda_2 \cdots \lambda_n. \tag{7.78}$$

The following properties further illustrate the behavior of determinants under scalar multiplication, transposition, and adjoints:

Theorem 7.35 Given an $n \times n$ square matrix A and a scalar k, the following properties hold:

$$\det(kA) = k^n \det(A), \tag{7.79}$$
$$\det(A^T) = \det(A), \tag{7.80}$$
$$\det(A^\dagger) = \det(A^*) = \det(A)^*. \tag{7.81}$$

Exercise 7.42 Using Def. 7.32, explain why the formulas in Theorem 7.35 hold. Consider the effects of scaling, transposition, and adjoints on the determinant computation.

Determinants are crucial in determining whether a matrix is invertible or singular:

Theorem 7.36 A square matrix A is singular if and only if $\det(A) = 0$. Conversely, A is invertible if and only if $\det(A) \neq 0$.

Theorem 7.36 follows intuitively from Theorems 7.32 and 7.33. If the inverse A^{-1} exists, then $AA^{-1} = I$, implying $\det(A)\det(A^{-1}) = \det(I) = 1$. Thus, if $\det(A) = 0$, it is impossible for $\det(A)\det(A^{-1})$ to equal 1, confirming that A is singular. This leads to:

Theorem 7.37 For an invertible matrix A, the determinant of its inverse is:

$$\det(A^{-1}) = \frac{1}{\det(A)}. \tag{7.82}$$

Exercise 7.43 Let $A = UBU^{-1}$, where A, B, U are square matrices and U is invertible. Prove that $\det(A) = \det(B)$.

3   Other Applications of Determinants

Determinants will be explored further in subsequent chapters. Below is an overview of their applications for reference:

- Geometric Interpretation: In § 8.1, matrices are discussed as linear operators on vectors. Determinants represent the ratio of change in area or volume induced by these operations.

- Eigenvalues: Determinants are essential in computing the eigenvalues of matrices. This will be examined in § 9.1.

- Unitary Operators: In quantum computing, unitary operators—key components of quantum gates—have determinants with absolute value 1. This property is explored in § 8.3.

- Advanced Properties: Further properties of determinants, such as those involving tensor products and matrix exponentials, will be discussed in § 10.2.3.5 and § 11.3.

Problem Set 7

7.1 A matrix A is defined to be **skew-symmetric** if $A^T = -A$. Provide one example of skew-symmetric matrices for each size: 1×1, 2×2, and 3×3.

7.2 A matrix A is defined to be **skew-Hermitian** if $A^\dagger = -A$. Provide one example of skew-Hermitian matrices for each size: 1×1, 2×2, and 3×3, with at least one non-real entry.

7.3 Determine which of the following sets are subspaces of $\mathbb{C}^{m \times n}$. If a set is a subspace, find a basis and determine its dimension.

(a) The set of all symmetric matrices of size $m \times n$.

(b) The set of all skew-symmetric matrices of size $m \times n$.

(c) The set of all Hermitian matrices of size $m \times n$.

(d) The set of matrices of size $m \times n$ where all entries are the same scalar $k \in \mathbb{C}$.

7.4 Prove that the set of all Hermitian matrices of size $m \times n$ forms a vector space over the field $\mathbb{R}$. Determine the dimension of this vector space.

7.5 Prove Eq. 7.24b using element-wise notation.

7.6 ✳ Prove Eq. 7.24c using element-wise notation.

7.7 Prove Eq. 7.25a using element-wise notation.

7.8 Given $n \times n$ matrices A, B, C and a scalar k, determine which of the following statements are true. Provide proofs or counterexamples for each.

(A) $(A - B)(A + B) = A^2 - B^2$.

(B) If $AB = 0$, then either $A = 0$ or $B = 0$.

(C) If $kA = 0$, then either $k = 0$ or $A = 0$.

(D) If $AB = AC$, then $B = C$.

7.9 Prove Theorem 7.9 using element-wise notation.

7.10 Given matrices $A_{m \times p}$ and $B_{p \times n}$ and their matrix product $C = AB$, demonstrate that the column vectors of C correspond to the products of A with the column vectors of B. Specifically, show that:
$$C = \begin{bmatrix} Ab_1 & Ab_2 & \cdots & Ab_n \end{bmatrix}.$$

7.11 Consider two matrices, $A_{m \times p}$ and $B_{p \times n}$, and their matrix product $C = AB$. Demonstrate that every column vector of C resides within the column space of A, which is defined as the span of the column vectors of A. (Hint: Use the result from Problem 7.10 and conduct a column-row expansion for each column vector of C.)

7.12 Provide the matrix representation for the following operators:

(a) $|+\rangle\langle-|$,

(b) $|-_i\rangle\langle+_i|$,

(c) $|+_i\rangle\langle+_i| + |-_i\rangle\langle-_i|$,

(d) $|+\rangle\langle+| + |-\rangle\langle-|$,

where $|+\rangle = \frac{1}{\sqrt{2}}(|0\rangle + |1\rangle)$, $|-\rangle = \frac{1}{\sqrt{2}}(|0\rangle - |1\rangle)$, $|+_i\rangle = \frac{1}{\sqrt{2}}(|0\rangle + i|1\rangle)$, and $|-_i\rangle = \frac{1}{\sqrt{2}}(|0\rangle - i|1\rangle)$.

7.13 Approve or disapprove each of the following statements:

(A) In a valid matrix product AB, if A has a row of zeros, then the product has a row of zeros.

(B) In a valid matrix product AB, if B has a row of zeros, then the product has a row of zeros.

(C) In a valid matrix product AB, if A has a column of zeros, then the product has a column of zeros.

(D) In a valid matrix product AB, if B has a column of zeros, then the product has a column of zeros.

7.14 Verify that for a general 2×2 matrix $A = \begin{bmatrix} a & b \\ c & d \end{bmatrix}$, its inverse is given by the formula $A^{-1} = \dfrac{1}{ad - bc} \begin{bmatrix} d & -b \\ -c & a \end{bmatrix}$ given $ad - bc \neq 0$.

7.15 Use Theorem 7.17 to show that the following matrix is singular:
$$\begin{bmatrix} 1 & 2 & 3 \\ 4 & 5 & 6 \\ 7 & 8 & 9 \end{bmatrix}$$

7.16 Prove Theorem 7.21 using both matrix representation and Dirac notation.

7.17 ✳ Demonstrate that there are infinitely many 2×2 matrices A such that $A^2 = I$.

7.18 Prove that if a matrix A is idempotent, then the matrix $I - A$ is also idempotent.

7.19 Given an $n \times n$ matrix A, a vector $|v\rangle$, and a scalar λ such that $A\,|v\rangle = \lambda\,|v\rangle$, prove that for any polynomial function $p(x)$, the equation $p(A)\,|v\rangle = p(\lambda)\,|v\rangle$ holds.

7.20 Prove Eqs. 7.59a to 7.59d.

7.21 Prove that $\mathrm{tr}(ABC) = \mathrm{tr}(CBA)$ if A, B, C are symmetric matrices, and $\mathrm{tr}(ABC) = \mathrm{tr}(CBA)^*$ if A, B, C are Hermitian matrices.

7.22 Given a general 2×2 matrix $A = \begin{bmatrix} a & b \\ c & d \end{bmatrix}$, demonstrate that A satisfies the equation $p(A) = 0$ where $p(z) = z^2 - \mathrm{tr}(A)z + \det(A)$.

7.23 Show that generally $\mathrm{tr}(ABC) \neq \mathrm{tr}(ACB)$ for square matrices A, B, C using element-wise notation.

7.24 Prove Theorem 7.29.

7.25 Show that if two pure states are orthogonal to each other, then their respective density matrices are also orthogonal to each other.

7.26 ✳ Apply Def. 7.30 to the general Cauchy-Schwarz inequality (Eq. 6.60) to establish the following inequalities for two square complex matrices of the same size:

$$0 \leq |\mathrm{tr}(A^\dagger B)|^2 \leq |\mathrm{tr}(A^\dagger A)||\mathrm{tr}(B^\dagger B)| \leq |\mathrm{tr}(A)|^2|\mathrm{tr}(B)|^2. \qquad (7.83)$$

7.27 Verify Theorem 7.32, which states that the determinant of the product of two matrices equals the product of their determinants, using 2×2 matrices as a general example.

8. Matrices as Linear Operators

Contents

Matrices are more than mere arrays of numbers; they encapsulate profound mathematical structures and play a central role in various applications. Each type of matrix possesses intrinsic properties that reflect specific functional or geometric characteristics. To fully appreciate their significance, it is essential to understand matrices as representations of linear transformations or operators acting between vector spaces. This chapter delves into the concepts, properties, and geometric interpretations of linear mappings as expressed through matrices.

In quantum computing, particular attention is given to unitary matrices in this chapter, as they play a fundamental role in representing quantum gates [1]. Hermitian matrices, which are equally significant in quantum mechanics, will be explored in detail in the next chapter. A key objective of this chapter is to deepen the understanding of unitary matrices by examining their unique properties and behaviors within the broader context of linear operators.

8.1 Introduction to Matrix Transformations

8.1.1 Linear Transformations

The concept of linearity is fundamental in linear algebra. One of the key aspects we have explored is the linear combination, which plays a crucial role in defining a vector space, its basis, coordinates, and dimensions. For matrices, the notion of linearity is intimately connected to their role as linear transformations, defined as follows:

> **Definition 8.1 — Linear Transformation.** Let V and W be vector spaces over the same field. A function $T : V \to W$ is termed a linear transformation if it satisfies the following conditions for all vectors $u, v \in V$ and any scalar k from the field:
>
> $$T(u + v) = T(u) + T(v), \qquad \text{(Additivity Property)} \qquad (8.1a)$$
> $$T(kv) = kT(v), \qquad \text{(Homogeneity Property)} \qquad (8.1b)$$

This definition implies that a transformation T is linear if it commutes with any linear combination of vectors from a vector space V over a field F:

$$T(\alpha u + \beta v) = \alpha T(u) + \beta T(v), \quad \text{for all } \alpha, \beta \in F \text{ and } u, v \in V. \qquad (8.2)$$

■ **Example 8.1** Demonstrate that the function $f : \mathbb{C} \to \mathbb{C}$ defined by $f(x) = ax$, where $a \in \mathbb{C}$, is a linear transformation.

Solution. To verify that f is a linear transformation, we need to confirm both additivity and homogeneity for any $u, v \in \mathbb{C}$ and any scalar $k \in \mathbb{C}$:

$$f(u) + f(v) = au + av = a(u + v) = f(u + v),$$
$$f(ku) = a(ku) = k(au) = kf(u).$$

Hence, f satisfies the criteria for a linear transformation according to Def. 8.1.   ■

■ **Example 8.2 — Affine Transformation.** Demonstrate that the function $f : \mathbb{C} \to \mathbb{C}$ defined by $f(x) = ax + b$, where $a, b \in \mathbb{C}$, is not a linear transformation if $b \neq 0$.

This transformation is referred to as an affine transformation (from the Latin word "affinis," meaning "connected" or "related"). Affine transformations preserve geometric structures such as straight lines and parallelism. Unlike linear transformations, affine transformations include translations, allowing them to combine rotations, scalings, shears, reflections, and translations.

Solution. To verify that f is not a linear transformation, consider the two properties required for linearity:

- Additivity: For any $u, v \in \mathbb{C}$,

$$f(u + v) = a(u + v) + b = au + av + b.$$

  Compare this to $f(u) + f(v)$:

$$f(u) + f(v) = (au + b) + (av + b) = au + av + 2b.$$

  Since $f(u + v) \neq f(u) + f(v)$ when $b \neq 0$, additivity is not satisfied.

- Homogeneity: For any scalar $k \in \mathbb{C}$ and any $u \in \mathbb{C}$,

$$f(ku) = a(ku) + b = k(au) + b.$$

  Compare this to $kf(u)$:

$$kf(u) = k(au + b) = k(au) + kb.$$

  Since $f(ku) \neq kf(u)$ when $b \neq 0$ and $k \neq 1$, homogeneity is not satisfied.

Therefore, $f(x) = ax + b$ is not a linear transformation if $b \neq 0$.   ∎

> (i) The above example demonstrates that within $\mathbb{C}$, the concept of a "linear transformation" differs from that of a "linear function." Specifically, a constant function $f(x) = b$, where $b \neq 0$, does not qualify as a linear transformation. This distinction arises because only linear functions that pass through the origin (with zero y-intercept) qualify as linear transformations.

> **Exercise 8.1** Demonstrate that the function $f : \mathbb{C} \to \mathbb{C}$ defined by $f(x) = x^2$ is not a linear transformation.

8.1.2   Matrix Transformations

Matrices can be understood as functions that map vectors from one vector space to another. Specifically, an $m \times n$ matrix A transforms a vector $|u\rangle$ in $\mathbb{C}^n$ into a vector $|v\rangle$ in $\mathbb{C}^m$ through matrix multiplication:

$$A_{m \times n} |u\rangle_{n \times 1} = |v\rangle_{m \times 1}.$$

Thus, for any vector $|u\rangle$ in $\mathbb{C}^n$, multiplication by A results in a new vector $|v\rangle$ in $\mathbb{C}^m$. Essentially, this corresponds to applying a function $f : \mathbb{C}^n \to \mathbb{C}^m$ to $|u\rangle$. In linear algebra, we refer to such a function as a **matrix transformation** governed by A.

> **Definition 8.2 — Matrix Transformation.** A matrix transformation is governed by an $m \times n$ matrix A, denoted as T_A, that maps $\mathbb{C}^n$ to $\mathbb{C}^m$ via the formula $A |u\rangle = |v\rangle$,

where $|u\rangle \in \mathbb{C}^n$ and $|v\rangle \in \mathbb{C}^m$. This transformation is formally expressed as:

$$T_A : \mathbb{C}^n \to \mathbb{C}^m, \quad \text{or} \quad |u\rangle \xrightarrow{T_A} |v\rangle . \tag{8.3}$$

When $m = n$, A is also known as a **matrix operator**.

■ **Example 8.3** Identify 2×2 matrices that perform the following transformations:

(a) Transform all vectors to the zero vector.

(b) Leave all vectors unchanged.

Solution.

(a) We require a 2×2 matrix A such that $A|u\rangle = \mathbf{0}$ for any $|u\rangle \in \mathbb{C}^2$. The only matrix that fulfills this requirement is the zero matrix, denoted $A = 0$.

Here, the zero vector $\mathbf{0}$ refers to a vector whose components are all zero, and the zero matrix 0 refers to a matrix with all entries equal to zero.

(b) We require a 2×2 matrix A such that $A|u\rangle = |u\rangle$ for any $|u\rangle \in \mathbb{C}^2$. The identity matrix I_2 meets this condition. To verify that I_2 is the sole solution, suppose another matrix B also satisfies this condition. Then:

$$I_2 |u\rangle = |u\rangle , \quad B|u\rangle = |u\rangle \quad \Rightarrow \quad (I_2 - B)|u\rangle = \mathbf{0} \text{ for any } |u\rangle .$$

Consequently, $I_2 - B = 0$, which implies $B = I_2$. Hence, I_2 is the only 2×2 matrix that meets the requirement.

■

Exercise 8.2 Identify a 2×2 matrix that converts all vectors in $\mathbb{C}^2$ to their opposites.

By definition, all matrix operators are square matrices. Given the prominence of square matrices in quantum computing, this section will primarily focus on such matrix operators. Discussions on matrix transformations where $n \neq m$ are deferred to § 8.5.

8.1.3 Linearity of Matrix Transformations

A fundamental theorem in linear algebra asserts that all matrix transformations are inherently linear, as demonstrated in the proof that follows.

Theorem 8.1 Every matrix transformation from $\mathbb{C}^n$ to $\mathbb{C}^m$ is a linear transformation.

Proof. Consider a matrix $A_{m \times n}$, vectors $|u\rangle , |v\rangle \in \mathbb{C}^n$, and a scalar k. Let the matrix transformation represented by A be denoted as T_A, defined by $T_A(|v\rangle) = A|v\rangle$. The linearity of T_A can be shown using basic matrix operations:

$$T_A(|u\rangle + |v\rangle) = A(|u\rangle + |v\rangle) = A|u\rangle + A|v\rangle = T_A(|u\rangle) + T_A(|v\rangle),$$
$$T_A(k|v\rangle) = A(k|v\rangle) = k(A|v\rangle) = kT_A(|v\rangle).$$

These properties satisfy both additivity and homogeneity, thus confirming that T_A is a linear transformation.

□

When $m = n$, A is specifically called a **linear operator**. An example of a linear operator is depicted in Fig. 8.1. Consider a vector $v = 3\hat{i} + 2\hat{j}$ in $\mathbb{R}^2$, and apply the linear operator $A = \begin{bmatrix} -1 & 2 \\ 1 & 1 \end{bmatrix}$. It demonstrates that the transformations of $v, \hat{i}, \hat{j}$ preserve the relationships dictated by their linear combination.

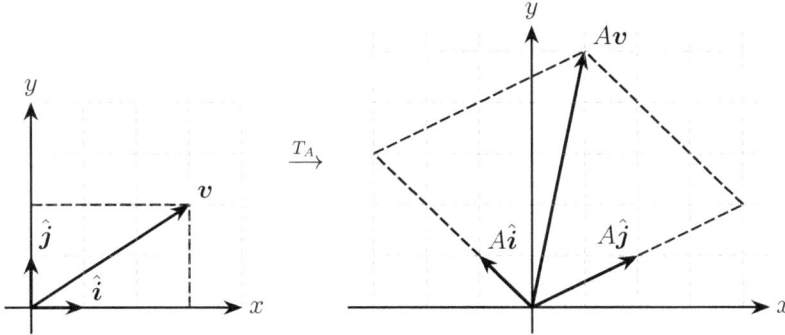

Figure 8.1: Geometric Example of a Linear Operator

Exercise 8.3 Calculate the transformed vectors $v, \hat{i}, \hat{j}$ by $A = \begin{bmatrix} -1 & 2 \\ 1 & 1 \end{bmatrix}$ and verify that $Av = 3A\hat{i} + 2A\hat{j}$, as shown in Fig. 8.1.

8.1.4   Matrix Transformations of Standard Basis Vectors

Due to the inherent linearity of matrix transformations, understanding their behavior can be effectively achieved by analyzing how they transform basis vectors, particularly the standard basis. This concept is captured in the theorem below.

Theorem 8.2 For any linear operator represented by an $n \times n$ matrix A, the columns of A represent the images of the standard basis vectors under the transformation induced by A:
$$A = \begin{bmatrix} Ae_1 & Ae_2 & \cdots & Ae_n \end{bmatrix}, \tag{8.4}$$
where $\{e_1, e_2, \ldots, e_n\}$ denotes the standard basis of $\mathbb{C}^n$.

Proof.   The identity matrix I_n is naturally represented by its column vectors, which form the standard basis:
$$I_n = \begin{bmatrix} e_1 & e_2 & \cdots & e_n \end{bmatrix}. \tag{8.5}$$
Multiplying A by I_n effectively evaluates the action of A on each of these basis vectors:
$$A = AI_n = A \begin{bmatrix} e_1 & e_2 & \cdots & e_n \end{bmatrix} = \begin{bmatrix} Ae_1 & Ae_2 & \cdots & Ae_n \end{bmatrix}.$$
Here, each column Ae_i is the image of the basis vector e_i under the transformation A, confirming the theorem's statement.   □

Using the transformation depicted in Fig. 8.1 as an example, $\hat{i}$ is transformed to $\begin{bmatrix} -1 & 1 \end{bmatrix}^T$ by T_A, which corresponds exactly to the first column of A. Similarly, $\hat{j}$ is transformed to $\begin{bmatrix} 2 & 1 \end{bmatrix}^T$, aligning with the second column of A.

In essence, a matrix is entirely defined by its column vectors. According to Theorem 8.2, these column vectors precisely dictate the transformation rules it embodies—there is nothing more or less to it. This insight unveils that a matrix not only acts as a linear transformation, but it also fundamentally *is* a linear transformation!

Theorem 8.3 Any linear transformation from $\mathbb{C}^n$ to $\mathbb{C}^m$ can be represented by a matrix $A \in \mathbb{C}^{m \times n}$.

Proof. By definition, any linear transformation $T : \mathbb{C}^n \to \mathbb{C}^m$ must satisfy both additivity and homogeneity. Consider a vector $v = [v_1 \ v_2 \ \cdots \ v_n]^T$. The action of T on v is expressed as:

$$T(v) = T(v_1 e_1 + v_2 e_2 + \cdots + v_n e_n) = v_1 T(e_1) + v_2 T(e_2) + \cdots + v_n T(e_n).$$

Construct the matrix

$$A = \begin{bmatrix} T(e_1) & T(e_2) & \cdots & T(e_n) \end{bmatrix}.$$

The matrix transformation $T_A(v) = Av$ then results in:

$$T_A(v) = Av = \begin{bmatrix} T(e_1) & T(e_2) & \cdots & T(e_n) \end{bmatrix} \begin{bmatrix} v_1 \\ v_2 \\ \vdots \\ v_n \end{bmatrix}$$

$$= v_1 T(e_1) + v_2 T(e_2) + \cdots + v_n T(e_n),$$

using the column-row expansion of matrix multiplication (Theorem 7.9). This shows that T_A and T define the same transformation, hence $T = T_A$, completing the proof. □

8.1.5 Examples of Matrix Operators

Theorem 8.2 simplifies the analysis of linear (matrix) operators by reducing it to examining their effects on standard basis vectors. In $\mathbb{R}^2$, such effects can be further visualized through geometric vectors. Subsequently, we will introduce several fundamental matrix operators in $\mathbb{R}^2$, which will aid in understanding the behavior of unitary and Hermitian matrices.

1 Uniform Scaling

Consider the matrix operator represented by:

$$\begin{bmatrix} 2 & 0 \\ 0 & 2 \end{bmatrix}.$$

This transformation converts $\hat{i} = (1,0)$ into $(2,0)$ and $\hat{j} = (0,1)$ into $(0,2)$, representing uniform scaling by a factor of two. Under this transformation, all vectors maintain their direction but double in length, as demonstrated by applying A to a general vector $v = (v_1, v_2)$:

$$\begin{bmatrix} 2 & 0 \\ 0 & 2 \end{bmatrix} \begin{bmatrix} v_1 \\ v_2 \end{bmatrix} = \begin{bmatrix} 2v_1 \\ 2v_2 \end{bmatrix}.$$

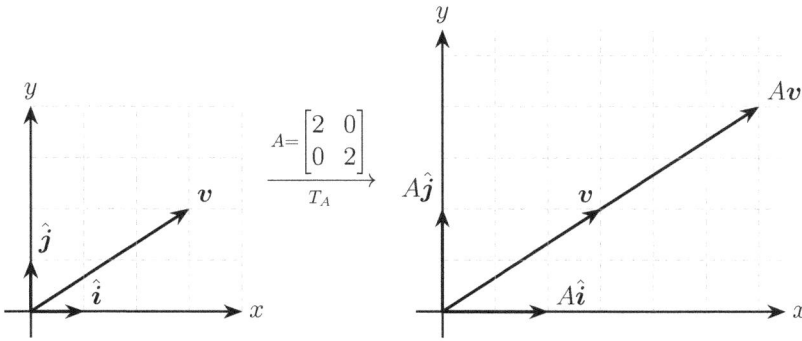

Figure 8.2: Example of Uniform Scaling Operator

A geometric illustration is provided in Fig. 8.2, showing both the original and transformed vectors.

In general, a matrix operator that uniformly scales any vector in $\mathbb{R}^2$ can be described by:

$$\begin{bmatrix} k & 0 \\ 0 & k \end{bmatrix},$$

where $k > 1$ indicates uniform expansion and $0 < k < 1$ indicates uniform contraction. When $k < 0$, the scaling factor is negative, indicating both a reflection about the origin and a change in magnitude.

2 Non-Uniform Scaling

Next, we examine a matrix operator represented by:

$$\begin{bmatrix} 1 & 0 \\ 0 & 2 \end{bmatrix}.$$

This operator transforms $\hat{i} = (1,0)$ to $(1,0)$ and $\hat{j} = (0,1)$ to $(0,2)$, scaling the y-component by a factor of two while retaining the x-component unchanged. This leads to non-uniform scaling, as depicted in Fig. 8.3. The direction of general vectors is not preserved by this transformation because the matrix acts differently on each axis, changing vector components independently.

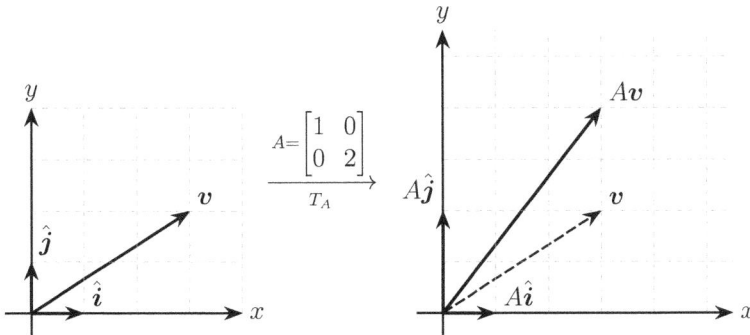

Figure 8.3: Example of Non-uniform Scaling Operator

The general form of a non-uniform scaling operator in $\mathbb{R}^2$ is:

$$\begin{bmatrix} k_x & 0 \\ 0 & k_y \end{bmatrix},$$

where $k_x \neq k_y$ distinguishes it from uniform scaling. This distinction is crucial due to the differing properties of their eigenvalues and eigenvectors, which will be explored further in § 9.1.

3　Rotation

Rotation operators hold particular significance in quantum computing. For visual context, consider a rotation operator in $\mathbb{R}^2$ represented by:

$$R_\theta = \begin{bmatrix} \cos\theta & -\sin\theta \\ \sin\theta & \cos\theta \end{bmatrix},$$

which rotates any vector counterclockwise by θ while preserving its length. Rotations are orthogonal transformations that maintain both the magnitude and angles between vectors. This transformation is illustrated in Fig. 8.4.

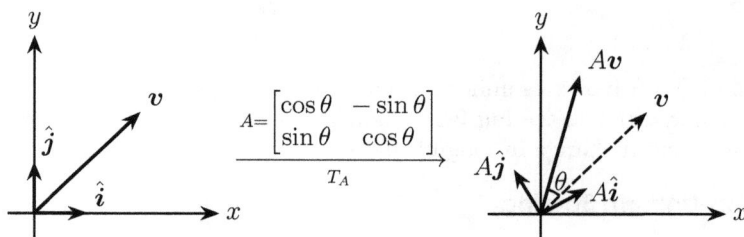

Figure 8.4: Example of Rotation Operator

■ **Example 8.4** Provide the matrix operator that rotates any vector clockwise by θ, while preserving their length.

Solution. Rotating clockwise by θ can be thought of as rotating counterclockwise by $-\theta$. Substituting θ with $-\theta$ in the rotation matrix R_θ yields:

$$R_{-\theta} = \begin{bmatrix} \cos(-\theta) & -\sin(-\theta) \\ \sin(-\theta) & \cos(-\theta) \end{bmatrix} = \begin{bmatrix} \cos\theta & \sin\theta \\ -\sin\theta & \cos\theta \end{bmatrix}.$$

■

Exercise 8.4 Demonstrate that the column vectors of the rotation matrix R_θ correspond to the vectors $\hat{i}$ and $\hat{j}$ rotated counterclockwise by θ.

Exercise 8.5 Demonstrate that the rotation operator R_θ preserves the length of any vector in $\mathbb{R}^2$.

4　Orthogonal Projection

A projection operator maps any vector from a vector space onto a specific subspace. Recall the projection formula we discussed in Theorem 6.18, which particularly addresses the projection of a vector onto the span of another vector—a scenario representing a projection onto a line. In $\mathbb{R}^2$, this scenario can be mathematically represented using the matrix operator:

$$P_\theta = \begin{bmatrix} \cos^2 \theta & \cos \theta \sin \theta \\ \cos \theta \sin \theta & \sin^2 \theta \end{bmatrix}.$$

This matrix operator orthogonally projects any vector in $\mathbb{R}^2$ onto the line defined by the direction vector $(\cos \theta, \sin \theta)$. The derivation of this property provides a practical application of theoretical concepts and is recommended as an exercise for further exploration (Problem 8.1).

For visual aid, an illustration of this projection in $\mathbb{R}^2$ is provided in Fig. 8.5. In Fig. 8.5, the line onto which the projection is made has a negative slope, indicating that θ corresponds to a negative angle.

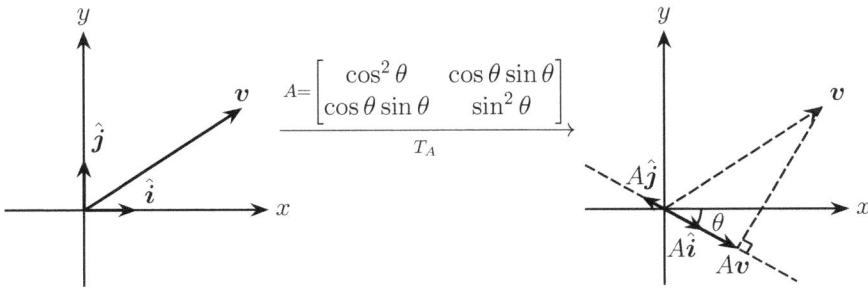

Figure 8.5: Example of Projection Operator

Exercise 8.6 Show that the determinant of P_θ is zero.

Exercise 8.7 Verify that P_θ is idempotent.

5   Reflection

Reflection operators are fundamental in the analysis of symmetrical properties across various dimensions. These operators can reflect a given vector about the origin, a line, or higher-dimensional hyperplanes. Consider the following transformations in $\mathbb{R}^2$:

(a) Reflection about the origin: The matrix $\begin{bmatrix} -1 & 0 \\ 0 & -1 \end{bmatrix}$ transforms the coordinates (x, y) to $(-x, -y)$.

(b) Reflection about the y-axis: The matrix $\begin{bmatrix} -1 & 0 \\ 0 & 1 \end{bmatrix}$ transforms the coordinates (x, y) to $(-x, y)$.

(c) Reflection about the line $y = x$: The matrix $\begin{bmatrix} 0 & 1 \\ 1 & 0 \end{bmatrix}$ transforms the coordinates (x, y) to (y, x).

The transformation about the line $y = x$ is further illustrated in Fig. 8.6.

For a general line at an angle θ from the x-axis, the reflection matrix is:

$$F_\theta = \begin{bmatrix} \cos 2\theta & \sin 2\theta \\ \sin 2\theta & -\cos 2\theta \end{bmatrix}, \tag{8.6}$$

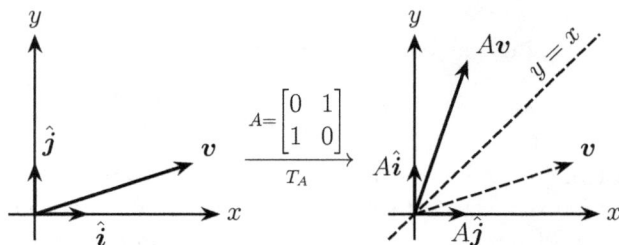

Figure 8.6: Example of Reflection Operator About the Line $y = x$

as shown in Fig. 8.7. The matrix F_θ reflects vectors across a general line at an angle θ with respect to the x-axis. The proof is left to the reader as an exercise (Problem 8.2).

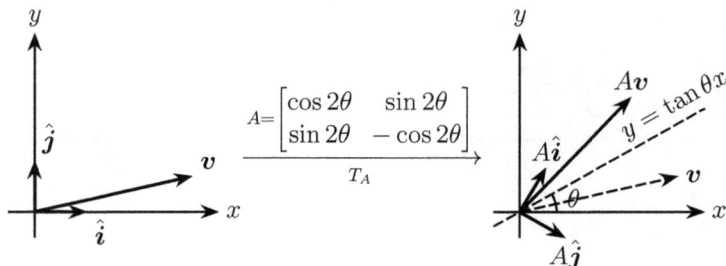

Figure 8.7: Example of Reflection Operator Across a General Line

Exercise 8.8 Geometrically interpret the matrix transformations represented by the following matrices in terms of reflection:

$$\text{(a) } \begin{bmatrix} 0 & 1 \\ -1 & 0 \end{bmatrix}, \text{ (b) } \begin{bmatrix} 1 & 0 \\ 0 & -1 \end{bmatrix}.$$

Exercise 8.9 Demonstrate that $F_\theta = 2P_\theta - I$ and that $F_\theta^2 = I$.

6 Shear

Shear operators modify the shape of objects by sliding one part parallel to a fixed line while keeping another direction fixed. In $\mathbb{R}^2$, a shear in the x-direction is described by:

$$\begin{bmatrix} 1 & k \\ 0 & 1 \end{bmatrix}.$$

This transformation shifts the x-component of a vector (x, y) to $(x + ky, y)$, effectively converting a rectangle into a parallelogram, as depicted in Fig. 8.8. Unlike scaling, which changes vector lengths, shear transformations primarily change the shape of objects while preserving areas.

For a shear transformation in the y-direction within $\mathbb{R}^2$, the transformation shifts the y-component of a vector (x, y) to $(x, y + kx)$. This transformation corresponds

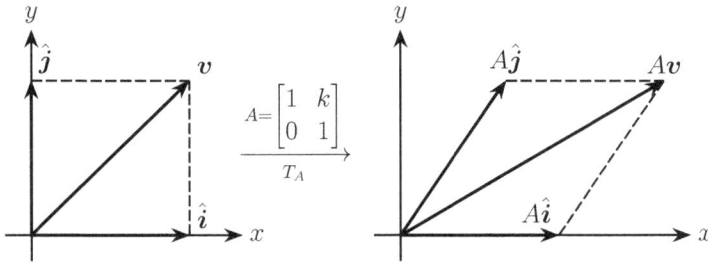

Figure 8.8: Example of Shear Operator in the x-direction

to the matrix:

$$\begin{bmatrix} 1 & 0 \\ k & 1 \end{bmatrix}.$$

Exercise 8.10 Illustrate geometrically the effect of a shear operator in the y-direction on a unit square formed by $\hat{i}$ and $\hat{j}$ within $\mathbb{R}^2$.

7 Summary

Table 8.1 provides a concise summary of the basic types of matrix operators and their exemplary geometric illustrations in $\mathbb{R}^2$. In the field of quantum computing, quantum gates are typically represented by unitary operators which preserve the norm of a vector. Key operators in this category include rotation and reflection operators, which will be further discussed in § 8.3.

Projection operators also play a crucial role in quantum computing, especially when computing probabilities of measurement outcomes or in spectral decomposition.

Conversely, scaling and shear operators, while less prevalent in direct quantum computing applications, offer valuable insights into the algebraic and geometric understanding of eigenvalues and eigenvectors, thus providing foundational knowledge essential for advanced studies.

8.2 Properties of Matrix Operators

8.2.1 Compositions of Matrix Transformations

The sequential execution of two or more matrix transformations creates a composite transformation. This concept is akin to the composition of functions in mathematics. We define a composite matrix transformation as follows:

Definition 8.3 — Composite Transformation. A composite transformation, involving a matrix transformation by $A_{p \times n}$ followed by $B_{m \times p}$, is denoted by:

$$(T_B \circ T_A)(\boldsymbol{v}) = T_B(T_A(\boldsymbol{v})), \tag{8.7}$$

facilitating a mapping from $\mathbb{C}^n$ through an intermediate space $\mathbb{C}^p$ to $\mathbb{C}^m$.

Operator Type	Matrix	Effect on Standard Basis Vectors
Uniform Scaling	$\begin{bmatrix} k & 0 \\ 0 & k \end{bmatrix}$, $k > 0$	
Non-Uniform Scaling	$\begin{bmatrix} k_x & 0 \\ 0 & k_y \end{bmatrix}$, $k_x, k_y > 0$	
Counterclockwise Rotation About the Origin	$\begin{bmatrix} \cos\theta & -\sin\theta \\ \sin\theta & \cos\theta \end{bmatrix}$	
Orthogonal Projection	$\begin{bmatrix} \cos^2\theta & \cos\theta\sin\theta \\ \cos\theta\sin\theta & \sin^2\theta \end{bmatrix}$	
Reflection About a Line Through the Origin	$\begin{bmatrix} \cos 2\theta & \sin 2\theta \\ \sin 2\theta & -\cos 2\theta \end{bmatrix}$	
Shear in the x-direction	$\begin{bmatrix} 1 & k \\ 0 & 1 \end{bmatrix}$	

Table 8.1: A Summary of Basic Matrix Operators in $\mathbb{R}^2$.

Theorem 8.4 The composite transformation $T_B \circ T_A$ corresponds to the matrix transformation represented by the product BA:

$$T_B \circ T_A = T_{BA}. \tag{8.8}$$

Proof. Consider any vector $v \in \mathbb{C}^n$:

$$(T_B \circ T_A)(v) = T_B(T_A(v)) = T_B(Av)$$
$$= B(Av) = (BA)v$$
$$= T_{BA}(v).$$

Thus, multiplying B and A in this order creates a matrix that encapsulates the effect of applying T_A followed by T_B, confirming that:

$$T_B \circ T_A = T_{BA}.$$

$\square$

Theorem 8.4 elucidates that matrix multiplication symbolizes composite transformations. When matrices A and B are square $(n \times n)$, their products (either AB or BA) act as composite operators within $\mathbb{C}^n$. This concept will be further explored through the following visual examples in $\mathbb{R}^2$.

1   Non-commutativity

Matrix multiplication is generally non-commutative, which implies $AB \neq BA$. This property can be observed in matrix transformations as follows:

Consider matrices $A = F_{\pi/4}$ and $B = F_{\pi/2}$ in $\mathbb{R}^2$ defined by Eq. 8.6. The transformation T_{BA} represents a sequence where a vector is first reflected by the line $y = x$ and then reflected about the y-axis. Conversely, T_{AB} performs the reflection by the y-axis first and $y = x$ second. Geometrically, reflecting first about the line $y = x$ changes the position of a vector relative to the y-axis, and the subsequent reflection about the y-axis does not return to the same original configuration that would have resulted from reversing the sequence of reflections. Figure 8.9 demonstrates that these transformations yield vectors in opposite directions.

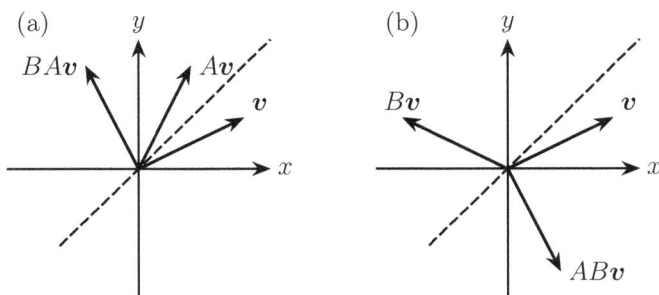

Figure 8.9: Example of Non-commutative Matrix Operators

Exercise 8.11 Verify that $AB = -BA$ for matrices $A = F_{\pi/4}$ and $B = F_{\pi/2}$.

Two operators A, B are commutative when the operations they represent can be performed in any order without affecting the outcome. The following example illustrates this concept.

■ **Example 8.5** Demonstrate that two rotation matrices in $\mathbb{R}^2$ are commutative.

Solution. Consider two rotation matrices R_θ and R_ϕ. Their product is computed as follows:

$$
R_\theta R_\phi = \begin{bmatrix} \cos\theta & -\sin\theta \\ \sin\theta & \cos\theta \end{bmatrix} \begin{bmatrix} \cos\phi & -\sin\phi \\ \sin\phi & \cos\phi \end{bmatrix}
$$

$$
= \begin{bmatrix} \cos\theta\cos\phi - \sin\theta\sin\phi & -(\cos\theta\sin\phi + \sin\theta\cos\phi) \\ \sin\theta\cos\phi + \cos\theta\sin\phi & \cos\theta\cos\phi - \sin\theta\sin\phi \end{bmatrix}
$$

$$
= \begin{bmatrix} \cos(\theta+\phi) & -\sin(\theta+\phi) \\ \sin(\theta+\phi) & \cos(\theta+\phi) \end{bmatrix} = R_{\theta+\phi}.
$$

Reversing the order of multiplication, we get:

$$
R_\phi R_\theta = \begin{bmatrix} \cos(\phi+\theta) & -\sin(\phi+\theta) \\ \sin(\phi+\theta) & \cos(\phi+\theta) \end{bmatrix} = R_{\phi+\theta}.
$$

Since $\theta + \phi = \phi + \theta$, it follows that $R_\theta R_\phi = R_\phi R_\theta = R_{\theta+\phi}$, confirming that the rotation matrices are commutative.

This result aligns with the geometric principle that a counterclockwise rotation by θ followed by ϕ is equivalent to a rotation by ϕ followed by θ, both culminating in a total rotation of $\theta + \phi$ counterclockwise.   ■

Although rotation operators are commutative in $\mathbb{R}^2$, this property does not extend to higher dimensions including $\mathbb{R}^3$. Further details are discussed in § 11.2.1.4.

2   Associativity

The principle of associativity in matrix multiplication is formally stated as $(AB)C = A(BC)$. Proving this property using matrix algebra, as introduced in § 7.2, might appear intricate; however, interpreting it through the lens of matrix transformations offers a clearer and more intuitive understanding.

Consider a vector v undergoing a sequence of transformations. If we apply $(AB)C$ to v, we first transform it with T_C, followed by the combined effect of T_B and then T_A, represented by T_{AB}. On the other hand, applying $A(BC)$ to v means we first transform it with the combined effect of T_C and then T_B, represented by T_{BC}, followed by T_A. Since the order of the individual transformations (T_C, T_B, and T_A) remains the same in both cases, we can conclude that $(AB)C$ and $A(BC)$ ultimately achieve the same result.

3   Inverse Operator

If a matrix A is invertible, there exists A^{-1} such that $AA^{-1} = A^{-1}A = I$. This relationship in the context of matrix transformations implies that for any vector v:

$$
(T_A \circ T_{A^{-1}})v = (T_{A^{-1}} \circ T_A)v = (AA^{-1})v = (A^{-1}A)v = Iv = v.
$$

In this case, the matrix operators A and A^{-1} effectively "cancel" each other out, restoring any vector to its original position. This relationship means that applying A followed by A^{-1} (or vice versa) is equivalent to performing no transformation at all, effectively leaving any vector unchanged. Hence, T_A and $T_{A^{-1}}$ are referred to as **inverse operators**.

This conceptual framework can also be applied intuitively to determine the inverses of matrices. For example, the inverse of a rotation matrix R_θ, which rotates vectors counterclockwise by θ, is $R_{-\theta}$, rotating clockwise by θ. Similarly, the inverse of a reflection operator F_θ is the operator itself, since reflecting about the same axis twice returns any vector to its original position.

Exercise 8.12 Explain the matrix identity $(AB)^{-1} = B^{-1}A^{-1}$ for invertible matrices $A, B \in \mathbb{C}^{n \times n}$ using the concept of matrix transformations.

4 Invertibility

The invertibility of a matrix is directly related to whether the corresponding operator can be "undone" or "reversed". Invertible matrices correspond to reversible transformations, such as rotation, reflection, and shear operations. An invertible matrix corresponds to an operator that provides a unique mapping from input vectors to output vectors, which means that each input is mapped to a distinct output without collapsing onto a lower-dimensional space.

In contrast, singular matrices correspond to non-invertible operations, such as projection operators. These operators reduce a vector from a higher-dimensional space to a lower-dimensional subspace and lack a one-to-one mapping, and thus are not reversible.

Exercise 8.13 Geometrically demonstrate that any projection operator P_θ in $\mathbb{R}^2$ is not bijective, and hence not reversible. (Hint: Illustrate that there are infinitely many vectors in $\mathbb{R}^2$ that result in the same projected vector when operated upon by P_θ).

8.2.2 Determinants and Matrix Operators

The determinant of a matrix provides significant insight into the behavior of the corresponding matrix operator. Specifically, in $\mathbb{R}^2$, the determinant quantifies the area of the transformed unit square, originally having an area of one, as outlined in the theorem below.

Theorem 8.5 For any matrix $A \in \mathbb{R}^{2 \times 2}$, its determinant represents the area of the parallelogram resulting from the transformation of the unit square by T_A. When $\det(A) = 0$, the transformation compresses the unit square onto a line or reduces it to the origin, indicating that the operator is not invertible.

An illustration of this theorem is provided in Fig. 8.10. Consider the matrix $A = \begin{bmatrix} a & b \\ c & d \end{bmatrix}$, which transforms $\hat{i}$ into (a, c) and $\hat{j}$ into (b, d). These transformed vectors form a parallelogram as depicted in the figure. Geometric analysis confirms that the area of this parallelogram is indeed $ad - bc$, which is precisely the determinant of A.

Exercise 8.14 Prove that the shaded area in Fig. 8.10 is equal to $ad - bc$.

When the vectors (a, c) and (b, d) are collinear, they fail to span a plane, resulting in the parallelogram collapsing to a line (or to the origin if both vectors are zero), which corresponds to an area of zero. Thus, $\det(A) = 0$ in these cases, reflecting the linear dependence of (a, c) and (b, d).

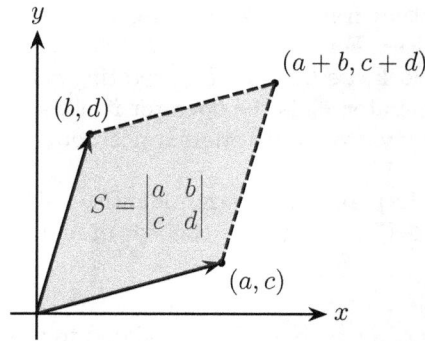

Figure 8.10: Illustration of Determinant as an Area

> **(i)** A negative value of det A indicates a transformation that reverses the orientation of $\hat{\imath}$ and $\hat{\jmath}$, typically involving a reflection or a similar operation that flips the coordinate system.

Theorem 8.5 further implies that if the determinant of a matrix is 1 or -1, the matrix operation preserves the area, which is essential in operations such as rotations, reflections, and shears. These properties can be verified by calculating the determinants of the respective transformation matrices.

When extending Theorem 8.5 to $\mathbb{R}^n$, the determinant of a matrix captures the transformed volume of a unit hypercube in higher dimensions. This extension introduces more complex mathematical considerations, which are omitted here for brevity.

8.2.3 Generalized Matrix Operators

Table 8.1 provides an overview of basic matrix operators in $\mathbb{R}^2$. This subsection extends those concepts to $\mathbb{C}^n$ and introduces their representations in Dirac notation, focusing on all operators except shear vectors, which are less pertinent to our discussion.

1 Scaling

General scaling operators in $\mathbb{C}^n$ are represented by diagonal matrices:

$$\Lambda = \begin{bmatrix} \lambda_0 & 0 & \cdots & 0 \\ 0 & \lambda_1 & \cdots & 0 \\ 0 & 0 & \ddots & 0 \\ 0 & 0 & \cdots & \lambda_{n-1} \end{bmatrix}, \tag{8.9}$$

where each diagonal element λ_i is a complex scalar that scales the corresponding component of a vector. This matrix can be expressed in Dirac notation as:

$$\Lambda = \sum_i \lambda_i |i\rangle\langle i|, \tag{8.10}$$

with $|i\rangle$ for $i = 0, 1, \ldots, n-1$ representing the standard basis vectors in $\mathbb{C}^n$.

Given a vector $|\psi\rangle = \sum_{i=0}^{n-1} \psi_i |i\rangle$, the action of Λ on $|\psi\rangle$ demonstrates the individual scaling of each component:

$$\Lambda |\psi\rangle = \sum_i \lambda_i |i\rangle\langle i| \sum_j \psi_j |j\rangle$$

$$= \sum_{i,j} \lambda_i \psi_j |i\rangle \langle i|j\rangle$$

$$= \sum_{i,j} \lambda_i \psi_j \delta_{ij} |i\rangle$$

$$= \sum_i \lambda_i \psi_i |i\rangle .$$

This expression illustrates that Λ scales each component $\psi_i |i\rangle$ of the vector by the corresponding scalar λ_i, effectively modifying each component independently based on its associated scaling factor.

2 Projection

The projection operator P_u, defined for any vector $|u\rangle \in \mathbb{C}^n$, projects a vector $|\psi\rangle$ orthogonally onto the one-dimensional subspace spanned by $|u\rangle$. It is given by:

$$P_u = |u\rangle \langle u| , \qquad (8.11)$$

which acts on $|\psi\rangle$ as:

$$P_u |\psi\rangle = \langle u|\psi\rangle |u\rangle ,$$

consistent with the unit vector projection formula from Eq. 6.58.

■ **Example 8.6** The projection operator P_θ in $\mathbb{R}^2$ is a specific instance of P_u, since P_θ projects any vector in $\mathbb{R}^2$ onto the line with an inclination angle θ (see Table 8.1). To demostrate this, choose $|u\rangle = \begin{bmatrix} \cos\theta \\ \sin\theta \end{bmatrix}$, the unit vector along the line of projection. The projection operator P_u can be calculated using the outer product:

$$P_u = |u\rangle \langle u| = \begin{bmatrix} \cos\theta \\ \sin\theta \end{bmatrix} \begin{bmatrix} \cos\theta & \sin\theta \end{bmatrix} = \begin{bmatrix} \cos^2\theta & \cos\theta\sin\theta \\ \cos\theta\sin\theta & \sin^2\theta \end{bmatrix} = P_\theta.$$

■

For projections onto a multidimensional subspace spanned by an orthonormal set $\{|u_1\rangle , |u_2\rangle , \ldots, |u_r\rangle\}$, the projection operator is:

$$P = \sum_{i=1}^{r} |u_i\rangle\langle u_i| , \qquad (8.12)$$

applied as:

$$P |\psi\rangle = \sum_{i=1}^{r} \langle u_i|\psi\rangle |u_i\rangle ,$$

which reproduces Eq. 6.72 in Dirac notation. Further discussions and applications of projections can be found in § 11.2.4.

Exercise 8.15 Confirm that both P_u and P are Hermitian operators.

3   Rotation and Reflection

Rotation and reflection operations, while distinct in their geometrical implications, share similar underlying matrix properties. A notable relationship in $\mathbb{R}^2$ involves the rotation matrix R_θ and the reflection matrix F_θ, which can be expressed as:

$$F_\theta = R_{2\theta} \begin{bmatrix} 1 & 0 \\ 0 & -1 \end{bmatrix} = R_{2\theta} F_0. \tag{8.13}$$

This equation indicates that applying F_θ is equivalent to first reflecting a vector about the x-axis and then rotating it counterclockwise by 2θ.

Exercise 8.16 Geometrically verify Eq. 8.13 by showing that reflecting any vector in $\mathbb{R}^2$ about the x-axis followed by a 2θ counterclockwise rotation results in a reflection about a line inclined by θ.

In the higher-dimensional space of $\mathbb{C}^n$, rotation and reflection operations fall under the category of **unitary operators**, which satisfy the condition $UU^\dagger = I$. These operators essentially function as basis-change matrices and are expressed in Dirac notation as:

$$U = \sum_{i=1}^{n} |u_i\rangle\langle v_i|, \tag{8.14}$$

where $\{u_i\}$ and $\{v_i\}$ are orthonormal bases of $\mathbb{C}^n$. Unitary matrices play a pivotal role in quantum computing due to their properties of preserving the norm of quantum states. This topic is thoroughly explored in the next two sections § 8.3 and § 8.4, emphasizing their fundamental importance in the field.

Exercise 8.17 Verify that the reflection operator F_θ and the rotation operator R_θ in $\mathbb{R}^2$ are unitary by showing that they satisfy the condition $UU^\dagger = I$.

8.3   Unitary Operators

Unitary matrices, often referred to as unitary operators, are foundational elements in the theory of quantum computing. They serve as the mathematical underpinnings for quantum gates, which are critical for controlling the evolution of quantum systems. Analogous to orthogonal matrices in real vector spaces, unitary matrices operate within complex vector spaces, extending the concept of orthogonality and norm-preservation to the complex field. By preserving both the norm and inner product of vectors, unitary matrices ensure the reversibility and probabilistic consistency of quantum operations. This section provides a detailed exposition on unitary matrices, elucidating their essential mathematical properties and commonly used forms in Dirac notation.

8.3.1   Definition

In complex vector spaces like $\mathbb{C}^n$, rotation and reflection operators are unitary matrices. These matrices are characterized by their ability to preserve the norm and inner products of vectors within these spaces.

Definition 8.4 — Unitary Matrix. A matrix $U \in \mathbb{C}^{n \times n}$ is defined as unitary if it preserves the norm of any vector $|v\rangle$ in $\mathbb{C}^n$, formally expressed as:

$$\| U |v\rangle \| = \|v\| \quad \text{for any } |v\rangle \in \mathbb{C}^n. \tag{8.15}$$

Definition 8.4 underscores the norm-preserving characteristic of unitary matrices, which is crucial for their application as quantum gates in quantum computing. The following theorem demonstrates that this property is equivalent to the familiar definition $U^\dagger = U^{-1}$, ensuring that unitary matrices preserve both norms and inner products.

Theorem 8.6 A matrix U is unitary if and only if:

$$U^\dagger U = U U^\dagger = I, \tag{8.16}$$

indicating that $U^{-1} = U^\dagger$.

Proof. First, if $U^\dagger = U^{-1}$, then $U^\dagger U = U U^\dagger = I$. For any vector $|v\rangle \in \mathbb{C}^n$:

$$\| U |v\rangle \|^2 = (U |v\rangle)^\dagger U |v\rangle = \langle v| U^\dagger U |v\rangle = \langle v|v\rangle = \|v\|^2.$$

Thus, U preserves the norm, proving that U is unitary by Def. 8.4.

Conversely, assume U is unitary. Then:

$$\| U |v\rangle \| = \|v\| \quad \Rightarrow \quad \| U |v\rangle \|^2 = \|v\|^2$$
$$\Rightarrow \quad \langle v|U^\dagger U|v\rangle = \langle v|v\rangle$$
$$\Rightarrow \quad \langle v|(U^\dagger U - I)|v\rangle = 0 \quad \text{for all } |v\rangle \in \mathbb{C}^n.$$

Since the quadratic form $\langle v|(U^\dagger U - I)|v\rangle$ is zero for all $|v\rangle$, the matrix $U^\dagger U - I$ must be the zero matrix. Thus:

$$U^\dagger U = I.$$

Similarly, $U U^\dagger = I$, implying $U^\dagger = U^{-1}$. This completes the proof. $\square$

8.3.2 Basic Properties

Unitary matrices also preserve the inner product, a fundamental property that ensures the orthogonality of vectors and the consistency of measurement probabilities throughout quantum transformations.

Theorem 8.7 — Preservation of Inner Product. If U is a unitary matrix, then for any vectors $|v\rangle, |w\rangle$ in $\mathbb{C}^n$:

$$\langle v|w\rangle = \langle Uv|Uw\rangle. \tag{8.17}$$

Proof. Applying the property $U^\dagger U = I$ for unitary matrices U to the right-hand side of Eq. 8.17 gives:

$$\langle Uv|Uw\rangle = \langle v| U^\dagger U |w\rangle = \langle v|I|w\rangle = \langle v|w\rangle.$$

$\square$

Since unitary matrices preserve norms and inner products, they are analogous to "rotation" and "reflection" operators in complex Hilbert spaces. Unitary matrices are always reversible, making them ideal for representing quantum gates, which implement reversible transformations of quantum states in quantum computing. In contrast, processes like quantum measurement or decoherence are inherently irreversible and are modeled by non-unitary operators.

Below are a few key properties of unitary matrices followed by their proofs.

Theorem 8.8 If U is unitary, then $U^\dagger$ is also unitary.

Proof. Since U is unitary, $U^\dagger U = I$. This indicates that

$$(U^\dagger)^{-1} = U = (U^\dagger)^\dagger.$$

By Theorem 8.6, $U^\dagger$ is also unitary. $\square$

Theorem 8.9 The determinant of a unitary matrix U has a modulus of one, i.e., $|\det(U)| = 1$.

Proof. Since $UU^\dagger = I$, using the property $\det(AB) = \det(A)\det(B)$:

$$\det(U)\det(U^\dagger) = \det(I) = 1.$$

Note that $\det(U^\dagger) = \det(U)^*$, thus:

$$\det(U)\det(U)^* = |\det(U)|^2 = 1 \quad \Rightarrow \quad |\det(U)| = 1.$$

$\square$

Theorem 8.10 The product of unitary matrices $U_1, U_2, \ldots, U_r$ is also unitary.

Proof. Let $U_1, U_2, \ldots, U_r$ be unitary matrices. Note that:

$$(U_1 U_2 \cdots U_r)^\dagger = U_r^\dagger \cdots U_2^\dagger U_1^\dagger.$$

Then:

$$\begin{aligned}
(U_1 U_2 \cdots U_r)^\dagger (U_1 U_2 \cdots U_r) &= U_r^\dagger \cdots U_2^\dagger U_1^\dagger U_1 U_2 \cdots U_r \\
&= (U_r^\dagger U_r) \cdots (U_2^\dagger U_2)(U_1^\dagger U_1) \\
&= I \cdots I \cdot I = I.
\end{aligned}$$

Since $(U_1 U_2 \cdots U_r)^\dagger (U_1 U_2 \cdots U_r) = I$, the product $U_1 U_2 \cdots U_r$ is unitary. $\square$

Theorem 8.11 — Orthonormality of Column Vectors. The column vectors of a unitary matrix form an orthonormal basis of $\mathbb{C}^n$.

Proof. Consider a unitary matrix $U \in \mathbb{C}^{n \times n}$. Partition U into its column vectors:

$$U = \begin{bmatrix} |u_1\rangle & |u_2\rangle & \cdots & |u_n\rangle \end{bmatrix}.$$

Similarly, partition $U^\dagger$ into its row vectors:

$$U^\dagger = \begin{bmatrix} \langle u_1| \\ \langle u_2| \\ \vdots \\ \langle u_n| \end{bmatrix}.$$

The matrix product $U^\dagger U$, using the definition of matrix multiplication, can be expanded as:

$$U^\dagger U = \begin{bmatrix} \langle u_1|u_1\rangle & \langle u_1|u_2\rangle & \cdots & \langle u_1|u_n\rangle \\ \langle u_2|u_1\rangle & \langle u_2|u_2\rangle & \cdots & \langle u_2|u_n\rangle \\ \vdots & \vdots & \ddots & \vdots \\ \langle u_n|u_1\rangle & \langle u_n|u_2\rangle & \cdots & \langle u_n|u_n\rangle \end{bmatrix}.$$

Given that $U^\dagger U = I$, each diagonal element $\langle u_i|u_i\rangle = 1$ and each off-diagonal element $\langle u_i|u_j\rangle = 0$ for $i \neq j$, $i, j = 1, 2, \ldots, n$. This ensures that the column vectors $|u_1\rangle, |u_2\rangle, \ldots, |u_n\rangle$ have unit norm and are orthogonal to each other, satisfying the orthonormality condition $\langle u_i|u_j\rangle = \delta_{ij}$. Therefore, the column vectors of U form an orthonormal basis of $\mathbb{C}^n$.   □

Exercise 8.18 Show that the row vectors of a unitary matrix also form an orthonormal basis of $\mathbb{C}^n$.

■ **Example 8.7** The identity matrix I is unitary since $I^\dagger I = II = I$. Its column vectors are the standard basis of $\mathbb{C}^n$.

The vectors $|+\rangle = \frac{1}{\sqrt{2}}(|0\rangle + |1\rangle)$ and $|-\rangle = \frac{1}{\sqrt{2}}(|0\rangle - |1\rangle)$ form an orthonormal basis of $\mathbb{C}^2$. Thus, the following 2×2 unitary matrices can be constructed by putting $|+\rangle, |-\rangle$ as column vectors:

$$\begin{bmatrix} \frac{1}{\sqrt{2}} & \frac{1}{\sqrt{2}} \\ \frac{1}{\sqrt{2}} & -\frac{1}{\sqrt{2}} \end{bmatrix} \quad \text{or} \quad \begin{bmatrix} \frac{1}{\sqrt{2}} & \frac{1}{\sqrt{2}} \\ -\frac{1}{\sqrt{2}} & \frac{1}{\sqrt{2}} \end{bmatrix}.$$

■

Exercise 8.19 Construct two unitary matrices based on the following orthonormal basis of $\mathbb{C}^2$: $\{|+_i\rangle, |-_i\rangle\}$ where $|+_i\rangle = \frac{1}{\sqrt{2}}(|0\rangle + i|1\rangle)$ and $|-_i\rangle = \frac{1}{\sqrt{2}}(|0\rangle - i|1\rangle)$.

8.3.3 Special Types of Unitary Operators

Unitary matrices (operators) are often represented as sums of outer products. This subsection introduces two essential forms of unitary operators: those that are both unitary and Hermitian, exemplified by the Pauli matrices in $\mathbb{C}^{2\times2}$, and those forming general basis transformation operators.

(i) When matrices are expressed in Dirac notation, they are commonly referred to as operators.

1 Simultaneous Unitary and Hermitian Operators

> **Theorem 8.12** Given $\{|\phi_i\rangle\}$, an orthonormal basis of $\mathbb{C}^n$, and $\lambda_i \in \{1, -1\}$, the following matrix is both unitary and Hermitian:
>
> $$S = \sum_i \lambda_i |\phi_i\rangle \langle \phi_i| . \tag{8.18}$$

Proof. To prove S is Hermitian, observe that

$$S^\dagger = \sum_i \lambda_i^* |\phi_i\rangle \langle \phi_i| = S,$$

since $\lambda_i = \lambda_i^*$ for $\lambda_i \in \{1, -1\}$.

To prove S is unitary, consider transforming an arbitrary vector $|\psi\rangle \in \mathbb{C}^n$ by S:

$$|\psi'\rangle = S|\psi\rangle = \sum_i \lambda_i |\phi_i\rangle \langle \phi_i| |\psi\rangle$$

$$= \sum_i (\lambda_i \langle \phi_i|\psi\rangle) |\phi_i\rangle .$$

Denoting $\psi_i' = \lambda_i \langle \phi_i|\psi\rangle$, the squared norm of $|\psi'\rangle$ is:

$$\|\psi'\|^2 = \sum_i |\psi_i'|^2 = \sum_i |\lambda_i|^2 |\langle \phi_i|\psi\rangle|^2 = \sum_i |\langle \phi_i|\psi\rangle|^2,$$

since $|\lambda_i|^2 = 1$. Using the decomposition $|\psi\rangle = \sum_i \langle \phi_i|\psi\rangle |\phi_i\rangle$, we find:

$$\|\psi\|^2 = \sum_i |\langle \phi_i|\psi\rangle|^2.$$

Therefore, $\|\psi'\|^2 = \|\psi\|^2$, confirming that S preserves norms and is unitary by Def. 8.4. □

Examples of matrices that fit Theorem 8.12 include the Pauli matrices, which play pivotal roles in quantum computing applications:

$$X = |+\rangle \langle +| - |-\rangle \langle -| , \tag{8.19a}$$
$$Y = |+_i\rangle \langle +_i| - |-_i\rangle \langle -_i| , \tag{8.19b}$$
$$Z = |0\rangle \langle 0| - |1\rangle \langle 1| . \tag{8.19c}$$

Chapter 12 provides an extensive exploration of Pauli matrices, delving into their unique properties.

> **Exercise 8.20** Derive the matrix representations of the Pauli matrices using Eq. 8.19 and confirm their unitary and Hermitian nature.

2 Basis Transformation Operators

Theorem 8.13 Given two sets of orthonormal bases of $\mathbb{C}^n$, $\{|u_i\rangle\}$ and $\{|v_i\rangle\}$ $(i = 1, 2, \ldots, n)$, and modulus-one complex numbers $\{c_i\}$ where $|c_i| = 1$, the matrix

$$U = \sum_i c_i |u_i\rangle \langle v_i| \tag{8.20}$$

is unitary.

The proof of Theorem 8.13 is similar to the methodology used in Theorem 8.12 and is left as an exercise to the reader (Problem 8.9).

Pauli matrices can also be expressed in a basis-transformation form. For example:

$$X = |0\rangle\langle 1| + |1\rangle\langle 0|, \tag{8.21a}$$
$$Y = -i|0\rangle\langle 1| + i|1\rangle\langle 0|. \tag{8.21b}$$

When $\{|u_i\rangle\}$ and $\{|v_i\rangle\}$ are distinct orthonormal bases or when $c_i \notin \{-1, 1\}$, the matrix U given by Theorem 8.13 may not be Hermitian. For instance, the matrix

$$|+_i\rangle\langle 0| + |-_i\rangle\langle 1|$$

is unitary but not Hermitian.

Exercise 8.21 Represent $|+_i\rangle\langle 0| + |-_i\rangle\langle 1|$ in matrix form. Verify that it is unitary but not Hermitian.

Unitary operators given by Eq. 8.20 are referred to as basis transformation operators and will be explored in detail in § 8.4.

8.3.4 Orthogonal Matrices

Orthogonal matrices are the real counterparts of unitary matrices in $\mathbb{R}^n$. They share many similar properties with unitary matrices and play a significant role in real matrix analysis. This section provides a brief overview of orthogonal matrices to systematically illustrate their analogy with unitary matrices.

1 Definition

Orthogonal matrices, which include rotation and reflection operators, preserve the norm of any vector subjected to their transformations in $\mathbb{R}^n$.

Definition 8.5 — Orthogonal Matrix. An $n \times n$ real matrix Q is termed an orthogonal matrix if it preserves the norm of any vector in $\mathbb{R}^n$, formulated as:

$$\|Qv\| = \|v\| \text{ for any } v \in \mathbb{R}^n. \tag{8.22}$$

Definition 8.5 aligns with the algebraic definition of orthogonal matrices, which states that $Q^T Q = I$. This is equivalent to saying that $Q^T = Q^{-1}$, meaning the transpose of Q is also its inverse.

Theorem 8.14 A matrix Q is orthogonal if and only if:

$$Q^T = Q^{-1}. \tag{8.23}$$

Exercise 8.22 Demonstrate that if Q is an orthogonal matrix, Q^T is also orthogonal.

2   Properties of Orthogonal Matrices

Orthogonal matrices exhibit several important properties, many of which mirror those of unitary matrices. These properties make them fundamental in geometry and the analysis of real linear transformations.

Orthogonal matrices preserve dot products between real vectors, as stated in the following theorem:

Theorem 8.15 — Preservation of Dot Products. If $Q \in \mathbb{R}^{n \times n}$ is an orthogonal matrix, then for any vectors $u, v \in \mathbb{R}^n$, the following identity holds:

$$u \cdot v = (Qu) \cdot (Qv). \tag{8.24}$$

Exercise 8.23 Prove Theorem 8.15 and further demonstrate that Q preserves the angle between any two vectors.

Orthogonal matrices are characterized by determinants of either $+1$ or -1. Intuitively, a determinant of $+1$ corresponds to transformations that preserve orientation, such as rotations, while -1 corresponds to transformations that reverse orientation, such as reflections, all while preserving volume or area.

Theorem 8.16 — Determinant of Orthogonal Matrices. If Q is an orthogonal matrix, then $\det Q = \pm 1$.

Theorem 8.17 — Product of Orthogonal Matrices. If $Q_1, Q_2, \ldots, Q_r$ are orthogonal matrices of the same size, their product $Q_1 Q_2 \cdots Q_r$ is also orthogonal.

Theorem 8.18 — Orthonormality of Column Vectors. The column vectors (or row vectors) of an orthogonal matrix Q form an orthonormal basis of $\mathbb{R}^n$.

The proofs of these theorems directly parallel the proofs of their unitary counterparts, and are omitted here.

Exercise 8.24 Verify the orthogonality of the matrix Q and the orthonormality of its column (row) vectors for:

$$Q = \begin{bmatrix} \frac{1}{3} & \frac{2}{3} & \frac{2}{3} \\ \frac{2}{3} & \frac{1}{3} & -\frac{2}{3} \\ -\frac{2}{3} & \frac{2}{3} & -\frac{1}{3} \end{bmatrix}.$$

8.4 Change of Basis

In quantum computing, the ability to represent qubit state vectors in different bases provides invaluable flexibility. Different bases offer diverse perspectives on the same quantum system, often simplifying calculations or revealing hidden structures. The crucial role of basis transformations in quantum computing necessitates a precise mathematical framework to describe them.

This section explores the mathematical formalism of basis changes. We begin by introducing basis transformation operators and their effect on state vectors. To build intuition, we first examine classical concepts of change of basis in $\mathbb{R}^2$, where state vectors are re-expressed in different bases. We then generalize these concepts to $\mathbb{C}^n$, the complex vector space relevant to quantum computing. Finally, we discuss basis-independent quantities and relations.

8.4.1 Introduction: Complementary Views of Change of Basis

1 Basis Change View

In the conventional view, a change of basis refers to a change in perspective for stationary vectors, re-expressed in a different basis. The vectors themselves remain unchanged, while their coordinates are modified relative to the new basis.

As an example, consider a vector $\boldsymbol{v} = (v_1, v_2) \in \mathbb{R}^2$ expressed in the standard basis $\{\hat{\boldsymbol{i}}, \hat{\boldsymbol{j}}\}$. When re-expressed in another orthonormal basis $\{\hat{\boldsymbol{i}}', \hat{\boldsymbol{j}}'\}$, its coordinates become (v_1', v_2'), as depicted in Fig. 8.11a.

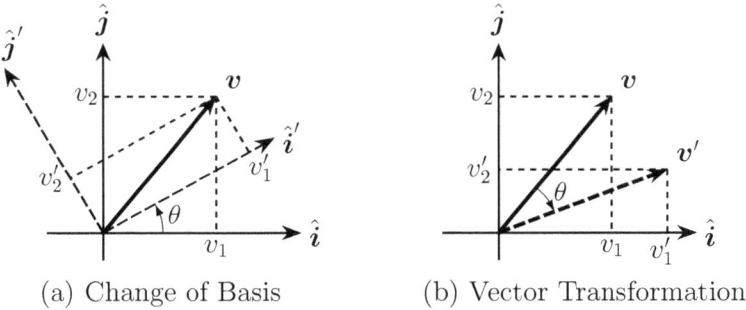

(a) Change of Basis   (b) Vector Transformation

Figure 8.11: Equivalence of Basis Change and Vector Transformation

This re-expression is facilitated by a matrix known as the **transition matrix** or **change-of-basis matrix**:

$$\begin{bmatrix} v_1' \\ v_2' \end{bmatrix} = T \begin{bmatrix} v_1 \\ v_2 \end{bmatrix}, \tag{8.25}$$

where T is the transition matrix from $\{\hat{\boldsymbol{i}}, \hat{\boldsymbol{j}}\}$ to $\{\hat{\boldsymbol{i}}', \hat{\boldsymbol{j}}'\}$.

In our example, the change of basis is a counterclockwise rotation by an angle θ. It is easy to verify that T is given by $R_{-\theta}$:

$$T = R_{-\theta} = \begin{bmatrix} \cos\theta & \sin\theta \\ -\sin\theta & \cos\theta \end{bmatrix}. \tag{8.26}$$

Note that the old basis vectors $\{\hat{\boldsymbol{i}}, \hat{\boldsymbol{j}}\}$ are transformed into the new basis vectors $\{\hat{\boldsymbol{i}}', \hat{\boldsymbol{j}}'\}$ via:

$$\hat{\boldsymbol{i}}' = U\hat{\boldsymbol{i}}, \quad \hat{\boldsymbol{j}}' = U\hat{\boldsymbol{j}}, \tag{8.27}$$

where $U \equiv T^{-1} = R_\theta$. (Refer to § 8.1.5.3 for the linear transformation by R_θ.)

The matrix U is commonly referred to as the **basis transformation matrix** (or operator). Thus, the transition matrix is the inverse $(T = U^{-1} = U^\dagger)$ of the basis transformation matrix.

> (i) In this section, we consider changes of basis between two orthonormal bases, unless specified otherwise. In such cases, T is a unitary matrix, as is $R_{-\theta}$. We use U to denote the basis transformation matrix, where $U = T^{-1}$, and $U^\dagger = T$.

2   Vector Transformation View

The change of basis depicted in Fig. 8.11a can also be viewed as a direct rotation of the vector v in the opposite direction (i.e., clockwise) by θ into v', while the coordinate system (i.e., basis) remains unchanged, as illustrated in Fig. 8.11b. Conversely, a unitary transformation of vectors can be viewed as a change of basis.

The relationship between v and v' is given by:

$$v' = Tv = R_{-\theta}v = U^\dagger v. \tag{8.28}$$

In this case, $T = U^\dagger$ functions as a linear transformation operator. (Refer to § 8.1 for details on such transformation operators.)

Change of Basis in Quantum Computing

Dual Perspectives of Change of Basis

In quantum computing, qubit state vectors are often represented in different bases, providing diverse perspectives on the same quantum system. This corresponds to the conventional view of change of basis illustrated in Fig. 8.11a.

During computations, qubit state vectors undergo physical transformations via unitary operations, such as quantum gates. These operations can also be interpreted as changes of basis, paralleling the vector transformation shown in Fig. 8.11b.

Notable Applications

Basis transformations play a crucial role in quantum computing, where different computational tasks may require different bases. For example, while the standard computational basis is frequently used, other bases, such as the Hadamard or Bell basis, are indispensable for operations involving superposition and entanglement.

A key application of basis transformations in quantum algorithms is the Quantum Fourier Transform (QFT), the quantum analogue of the classical discrete Fourier transform. The QFT transforms states from the computational basis to the Fourier basis (§ 11.2.5), leveraging quantum parallelism to process multiple states simultaneously. This ability underpins algorithms such as Shor's algorithm for integer factorization.

Unitary Evolution of Quantum States

The evolution of a quantum system is described by a unitary operator:

$$U(t) = e^{-\frac{i}{\hbar}Ht},$$

where $\hbar$ is the reduced Planck constant and H is the Hamiltonian of the system. For an initial state $\psi(0)\rangle$, the state at time t is:

$$|\psi(t)\rangle = U(t)|\psi(0)\rangle.$$

This evolution exemplifies a basis transformation in quantum mechanics.

8.4.2   A General Framework for Change of Basis

1   Complete Orthonormal Basis

As introduced in § 5.4.2, a basis is a minimal set of vectors that spans a vector space.

Definition 8.6 A basis $S = \{|b_i\rangle\} \equiv \{|b_1\rangle, |b_2\rangle, \ldots\}$ in $\mathbb{C}^n$ is orthonormal if:

$$\langle b_i | b_j \rangle = \delta_{ij} \tag{8.29}$$

where $i, j = 1, 2, \ldots, n$, and δ_{ij} is the Kronecker delta function, defined as $\delta_{ij} = 1$ if $i = j$ and $\delta_{ij} = 0$ if $i \neq j$.

The basis is complete if:

$$\sum_{i=1}^{n} |b_i\rangle\langle b_i| = I, \tag{8.30}$$

where I is the identity operator.

Theorem 8.19 An orthonormal basis with n basis vectors in $\mathbb{C}^n$ is complete.

Proof.   Since $S = \{|b_i\rangle\}$ is an orthonormal basis in $\mathbb{C}^n$, by definition, S spans the entire vector space $\mathbb{C}^n$. Thus, any vector $|v\rangle \in \mathbb{C}^n$ can be expressed as:

$$|v\rangle = \sum_{i=1}^{n} c_i |b_i\rangle,$$

where the coefficients c_i are given by the inner product:

$$c_i = \langle b_i | v \rangle.$$

Now consider the operator:

$$A = \sum_{j=1}^{n} |b_j\rangle\langle b_j|.$$

Applying this operator to any vector $|v\rangle \in \mathbb{C}^n$, we obtain:

$$A |v\rangle = \left(\sum_{j=1}^{n} |b_j\rangle\langle b_j| \right) \left(\sum_{i=1}^{n} c_i |b_i\rangle \right)$$

$$= \sum_{j=1}^{n} \sum_{i=1}^{n} c_i |b_j\rangle \langle b_j | b_i \rangle.$$

By using the orthonormal condition Eq. 8.29, where $\langle b_j | b_i \rangle = \delta_{ij}$, this reduces to:

$$A |v\rangle = \sum_{j=1}^{n} \sum_{i=1}^{n} c_i |b_j\rangle \, \delta_{ij}$$

$$= \sum_{i=1}^{n} c_i |b_i\rangle$$

$$= |v\rangle.$$

This shows that A acts as the identity operator I on $\mathbb{C}^n$, which implies that:

$$\sum_{i=1}^{n} |b_i\rangle\langle b_i| = I,$$

which is the completeness condition Eq. 8.30. $\qquad\qquad\qquad\qquad\qquad\qquad\qquad$ ◻

Completeness Condition in Hilbert Spaces

The completeness condition states that any vector in a given Hilbert space (§ 6.3.4) can be expressed as a linear combination of vectors from a specific orthonormal basis. In finite-dimensional Hilbert spaces, such as $\mathbb{C}^n$, this concept is straightforward because orthonormal bases are automatically complete. However, in infinite-dimensional Hilbert spaces, the completeness condition becomes significantly more complex and essential, as Hilbert spaces are rigorously defined as complete complex inner product spaces.

To clarify the terminology, we have so far used "orthonormal basis" and "complete orthonormal basis" interchangeably, assuming that an orthonormal basis is inherently complete in the context of finite sets of vectors. In infinite-dimensional Hilbert spaces, however, the term "completeness" explicitly refers to the basis's ability to span the entire space. This distinction is crucial for fully understanding what constitutes a basis, particularly when extending beyond the confines of finite-dimensional vector spaces.

Additionally, an orthonormal basis remains orthonormal when each basis vector is scaled with a phase factor.

Theorem 8.20 Let $S = \{|b_j\rangle\}$ be an orthonormal basis in $\mathbb{C}^n$. Then $S' = \{e^{i\theta_j}\,|b_j\rangle\}$, where each θ_j is a real number, is also an orthonormal basis in $\mathbb{C}^n$.

Proof. Let $|b_j'\rangle = e^{i\theta_j}\,|b_j\rangle$. Then,

$$\langle b_k'|b_j'\rangle = \langle e^{-i\theta_k} b_k|e^{i\theta_j} b_j\rangle = e^{i(\theta_j - \theta_k)}\,\langle b_k|b_j\rangle = e^{i(\theta_j - \theta_k)}\delta_{kj} = \delta_{kj}.$$

In the last step, when $k = j$, the factor $e^{i(\theta_j - \theta_k)} = 1$. When $k \neq j$, the product is zero due to $\delta_{kj} = 0$.

Thus, $S' = \{|b_j'\rangle\}$ is also an orthonormal basis. $\qquad\qquad\qquad\qquad\qquad\qquad$ ◻

2 The Basis Transformation Operator

Generalizing our intuition of basis change in $\mathbb{R}^2$, now let's consider the change from orthonormal basis $S = \{|b_i\rangle\}$ to another orthonormal basis $S' = \{|b_i'\rangle\}$, both in $\mathbb{C}^n$.

The orthonormality and completeness of S and S' require:

$$\langle b_i|b_j\rangle = \langle b_i'|b_j'\rangle = \delta_{ij},$$

$$\sum_{i=1}^{n} |b_i\rangle\langle b_i| = \sum_{i=1}^{n} |b_i'\rangle\langle b_i'| = I,$$

where $i, j = 1, 2, \ldots, n$.

For example, the bases depicted in Fig. 8.11 are given by:

$$|b_1\rangle = \hat{i} = \begin{bmatrix} 0 \\ 1 \end{bmatrix} = |0\rangle, \qquad\qquad |b_2\rangle = \hat{j} = \begin{bmatrix} 1 \\ 0 \end{bmatrix} = |1\rangle, \qquad (8.31a)$$

$$|b_1'\rangle = \hat{i}' = \begin{bmatrix} \cos\theta \\ \sin\theta \end{bmatrix}, \qquad\qquad |b_2'\rangle = \hat{j}' = \begin{bmatrix} -\sin\theta \\ \cos\theta \end{bmatrix}. \qquad (8.31b)$$

Mirroring Eq. 8.27, we define the basis transformation operator U as:

$$|b_i'\rangle = U |b_i\rangle. \qquad (8.32)$$

This relationship can be viewed as a general coordinate rotation, illustrated in Fig. 8.12a.

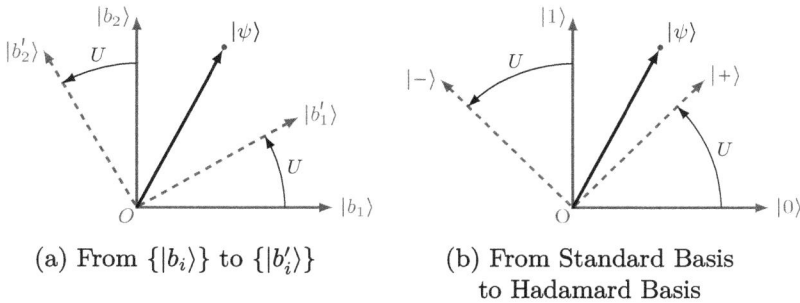

(a) From $\{|b_i\rangle\}$ to $\{|b_i'\rangle\}$ (b) From Standard Basis to Hadamard Basis

Figure 8.12: Illustration of Basis Rotation

Since U transforms $|b_i\rangle$ to $|b_i'\rangle$, it consists of terms like $|b_i'\rangle\langle b_i|$, hence the following theorem:

Theorem 8.21 — Basis Transformation Operator. The basis transformation operator U defined by Eq. 8.32 is a unitary operator given by:

$$U = \sum_{i=1}^{n} |b_i'\rangle\langle b_i|. \qquad (8.33)$$

Proof.

Unitarity

Consider:

$$UU^\dagger = \left(\sum_{i=1}^{n} |b_i'\rangle\langle b_i| \right) \left(\sum_{j=1}^{n} |b_j\rangle\langle b_j'| \right).$$

Expanding this product:

$$UU^\dagger = \sum_{i=1}^{n}\sum_{j=1}^{n} |b_i'\rangle \langle b_i|b_j\rangle \langle b_j'| = \sum_{i=1}^{n} |b_i'\rangle \langle b_i'| = I,$$

where we used the orthonormality condition $\langle b_i|b_j\rangle = \delta_{ij}$ and the completeness relation for the basis $\{|b_i'\rangle\}$.

Similarly, we can show that $U^\dagger U = I$. Thus, U is unitary.

Basis Transformation

Applying U to the vector $|b_i\rangle$ gives:

$$U\,|b_i\rangle = \left(\sum_{j=1}^{n} |b'_j\rangle\langle b_j|\right)|b_i\rangle$$

$$= \sum_{j=1}^{n} |b'_j\rangle\,\langle b_j|b_i\rangle\,.$$

Using the orthonormality condition $\langle b_j|b_i\rangle = \delta_{ji}$, this reduces to $U\,|b_i\rangle = |b'_i\rangle$.

Therefore, we have demonstrated that U is a unitary operator and that it correctly transforms the basis vectors $|b_i\rangle$ to $|b'_i\rangle$, completing the proof.   □

> ⓘ If we introduce a phase factor, $e^{i\theta_j}$ where θ_j is a real angle, to each basis vector, the basis remains orthonormal. Thus, we can also define the basis transformation operator as $U = \sum_{j=1}^{n} e^{i\theta_j}\,|b'_j\rangle\langle b_j|$, which was investigated in the context of unitary operators in Theorem 8.13.

3   The Transition Matrix

Recall that from § 8.4.1, the transition matrix converts the coordinates of a vector from one basis to another. It is the inverse ($U^{-1} = U^\dagger$) of the basis transformation matrix U. This concept is formalized by the following theorem:

Theorem 8.22 Consider a vector $|\psi\rangle$ represented in two different bases:

$$|\psi\rangle = \sum_j c_j\,|b_j\rangle = \sum_j c'_j\,|b'_j\rangle\,. \tag{8.34}$$

(a) The coordinates in each basis are related by the following equations:

$$c_i = \sum_j c'_j\,\langle b_i|b'_j\rangle\,,\quad c'_i = \sum_j c_j\,\langle b'_i|b_j\rangle\,. \tag{8.35}$$

(b) Let $|c\rangle$ be the coordinate vector formed by (c_i), and let $|c'\rangle$ be the coordinate vector formed by (c'_i). These vectors represent $|\psi\rangle$ in the two bases. The relationship between these coordinate vectors is given by:

$$|c'\rangle = U^\dagger\,|c\rangle\,, \tag{8.36}$$

where U is the basis transformation operator defined by Eq. 8.32.

> ⓘ Note that U is the basis transformation operator that maps the original basis vectors to the new basis, as in $|b'_i\rangle = U\,|b_i\rangle$. In contrast, $U^\dagger$ is used to transform the coordinate vectors, as in $|c'\rangle = U^\dagger\,|c\rangle$. The transformation by U changes the basis itself, while the transformation by $U^\dagger$ changes the representation of a vector in the new basis.

Proof. For part (a), given $|\psi\rangle = \sum_j c'_j |b'_j\rangle$, we have:

$$c_i = \langle b_i|\psi\rangle = \sum_j c'_j \langle b_i|b'_j\rangle.$$

Similarly, given $|\psi\rangle = \sum_j c_j |b_j\rangle$, we get:

$$c'_i = \langle b'_i|\psi\rangle = \sum_j c_j \langle b'_i|b_j\rangle.$$

For part (b), substituting $|b'_j\rangle = U |b_j\rangle$ into the above expression for c_i, we obtain:

$$c_i = \sum_j c'_j \langle b_i|U|b_j\rangle.$$

Note that $\langle b_i|U|b_j\rangle$ represents the (i, j) element of U, denoted by U_{ij} (see Eq. 8.39). Therefore,

$$c_i = \sum_j U_{ij}c'_j,$$

which is equivalent to $|c\rangle = U |c'\rangle$ in matrix form. The inverse of this relation leads to $|c'\rangle = U^\dagger |c\rangle$.   □

(i) | **An Advantage of Dirac Notation:** As illustrated in Theorem 8.22, a key advantage of Dirac notation is its ability to distinguish between abstract vectors, such as $|\psi\rangle$, and their specific representations in different bases, like $|c\rangle$ and $|c'\rangle$.

By working with abstract vectors, Dirac notation simplifies the calculation of vector coordinates under a change of basis, treating it as an ordinary algebraic operation. This feature is particularly useful, as demonstrated in the following example.

■ **Example 8.8** Compute the basis transformation operator from the computational basis $\{|0\rangle, |1\rangle\}$ to the Hadamard basis $\{|+\rangle, |-\rangle\}$ (illustrated in Fig. 8.12b) and determine the transformed coordinates of $|+_i\rangle = \frac{1}{\sqrt{2}}(|0\rangle + i|1\rangle)$.

Computing the Transformed Coordinates via $U^\dagger$

According to Theorem 8.21, the basis transformation operator is:

$$U = |+\rangle\langle 0| + |-\rangle\langle 1|,$$

which maps $|0\rangle \mapsto |+\rangle$ and $|1\rangle \mapsto |-\rangle$. The inverse transformation is given by

$$U^\dagger = |0\rangle\langle +| + |1\rangle\langle -|,$$

Applying $U^\dagger$ to $|+_i\rangle$ yields its transformed coordinates in the new basis:

$$U^\dagger |+_i\rangle = \frac{1}{\sqrt{2}} (|0\rangle\langle +| + |1\rangle\langle -|) (|0\rangle + i|1\rangle)$$

$$= \frac{1}{\sqrt{2}} (|0\rangle \langle +|0\rangle + |1\rangle \langle -|0\rangle + i|0\rangle \langle +|1\rangle + i|1\rangle \langle -|1\rangle)$$

$$= \frac{1+i}{2} |0\rangle + \frac{1-i}{2} |1\rangle.$$

Thus, the coordinates of $|+_i\rangle$ in the new basis are $(\frac{1+i}{2}, \frac{1-i}{2})$. Here, we have used the relationships $\langle +|0\rangle = \langle +|1\rangle = \langle -|0\rangle = \frac{1}{\sqrt{2}}$ and $\langle -|1\rangle = -\frac{1}{\sqrt{2}}$.

Obtaining the Transformed Coordinates Directly

The Dirac notation allows us to treat $|+_i\rangle$ as an abstract vector and calculate these coordinates algebraically without going through $U^\dagger$. To achieve this, we substitute $|0\rangle = \frac{1}{\sqrt{2}}(|+\rangle + |-\rangle)$ and $|1\rangle = \frac{1}{\sqrt{2}}(|+\rangle - |-\rangle)$, and obtain:

$$|-_i\rangle = \frac{1}{\sqrt{2}}(|0\rangle + i\,|1\rangle)$$

$$= \frac{1}{\sqrt{2}}\left(\frac{|+\rangle + |-\rangle}{\sqrt{2}} + i\frac{|+\rangle - |-\rangle}{\sqrt{2}}\right)$$

$$= \frac{1+i}{2}|+\rangle + \frac{1-i}{2}|-\rangle.$$

This confirms that the coordinates of $|+_i\rangle$ in the $\{|+\rangle, |-\rangle\}$ basis are $(\frac{1+i}{2}, \frac{1-i}{2})$. ∎

Exercise 8.25 Derive the basis transformation operator from the Hadamard basis $\{|+\rangle, |-\rangle\}$ back to the computational basis $\{|0\rangle, |1\rangle\}$, and determine the coordinates of $|\psi\rangle = \frac{1}{\sqrt{2}}(|+\rangle - i\,|-\rangle)$ in the computational basis.

4   Constructing the Basis Transformation Matrix

Next, we explore how U can be constructed.

Theorem 8.23 The basis transformation matrix U defined by Eq. 8.32 can be expressed as:

$$U = \sum_{i,j} \langle b_i | b'_j \rangle\, |b_i\rangle\langle b_j|. \tag{8.37}$$

If $\{|b_i\rangle\}$ is the standard (computational) basis, then $|b_i\rangle\langle b_j|$ represents the (i,j) element of U, denoted as U_{ij}. Thus, Eq. 8.37 implies that $U_{ij} = \langle b_i | b'_j \rangle$, leading to:

$$U = \begin{bmatrix} \langle b_1 | b'_1 \rangle & \langle b_1 | b'_2 \rangle & \cdots \\ \langle b_2 | b'_1 \rangle & \langle b_2 | b'_2 \rangle & \cdots \\ \vdots & & \ddots \end{bmatrix}. \tag{8.38}$$

More generally, $\langle b_i | b'_j \rangle$ is the matrix element U_{ij} in the basis $\{|b_i\rangle\}$:

$$U_{ij} = \langle b_i | U | b_j \rangle = \langle b_i | b'_j \rangle. \tag{8.39}$$

Exercise 8.26 Before proceeding with a proof, verify that Eqs. 8.31 and 8.38 yield $U = R_\theta = \begin{bmatrix} \cos\theta & -\sin\theta \\ \sin\theta & \cos\theta \end{bmatrix}$, as expected from Eq. 8.26.

Proof.   According to Theorem 8.21, U is given by:

$$U = \sum_j |b'_j\rangle\langle b_j|.$$

By the completeness identity for the basis $\{|b_i\rangle\}$:

$$I = \sum_i |b_i\rangle\langle b_i|.$$

Therefore,

$$U = \sum_j I \, |b'_j\rangle\langle b_j|$$

$$= \sum_j \sum_i |b_i\rangle\langle b_i| \, |b'_j\rangle\langle b_j|$$

$$= \sum_{i,j} |b_i\rangle \, (\langle b_i| \, |b'_j\rangle) \, \langle b_j|$$

$$= \sum_{i,j} \langle b_i|b'_j\rangle \, |b_i\rangle \, \langle b_j| \, .$$

In the last step, we can move $\langle b_i|b'_j\rangle$ to the front because it is a scalar. □

Exercise 8.27 Prove that Eq. 8.37 can also be written as

$$U = \sum_{i,j} \langle b_i|b'_j\rangle \, |b'_i\rangle\langle b'_j| \, . \tag{8.40}$$

Note that the coefficient $\langle b_i|b'_j\rangle$ is the same in both Eqs. 8.37 and 8.40.

This result shows $\langle b_i|b'_j\rangle$ is also the matrix element U_{ij} in the basis $\{|b'_i\rangle\}$.

$$U_{ij} = \langle b'_i|U|b'_j\rangle = \langle b_i|b'_j\rangle \, .$$

Equivalently, Eq. 8.38 can be expressed as follows:

Theorem 8.24 — Basis Transformation Matrix From Standard Basis. If U is the basis transformation matrix from the standard basis $S = \{|0\rangle, |1\rangle, \ldots, |n-1\rangle\}$ to an orthonormal basis $S' = \{|\phi_1\rangle, |\phi_2\rangle, \ldots, |\phi_n\rangle\}$ in $\mathbb{C}^n$, then U can be constructed by using the basis vectors of S' as its column vectors:

$$U = \begin{bmatrix} |\phi_1\rangle & |\phi_2\rangle & \cdots & |\phi_n\rangle \end{bmatrix} . \tag{8.41}$$

Proof. According to Eq. 8.38, $U^\dagger$ is given by:

$$\begin{bmatrix} \langle b_1|b'_1\rangle & \langle b_1|b'_2\rangle & \cdots \\ \langle b_2|b'_1\rangle & \langle b_2|b'_2\rangle & \cdots \\ \vdots & & \ddots \end{bmatrix} = \begin{bmatrix} \langle 0|\phi_1\rangle & \langle 0|\phi_2\rangle & \cdots \\ \langle 1|\phi_1\rangle & \langle 1|\phi_2\rangle & \\ \vdots & & \ddots \end{bmatrix} . \tag{8.42}$$

Apparently, the i-the column of U is given by the components of $|\phi_i\rangle$. □

Exercise 8.28 Calculate the transition matrix from the standard basis to S', where

$$S' = \left\{ \left(\frac{1}{2}, \frac{\sqrt{3}}{2} \right), \left(-\frac{\sqrt{3}}{2}, \frac{1}{2} \right) \right\} .$$

Then, determine the coordinates of $\left(\frac{1}{\sqrt{2}}, \frac{1}{\sqrt{2}} \right)$ using this matrix.

Theorem 8.24 provides a method to construct transition matrices involving the standard basis. For transitions between two arbitrary bases, the process can be simplified by using the standard basis as an intermediary:

$$\text{Basis } S_1 \quad \Rightarrow \quad \text{Standard Basis} \quad \Rightarrow \quad \text{Basis } S_2.$$

This approach leads to the following result:

> **Theorem 8.25 — Basis Transformation Matrix Between Arbitrary Bases.** Given two orthonormal bases S_1 and S_2 in $\mathbb{C}^n$, the basis transformation matrix from S_1 to S_2 is given by
>
> $$U = U_2 U_1^\dagger, \tag{8.43}$$
>
> where U_1 and U_2 are the basis transformation matrices from the standard basis to S_1 and S_2, respectively, which can be constructed using Theorem 8.24.

Exercise 8.29 Prove Theorem 8.25.

Exercise 8.30 Compute the basis transformation matrix from
$\left\{ \left(\frac{1}{\sqrt{2}}, \frac{1}{\sqrt{2}} \right), \left(\frac{1}{\sqrt{2}}, -\frac{1}{\sqrt{2}} \right) \right\}$ to $\left\{ \left(\frac{1}{2}, \frac{\sqrt{3}}{2} \right), \left(-\frac{\sqrt{3}}{2}, \frac{1}{2} \right) \right\}$.

i) Theorems 8.24 and 8.25 are applicable even when S_1 or S_2 are not orthonormal bases. In such cases, Hermitian adjoint should be replaced by inverse, e.g., $U^\dagger$ by U^{-1}. This distinction is necessary because, for non-orthonormal bases, the transformation matrices are no longer guaranteed to be unitary, and the inverse must be used to properly account for the basis transformation.

Exercise 8.31 Find the basis transformation matrix for changing a vector given in the standard basis with basis vectors

$$|u_1\rangle = [1\ 0\ 0\ 0]^T, \quad |u_2\rangle = [0\ 1\ 0\ 0]^T, \quad |u_3\rangle = [0\ 0\ 1\ 0]^T, \quad |u_4\rangle = [0\ 0\ 0\ 1]^T$$

to a vector given in the v basis with basis vectors

$$|v_1\rangle = \frac{1}{2}[1\ 1\ 1\ 1]^T, \quad |v_2\rangle = \frac{1}{\sqrt{2}}[0\ 1\ 0\ 1]^T, \quad |v_3\rangle = [0\ 0\ 0\ 1]^T, \quad |v_4\rangle = [0\ 0\ 1\ 0]^T.$$

8.4.3   Matrices Under Change of Basis

1   Matrix Representation Under a Change of Basis

The transformation properties of a matrix must remain consistent under any change of basis. This requirement leads to the following definition.

> **Definition 8.7 — Matrix Representation Under a Change of Basis.** Let $A \in \mathbb{C}^{n \times n}$ be a matrix representing a linear operator relative to an orthonormal basis S. Let U be the basis transformation operator from S to another orthonormal basis S'. Let a vector in $\mathbb{C}^n$ be represented by $|c\rangle$ in S and by $|c'\rangle$ in S'.
>
> Then the matrix representation of A relative to S', denoted by A', satisfies the following relationship, consistent with Theorem 8.22:
>
> $$A' |c'\rangle = U^\dagger A |c\rangle. \tag{8.44}$$

Equation 8.44 implies that for any vector, applying the transformation A in basis S and applying A' in basis S' yield equivalent results. This equivalence leads to a practical formula for computing A':

Theorem 8.26 Let U be the basis transformation operator from an orthonormal basis S to another orthonormal basis S'. If A is a matrix representing a linear transformation relative to S, then its matrix representation A' relative to S' is given by:

$$A' = U^\dagger A U. \tag{8.45}$$

Proof. From Eq. 8.44, and using $|c'\rangle = U^\dagger |c\rangle$, we have:

$$A'U^\dagger |c\rangle = U^\dagger A |c\rangle \quad \text{for all } |c\rangle \in \mathbb{C}^n.$$

This implies that $A'U^\dagger = U^\dagger A$.

Multiplying both sides by U, and using the fact that $U^\dagger U = UU^\dagger = I$ (since U is unitary), we obtain:

$$A' = U^\dagger A U.$$

$\square$

Equation 8.45 describes a relationship known as **matrix similarity**, specifically unitary similarity in this context. Matrices A and A' are said to be unitarily similar, meaning they represent the same transformation under different bases.

■ **Example 8.9** Given $A = \dfrac{1}{2}\begin{bmatrix} 1 & 1 \\ 1 & 1 \end{bmatrix}$ in the computational basis, compute A' under the basis $S = \{|+\rangle, |-\rangle\}$.

Solution. As shown in Example 8.8, the transformation matrix from the computational basis to S is given by:

$$U = |+\rangle\langle 0| + |-\rangle\langle 1|$$

$$= \frac{1}{\sqrt{2}} \left(|0\rangle + |1\rangle \right) \langle 0| + \frac{1}{\sqrt{2}} \left(|0\rangle - |1\rangle \right) \langle 1|$$

$$= \frac{1}{\sqrt{2}} \left(|0\rangle\langle 0| + |1\rangle\langle 0| + |0\rangle\langle 1| - |1\rangle\langle 1| \right).$$

In matrix form, U corresponds to the Hadamard matrix:

$$U = H = \frac{1}{\sqrt{2}} \begin{bmatrix} 1 & 1 \\ 1 & -1 \end{bmatrix}. \tag{8.46}$$

Using Eq. 8.45, the transformed matrix A' in the new basis is calculated as:

$$A' = H^\dagger A H = \frac{1}{4} \begin{bmatrix} 1 & 1 \\ 1 & -1 \end{bmatrix} \begin{bmatrix} 1 & 1 \\ 1 & 1 \end{bmatrix} \begin{bmatrix} 1 & 1 \\ 1 & -1 \end{bmatrix} = \begin{bmatrix} 1 & 0 \\ 0 & 0 \end{bmatrix}.$$

■

Exercise 8.32 Prove that the identity matrix I retains the same matrix representation under any unitary transformation.

8.4.4 Basis-independent Quantities and Relationships

Bases can be viewed as different perspectives for analyzing the same system. While selecting an appropriate basis is often crucial for simplifying specific problems, the fundamental characteristics of a system must remain invariant under any change of basis. Mathematically, this means that certain scalar quantities and relationships are preserved, irrespective of the basis chosen.

We begin by examining the sesquilinear form $\langle u|A|v \rangle$, which generalizes the inner product. Here, $|v\rangle$, $|v'\rangle$, A, and similar terms represent vectors and matrices in different bases, analogous to $|c\rangle$ in Theorem 8.22.

Theorem 8.27 For any vectors $|u\rangle$, $|v\rangle$ in an orthonormal basis S and their corresponding vectors $|u'\rangle$, $|v'\rangle$ in another orthonormal basis S', if A is an operator relative to S and A' corresponds to A relative to S', then the following holds:

$$\langle u|A|v \rangle = \langle u'|A'|v' \rangle. \tag{8.47}$$

Proof. Using $|u'\rangle = U^\dagger |u\rangle$, $|v'\rangle = U^\dagger |v\rangle$, and $A' = U^\dagger AU$, we have:

$$\langle u'|A'|v' \rangle = \langle U^\dagger u|U^\dagger AU|U^\dagger v \rangle = \langle u|UU^\dagger AUU^\dagger|v \rangle = \langle u|A|v \rangle,$$

where we used the unitarity condition $UU^\dagger = I$.   □

When $|u\rangle = |v\rangle$, the sesquilinear form reduces to a quadratic form $\langle v|A|v \rangle$, which is similarly invariant under a change of basis:

$$\langle u|A|u \rangle = \langle u'|A'|u' \rangle. \tag{8.48}$$

Setting $A = I$ yields the invariance of the inner product:

$$\langle u|v \rangle = \langle u'|v' \rangle. \tag{8.49}$$

Furthermore, setting $|u\rangle = |v\rangle$ yields the invariance of the vector norm:

$$\|v\| = \|v'\|. \tag{8.50}$$

■ **Example 8.10** Demonstrate Eq. 8.48 using A and $|+_i\rangle$ given in Examples 8.8 and 8.9.

Solution. As shown in Example 8.9, the representations of the matrix A in the computational basis and in the S basis are:

$$A = \frac{1}{2}\begin{bmatrix} 1 & 1 \\ 1 & 1 \end{bmatrix}, \quad A' = H^\dagger AH = \begin{bmatrix} 1 & 0 \\ 0 & 0 \end{bmatrix}.$$

Given the vector:

$$|+_i\rangle = \frac{1}{\sqrt{2}}(|0\rangle + i|1\rangle) = \frac{1}{\sqrt{2}}\begin{bmatrix} 1 \\ i \end{bmatrix},$$

its representation in the S basis is (see Example 8.8):

$$|+_i'\rangle = H^\dagger |+_i\rangle = \frac{1}{2}\begin{bmatrix} 1+i \\ 1-i \end{bmatrix}.$$

At this point, it is straightforward to verify through matrix multiplication that:

$$\langle +_i|A|+_i \rangle = \langle +_i'|A'|+_i' \rangle = \frac{1}{2}.$$

■

Exercise 8.33 Prove that the inner product $\langle u|v \rangle$ is basis invariant directly using Theorem 8.22 directly.

Theorem 8.28 summarizes several important quantities and relationships that are invariant under a basis change. In quantum computing, these invariances also extend to advanced metrics like measurement fidelity and quantum entropy.

Theorem 8.28 For any vectors $|u\rangle, |v\rangle \in \mathbb{C}^n$, and matrices $A, B \in \mathbb{C}^{n \times n}$, along with their respective transformed representations $|u'\rangle, |v'\rangle, A', B'$ under any unitary transformation U, the following relationships hold:

- Norm: $\|v\| = \|v'\|$.

- Inner Product: $\langle u|v \rangle = \langle u'|v' \rangle$.

- Sesquilinear Form: $\langle u|A|v \rangle = \langle u'|A'|v' \rangle$.

- Trace: $\operatorname{tr} A = \operatorname{tr} A'$.

- Determinant: $\det A = \det A'$.

- Inner Product of Matrices: $\operatorname{tr}(A^\dagger B) = \operatorname{tr}(A'^\dagger B')$.

- Unitarity: If $AA^\dagger = I$, then $A'A'^\dagger = I$.

- Hermiticity: If $A = A^\dagger$, then $A' = A'^\dagger$.

- Commutativity: If $[A, B] = 0$, then $[A', B'] = 0$.

Exercise 8.34 Prove the last four items in Theorem 8.28.

8.5 Structure and Solutions of Linear Systems

Solving a system of linear equations, commonly known as a linear system, is a foundational problem in applied mathematics. Typical linear algebra textbooks introduce linear systems early on, often alongside matrix algebra. We have deferred this introduction until now, allowing us to build on the foundational concepts established so far in order to present deeper insights into solution spaces, fundamental spaces of matrices, and the distinction between underdetermined and overdetermined systems.

While the introduction is delayed, the importance of solving linear systems remains central, both in classical and quantum contexts. For example, in quantum computing, the Harrow-Hassidim-Lloyd (HHL) algorithm is specifically designed to solve linear systems using quantum techniques, highlighting the relevance of these concepts in advanced applications.

This section introduces linear systems in a way that leverages our previous discussions on vector spaces and matrix operators. We will cover essential properties, such as rank, nullity, and the structure of solution spaces. Moreover, we will analyze underdetermined and overdetermined systems to understand how their solutions can vary from unique, to infinite, or none.

Throughout, our emphasis will be on general matrices, which may not necessarily be square, and we will use standard vector representations instead of Dirac notation to underscore our focus on general linear algebra concepts, rather than quantum-specific formalism.

8.5.1 Basics of Linear Systems

1 Definition

A linear system consists of a set of linear equations concerning unknown variables. It can be efficiently expressed using matrix multiplication, as defined below.

> **Definition 8.8 — Linear System.** A linear system with m equations and n unknowns is generally represented as:
>
> $$\begin{cases} a_{11}x_1 + a_{12}x_2 + \cdots + a_{1n}x_n = b_1 \\ a_{21}x_1 + a_{22}x_2 + \cdots + a_{2n}x_n = b_2 \\ \quad \vdots \qquad \quad \vdots \qquad \qquad \vdots \qquad \quad \vdots \\ a_{m1}x_1 + a_{m2}x_2 + \cdots + a_{mn}x_n = b_m \end{cases}, \tag{8.51}$$
>
> where a_{ij} and b_i for $i = 1, 2, \ldots, m$ and $j = 1, 2, \ldots, n$ are known scalars, and x_j are the unknown variables. This system can be expressed in matrix form as:
>
> $$\begin{bmatrix} a_{11} & a_{12} & \cdots & a_{1n} \\ a_{21} & a_{22} & \cdots & a_{2n} \\ \vdots & \vdots & \ddots & \vdots \\ a_{m1} & a_{m2} & \cdots & a_{mn} \end{bmatrix} \begin{bmatrix} x_1 \\ x_2 \\ \vdots \\ x_n \end{bmatrix} = \begin{bmatrix} b_1 \\ b_2 \\ \vdots \\ b_m \end{bmatrix}, \tag{8.52}$$
>
> or more succinctly as
>
> $$Ax = b, \tag{8.53}$$
>
> where $A = [a_{ij}]$ is termed the **coefficient matrix**, $x = \begin{bmatrix} x_1 & x_2 & \cdots & x_n \end{bmatrix}^T$ is the vector of unknowns, and $b = \begin{bmatrix} b_1 & b_2 & \cdots & b_m \end{bmatrix}^T$ is the vector of constants.

Just like vectors and matrices, a linear system can either be real or complex, depending on whether the matrix A and vector b are composed of real or complex numbers. Correspondingly, the solutions to these systems are real for real linear systems and complex for complex systems.

A linear system is considered **consistent** if it has at least one solution; otherwise, it is deemed **inconsistent**.

2 Number of Solutions

A fundamental aspect of linear systems is determining the number of possible solutions. The following theorem categorizes all possible scenarios:

> **Theorem 8.29** Any linear system may have no solution, exactly one solution, or infinitely many solutions.

Proof. Assume that a linear system has two distinct solutions, x_1 and x_2. Given $Ax_1 = b$ and $Ax_2 = b$, with $x_1 \neq x_2$, it follows that:

$$A(x_1 - x_2) = Ax_1 - Ax_2 = b - b = 0.$$

Defining $u = x_1 - x_2$ (a nonzero vector), we have $Au = 0$. Constructing a vector $v = x_1 + ku$, where k is any scalar, we find:

$$Av = A(x_1 + ku) = Ax_1 + kAu = b + k0 = b.$$

This shows that if a linear system has two distinct solutions, it must have infinitely many solutions. Hence, the only possible scenarios for the number of solutions are none, one, or infinitely many. □

To gain an intuitive understanding of Theorem 8.29, consider linear systems in $\mathbb{R}^2$ or $\mathbb{R}^3$ (real systems with two or three unknowns). Visualizing these systems can be particularly enlightening:

$$\begin{cases} a_1 x + b_1 y = c_1 \\ a_2 x + b_2 y = c_2 \end{cases}. \tag{8.54}$$

Each equation represents a line in a two-dimensional plane, and the solutions are the points (x, y) where these lines intersect. There are three possible outcomes, as depicted in Fig. 8.13: (a) the lines are parallel and never intersect, indicating no solution; (b) the lines intersect at a single point, yielding a unique solution; (c) the lines coincide, resulting in infinitely many solutions.

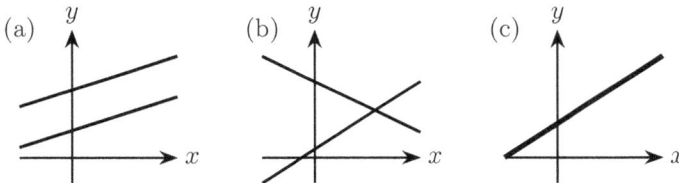

Figure 8.13: Geometric Illustration of Linear Systems in 2D

Exercise 8.35 Geometrically demonstrate how a linear system with three equations and two unknowns could have no, one, or infinitely many solutions.

For a linear system with three unknowns, each equation represents a *plane* in three-dimensional space, and the solution is defined as the set of points where all these planes intersect. Consider a system defined by three equations:

$$\begin{cases} a_1 x + b_1 y + c_1 z = d_1 \\ a_2 x + b_2 y + c_2 z = d_2 \\ a_3 x + b_3 y + c_3 z = d_3 \end{cases}. \tag{8.55}$$

Figure 8.14 illustrates scenarios where there are no solutions, exactly one solution, or infinitely many solutions. However, the complexity of three-dimensional geometry can produce additional configurations. Readers are encouraged to explore all potential configurations of three planes.

Exercise 8.36 Sketch a configuration of three planes that are not parallel yet do not share a common intersection point.

ⓘ When a linear system has infinitely many solutions, these solutions can form various geometric structures such as a line, a plane, or even a higher-dimensional

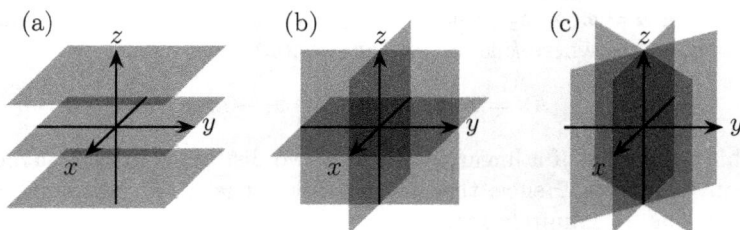

Figure 8.14: Geometric Illustration of Linear Systems in 3D

hyperplane, depending on the number of free variables and the relationships between the equations.

■ **Example 8.11** Consider the following linear system with three equations and three unknowns:

$$\begin{cases} x + y + z & = 1, \\ 2x + 2y + 2z & = 2, \\ 3x + 3y + 3z & = 3. \end{cases}$$

All three equations are scalar multiples of $x + y + z = 1$, making them linearly dependent. As a result, any (x, y, z) satisfying the equation $x + y + z = 1$ is a solution to this linear system.

Therefore, the system has infinitely many solutions, which form a two-dimensional plane in $\mathbb{R}^3$.   ■

3   General Solution to a Linear System

When a linear system admits infinitely many solutions, it is possible to express these solutions through a general parametric formula. This formula is termed the **general solution** of the linear system.

■ **Example 8.12** Find the general solution for the linear system:

$$\begin{cases} x + 2y - 3z = 1 \\ 2x - y + z = 2 \end{cases}.$$

Solution. To eliminate y, we multiply the second equation by 2 and add it to the first, yielding:

$$5x - z = 5.$$

With two unknowns and one equation, we introduce a free parameter t, setting $x = t$. This substitution gives $z = 5t - 5$. Substituting $x = t$ and $z = 5t - 5$ back into the second equation, we derive $y = 7t - 7$. Thus, the general solution is:

$$\begin{cases} x = t \\ y = 7t - 7 \\ z = 5t - 5 \end{cases}.$$

This represents the parametric equation of a line in three-dimensional space.   ■

The example above illustrates that any value of t generates a specific solution for the system. For example, setting $t = 1$ produces the solution $(1, 0, 0)$, which can be verified directly. Rewriting the general solution in vector form can often be insightful:

$$\begin{bmatrix} x \\ y \\ z \end{bmatrix} = t \begin{bmatrix} 1 \\ 7 \\ 5 \end{bmatrix} + \begin{bmatrix} 0 \\ -7 \\ -5 \end{bmatrix}.$$

We will explore why this vector form is particularly useful later in this section.

Exercise 8.37 Determine the general solution for the following linear system:

$$\begin{cases} x + y - z + w = 2 \\ x - 2y + z - w = 1 \\ 3x - 3y + z - w = 4 \end{cases}.$$

Analyze its geometric representation and discuss the nature of the solution set.

8.5.2 Homogeneous Linear Systems

Homogeneous linear systems form a fundamental class in linear algebra due to their simplicity and the insights they provide about the structure of solutions.

Definition 8.9 — Homogeneous Linear System. A homogeneous linear system is characterized by having all zeros on the right-hand side of the equation, expressed as: $A\boldsymbol{x} = \boldsymbol{0}$.

For every homogeneous linear system, the trivial solution $\boldsymbol{x} = \boldsymbol{0}$ always exists, meaning homogeneous systems are always consistent and have at least one solution.

An important feature of homogeneous systems is captured in the following theorem:

Theorem 8.30 The set of all solutions to a homogeneous linear system $A_{m \times n}\boldsymbol{x} = \boldsymbol{0}$ forms a subspace of $\mathbb{C}^n$ (or $\mathbb{R}^n$), known as the **solution space** of the system.

Proof.   Let $\boldsymbol{x}_1, \boldsymbol{x}_2 \in \mathbb{C}^n$ be two solutions to $A\boldsymbol{x} = \boldsymbol{0}$, and let $k \in \mathbb{C}$ be a scalar. Then,

$$A(\boldsymbol{x}_1 + \boldsymbol{x}_2) = A\boldsymbol{x}_1 + A\boldsymbol{x}_2 = \boldsymbol{0} + \boldsymbol{0} = \boldsymbol{0},$$
$$A(k\boldsymbol{x}_1) = k(A\boldsymbol{x}_1) = k\boldsymbol{0} = \boldsymbol{0}.$$

Thus, the solution set is closed under addition and scalar multiplication. By Theorem 5.3, this set is a subspace of $\mathbb{C}^n$.   □

While solutions of homogeneous systems always form a vector space, non-homogeneous systems characterized by $A\boldsymbol{x} = \boldsymbol{b}$ do not generally form a vector space. However, there is a relationship between the solutions of non-homogeneous and homogeneous systems, articulated in the following theorem:

Theorem 8.31 — General Solution to a Non-homogeneous Linear System. The general solution to a non-homogeneous linear system $A\boldsymbol{x} = \boldsymbol{b}$ consists of any particular solution to the system plus the general solution of the corresponding homogeneous system $A\boldsymbol{x} = \boldsymbol{0}$.

The proof of this theorem is left to the reader as an exercise (Problem 8.13). Here, we provide a geometric interpretation to gain further insight. Essentially, this theorem indicates that the solution set to $A\boldsymbol{x} = \boldsymbol{b}$ forms a line, plane, or hyperplane that is *parallel* to the solution space of the corresponding homogeneous system $A\boldsymbol{x} = \boldsymbol{0}$. Any particular solution to $A\boldsymbol{x} = \boldsymbol{b}$ can be thought of as a "shift" that defines the position of these parallel structures.

> ⓘ For a non-homogeneous system, the set of solutions does not form a vector space and therefore cannot be called a "solution space." Instead, such sets of solutions are formally known as "affine subspaces."

Consider the situation in $\mathbb{R}^2$ as illustrated in Fig. 8.15. If the solution space of the homogeneous system $A\boldsymbol{x} = \boldsymbol{0}$ is a line passing through the origin, then the general solution set to the non-homogeneous system $A\boldsymbol{x} = \boldsymbol{b}$ will be a line parallel to the original. Adding any solution $\boldsymbol{x}_h$ from the homogeneous system to a particular solution $\boldsymbol{x}_p$ of the non-homogeneous system yields another valid solution to $A\boldsymbol{x} = \boldsymbol{b}$.

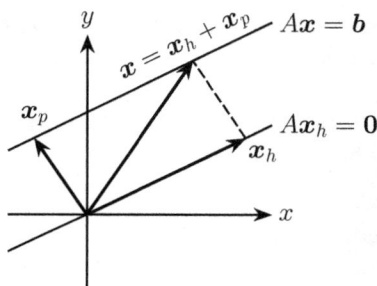

Figure 8.15: General Solution to a Non-homogeneous System

As shown in Example 8.12, the general solution for a given non-homogeneous linear system can be expressed as:

$$\begin{bmatrix} x \\ y \\ z \end{bmatrix} = t \begin{bmatrix} 1 \\ 7 \\ 5 \end{bmatrix} + \begin{bmatrix} 0 \\ -7 \\ -5 \end{bmatrix}.$$

Readers are encouraged to verify that the first term represents the general solution of the corresponding homogeneous system, while the second term provides a particular solution to the non-homogeneous system.

> **Exercise 8.38** Verify that the general solution derived in Exercise 8.37 follows the structure described in Theorem 8.31.

8.5.3   Fundamental Spaces and Orthogonal Complements

The fundamental spaces associated with a matrix play a central role in analyzing the behavior of linear systems and the linear transformations they represent. This subsection introduces these spaces and examines their interrelationships. Our focus will be on complex matrices, as their properties are particularly significant in the context of quantum computing.

Throughout this subsection, a vector may refer to an $n \times 1$ column vector when it is multiplied on the right by an $m \times n$ matrix A, or a $1 \times m$ row vector when it is multiplied on the left by A.

1 Column Space and Row Space of a Matrix

While the homogeneous linear system $Ax = 0$ is always consistent, the consistency of the non-homogeneous system $Ax = b$ depends on the matrix A and the vector b. To fully understand the behavior of these systems, it is instructive to consider matrix transformations, leading us to the concept of the fundamental spaces of a matrix.

Consider an $m \times n$ matrix A and an $n \times 1$ column vector x:

$$Ax = x_1 c_1 + x_2 c_2 + \cdots + x_n c_n, \tag{8.56}$$

where A is partitioned into its column vectors $A = \begin{bmatrix} c_1 & c_2 & \cdots & c_n \end{bmatrix}$ and $x = \begin{bmatrix} x_1 & x_2 & \cdots & x_n \end{bmatrix}^T$.

Equation (8.56) represents a linear combination of A's column vectors. Thus, $Ax = b$ is consistent if and only if b can be expressed as a linear combination of the column vectors $\{c_1, c_2, \ldots, c_n\}$:

$$x_1 c_1 + x_2 c_2 + \cdots + x_n c_n = b.$$

This expression clearly shows that $Ax = b$ is consistent if and only if b resides within the span of these column vectors, introducing the concept of the column space.

Definition 8.10 — Column Space. The column space of an $m \times n$ matrix A, denoted as $\mathrm{col}(A)$ or $C(A)$, is the span of A's column vectors:

$$\mathrm{col}(A) \equiv C(A) = \{Ax \mid x \in \mathbb{C}^n\}. \tag{8.57}$$

$\mathrm{col}(A)$ is a subspace of $\mathbb{C}^m$.

The column space is synonymous with the **range** of A, denoted as $\mathrm{range}(A)$, as it represents all possible outputs $T_A(x) = Ax$ for input vectors x.

Theorem 8.32 A linear system $Ax = b$ is consistent if and only if b is in the column space (range) of A.

The row space of A pertains to the linear combinations of the adjoints of A's row vectors. The adjoint is necessary here because the row space is defined as a subspace of column vectors.

Definition 8.11 — Row Space. The row space of an $m \times n$ matrix A, denoted as $\mathrm{row}(A)$ or $R(A)$, is the span of all its row vectors adjoined:

$$\mathrm{row}(A) \equiv R(A) = \{A^\dagger x \mid x \in \mathbb{C}^m\}. \tag{8.58}$$

$\mathrm{row}(A)$ is a subspace of $\mathbb{C}^n$.

The row space of A is equivalent to the column space of $A^\dagger$, i.e., $\mathrm{row}(A) = \mathrm{col}(A^\dagger)$. Note that in this convention, $\mathrm{row}(A) = \mathrm{col}(A^T)$ may not be true for complex matrices.

2   Null Space of a Matrix and Orthogonal Complements

A crucial aspect of understanding matrices involves the relationship between the row space of a matrix and the solutions to the homogeneous system $A\boldsymbol{x} = \boldsymbol{0}$, known as the null space of A.

> **Definition 8.12 — Null Space.** The null space of a matrix A, denoted as $\mathrm{null}(A)$ or $N(A)$, encompasses all solutions to the linear system $A\boldsymbol{x} = \boldsymbol{0}$.

Exploring the relationship between the $\mathrm{row}(A)$ and $\mathrm{null}(A)$ requires considering a matrix A partitioned into row vectors:

$$A = \begin{bmatrix} \boldsymbol{r}_1 \\ \boldsymbol{r}_2 \\ \vdots \\ \boldsymbol{r}_m \end{bmatrix},$$

where the product $A\boldsymbol{x}$ can be expressed as:

$$A\boldsymbol{x} = \begin{bmatrix} \boldsymbol{r}_1\boldsymbol{x} \\ \boldsymbol{r}_2\boldsymbol{x} \\ \vdots \\ \boldsymbol{r}_m\boldsymbol{x} \end{bmatrix}.$$

The condition $A\boldsymbol{x} = \boldsymbol{0}$ implies:

$$\boldsymbol{r}_1\boldsymbol{x} = \boldsymbol{r}_2\boldsymbol{x} = \cdots = \boldsymbol{r}_m\boldsymbol{x} = 0,$$

or equivalently, in terms of the complex dot product (defined between two column vectors):

$$\boldsymbol{r}_1^{\dagger} \cdot \boldsymbol{x} = \boldsymbol{r}_2^{\dagger} \cdot \boldsymbol{x} = \cdots = \boldsymbol{r}_m^{\dagger} \cdot \boldsymbol{x} = 0.$$

This demonstrates that any solution $\boldsymbol{x}$ must be orthogonal to the adjoint (complex conjugate) of every row vector of A. In other words, the null space of A consists of all vectors orthogonal to the adjoints of the row vectors of A. Thus, the row space of A and the null space of A are said to be orthogonal complements of each other:

$$\mathrm{null}(A) = \mathrm{row}(A)^{\perp}. \tag{8.59}$$

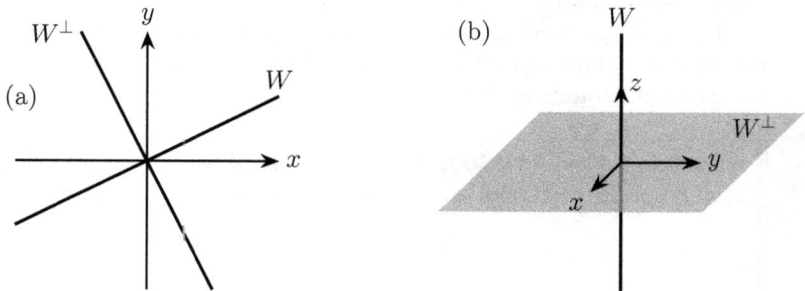

Figure 8.16: Orthogonal Complements in $\mathbb{R}^2$ and $\mathbb{R}^3$

> **Definition 8.13 — Orthogonal Complement.** The orthogonal complement of a subspace W within an inner product space V, denoted $W^\perp$, is the set of all vectors in V that are orthogonal to every vector in W.

Figure 8.16 illustrates orthogonal complements in $\mathbb{R}^2$ and $\mathbb{R}^3$. In $\mathbb{R}^2$, this typically manifests as two perpendicular lines through the origin. In $\mathbb{R}^3$, it often involves a plane and a line normal to that plane.

> **Exercise 8.39** Why are two perpendicular lines in $\mathbb{R}^3$ not considered orthogonal complements?

Orthogonal complements satisfy the following general properties, with their proofs left as exercises for the reader (Problem 8.14).

> **Theorem 8.33** Let W and $W^\perp$ be subspaces of $\mathbb{C}^n$. The following properties hold:
> (a) $W \cap W^\perp = \{\mathbf{0}\}$,
> (b) $\dim(W) + \dim(W^\perp) = n$,
> (c) If S is a basis for W and S' is a basis for $W^\perp$, then $S \cup S'$ forms a basis for $\mathbb{C}^n$.

For a complex matrix A, the orthogonal complement of $\text{null}(A)$ is the row space of A, or equivalently, the column space of $A^\dagger$, as illustrated in the following example.

■ **Example 8.13** Consider the 3×3 complex matrix

$$A = \begin{bmatrix} 1 & i & 1 \\ 0 & 1 & -i \\ i & 0 & 0 \end{bmatrix}.$$

Without showing details, it can be verified that:

$$\text{null}(A) = \text{span} \left\{ \begin{bmatrix} 0 \\ i \\ 1 \end{bmatrix} \right\}.$$

Meanwhile, the row space of A, equivalent to the column space of $A^\dagger$, is given by:

$$\text{row}(A) = \text{span} \left\{ \begin{bmatrix} 1 \\ -i \\ 1 \end{bmatrix}, \begin{bmatrix} 0 \\ 1 \\ i \end{bmatrix} \right\}.$$

The vector in $\text{null}(A)$ is orthogonal to every vector in $\text{row}(A)$. For example,

$$(0, i, 1) \cdot (1, -i, 1) = 0 \cdot 1 + (-i) \cdot (-i) + 1 \cdot 1 = -1 + 1 = 0,$$

and

$$(0, i, 1) \cdot (0, 1, i) = 0 \cdot 0 + (-i) \cdot 1 + 1 \cdot (i) = -i + i = 0.$$

Moreover,

$$\dim(\text{null}(A)) + \dim(\text{row}(A)) = 1 + 2 = 3 = \dim(\mathbb{C}^3).$$

Thus, $\text{null}(A)$ and $\text{row}(A)$ form orthogonal complements in $\mathbb{C}^3$. Combining their bases $\{(0, i, 1), (1, -i, 1), (0, 1, i)\}$ gives a complete basis of $\mathbb{C}^3$. ■

3 The Four Fundamental Spaces of Matrices

Every complex matrix A and its conjugate transpose are associated with four **fundamental spaces**, which are important from a theoretical perspective:

(a) the column space of A, also known as the **range** of A;

(b) the row space of A (equivalently the column space of $A^\dagger$);

(c) the null space of A, also known as the **kernel** of A;

(d) the null space of $A^\dagger$, also known as the **left null space** of A.

These spaces are organized into two pairs of orthogonal complements.

Theorem 8.34 For any matrix $A \in \mathbb{C}^{m \times n}$:

(a) row(A) and null(A), both subspaces of $\mathbb{C}^n$, are orthogonal complements of each other;

(b) col(A) and null$(A^\dagger)$, both subspaces of $\mathbb{C}^m$, are orthogonal complements of each other.

Exercise 8.40 Define the left null space of a matrix $A \in \mathbb{C}^{m \times n}$ as the set of vectors $\{y \mid y^\dagger A = 0, \ y \in \mathbb{C}^m\}$. Show that if $y^\dagger A = 0$, then $(Ax) \cdot y = 0$ for any $x \in \mathbb{C}^n$, implying that y is orthogonal to every vector in the column space of A.

8.5.4 Rank and Nullity of a Matrix

1 Rank

The concept of rank is foundational in linear algebra, particularly in relation to the dimensional properties of the column and row spaces of a matrix.

Theorem 8.35 For any matrix $A \in \mathbb{C}^{m \times n}$, the dimensions of the column space and the row space are equal: $\dim(\mathrm{col}(A)) = \dim(\mathrm{row}(A))$.

Definition 8.14 — Rank. The rank of a matrix A, denoted rank(A), is defined as the dimension of col(A) or equivalently row(A).

While the proof of Theorem 8.35 often involves Gaussian elimination methods beyond the scope of this text, interested readers are encouraged to explore this topic further through self-study (Problem 8.16).

Exercise 8.41 Prove that $\mathrm{rank}(A) = \mathrm{rank}(A^T) = \mathrm{rank}(A^*) = \mathrm{rank}(A^\dagger)$.

The rank of a matrix can be determined by analyzing the linear independence of its column or row vectors. Consider the following example:

■ **Example 8.14** Evaluate the ranks of the matrices below:

$$\text{(a)} \ \begin{bmatrix} 0 & 0 \\ 0 & 0 \end{bmatrix}, \quad \text{(b)} \ \begin{bmatrix} 1 & i \\ -i & 1 \end{bmatrix}, \quad \text{(c)} \ \begin{bmatrix} 1 & i \\ i & 1 \end{bmatrix}.$$

Solution.

(a) The zero matrix has a rank of zero, as its column space is the zero subspace with $\dim(\{0\}) = 0$.

(b) The two row vectors are linearly dependent since $(1, i) = i(-i, 1)$. Therefore, there is only one linearly independent row vector, and the rank is one.

(c) To determine the linear independence of $(1, i)$ and $(i, 1)$, compute the determinant:

$$\begin{vmatrix} 1 & i \\ i & 1 \end{vmatrix} = 1 - i^2 = 2 \neq 0.$$

This non-zero determinant indicates linear independence; hence, the rank is two.

∎

Exercise 8.42 Evaluate the rank of the following matrices:

$$\text{(a)} \begin{bmatrix} 1 & 2 & 3 \\ 2 & 4 & 6 \\ 0 & 0 & 0 \end{bmatrix}; \quad \text{(b)} \begin{bmatrix} 1 & 2 & 3 \\ 3 & 4 & 5 \\ 5 & 6 & 7 \end{bmatrix}.$$

Rank is a fundamental property of a matrix, indicating the number of linearly independent row or column vectors, and effectively measures the amount of unique information the matrix can encode. Consider the **projection matrix** below, which always has a rank of one:

$$P = |\psi\rangle \langle\psi|.$$

For instance, let $|\psi\rangle = \begin{bmatrix} 1 \\ i \end{bmatrix}$:

$$|\psi\rangle \langle\psi| = \begin{bmatrix} 1 \\ i \end{bmatrix} \begin{bmatrix} 1 & -i \end{bmatrix} = \begin{bmatrix} 1 & -i \\ i & 1 \end{bmatrix},$$

where the columns $c_1 = \begin{bmatrix} 1 & i \end{bmatrix}^T$ and $c_2 = \begin{bmatrix} -i & 1 \end{bmatrix}^T$ are clearly linearly dependent since $c_2 = -ic_1$. Thus, the rank of P is one, reflecting that $|\psi\rangle \langle\psi|$ projects any vector onto the span of $|\psi\rangle$, a one-dimensional subspace of $\mathbb{C}^n$.

Exercise 8.43 Prove that a matrix of the form $|\psi\rangle \langle\phi|$ is of rank one in $\mathbb{C}^{n \times n}$ for any non-zero $|\psi\rangle, |\phi\rangle \in \mathbb{C}^n$.

Theorem 8.36 For any matrix $A \in \mathbb{C}^{m \times n}$, the rank of A is capped by the following inequality:

$$\text{rank}(A) \leq \min(m, n).$$

This theorem follows from the fact that the dimension of the column space (span of column vectors) of an $m \times n$ matrix cannot exceed n, and similarly, the dimension of the row space (span of row vectors) cannot exceed m. Hence, the rank, which counts the maximum number of linearly independent column or row vectors, is bounded by the smaller of m and n.

Definition 8.15 — Full-Rank Matrix. A matrix A is described as a full-rank matrix if its rank achieves the maximal dimension possible for its size, specifically:

$$A_{m\times n} \text{ is full-rank if } \text{rank}(A) = \min(m, n).$$

For square matrices $(m = n)$, a full-rank matrix is equivalently an invertible matrix, a direct corollary of Theorem 7.17. This relationship is elucidated by the following theorem:

Theorem 8.37 A square matrix $A \in \mathbb{C}^{n\times n}$ is invertible if and only if it is full-rank, meaning:

$$\text{rank}(A) = n.$$

Theorem 8.37 indicates that if $\det(A) \neq 0$, then A is invertible and $\text{rank}(A) = n$. Conversely, if $\det(A) = 0$, then A is singular, with $0 \leq \text{rank}(A) \leq n - 1$. In such cases, the exact rank of A is linked to another property—the nullity of the matrix.

2 Nullity

The nullity of a matrix measures the dimension of its null space.

Definition 8.16 — Nullity. The nullity of a matrix A, denoted as $\text{nullity}(A)$, is defined as the dimension of the null space of A.

The null space and the row space of a matrix are orthogonal complements. This relationship leads to an important theorem in linear algebra known as the rank-nullity theorem.

Theorem 8.38 — Rank-Nullity Theorem. For any matrix $A \in \mathbb{C}^{m\times n}$, the sum of the rank and the nullity of A equals the number of columns in A, expressed as:

$$\text{rank}(A) + \text{nullity}(A) = n. \tag{8.60}$$

Exercise 8.44 Show that for complex matrices, $\text{rank}(A) + \text{nullity}(A^\dagger) = m$.

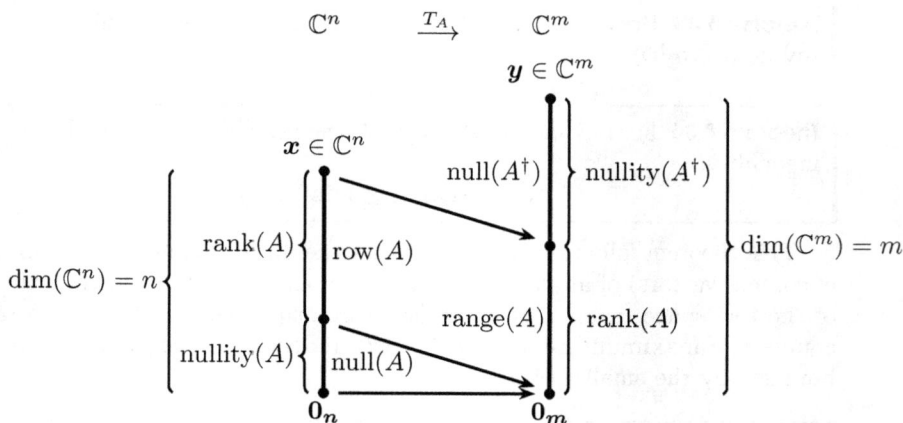

Figure 8.17: Conceptual Illustration of the Rank-Nullity Theorem

The rank-nullity theorem provides powerful insight into the structure of linear transformations. As depicted in Fig. 8.17, an $m \times n$ matrix A maps vectors from $\mathbb{C}^n$ to $\mathbb{C}^m$. The vectors in the null space of A are transformed to the zero vector, with their dimension given by nullity(A). Conversely, vectors not in the null space, which have non-zero projections onto the row space of A, are mapped by A onto the range of A in $\mathbb{C}^m$.

Upon transformation by A, the range does not necessarily span the entirety of $\mathbb{C}^m$. Instead, it covers a subspace whose dimension equals the rank of A. The complementary dimensions in $\mathbb{C}^m$ are spanned by the null space of $A^\dagger$, which forms the orthogonal complement to the column space of A.

Exercise 8.45 Is it true or false that the column space of a matrix $A_{m \times n}$ is equivalent to the set of vectors $\{x \mid A^\dagger x \neq 0, x \in \mathbb{C}^m\}$?

8.5.5 ✳ Underdetermined and Overdetermined Linear Systems

Basic algebra teaches us that typically, the number of equations should match the number of unknowns to completely determine a system—two equations for two unknowns, three for three, and so on. This principle extends to linear systems as well. For a linear system represented by $A_{m \times n} x = b$, where m is the number of equations and n is the number of unknowns, the system typically exhibits:

(a) a unique solution when $m = n$ (square system);

(b) no solution or infinitely many solutions when $m > n$ (overdetermined);

(c) infinitely many solutions when $m < n$ (underdetermined).

In practice, each case has significant implications in various fields, from engineering to economics, and understanding them is vital for effectively applying linear algebra.

1 Linear Systems with a Unique Solution

A linear system has a unique solution typically when the matrix A is square ($m = n$) and invertible. The solution can be directly computed if the inverse of A exists:

$$Ax = b \quad \text{and} \quad A \text{ is invertible} \quad \Rightarrow \quad x = A^{-1}b. \tag{8.61}$$

Such systems frequently arise in numerical simulations where physical phenomena, governed by ordinary or partial differential equations, are modeled. The matrix A often represents system parameters or interactions within networks, while the vector b typically encapsulates external forces or boundary conditions.

i | When a square matrix A is invertible, the homogeneous linear system $Ax = 0$ possesses only the unique solution $x = 0$, known as the **trivial solution**. Conversely, for $Ax = 0$ to admit non-trivial solutions, the matrix A must be singular (not invertible).

Exercise 8.46 Solve the linear system by finding the inverse of the coefficient matrix:
$$\begin{bmatrix} 1 & i \\ i & 1 \end{bmatrix} \begin{bmatrix} x \\ y \end{bmatrix} = \begin{bmatrix} 1 + i \\ 1 - i \end{bmatrix}.$$

2 Overdetermined Linear Systems

In cases where $m > n$, the system has more equations than unknowns, typically resulting in an overdetermined condition. Such systems often lack an exact solution due to excessive constraints. A common application of overdetermined systems is in data fitting, particularly when attempting to fit multiple data points to a simple linear model.

Consider m data points (x_i, y_i) for $i = 1, 2, \ldots, m$, and the objective is to fit these points to a linear model of the form $y = \beta_0 + \beta_1 x$. Here, β_0 and β_1 are the coefficients to be determined. Typically, with $m > 2$, this setup leads to an overdetermined linear system:

$$\begin{cases} \beta_0 + x_1\beta_1 & = y_1, \\ \beta_0 + x_2\beta_1 & = y_2, \\ \quad \cdots \\ \beta_0 + x_m\beta_1 & = y_m. \end{cases}$$

Expressed in matrix form, this system is:

$$\begin{bmatrix} 1 & x_1 \\ 1 & x_2 \\ \vdots & \vdots \\ 1 & x_m \end{bmatrix} \begin{bmatrix} \beta_0 \\ \beta_1 \end{bmatrix} = \begin{bmatrix} y_1 \\ y_2 \\ \vdots \\ y_m \end{bmatrix}.$$

Given the absence of an exact solution for such a system, the focus shifts to identifying the **best-approximation** solution, which is typically achieved by minimizing the norm of the error vector.

Definition 8.17 — Least Squares Solution. For any linear system $Ax = b$, the vector $b - Ax$ is referred to as the **error vector**. The solution x that minimizes the square of the norm of this error vector, $\|b - Ax\|^2$, is known as the least squares solution to the linear system, denoted as $\hat{x}$:

$$\hat{x} = \arg\min_x \|b - Ax\|^2. \qquad (8.62)$$

This approach is termed "least squares" because it minimizes the sum of the squares of the errors in predictions made on the dependent variable. The minimized error is called the **least squares error**.

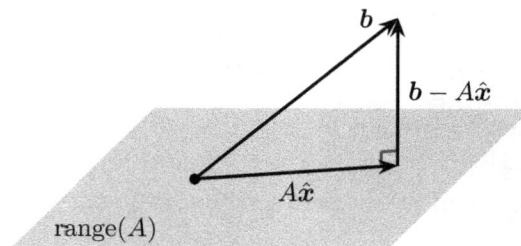

Figure 8.18: Conceptual Illustration of the Least Squares Solution

Determining $\hat{x}$ may initially seem challenging. However, leveraging our understanding of fundamental spaces reveals a systematic approach for deriving $\hat{x}$. If an exact solution x exists, it inherently serves as the least squares solution. If no exact solution exists, this implies that b lies outside the column space of A, as per Theorem 8.32. This scenario can be geometrically visualized as shown in Fig. 8.18.

From Fig. 8.18, it is evident that to minimize the error vector's norm, $A\hat{x}$ should represent the orthogonal projection of b onto the range of A. Consequently, the minimized error vector $b - A\hat{x}$ must be orthogonal to the range of A, placing it within its orthogonal complement:

(a) For a real matrix A, $\mathrm{col}(A)^\perp$ is the null space of A^T. To ensure $b - A\hat{x}$ resides in $\mathrm{null}(A^T)$, it must satisfy:

$$A^T(b - A\hat{x}) = 0 \quad \Rightarrow \quad A^T A\hat{x} = A^T b. \tag{8.63}$$

(b) For a complex matrix A, $\mathrm{col}(A)^\perp$ is the null space of $A^\dagger$. Thus, we derive:

$$A^\dagger(b - A\hat{x}) = 0 \quad \Rightarrow \quad A^\dagger A\hat{x} = A^\dagger b. \tag{8.64}$$

Equations 8.63 and 8.64 are known as the **normal equations** associated with the real or complex linear system $Ax = b$. It is well-established that these equations are consistently solvable. Moreover, $A^T A$ (or $A^\dagger A$) is almost always invertible for overdetermined systems in practice. This observation leads to the following formulation for $\hat{x}$:

Theorem 8.39 There exists a unique least squares solution to $Ax = b$:

$$\hat{x} = \begin{cases} (A^T A)^{-1} A^T b, & \text{if } A \text{ is real,} \\ (A^\dagger A)^{-1} A^\dagger b, & \text{if } A \text{ is complex.} \end{cases} \tag{8.65}$$

assuming $A^T A$ (or $A^\dagger A$) is invertible.

This theorem also yields the formula for the projection of b onto the column space of A, which is $A\hat{x}$:

Theorem 8.40 When $A^T A$ (or $A^\dagger A$) is invertible, the projection of b onto the column space of A is calculated as:

$$\mathrm{proj}_{\mathrm{col}(A)} b = A\hat{x} = \begin{cases} A(A^T A)^{-1} A^T b, & \text{for } A \text{ is real,} \\ A(A^\dagger A)^{-1} A^\dagger b, & \text{for } A \text{ is complex.} \end{cases} \tag{8.66}$$

Exercise 8.47 Write out the associated normal equation for the following inconsistent linear system with a single unknown x:

$$\begin{cases} x = x_1 \\ x = x_2 \\ x = x_3. \end{cases}$$

Calculate $\hat{x}$, the least squares solution to this system. What observations can you make from the result?

The foundational principles of least-squares problem-solving lay the groundwork for regression theory. These concepts will be revisited and expanded upon in § 11.1.5 and § 11.2.4, employing Dirac notation to enhance clarity and precision in theoretical developments.

3   Underdetermined Linear Systems

When $n > m$, the linear system becomes underdetermined. As illustrated in Example 8.12, such systems have free parameters in their general solutions, indicating more degrees of freedom than constraints. This scenario often arises in data modeling when the number of parameters exceeds the available data samples, a common occurrence in fields like machine learning with neural networks or gene expression analysis in bioinformatics.

To stabilize solutions in these contexts, regularization techniques such as Lasso or ridge regression are frequently employed. Below, we briefly introduce ridge regression, which adjusts solutions by incorporating a penalty proportional to the square of the magnitude of the coefficients.

For an underdetermined system where $Ax = b$ results in a typically singular $A^T A$ (or $A^\dagger A$), the standard least squares solution from Eq. 8.65 becomes inapplicable due to the existence of infinitely many solutions. Ridge regression, also known as **Tikhonov regularization**, optimizes a modified objective:

$$\hat{x} = \arg\min_{x} \|b - Ax\|^2 + \|\Gamma x\|^2, \tag{8.67}$$

where Γ is often chosen as $\sqrt{\lambda}I$, where λ is a regularization parameter. This approach is commonly referred to as L_2 regularization. The resultant formulation for $\hat{x}$ is:

$$\hat{x} = \begin{cases} (A^T A + \lambda I)^{-1} A^T b, & \text{if } A \text{ is real,} \\ (A^\dagger A + \lambda I)^{-1} A^\dagger b, & \text{if } A \text{ is complex.} \end{cases} \tag{8.68}$$

This regularization is especially useful for mitigating overfitting in regression and machine learning scenarios.

> i The selection of the regularization parameter λ is crucial and is typically adjusted using cross-validation techniques to balance the bias-variance tradeoff. A larger λ increases model bias while reducing variance, helping to prevent overfitting in small or noisy datasets. Conversely, a smaller λ may enable the model to capture more complex patterns but at an increased risk of overfitting.

Problem Set 8

8.1 Prove that the matrix

$$\begin{bmatrix} \cos^2 \theta & \cos \theta \sin \theta \\ \cos \theta \sin \theta & \sin^2 \theta \end{bmatrix} \tag{8.69}$$

orthogonally projects any vector in $\mathbb{R}^2$ onto the line spanned by the vector $(\cos \theta, \sin \theta)$ using Theorem 6.18.

8.2 Prove that the matrix operator, which reflects any vector in $\mathbb{R}^2$ about the line inclined at an angle θ, is given by:

$$\begin{bmatrix} \cos 2\theta & \sin 2\theta \\ \sin 2\theta & -\cos 2\theta \end{bmatrix}. \tag{8.70}$$

Hint: Analyze the scenario by considering the vector average of the original vector and its transformed counterpart after reflection.

8.3 Prove that any two rotation operators in $\mathbb{R}^2$ are commutative by verifying that $[R_\theta, R_\phi] = 0$ for any angles θ and ϕ.

8.4 For $n \times n$ matrices A and B, it is true that if either the product AB or BA is invertible, then both A and B must also be invertible. Explain this theorem from the perspective of matrix transformations.

8.5 Utilize your understanding of matrix operators to identify four distinct 2×2 matrices A such that:

$$A^4 = R_\pi.$$

8.6 Verify that the following basic rotation matrices in 3D are orthogonal matrices:

$$R_x(\theta) = \begin{bmatrix} 1 & 0 & 0 \\ 0 & \cos\theta & -\sin\theta \\ 0 & \sin\theta & \cos\theta \end{bmatrix}, \quad R_y(\theta) = \begin{bmatrix} \cos\theta & 0 & \sin\theta \\ 0 & 1 & 0 \\ -\sin\theta & 0 & \cos\theta \end{bmatrix},$$

$$R_z(\theta) = \begin{bmatrix} \cos\theta & -\sin\theta & 0 \\ \sin\theta & \cos\theta & 0 \\ 0 & 0 & 1 \end{bmatrix}.$$

Additionally, compute the general 3D rotation matrix $R = R_z(\alpha)R_y(\beta)R_x(\gamma)$ in complete matrix form and verify that it is also an orthogonal matrix. Here α, β, γ represents yaw, pitch and roll angles, respectively.

8.7 The identity matrix I is both Hermitian and unitary. Express the 2×2 identity matrix I_2 in the form prescribed by Theorem 8.12 using the following basis sets:

(a) The computational basis $\{|0\rangle, |1\rangle\}$,

(b) The plus-minus basis $\{|+\rangle, |-\rangle\}$,

(c) The plus-minus i basis $\{|+_i\rangle, |-_i\rangle\}$,

where $|+\rangle = \frac{1}{\sqrt{2}}(|0\rangle + |1\rangle)$, $|-\rangle = \frac{1}{\sqrt{2}}(|0\rangle - |1\rangle)$, $|+_i\rangle = \frac{1}{\sqrt{2}}(|0\rangle + i|1\rangle)$, and $|-_i\rangle = \frac{1}{\sqrt{2}}(|0\rangle - i|1\rangle)$. What do you find from your results?

8.8 Construct a unitary matrix in $\mathbb{C}^4$ with the first column being $\begin{bmatrix} \frac{1}{2} & \frac{1}{2} & \frac{1}{2} & \frac{1}{2} \end{bmatrix}^T$.

8.9 Prove Theorem 8.13 using Def. 8.4.

8.10 Let U be the basis transformation operator from a basis of $\{|b\rangle\}$ to another basis of $\{|b'\rangle\}$. Find the matrix representation of U under the computational basis, basis of $\{|b\rangle\}$, and basis of $\{|b'\rangle\}$, respectively.

8.11 Find the matrix representations of all Pauli matrices relative to each of the folllowing bases:

(a) $\{|+\rangle, |-\rangle\}$,

(b) $\{|+_i\rangle, |-_i\rangle\}$.

8.12 Prove all the identities in Theorem 8.28.

8.13 Prove Theorem 8.31.

8.14 Prove Theorem 8.33.

8.15 Determine the number of distinct subspaces among the row spaces, column spaces, and null spaces of the matrices A, A^T, A^*, and $A^\dagger$. Additionally, identify all pairs of these subspaces that are orthogonal complements of each other.

8.16 $*$ Do a self-study on Gaussian Elimination method and provide a proof to Theorem 8.35.

8.17 $*$ Prove that for a matrix $A \in \mathbb{C}^{m \times n}$, $\mathrm{rank}(AA^T) = \mathrm{rank}(A^T A) \leq \min(m, n)$.

8.18 $*$ Prove that for any matrices A and B, provided that the product AB is well-defined, the rank of the product matrix AB satisfies the inequality:

$$\mathrm{rank}(AB) \leq \min(\mathrm{rank}(A), \mathrm{rank}(B)).$$

9. Spectral Decomposition of Matrices

Contents

In the previous chapter, we explored matrices as operators on vectors, with particular attention to how unitary matrices enable changes of basis within vector spaces. However, when dealing with more general matrices, such as non-unitary or non-square matrices, the analysis becomes more complex. For instance, consider the linear operator in $\mathbb{R}^2$:

$$A = \begin{bmatrix} 2 & 1 \\ 1 & 3 \end{bmatrix}.$$

The transformations performed by this matrix—whether they involve scaling, rotation, or other geometric effects—are not immediately obvious. To fully understand the behavior of such matrices, more advanced tools are required.

This chapter establishes a foundation of those tools through the study of eigenvalues and eigenvectors. The term "eigen" is derived from the German word meaning "own" or "inherent," reflecting how these concepts capture the intrinsic properties of a matrix. Building on this foundation, we will explore critical topics such as diagonalization, the spectral theorem, and normal and Hermitian matrices, each of which plays a pivotal role in matrix analysis. The tools introduced here will equip us to analyze matrices more effectively and provide the groundwork for exploring complex quantum systems.

Many interconnected concepts and relationships are covered in this chapter. Table 9.1 provides a brief summary for easy reference.

9.1 Eigenvalues and Eigenvectors

Eigenvectors are special "directions" associated with a square matrix, characterized by the property that any vector aligned with these directions remains parallel upon transformation by the matrix, though scaled by a scalar factor known as the eigenvalue. For an $n \times n$ matrix, up to n linearly independent such directions may exist, forming a basis referred to as an eigenbasis.

Eigenbasis analysis involves decomposing any vector into components aligned with the eigenvectors of the matrix. This simplifies the transformation by applying straightforward scaling rules to each component, which can then be reassembled to form the transformed vector. The utility of eigenbasis analysis lies in its reliance on linearity and the strategic selection of the basis, making it a powerful tool extensively employed in engineering, physics, and other scientific disciplines.

The purpose of this section is to provide readers with a thorough understanding of eigenbasis analysis, supported by detailed explanations and practical examples. This will lay the foundation for exploring the broader concept of spectral decomposition.

9.1.1 Definition and Basic Properties

1 Definition

A square matrix serves as a linear operator in $\mathbb{C}^n$. As discussed in § 8.1, a vector typically alters its direction upon transformation. However, exceptions exist, such as with non-uniform scaling operators like:

$$A = \begin{bmatrix} 1 & 0 \\ 0 & 2 \end{bmatrix},$$

which maintains the direction for vectors along the x-axis or y-axis. Conversely, it modifies the direction of all other vectors, as depicted in Fig. 9.1.

> ⓘ In this section, the term "direction" specifically denotes a one-dimensional subspace or a line through the origin. The phrase "along the same direction" implies being "parallel to the original direction". Thus, $k\boldsymbol{v}$ aligns with the direction of $\boldsymbol{v}$ for any scalar k. We use "direction" here for its intuitive alignment with geometric representations in $\mathbb{R}^2$.

Concept	Description and Relationships
Eigenvalues and Eigenvectors	For a matrix A, a scalar λ is an eigenvalue if there exists a non-zero vector v such that $Av = \lambda v$. The vector v is the corresponding eigenvector.
Characteristic Equation	The equation $\det(\lambda I - A) = 0$ is the characteristic equation of a matrix, a polynomial equation in λ. Solving it gives the eigenvalues λ. An $n \times n$ matrix has n eigenvalues (counting multiplicities).
Degeneracy and Eigenspace	If an eigenvalue λ has multiplicity greater than 1, it is called degenerate. The corresponding eigenspace is the set of all eigenvectors associated with that eigenvalue.
Eigenbasis	A matrix is diagonalizable if it has a complete set of n linearly independent eigenvectors. These eigenvectors form an eigenbasis for the vector space.
Gram-Schmidt Process	An algorithm to orthogonalize a set of linearly independent vectors, often used to generate an orthonormal basis for the eigenspace associated with a degenerate eigenvalue.
Similar Matrices	Two matrices A and B are similar if $B = P^{-1}AP$ for some invertible matrix P. Similar matrices have the same eigenvalues and represent the same linear transformation in different bases.
Diagonalization (Eigendecomposition)	A matrix A is diagonalizable if it can be written as $A = P\Lambda P^{-1}$, where Λ is a diagonal matrix of eigenvalues and P is the matrix of eigenvectors.
Spectral Decomposition	A special case of eigendecomposition for normal matrices, where A is written as $A = U\Lambda U^\dagger$, with U being a unitary matrix. Alternatively, A can be expressed as $A = \sum_i \lambda_i \lvert\phi_i\rangle \langle\phi_i\rvert$, where $\{\lvert\phi_i\rangle\}$ is an orthonormal eigenbasis.
Normal Matrices	A matrix A is normal if it satisfies $AA^\dagger = A^\dagger A$. Normal matrices are always unitarily diagonalizable, allowing for a spectral decomposition.
Hermitian Matrices	A matrix A is Hermitian if $A = A^\dagger$. Hermitian matrices have real eigenvalues and are a subset of normal matrices.

Table 9.1: Summary of Key Concepts and Relationships in Spectral Decomposition

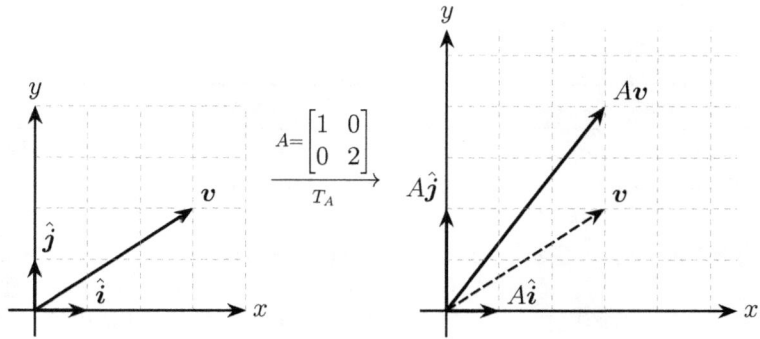

Figure 9.1: Illustration of Eigenvectors for Non-uniform Scaling

In Fig. 9.1, vectors $\hat{i}$ and $\hat{j}$ are identified as eigenvectors of matrix A. Given $A\hat{i} = \hat{i}$ and $A\hat{j} = 2\hat{j}$, the eigenvalues are respectively one for $\hat{i}$, and two for $\hat{j}$. Generally, eigenvectors and eigenvalues are formally defined as follows:

Definition 9.1 — Eigenvalue and Eigenvector. For a matrix $A \in \mathbb{C}^{n \times n}$, if there exists any nonzero vector $|v\rangle \in \mathbb{C}^n$ and any scalar $\lambda \in \mathbb{C}$ satisfying

$$A\,|v\rangle = \lambda\,|v\rangle, \tag{9.1}$$

then $|v\rangle$ qualifies as an **eigenvector** of A, and λ is the associated **eigenvalue**.

In Def. 9.1, the equation $A\,|v\rangle = \lambda\,|v\rangle$ indicates that $|v\rangle$ remains parallel to itself after being transformed by A. Notably, if $|v\rangle$ is an eigenvector associated with the eigenvalue λ, then any nonzero scalar multiple of $|v\rangle$ also serves as an eigenvector for the same eigenvalue:

$$A\,|v\rangle = \lambda\,|v\rangle \quad \Rightarrow \quad A(k\,|v\rangle) = \lambda(k\,|v\rangle) \text{ for any nonzero } k \in \mathbb{C}.$$

It is important to recognize that while eigenvalues can be zero, eigenvectors must be non-zero vectors. When $\lambda = 0$, $|v\rangle$ belongs to the null space of A, since $A\,|v\rangle = \mathbf{0}$.

Exercise 9.1 Apply Def. 9.1 to determine the eigenvalues and corresponding eigenvectors of the following matrices, analyzing the transformation rules they represent:

$$\text{(a) } \begin{bmatrix} 1 & 0 \\ 0 & 0 \end{bmatrix}, \text{ (b) } \begin{bmatrix} 1 & 1 \\ 1 & 1 \end{bmatrix}, \text{ (c) } \begin{bmatrix} 2 & 0 \\ 0 & 2 \end{bmatrix}$$

2 Basic Properties

Theorem 9.1 If λ is an eigenvalue of a matrix A, then:

(a) λ^k is an eigenvalue of A^k for any positive integer k.

(b) λ^{-1} is an eigenvalue of A^{-1} provided that A is invertible.

Proof.

(a) Given that $A\boldsymbol{v} = \lambda\boldsymbol{v}$, applying A repeatedly to both sides yields:

$$A^2\boldsymbol{v} = A(A\boldsymbol{v}) = A(\lambda\boldsymbol{v}) = \lambda A\boldsymbol{v} = \lambda^2\boldsymbol{v}.$$

Similarly,
$$A^3\boldsymbol{v} = A(A^2\boldsymbol{v}) = A(\lambda^2\boldsymbol{v}) = \lambda^2 A\boldsymbol{v} = \lambda^3\boldsymbol{v}.$$

Continuing this process, we see that:

$$A^k\boldsymbol{v} = \lambda^k\boldsymbol{v} \quad \text{for any positive integer } k.$$

Thus, λ^k is an eigenvalue of A^k with the same eigenvector $\boldsymbol{v}$.

(b) If A is invertible, multiplying A^{-1} to both sides of $A\boldsymbol{v} = \lambda\boldsymbol{v}$ yields:

$$\boldsymbol{v} = \lambda A^{-1}\boldsymbol{v}.$$

Therefore, dividing both sides by λ, we get:

$$A^{-1}\boldsymbol{v} = \lambda^{-1}\boldsymbol{v}.$$

This shows that λ^{-1} is an eigenvalue of A^{-1}, provided A is invertible.

$\square$

Theorem 9.2 All eigenvalues λ_i of a unitary matrix U have a magnitude of one, that is, $|\lambda_i| = 1$.

Proof. Not losing generality, suppose $|v_i\rangle$ is a normalized eigenvector associated with the eigenvalue λ_i of the unitary matrix U. By definition, we have:

$$U|v_i\rangle = \lambda_i|v_i\rangle.$$

Applying the conjugate transpose to both sides and then multiplying by the original, we derive:

$$(U|v_i\rangle)^\dagger(U|v_i\rangle) = (\lambda_i|v_i\rangle)^\dagger(\lambda_i|v_i\rangle).$$

The left-hand side simplifies using the unitarity of U and the normalization of $|v_i\rangle$:

$$(U|v_i\rangle)^\dagger(U|v_i\rangle) = \langle v_i|U^\dagger U|v_i\rangle = \langle v_i|v_i\rangle = 1.$$

The right-hand side becomes:

$$(\lambda_i|v_i\rangle)^\dagger(\lambda_i|v_i\rangle) = \lambda_i^*\lambda_i\langle v_i|v_i\rangle = |\lambda_i|^2.$$

Equating the two results, we conclude:

$$|\lambda_i|^2 = 1 \quad \Rightarrow \quad |\lambda_i| = 1.$$

This confirms that the magnitude of any eigenvalue λ_i of a unitary matrix U is 1. $\square$

Exercise 9.2 Determine the eigenvalues and corresponding eigenvectors for $n \times n$ matrices in the following form:

$$S = \sum_{i=1}^{n} \lambda_i|\phi_i\rangle\langle\phi_i| \quad \text{where } \{|\phi_i\rangle\} \text{ is an orthonormal basis of } \mathbb{C}^n.$$

Even for real matrices, eigenvalues and eigenvectors can be complex. Consider the rotation matrix:

$$R_\theta = \begin{bmatrix} \cos\theta & -\sin\theta \\ \sin\theta & \cos\theta \end{bmatrix}.$$

For $\theta \neq k\pi$, $k \in \mathbb{Z}$, no real eigenvector exists in $\mathbb{R}^2$ because every nonzero vector is rotated counterclockwise by the angle θ rather than being scaled. Does this imply the absence of eigenvalues or eigenvectors? Interestingly, they still exist but are complex.

■ **Example 9.1** Confirm that for R_θ, the eigenvalues are $\lambda_1 = e^{-i\theta}$ and $\lambda_2 = e^{i\theta}$, with the corresponding eigenvectors $|+_i\rangle$ and $|-_i\rangle$, respectively.

Solution. The matrix representation of $|+_i\rangle$ is $\frac{1}{\sqrt{2}} \begin{bmatrix} 1 \\ i \end{bmatrix}$. The scalar factor $\frac{1}{\sqrt{2}}$ is irrelevant for determining eigenvectors. Thus, we compute:

$$\begin{bmatrix} \cos\theta & -\sin\theta \\ \sin\theta & \cos\theta \end{bmatrix} \begin{bmatrix} 1 \\ i \end{bmatrix} = \begin{bmatrix} \cos\theta - i\sin\theta \\ \sin\theta + i\cos\theta \end{bmatrix} = (\cos\theta - i\sin\theta) \begin{bmatrix} 1 \\ i \end{bmatrix}.$$

Hence, $|+_i\rangle$ is an eigenvector of R_θ associated with the eigenvalue $\cos\theta - i\sin\theta = e^{-i\theta}$.

Similarly, it can be verified that $R_\theta |-_i\rangle = e^{i\theta} |-_i\rangle$, confirming that $|-_i\rangle$ is an eigenvector of R_θ corresponding to the eigenvalue $e^{i\theta}$.   ■

Why do some real matrices have complex eigenvalues and eigenvectors? In the next subsection, we will introduce a systematic method to determine eigenvalues and eigenvectors for any matrix.

Left Eigenvectors

Typically, the term "eigenvector," as outlined in Def. 9.1, refers to what are technically known as **right eigenvectors**. However, matrices also have **left eigenvectors**, which are defined distinctly. A left eigenvector is any *row vector u* that satisfies the following condition:

$$uA = \kappa u, \tag{9.2}$$

where κ is the eigenvalue associated with u.

Notably, left eigenvectors of a matrix A are equivalent to the transpose of right eigenvectors of the transpose matrix A^T, and they share the same eigenvalues (Problem 9.5a). Beyond this, there is generally no direct relationship between the left and right eigenvectors of a non-symmetric matrix, highlighting the unique properties of each type of eigenvector.

9.1.2   Finding Eigenvalues and Eigenvectors

Determining eigenvalues and eigenvectors for a matrix, even as simple as R_θ, can be impractical by mere observation. A more systematic method is required, which is facilitated by the characteristic equation of a matrix.

1 The Characteristic Equation

Theorem 9.3 Given a matrix $A \in \mathbb{C}^{n \times n}$, a scalar λ is an eigenvalue of A if and only if it satisfies:
$$\det(\lambda I - A) = 0, \tag{9.3}$$
where I is the identity matrix of size $n \times n$. This equation is called the **characteristic equation** of A.

Proof. Assume λ is an eigenvalue of A with a corresponding eigenvector $|v\rangle$. According to the definition of eigenvalues and eigenvectors, we have:
$$A |v\rangle = \lambda |v\rangle = \lambda I |v\rangle .$$

Rewriting this, we get:
$$(\lambda I - A) |v\rangle = \mathbf{0}.$$

Since $|v\rangle \neq \mathbf{0}$ (the eigenvector is nonzero), this implies that the matrix $\lambda I - A$ is singular, meaning:
$$\det(\lambda I - A) = 0.$$

Conversely, if $\det(\lambda I - A) = 0$, then $\lambda I - A$ is singular, implying the existence of a non-trivial solution $|v\rangle \neq \mathbf{0}$ such that:
$$(\lambda I - A) |v\rangle = \mathbf{0} \quad \Rightarrow \quad A |v\rangle = \lambda |v\rangle ,$$

confirming that λ is an eigenvalue of A.   □

■ **Example 9.2** Identify the eigenvalues of the matrix:
$$A = \begin{bmatrix} 0 & 1 \\ -1 & 0 \end{bmatrix}.$$

Solution. The characteristic equation for A is:
$$\det(\lambda I - A) = \begin{vmatrix} \lambda & -1 \\ 1 & \lambda \end{vmatrix} = \lambda^2 + 1 = (\lambda - i)(\lambda + i) = 0.$$

Therefore, the eigenvalues are $\lambda = \pm i$.   ■

2 Number of Eigenvalues

The following theorem asserts that a matrix $A \in \mathbb{C}^{n \times n}$ has at least one, and at most n, distinct eigenvalues (and associated eigenvectors).

Theorem 9.4 The characteristic equation of an $n \times n$ matrix, $\det(\lambda I - A) = 0$, is a polynomial equation of degree n. This equation has at least one, and up to n, distinct real or complex roots.

Proof. We first observe that $p_A(\lambda) = \det(\lambda I - A)$ is a polynomial in λ of degree n, known as the **characteristic polynomial** of A. Specifically, the matrix $\lambda I - A$ takes the form:
$$\lambda I - A = \begin{bmatrix} \lambda - a_{11} & -a_{12} & \cdots & -a_{1n} \\ -a_{21} & \lambda - a_{22} & \cdots & -a_{2n} \\ \vdots & \vdots & \ddots & \vdots \\ -a_{n1} & -a_{n2} & \cdots & \lambda - a_{nn} \end{bmatrix}.$$

When expanding the determinant $\det(\lambda I - A)$, the highest power of λ arises from the product of the diagonal elements, resulting in the term λ^n. Therefore, the characteristic polynomial $p_A(\lambda)$ is a degree-n polynomial.

By the Fundamental Theorem of Algebra, a degree-n polynomial has exactly n roots in $\mathbb{C}$ (counting multiplicities), although it may have fewer than n distinct roots.   □

The occurrence of complex eigenvalues in real matrices is due to the fact that polynomial equations with real coefficients can have complex roots. This is demonstrated in Example 9.2.

3   Procedure for Finding Eigenvalues and Eigenvectors

Once an eigenvalue λ is found by solving the characteristic equation $\det(\lambda I - A) = 0$, the corresponding eigenvector(s) can be determined by solving the linear system $(\lambda I - A)\,|v\rangle = \mathbf{0}$.

■ **Example 9.3** Determine the eigenvalues and eigenvectors of the matrix:

$$A = \begin{bmatrix} 0 & 1 \\ 1 & 0 \end{bmatrix}.$$

Solution. The characteristic equation of A is given by:

$$\det(\lambda I - A) = \begin{vmatrix} \lambda & -1 \\ -1 & \lambda \end{vmatrix} = \lambda^2 - 1 = (\lambda - 1)(\lambda + 1) = 0.$$

This gives the eigenvalues $\lambda_1 = 1$ and $\lambda_2 = -1$. Substituting each eigenvalue into the system $(\lambda I - A)\,|v\rangle = \mathbf{0}$, where $|v\rangle = \begin{bmatrix} x \\ y \end{bmatrix}$, we obtain:

$$\lambda_1 = 1: \quad \begin{bmatrix} 1 & -1 \\ -1 & 1 \end{bmatrix} \begin{bmatrix} x \\ y \end{bmatrix} = \begin{bmatrix} 0 \\ 0 \end{bmatrix} \quad \Rightarrow \quad x = y \quad \Rightarrow \quad |v_1\rangle = t \begin{bmatrix} 1 \\ 1 \end{bmatrix}.$$

$$\lambda_2 = -1: \quad \begin{bmatrix} -1 & -1 \\ -1 & -1 \end{bmatrix} \begin{bmatrix} x \\ y \end{bmatrix} = \begin{bmatrix} 0 \\ 0 \end{bmatrix} \quad \Rightarrow \quad x = -y \quad \Rightarrow \quad |v_2\rangle = t \begin{bmatrix} 1 \\ -1 \end{bmatrix}.$$

Normalizing the eigenvectors, we get

$$\lambda_1 = 1: \quad |v_1\rangle = \frac{1}{\sqrt{2}} \begin{bmatrix} 1 \\ 1 \end{bmatrix}$$

$$\lambda_2 = -1: \quad |v_2\rangle = \frac{1}{\sqrt{2}} \begin{bmatrix} 1 \\ -1 \end{bmatrix}.$$

■

Exercise 9.3 Calculate the eigenvectors for A corresponding to each eigenvalue found in Example 9.2.

Exercise 9.4 Determine the eigenvalues and corresponding eigenvectors for the matrix:

$$\begin{bmatrix} 1 & 2 \\ 3 & 2 \end{bmatrix}.$$

Cayley-Hamilton theorem

An important theorem in matrix theory is the Cayley-Hamilton theorem, which asserts that every matrix satisfies its own characteristic equation. Specifically, if the characteristic equation of a matrix A is defined as

$$\det(A - \lambda I) = \lambda^n + c_1\lambda^{n-1} + c_2\lambda^{n-2} + \ldots + c_{n-1}\lambda + c_n = 0, \qquad (9.4)$$

then matrix A itself will fulfill the corresponding matrix polynomial equation:

$$A^n + c_1 A^{n-1} + c_2 A^{n-2} + \ldots + c_{n-1}A + c_n I = 0. \qquad (9.5)$$

For more discussions on the Cayley-Hamilton theorem and its practical implications, please refer to § 11.3.1.5.

9.1.3 Degenerate Eigenvalues and Eigenspaces

As illustrated in Fig. 9.1, an eigenvector of a matrix A defines a direction in which the transformation by A results in a scaled version of the vector, with the scaling factor given by the corresponding eigenvalue.

When multiple eigenvectors correspond to the same eigenvalue, they span a subspace known as the eigenspace associated with that eigenvalue. In such cases, the eigenvalue is termed *degenerate*. Eigenspaces associated with degenerate eigenvalues possess unique properties and play a central role in understanding the structure of matrices and their transformations, which we will explore in this subsection.

1 Degenerate Eigenvalue and Algebraic Multiplicity

The characteristic equation of an $n \times n$ matrix A can be expressed in fully factorized form as:

$$(\lambda - \lambda_1)(\lambda - \lambda_2) \cdots (\lambda - \lambda_n) = 0,$$

where $\lambda_1, \lambda_2, \ldots, \lambda_n$ are the eigenvalues of A. Sometimes, certain eigenvalues appear as repeated roots of the characteristic equation.

For example, consider the characteristic equation of the identity matrix I:

$$\det(\lambda I - I) = \det((\lambda - 1)I) = (\lambda - 1)^n = 0.$$

Here, $\lambda = 1$ is an eigenvalue of I with an algebraic multiplicity of n, reflecting the repeated factor $(\lambda - 1)$ in the characteristic equation.

Definition 9.2 — Algebraic Multiplicity of Eigenvalue. The algebraic multiplicity of an eigenvalue λ_0 is the number of times the factor $(\lambda - \lambda_0)$ appears in the fully factorized characteristic equation.

Definition 9.3 — Degenerate Eigenvalue. An eigenvalue with an algebraic multiplicity greater than one is called a degenerate eigenvalue.

Exercise 9.5 Examine the operators introduced in § 8.1.5 for $\mathbb{R}^2$ and determine which exhibit degenerate eigenvalues.

2   Eigenspace and Geometric Multiplicity

Identifying eigenvectors associated with degenerate eigenvalues requires careful consideration. Take, for instance, the identity matrix I_2; every nonzero vector in $\mathbb{C}^2$ is an eigenvector corresponding to $\lambda = 1$, as demonstrated by the equation $I_2 |v\rangle = |v\rangle$ for any vector $|v\rangle$. This means every direction in the vector space is represented by an eigenvector. How should we understand this?

To find eigenvectors for a given eigenvalue λ, we solve the linear system $(\lambda I - A) |v\rangle = \mathbf{0}$. For $A = I_2$ and $\lambda = 1$, the coefficient matrix reduces to $\lambda I - A = \mathbf{0}$, a zero matrix with rank zero. By the rank-nullity theorem (Theorem 8.38), the nullity is two, meaning the entire space $\mathbb{R}^2$ forms the solution space. This is the eigenspace corresponding to $\lambda = 1$.

> **Definition 9.4 — Eigenspace and Geometric Multiplicity.** The eigenspace of a matrix A associated with an eigenvalue λ is the null space of $(\lambda I - A)$, which solves the homogeneous system $(\lambda I - A) |v\rangle = \mathbf{0}$. The dimension of this eigenspace is called the geometric multiplicity of λ.

For I_2, the single eigenvalue $\lambda = 1$ has both an algebraic and geometric multiplicity of two. The following theorem relates the algebraic and geometric multiplicities of an eigenvalue.

> **Theorem 9.5** The geometric multiplicity of any eigenvalue of a square matrix is always less than or equal to its algebraic multiplicity.

The proof of this theorem is somewhat involved, and we will not explore it here. However, interested readers are encouraged to investigate it as a self-study exercise. Intuitively, the geometric multiplicity cannot exceed the algebraic multiplicity because the algebraic multiplicity sets an upper limit on the number of linearly independent eigenvectors that *could* exist for a given eigenvalue, while the geometric multiplicity represents the actual number of linearly independent eigenvectors that do exist.

Below, we give an example illustrating a situation where geometric multiplicity is less than algebraic multiplicity.

■ **Example 9.4** Find the eigenvalues and eigenspaces of the matrix:

$$A = \begin{bmatrix} 1 & 1 \\ 0 & 1 \end{bmatrix}.$$

Solution. The characteristic equation is:

$$\det(\lambda I - A) = \begin{vmatrix} \lambda - 1 & -1 \\ 0 & \lambda - 1 \end{vmatrix} = (\lambda - 1)^2 = 0.$$

Thus, $\lambda = 1$ is the sole eigenvalue with an algebraic multiplicity of two. Solving the system:

$$(\lambda I - A) |v\rangle = \mathbf{0} \quad \Rightarrow \quad \begin{bmatrix} 0 & -1 \\ 0 & 0 \end{bmatrix} \begin{bmatrix} x \\ y \end{bmatrix} = \begin{bmatrix} 0 \\ 0 \end{bmatrix} \quad \Rightarrow \quad |v\rangle = t \begin{bmatrix} 1 \\ 0 \end{bmatrix}.$$

Hence, the eigenspace for $\lambda = 1$ is spanned by $\{(1,0)\}$, so its geometric multiplicity is one.   ■

For non-degenerate eigenvalues, the algebraic and geometric multiplicities are both one, yielding a one-dimensional eigenspace spanned by a single eigenvector.

In contrast, degenerate eigenvalues may have eigenspaces spanned by two or more linearly independent eigenvectors, making "eigenspace" a more suitable term than "eigenvector" for describing the set of vectors associated with such an eigenvalue.

3 Eigenvalue Multiplicity and Shear Transformations

From Example 9.4, we observe that the matrix A acts as a shear operator in the x-direction. The geometric interpretation of a shear operator, illustrated in Fig. 8.8, shows that only the x-unit vector maintains its direction during the transformation, identifying it as the sole unit eigenvector of this operator. Other directions, such as the y-direction in the example, change under the transformation. As a result, most vectors fail to remain invariant in direction, explaining why a shear operator has fewer eigenvectors than the dimension of the space.

It can be shown that matrices with eigenvalues whose geometric multiplicity is less than their algebraic multiplicity often represent shear transformations in certain directions. This concept is explored further in the theory of Jordan forms, which we will cover in § 13.1.3.

Exercise 9.6 Find the eigenvalues and their algebraic and geometric multiplicities for the matrix:
$$\begin{bmatrix} 1 & 1 & 0 \\ 0 & 1 & 0 \\ 0 & 0 & 1 \end{bmatrix}.$$

Degeneracy in Quantum Computing

In quantum mechanics, the quantized energy levels of a system are represented by the eigenvalues of the system's Hamiltonian, the operator that determines the total energy of the system. Degenerate eigenvalues occur when multiple orthogonal eigenstates correspond to the same energy level. This phenomenon, known as degeneracy, often arises from the system's symmetries.

Degenerate states play a critical role in quantum error correction, which enhances the robustness of quantum computing. By encoding information across entangled states with the same energy, degenerate eigenvalues help protect information against errors, enabling advantages that classical systems cannot achieve. Understanding and utilizing degeneracy is essential for developing complex quantum computations and improving the error resilience of quantum algorithms. Degenerate eigenspaces are thus key to advancing quantum computing technologies.

9.1.4 Eigenbasis

Now let's focus on the case where an $n \times n$ matrix A has n linearly independent eigenvectors. (The exact conditions for this will be discussed in Theorem 9.7, § 9.2.2, and § 9.3 from different angles.) In this situation, the eigenvectors of A form a basis for $\mathbb{C}^n$, known as an eigenbasis.

Definition 9.5 — Eigenbasis. An eigenbasis of a matrix $A \in \mathbb{C}^{n \times n}$ is a basis for $\mathbb{C}^n$ consisting entirely of eigenvectors of A.

When does a matrix possess an eigenbasis, and when does it not? We will explore this question across two distinct scenarios.

1 Eigenbasis with No Degenerate Eigenvalues

In cases where matrix A does not possess any degenerate eigenvalues, it must have n distinct eigenvalues. The following theorem guarantees that in such scenarios, n linearly independent eigenvectors can always be found to form an eigenbasis.

Theorem 9.6 Eigenvectors corresponding to distinct eigenvalues of a matrix are linearly independent.

Proof.   We begin the proof using mathematical induction. Assume λ_1 is an eigenvalue of $A_{n \times n}$ with corresponding eigenvector $|v_1\rangle$. Clearly, the set $\{|v_1\rangle\}$ is linearly independent, as $|v_1\rangle$ is nonzero.

Introduce another eigenvalue λ_2 and its eigenvector $|v_2\rangle$. If the set $\{|v_1\rangle, |v_2\rangle\}$ were linearly dependent, there would exist a nonzero scalar k such that $|v_2\rangle = k|v_1\rangle$. Applying the eigenvalue-eigenvector relationship:

$$A|v_2\rangle = \lambda_2|v_2\rangle \quad \Rightarrow \quad A(k|v_1\rangle) = \lambda_2(k|v_1\rangle) \quad \Rightarrow \quad A|v_1\rangle = \lambda_2|v_1\rangle.$$

This result contradicts the initial condition $A|v_1\rangle = \lambda_1|v_1\rangle$, given $\lambda_1 \neq \lambda_2$, affirming that $\{|v_1\rangle, |v_2\rangle\}$ must be linearly independent.

Now consider $\lambda_1, \lambda_2, \ldots, \lambda_r$ as r distinct eigenvalues of A, where $r < n$, with corresponding linearly independent eigenvectors $\{|v_1\rangle, |v_2\rangle, \ldots, |v_r\rangle\}$. For another distinct eigenvalue λ_{r+1} and its eigenvector $|v_{r+1}\rangle$, assume it can be expressed as a linear combination of the preceding eigenvectors:

$$|v_{r+1}\rangle = \sum_i c_i|v_i\rangle,$$

where c_i are scalar coefficients.

Inserting this expression into the eigenvalue-eigenvector relation for λ_{r+1}, we deduce:

$$A\sum_i c_i|v_i\rangle = \lambda_{r+1}\sum_i c_i|v_i\rangle$$

$$\Rightarrow \quad \sum_i c_i(A|v_i\rangle) = \sum_i c_i\lambda_{r+1}|v_i\rangle$$

$$\Rightarrow \quad \sum_i c_i(\lambda_i|v_i\rangle) = \sum_i c_i\lambda_{r+1}|v_i\rangle$$

$$\Rightarrow \quad \sum_i c_i(\lambda_i - \lambda_{r+1})|v_i\rangle = \mathbf{0}.$$

By the linear independence of $\{|v_i\rangle\}$, all coefficients $c_i = 0$, since $\lambda_i - \lambda_{r+1} \neq 0$. This leads to $|v_{r+1}\rangle = \mathbf{0}$, a contradiction because $|v_{r+1}\rangle$ is an eigenvector.

Therefore, $|v_{r+1}\rangle$ cannot be a linear combination of $\{|v_i\rangle\}$, concluding that all considered eigenvectors are linearly independent, completing the proof.   □

■ **Example 9.5** Determine an eigenbasis for the matrix $A = \begin{bmatrix} 1 & 2 \\ 2 & 1 \end{bmatrix}$.

Solution. The characteristic equation of A is calculated as follows:

$$\det(\lambda I - A) = \begin{vmatrix} \lambda - 1 & -2 \\ -2 & \lambda - 1 \end{vmatrix} = (\lambda - 1)^2 - 4 = (\lambda + 1)(\lambda - 3) = 0.$$

Solving for the eigenvectors for each eigenvalue, we find:

$$\lambda_1 = -1 : \quad \begin{bmatrix} -2 & -2 \\ -2 & -2 \end{bmatrix} \begin{bmatrix} x \\ y \end{bmatrix} = \begin{bmatrix} 0 \\ 0 \end{bmatrix} \quad \Rightarrow \quad x = -y \quad \Rightarrow \quad |v_1\rangle = t \begin{bmatrix} 1 \\ -1 \end{bmatrix}.$$

$$\lambda_2 = 3 : \quad \begin{bmatrix} 2 & -2 \\ -2 & 2 \end{bmatrix} \begin{bmatrix} x \\ y \end{bmatrix} = \begin{bmatrix} 0 \\ 0 \end{bmatrix} \quad \Rightarrow \quad x = y \quad \Rightarrow \quad |v_2\rangle = t \begin{bmatrix} 1 \\ 1 \end{bmatrix}.$$

Thus, the set $\{(1, 1), (1, -1)\}$ forms an eigenbasis for A. ■

Exercise 9.7 Determine an eigenbasis for the matrix:

$$\begin{bmatrix} 1 & 1 & 0 \\ 1 & 1 & 0 \\ 0 & 0 & 1 \end{bmatrix}.$$

2 Eigenbasis with Degenerate Eigenvalues

The presence of degenerate eigenvalues imposes specific conditions on the existence of an eigenbasis. The following theorem clarifies these conditions:

> **Theorem 9.7** An eigenbasis of a matrix exists if and only if the geometric multiplicity of every eigenvalue equals its algebraic multiplicity.

Proof. Consider $A \in \mathbb{C}^{n \times n}$ with eigenvalues $\lambda_1, \lambda_2, \ldots, \lambda_r$, where each eigenvalue λ_i has algebraic and geometric multiplicities of n_i, respectively, and:

$$n_1 + n_2 + \cdots + n_r = n,$$

reflecting the sum of the algebraic multiplicities equaling the degree of the characteristic polynomial.

The equality of the geometric and algebraic multiplicities implies that each eigenvalue λ_i has an eigenspace dimension of n_i, spanned by n_i linearly independent eigenvectors. Given that Theorem 9.6 establishes the linear independence of eigenvectors corresponding to distinct eigenvalues, the aggregation of these eigenvectors:

$$n_1 + n_2 + \cdots + n_r = n,$$

provides n linearly independent vectors that form an eigenbasis for $\mathbb{C}^n$.

Conversely, if any degenerate eigenvalue has a geometric multiplicity that is less than its algebraic multiplicity, the total number of linearly independent eigenvectors will be less than n. Consequently, it will not be possible to form an eigenbasis for $\mathbb{C}^n$, as the required condition of having n linearly independent vectors is not satisfied. □

As we will explore in § 9.2.2, a matrix that possesses an eigenbasis is a **diagonalizable matrix**. In contrast, matrices that lack an eigenbasis are known as **defective matrices**. In quantum computing, most matrices of interest are diagonalizable, which will be the primary focus of our study.

3 Applying an Eigenbasis: A Dynamic System Example

Understanding the behavior of linear operators through their eigenbasis provides significant insights, particularly in dynamic systems where the operator is repeatedly applied. This iterative application simplifies predictions about the system's long-term behavior. We illustrate this through an example involving population dynamics.

■ **Example 9.6** Consider a dynamic system modeled by the matrix

$$A = \begin{bmatrix} 0.7 & 0.2 \\ 0.3 & 0.8 \end{bmatrix},$$

which represents the annual exchange rates between two population groups. Initially, group one contains 8000 members, and group two has 2000. Each year, 30% of group one's population transfers to group two, and 20% of group two's population moves to group one. The goal is to determine the population distribution between the two groups after k years using the matrix A.

Solution. To compute the population distribution after k years, we apply the recursive formula $A\boldsymbol{x}_{k-1} = \boldsymbol{x}_k$, where $\boldsymbol{x}_k$ represents the population vector in year k. This formula updates the population distribution based on the previous year's values. After k years, the desired distribution can be obtained by computing $A^k \boldsymbol{x}_0$, where $\boldsymbol{x}_0 = \begin{bmatrix} 8000 \\ 2000 \end{bmatrix}$.

The matrix A has eigenvalues $\lambda_1 = 0.5$ and $\lambda_2 = 1$, with corresponding eigenvectors

$$\boldsymbol{v}_1 = \begin{bmatrix} -1 \\ 1 \end{bmatrix} \quad \text{and} \quad \boldsymbol{v}_2 = \begin{bmatrix} 2 \\ 3 \end{bmatrix}.$$

Decomposing $\boldsymbol{x}_0$ in terms of the eigenbasis, we get:

$$\boldsymbol{x}_0 = \begin{bmatrix} 8000 \\ 2000 \end{bmatrix} = -4000 \begin{bmatrix} -1 \\ 1 \end{bmatrix} + 2000 \begin{bmatrix} 2 \\ 3 \end{bmatrix}. \tag{9.6}$$

The key insight is that for any vector $\boldsymbol{w} = c_1 \boldsymbol{v}_1 + c_2 \boldsymbol{v}_2$, the operation $A^k \boldsymbol{w}$ simplifies to:

$$A^k \boldsymbol{w} = c_1 A^k \boldsymbol{v}_1 + c_2 A^k \boldsymbol{v}_2 = c_1 \lambda_1^k \boldsymbol{v}_1 + c_2 \lambda_2^k \boldsymbol{v}_2. \tag{9.7}$$

Applying A^k to this decomposition results in:

$$A^k \begin{bmatrix} 8000 \\ 2000 \end{bmatrix} = -4000 \cdot 0.5^k \begin{bmatrix} -1 \\ 1 \end{bmatrix} + 2000 \cdot 1^k \begin{bmatrix} 2 \\ 3 \end{bmatrix} = 0.5^k \begin{bmatrix} 4000 \\ -4000 \end{bmatrix} + \begin{bmatrix} 4000 \\ 6000 \end{bmatrix}.$$

The component along $\boldsymbol{v}_1$ diminishes over time because $\lambda_1 = 0.5$, while the component along $\boldsymbol{v}_2$ remains constant since $\lambda_2 = 1$. Consequently, as k increases, the population stabilizes to the configuration associated with $\boldsymbol{v}_2$, offering a clear view of the long-term population dynamics.

Thus, as k increases, the population vector converges to $\tilde{\boldsymbol{x}} = \begin{bmatrix} 4000 \\ 6000 \end{bmatrix}$, indicating that the population will stabilize at these values.   ■

4 Expansion in the Eigenbasis

The expansion of a vector in the eigenbasis refers to the process of expressing a vector as a linear combination of the eigenvectors of a linear operator. When the operator is applied, each component of the vector is scaled by its corresponding eigenvalue.

In Fig. 9.2, using the matrix A from the earlier dynamic system example (Eq. 9.6), we observe that the component along v_2 remains unchanged since $\lambda_2 = 1$. In contrast, the component along v_1 is halved in magnitude with each application of A, as $\lambda_1 = 0.5$. With repeated applications, the influence of v_1 diminishes and eventually vanishes, leaving the system's behavior determined solely by v_2, leading to long-term stability.

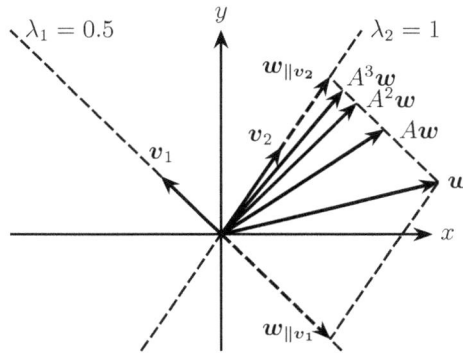

Figure 9.2: Illustration of Vector Expansion in the Eigenbasis

---

Singular Values of Non-square Matrices

While eigenvalues, eigenvectors, eigenspaces, and eigenbases are instrumental in analyzing square matrices through eigen-analysis, what about non-square matrices? Indeed, there is a parallel system for non-square matrices involving singular values and singular value decomposition (SVD). Detailed exploration of SVD will be covered in a separate section (§ 13.2) in a later chapter, providing a comprehensive introduction to this essential method.

9.2 Matrix Similarity and Diagonalization

The previous section showed that when a linear operator has an eigenbasis, it behaves like a non-uniform scaling operator under this basis, with each scaling factor corresponding to an eigenvalue. This insight establishes a clear analogy between such matrices and diagonal matrices, since diagonal matrices act as non-uniform scaling operators under the standard basis.

In the language of linear algebra, a matrix with an eigenbasis is called diagonalizable, meaning it is similar to a diagonal matrix. This section explores the elegant theory of diagonalization and matrix similarity, setting the stage for further discussions on spectral decomposition and Hermitian matrices, which are central to this chapter and crucial for understanding the theoretical foundations of quantum computing.

9.2.1   Matrix Similarity

1   Definition

In § 8.4.3, we explored how the representation of a linear operator changes under a unitary basis transformation. This concept extends to matrix similarity, a fundamental idea for understanding how matrices relate under different bases.

> i | Any invertible matrix P inherently represents a basis transformation between two bases of $\mathbb{C}^n$. The proof of this property is left as an exercise for the reader (Problem 9.8). Furthermore, if both bases are orthonormal, then P is not only invertible but also unitary (discussed in Theorem 8.21).

> **Definition 9.6 — Matrix Similarity.** Two square matrices $A, B \in \mathbb{C}^{n \times n}$ are said to be similar if there exists an invertible matrix P such that:
>
> $$B = P^{-1}AP. \tag{9.8}$$

This similarity indicates that matrices A and B represent the same linear transformation under two different bases of $\mathbb{C}^n$, with P acting as the change of basis matrix between these two bases.

To illustrate, consider rearranging Eq. 9.8 as $PB = AP$, and apply it to an eigenvector $|v\rangle \in \mathbb{C}^n$ of B, where $B|v\rangle = \lambda|v\rangle$ for some eigenvalue λ:

$$PB = AP \quad \Rightarrow \quad PB|v\rangle = AP|v\rangle \quad \Rightarrow \quad \lambda(P|v\rangle) = A(P|v\rangle). \tag{9.9}$$

Thus, $P|v\rangle$ is an eigenvector of A corresponding to the same eigenvalue λ. This shows that P transforms an eigenvector of B into an eigenvector of A, preserving the same eigenvalue. In other words, P acts as a change of basis matrix, converting vectors from the basis associated with B to the basis associated with A. This equivalence demonstrates that similar matrices are simply different representations of the same linear transformation under a specific change of basis.

2   Matrix Similarity and Invariants

Similar matrices represent the same linear transformation expressed in different bases, thereby preserving several intrinsic properties. These invariants are crucial for understanding the essence of matrix transformations and their theoretical implications.

> **Theorem 9.8** The following attributes are invariant under similarity transformations:
>
> (a) Determinant and invertibility
>
> (b) Trace
>
> (c) Rank and nullity (see § 8.5.4 for definitions)
>
> (d) Characteristic polynomial
>
> (e) Eigenvalues, along with their algebraic and geometric multiplicities (excluding eigenvectors)

Proof. Let A and B be two similar matrices, meaning there exists an invertible matrix P such that $B = P^{-1}AP$. We will demonstrate that each of the listed attributes remains invariant under this similarity transformation.

(a) Determinant and Invertibility: Using the property of determinants,

$$\det(P^{-1}AP) = \det(P^{-1})\det(A)\det(P).$$

Since $\det P^{-1} \det P = 1$, we have:

$$\det B = \det(P^{-1}AP) = \det A.$$

Thus, the determinant is invariant. As a result, if A is invertible (i.e., $\det A \neq 0$), then B is also invertible, and vice versa.

(b) Trace: Using the cyclic property of the trace, we get:

$$\operatorname{tr} B = \operatorname{tr}(P^{-1}AP) = \operatorname{tr}(APP^{-1}) = \operatorname{tr} A.$$

Hence, we have used the cyclic property of trace.

(c) Rank and Nullity: We first prove that a similarity transform preserves the nullity (the dimension of the null space or kernel) of a matrix.

Let $x \in \mathbb{C}^n$. Then $x \in \ker(A)$ if and only if $Ax = 0$. Consider $B = P^{-1}AP$ for some invertible matrix P.

Suppose $y \in \ker(B)$. Then

$$By = 0 \implies P^{-1}APy = 0 \implies A(Py) = 0.$$

Let $x = Py$, so $x \in \ker(A)$. Conversely, if $x \in \ker(A)$, then $Ax = 0$, and for $y = P^{-1}x$, we have

$$By = P^{-1}APy = P^{-1}Ax = P^{-1} \cdot 0 = 0,$$

so $y \in \ker(B)$.

Thus, the map $x \mapsto P^{-1}x$ establishes a bijection between $\ker(A)$ and $\ker(B)$, implying $\dim \ker(A) = \dim \ker(B)$. Therefore, $\operatorname{null}(A) = \operatorname{null}(B)$.

The rank of a matrix (the dimension of its column space) is related to the nullity by the rank-nullity theorem:

$$\operatorname{rank}(A) + \operatorname{nullity}(A) = n.$$

Since the nullity is invariant under similarity transformations, the rank must also be invariant. Therefore, similarity transforms preserve both rank and nullity.

(d) Characteristic Polynomial: Since determinant is invariant under similarity transformations, we have:

$$\det(\lambda I - B) = \det(\lambda I - P^{-1}AP) = \det(P^{-1}(\lambda I - A)P) = \det(\lambda I - A).$$

Thus, the characteristic polynomial is the same for A and B, meaning it is invariant under similarity transformations.

(e) Eigenvalues, Algebraic and Geometric Multiplicities: Eigenvalues are the solutions to the characteristic equation $\det(\lambda I - A) = 0$. Since the characteristic polynomial is invariant under similarity transformations, the eigenvalues of A and B are the same.

The algebraic multiplicity (the number of times an eigenvalue appears as a root of the characteristic equation) is also preserved since the characteristic polynomial is unchanged.

The geometric multiplicity (the dimension of the eigenspace corresponding to each eigenvalue) is determined by the null space of $\lambda I - A$. As shown above, the nullity (and therefore geometric multiplicity) is invariant under similarity transformations.

While eigenvalues and their multiplicities are invariant, eigenvectors themselves are not necessarily preserved, as the similarity transformation affects the eigenbasis.

□

These invariants are instrumental in confirming or refuting claims of matrix similarity. For instance, consider the matrices:

$$\begin{bmatrix} 1 & 0 \\ 0 & 0 \end{bmatrix} \quad \text{and} \quad \begin{bmatrix} 0 & 0 \\ 1 & 0 \end{bmatrix}.$$

These matrices are not similar, as evidenced by their differing traces.

To determine whether two matrices are similar, particularly when they are diagonalizable, the following theorems are useful:

Theorem 9.9 Matrix similarity is transitive. That is, if matrices A, B, and C are such that A is similar to both B and C, then B and C are similar to each other.

Proof. Since A is similar to both B and C, there exist invertible matrices P and Q such that:
$$B = P^{-1}AP, \quad \text{and} \quad A = Q^{-1}CQ.$$

Substituting the second equation into the first, we get:

$$B = P^{-1}(Q^{-1}CQ)P = (P^{-1}Q^{-1})C(QP) = (QP)^{-1}CQP.$$

This shows that B is similar to C, with QP acting as the similarity transformation, confirming the transitivity of matrix similarity.   □

Theorem 9.10 Two diagonalizable matrices A and B, each in $\mathbb{C}^{n \times n}$, are similar if and only if they have identical characteristic polynomials.

Readers are encouraged to construct the proof of Theorem 9.10 after studying the next section on diagonalization. This exercise will help reinforce the understanding of these underlying principles.

Exercise 9.8 Determine whether the following matrices are similar to $\begin{bmatrix} 1 & 0 \\ 1 & 0 \end{bmatrix}$:

(a) $\begin{bmatrix} 1 & 1 \\ 0 & 0 \end{bmatrix}$, (b) $\begin{bmatrix} 0 & 0 \\ 1 & 1 \end{bmatrix}$, (c) $\begin{bmatrix} 1 & 0 \\ 0 & 0 \end{bmatrix}$.

9.2.2 Diagonalization

1 Definition

Similar matrices share many key characteristics, leading to the question: among matrices similar to a given matrix, which form is the simplest? As demonstrated in Example 9.6 from the previous section, a diagonal matrix represents the simplest form when the original matrix has a complete eigenbasis in $\mathbb{C}^n$. This observation introduces the concept of diagonalization, also known as eigendecomposition.

Definition 9.7 — Diagonalizable Matrix. A square matrix A is diagonalizable if it is similar to a diagonal matrix Λ. Specifically, there exists an invertible matrix P such that:
$$\Lambda = P^{-1}AP. \tag{9.10}$$
The process of converting A into Λ is referred to as diagonalization, or eigendecomposition.

■ **Example 9.7** To deepen understanding of Def. 9.7 and to demonstrate how to derive P and Λ for a specific matrix A, consider the matrix from Example 9.6:
$$A = \begin{bmatrix} 0.7 & 0.2 \\ 0.3 & 0.8 \end{bmatrix}.$$
The eigenvalues of A are $\lambda_1 = 0.5$ and $\lambda_2 = 1$, with corresponding eigenvectors $v_1 = \begin{bmatrix} -1 \\ 1 \end{bmatrix}$ and $v_2 = \begin{bmatrix} 2 \\ 3 \end{bmatrix}$.

This configuration suggests that the linear transformation induced by A scales vectors along v_1 by a factor of 0.5 and leaves vectors along v_2 unchanged. This behavior corresponds to non-uniform scaling, as represented by the diagonal matrix Λ:
$$\Lambda = \begin{bmatrix} 0.5 & 0 \\ 0 & 1 \end{bmatrix}.$$
Thus, we expect A to be similar to Λ.

Indeed, if we form $P = \begin{bmatrix} v_1 & v_2 \end{bmatrix} = \begin{bmatrix} -1 & 2 \\ 1 & 3 \end{bmatrix}$, where the columns are the eigenvectors of A, P will diagonalize A to Λ:
$$P^{-1}AP = \begin{bmatrix} -1 & 2 \\ 1 & 3 \end{bmatrix}^{-1} \begin{bmatrix} 0.7 & 0.2 \\ 0.3 & 0.8 \end{bmatrix} \begin{bmatrix} -1 & 2 \\ 1 & 3 \end{bmatrix} = \begin{bmatrix} 0.5 & 0 \\ 0 & 1 \end{bmatrix}.$$
This confirms that A is similar to Λ. ∎

2 Theorem of Diagonalizability

Some matrices are not diagonalizable. An example is $A = \begin{bmatrix} 1 & 1 \\ 0 & 1 \end{bmatrix}$, as seen in Example 9.4. This matrix represents a shear transformation, which shifts vectors in one direction without scaling them uniformly in two directions, as a diagonal matrix would. Because it lacks enough distinct directions (eigenvectors) for scaling, it is not diagonalizable.

The following theorem provides a concise characterization of diagonalizability based on the presence of a complete set of linearly independent eigenvectors.

Theorem 9.11 A square matrix $A \in \mathbb{C}^{n \times n}$ is diagonalizable if and only if it possesses n linearly independent eigenvectors. In such cases, A can be diagonalized using the formula:

$$P^{-1}AP = \Lambda \equiv \begin{bmatrix} \lambda_1 & 0 & \cdots & 0 \\ 0 & \lambda_2 & \cdots & 0 \\ 0 & 0 & \ddots & 0 \\ 0 & 0 & \cdots & \lambda_n \end{bmatrix}, \tag{9.11}$$

where λ_i are the eigenvalues of A, and P is defined as:

$$P = \begin{bmatrix} |v_1\rangle & |v_2\rangle & \cdots & |v_n\rangle \end{bmatrix},$$

with each column vector $|v_i\rangle$ being an eigenvector of A corresponding to λ_i.

Proof. Assume A has n linearly independent eigenvectors $\{|v_i\rangle\}$, corresponding to eigenvalues $\{\lambda_i\}$, not necessarily distinct.

Since the set $\{|v_i\rangle\}$ is linearly independent, it forms an eigenbasis of $\mathbb{C}^n$ associated with A, and thus P is invertible. To prove Eq. 9.11, it suffices to demonstrate that

$$P\Lambda = AP, \tag{9.12}$$

where $\Lambda = \text{diag}(\lambda_1, \lambda_2, \ldots, \lambda_n)$.

Examining this equation column by column, the ith column of AP is $A|v_i\rangle$, which is the product of A and the ith column of P. On the other hand, the ith column of $P\Lambda$ is $\lambda_i |v_i\rangle$ according to matrix multiplication.

Since $|v_i\rangle$ is an eigenvector associated with λ_i, we have $A|v_i\rangle = \lambda_i |v_i\rangle$. Therefore, each column of $P\Lambda$ matches the respective column of AP, confirming that:

$$P\Lambda = AP,$$

leading to:

$$P^{-1}AP = \Lambda,$$

which proves that A is diagonalizable.

Conversely, if A is diagonalizable, there exists an invertible matrix P such that $P^{-1}AP = \Lambda$, where Λ is diagonal. Each column of P corresponds to an eigenvector of A, implying that A must have n linearly independent eigenvectors.   □

In essence, Theorem 9.11 asserts that any matrix A with a complete eigenbasis can be represented as a diagonal matrix under this basis. Conversely, if A lacks a complete eigenbasis, it is not diagonalizable.

Note that the diagonalization of a matrix is not unique. First, the eigenvectors can be scaled. Furthermore, the diagonal matrix Λ is not unique, as any permutation of its eigenvalues—along with the corresponding reordering of the eigenvectors in P—remains a valid configuration.

Exercise 9.9 Determine a matrix P such that

$$P^{-1} \begin{bmatrix} 2 & 0 \\ 0 & 3 \end{bmatrix} P = \begin{bmatrix} 3 & 0 \\ 0 & 2 \end{bmatrix}.$$

Describe the steps and considerations involved in rearranging the eigenvalues and corresponding eigenvectors.

Exercise 9.10 Devise an alternate diagonalization of the matrix A in Example 9.7.

Exercise 9.11 Derive a diagonalization for the matrix A, where

$$A = \begin{bmatrix} 2 & 1 \\ 1 & 2 \end{bmatrix}.$$

Detail the process of finding the eigenvalues, eigenvectors, and constructing the matrix P.

3   Properties of Diagonalization

Combining Theorem 9.8 and Theorem 9.11 leads to several convenient formulas for evaluating the determinant, trace, and rank of a matrix.

Theorem 9.12 For a diagonalizable matrix A with eigenvalues $\lambda_1, \lambda_2, \ldots, \lambda_n$ (not necessarily distinct), the following properties hold:

(a) $\det A = \prod_{i=1}^{n} \lambda_i,$ (9.13a)

(b) $\operatorname{tr} A = \sum_{i=1}^{n} \lambda_i,$ (9.13b)

(c) $\operatorname{rank}(A)$ is the sum of the algebraic multiplicities of its nonzero eigenvalues.
                                                                          (9.13c)

These properties are straightforward to verify since the right-hand side corresponds directly to the properties of the diagonal matrix $\Lambda = \operatorname{diag}(\lambda_1, \lambda_2, \ldots, \lambda_n)$, which shares the same characteristics as A.

ⓘ Equations 9.13a and 9.13b hold true for non-diagonalizable matrices as well, while Eq. 9.13c may not apply in the case of non-diagonalizable matrices.

Powers of a Diagonalizable Matrix

Diagonalization simplifies the computation of A^k for a diagonalizable matrix A and a positive integer k.

Theorem 9.13 Given a diagonalizable matrix A such that $\Lambda = P^{-1}AP$ for some invertible matrix P and diagonal matrix Λ, then for any positive integer k:

$$A^k = P\Lambda^k P^{-1}.$$ (9.14)

Proof. Since $\Lambda = P^{-1}AP$, we can express A as $A = P\Lambda P^{-1}$. Then, for any positive integer k,

$$
\begin{aligned}
A^k &= (P\Lambda P^{-1})(P\Lambda P^{-1})\cdots(P\Lambda P^{-1}) \quad \text{(repeated k times)} \\
&= P\Lambda(P^{-1}P)\Lambda(P^{-1}P)\cdots(P^{-1}P)\Lambda P^{-1} \\
&= P\Lambda^k P^{-1}.
\end{aligned}
$$

□

Exercise 9.12 Compute A^k for the matrix A, where k is a positive integer and

$$
A = \begin{bmatrix} 0.7 & 0.2 \\ 0.3 & 0.8 \end{bmatrix}.
$$

Inverse of a Diagonalizable Matrix

Theorem 9.14 Given a diagonalizable matrix A such that $\Lambda = P^{-1}AP$ for some invertible matrix P and a diagonal matrix Λ with non-zero diagonal elements, A is invertible, and:

$$
A^{-1} = P\Lambda^{-1}P^{-1}. \tag{9.15}
$$

9.2.3 $*$ Commutative Matrices and Shared Eigenbasis

Eigenbases, introduced in § 9.1.4, play a key role in the formulation of the uncertainty principle, a cornerstone of quantum mechanics. Mathematically, the uncertainty principle builds on the following theorem.

Theorem 9.15 Given two diagonalizable matrices $A, B \in \mathbb{C}^{n \times n}$, they commute if and only if there exists a common eigenbasis for $\mathbb{C}^n$ shared by both A and B.

Proof. First, assume that $\{|v_i\rangle\}$ is a common eigenbasis for both A and B, corresponding to eigenvalues a_i for A and b_i for B. If A and B share this eigenbasis, they can be simultaneously diagonalized by the same matrix $P = [|v_1\rangle \; |v_2\rangle \; \cdots \; |v_n\rangle]$, giving

$$
D_A = P^{-1}AP, \quad D_B = P^{-1}BP,
$$

and thus

$$
A = PD_AP^{-1}, \quad B = PD_BP^{-1},
$$

where $D_A = \text{diag}(a_1, a_2, \ldots, a_n)$ and $D_B = \text{diag}(b_1, b_2, \ldots, b_n)$. Hence,

$$
AB = PD_AD_BP^{-1}, \quad BA = PD_BD_AP^{-1}.
$$

Since diagonal matrices commute, $D_AD_B = D_BD_A$, leading to $AB = BA$.

Conversely, suppose $AB = BA$. Since A is diagonalizable, there exists a basis of eigenvectors $\{|v_i\rangle\}$ such that $A|v_i\rangle = a_i|v_i\rangle$. Applying B to this equation gives

$$
B(A|v_i\rangle) = a_iB|v_i\rangle.
$$

Since $AB = BA$, we have

$$
A(B|v_i\rangle) = a_iB|v_i\rangle,
$$

implying that $B|v_i\rangle$ is also an eigenvector of A with eigenvalue a_i.

There are two cases:

(a) If a_i has algebraic multiplicity one, $B|v_i\rangle$ must be a scalar multiple of $|v_i\rangle$, meaning $|v_i\rangle$ is also an eigenvector of B.

(b) If a_i has algebraic multiplicity greater than one, $B|v_i\rangle$ remains in the eigenspace of A corresponding to a_i, so B preserves the eigenspaces of A. Thus, we can choose a basis for each eigenspace of A that consists of eigenvectors of B.

Therefore, a common eigenbasis for both A and B can always be found, provided they are both diagonalizable and commute. $\qquad\square$

Exercise 9.13 Consider the following questions related to Theorem 9.15:

(a) The identity matrix I_n commutes with any $n \times n$ matrix A. Explain why this does not contradict Theorem 9.15, even though I_n commutes with any matrix regardless of a shared eigenbasis.

(b) Provide an example of two non-diagonalizable (deficient) matrices A and B that commute but do not share a common eigenbasis, showing that Theorem 9.15 does not always apply in such cases.

(c) Provide a counterexample where at least one of the matrices A or B is not diagonalizable, and explain why this invalidates Theorem 9.15.

9.3 Spectral Decomposition of Normal Matrices

This section focuses on normal matrices, which are distinguished by their ability to be diagonalized by unitary matrices. Normal matrices are fundamental in both linear algebra and quantum computing, as they enable spectral decomposition. This powerful theorem reveals the underlying structure of operators in complex vector spaces, providing deep insights into their behavior.

Eigendecomposition vs. Spectral Decomposition

Eigendecomposition applies to diagonalizable matrices and expresses the matrix in diagonal form using its eigenvectors. It can apply to any matrix that has a full set of linearly independent eigenvectors. The decomposition is expressed as $A = P\Lambda P^{-1}$, where P is the matrix of eigenvectors, and Λ is the diagonal matrix of eigenvalues.

Spectral decomposition is a special case of eigendecomposition that applies specifically to normal matrices (including diagonal, Hermitian, unitary, and real symmetric matrices). Normal matrices are unitarily diagonalizable, meaning they can be written as $A = U\Lambda U^\dagger$, where U is unitary. Consequently, A can also be expressed as $A = \sum_i \lambda_i |\phi_i\rangle \langle\phi_i|$, where $\{|\phi_i\rangle\}$ is an orthonormal eigenbasis of A, and λ_i are the corresponding eigenvalues.

9.3.1 Normal Matrices

Normal matrices are square matrices that commute with their conjugate transpose:

Definition 9.8 — Normal Matrix. A normal matrix A is a square matrix that satisfies the commutation relation:

$$AA^\dagger = A^\dagger A.$$

Theorem 9.16 Unitary matrices, which satisfy $UU^\dagger = U^\dagger U = I$, Hermitian matrices, which satisfy $H = H^\dagger$, and skew-Hermitian matrices, which satisfy $A^\dagger = -A$, are normal matrices.

Proof.   Unitary matrices are normal because $UU^\dagger = U^\dagger U = I$.

Hermitian matrices are normal because $HH^\dagger = H^\dagger H = H^2$. Similarly, skew-Hermitian matrices are normal because $AA^\dagger = A^\dagger A = -A^2$.   □

Exercise 9.14 Demonstrate that a 2×2 complex matrix of the form

$$\begin{bmatrix} a & b \\ b & a \end{bmatrix},$$

where a is a real number ($a \in \mathbb{R}$) and b is a complex number ($b \in \mathbb{C}$), is a normal matrix.

9.3.2 Spectral Decomposition

Normal matrices are distinguished by their ability to be diagonalized by unitary matrices, which is formally expressed by the following theorem:

Theorem 9.17 Let A be a normal matrix, i.e., a square matrix that satisfies $AA^\dagger = A^\dagger A$. Then, there exists a unitary matrix U and a diagonal matrix Λ such that:

$$A = U\Lambda U^\dagger. \tag{9.16}$$

The proof of this theorem is deferred until § 9.3.4. We will first explore its key implications and applications.

Theorem 9.18 In Theorem 9.17, Λ is a diagonal matrix containing the eigenvalues of A:

$$\Lambda = \text{diag}(\lambda_1, \lambda_2, \ldots, \lambda_n), \tag{9.17}$$

and the columns of U are the eigenvectors of A:

$$U = \begin{bmatrix} |\phi_1\rangle & |\phi_2\rangle & \cdots & |\phi_n\rangle \end{bmatrix}, \tag{9.18}$$

where $A|\phi_i\rangle = \lambda_i |\phi_i\rangle$.

Proof.   We can rewrite Eq. 9.16 as

$$AU = U\Lambda.$$

This equation is precisely $A|\phi_i\rangle = \lambda_i |\phi_i\rangle$ for each column of U.   □

1 Spectral Decomposition

Theorem 9.19 A normal matrix $A \in \mathbb{C}^{n \times n}$ can be represented as a sum of projection operators weighted by its eigenvalues, known as the spectral decomposition of the matrix:

$$A = \sum_{i=1}^{n} \lambda_i \, |\phi_i\rangle \, \langle\phi_i| , \qquad (9.19)$$

where λ_i are the eigenvalues of A, and $|\phi_i\rangle$ are the corresponding normalized eigenvectors.

Proof. Equation 9.19 follows directly from Eq. 9.16. To confirm this, we show that Eq. 9.19 correctly yields the eigenvalues $\{\lambda_i\}$ and the corresponding eigenvectors $\{|\phi_i\rangle\}$:

$$A \, |\phi_i\rangle = \sum_{j=1}^{n} \lambda_j \, |\phi_j\rangle\langle\phi_j| \, |\phi_i\rangle$$

$$= \sum_{j=1}^{n} \lambda_j \, |\phi_j\rangle \, \delta_{ji}$$

$$= \lambda_i \, |\phi_i\rangle .$$

Therefore, A satisfies $A \, |\phi_i\rangle = \lambda_i \, |\phi_i\rangle$ for each i, confirming the spectral decomposition representation. $\qquad \square$

Exercise 9.15 Confirm that Eq. 9.19 accurately represents the matrix given below by finding its eigenvalues and an orthonormal eigenbasis:

$$\begin{bmatrix} 1 & i \\ -i & -1 \end{bmatrix}.$$

2 Completeness of Orthonormal Basis

Theorem 9.20 A normal matrix has a special property: it possesses a set of orthonormal eigenvectors, $\{|\phi_i\rangle\}$, which form a complete orthonormal eigenbasis.

The orthonormal condition is expressed as

$$\langle\phi_i|\phi_j\rangle = \delta_{ij}, \qquad (9.20)$$

where δ_{ij} is the Kronecker delta function, defined as $\delta_{ij} = 1$ if $i = j$ and $\delta_{ij} = 0$ if $i \neq j$.

The completeness condition is expressed as

$$\sum_{i=1}^{n} |\phi_i\rangle\langle\phi_i| = I, \qquad (9.21)$$

where I is the identity matrix.

Proof. The orthonormal condition follows from Theorem 9.17, where $\{|\phi_i\rangle\}$ are given by the columns of U. Since U is unitary, its columns form an orthonormal basis.

The completeness condition is precisely the spectral decomposition of the identity matrix I. For I, the sole eigenvalue is one, and every nonzero vector in $\mathbb{C}^n$ is an eigenvector. Consequently, every orthonormal basis serves as a set of orthonormal eigenvectors of I.   □

9.3.3 Applications

1 Eigenvalues of Special Matrices

Given Theorems 9.17 and 9.19, it is instructive to classify normal matrices into unitary, Hermitian, or skew-Hermitian types based on the characteristics of their eigenvalues:

Theorem 9.21 Normal matrices reduce to the following special types under conditions applied uniformly to all eigenvalues λ_i for $i = 1, 2, \ldots, n$:

(a) A normal matrix is **unitary** ($U^\dagger U = I$) if and only if all its eigenvalues have modulus one, $|\lambda_i| = 1$.

(b) A normal matrix is **Hermitian** ($A^\dagger = A$) if and only if all its eigenvalues are real, $\lambda_i \in \mathbb{R}$.

(c) A normal matrix is **skew-Hermitian** ($A^\dagger = -A$) if and only if all its eigenvalues are purely imaginary, $i\lambda_i \in \mathbb{R}$.

The proof of Theorem 9.21 is proposed as an exercise for the reader (Problem 9.20).

An intuition for this theorem is that unitary and Hermitian matrices, as subsets of normal matrices, are distinguished by their eigenvalues. Drawing an analogy with the complex number plane, consider the behavior of A in its action on each eigenvector, $A\,|v_i\rangle = \lambda_i\,|v_i\rangle$. Hermitian matrices can be likened to real numbers ($\lambda \in \mathbb{R}$) as they scale vectors in each eigenspace by a real factor, whereas unitary matrices are analogous to complex numbers on the unit circle ($e^{i\theta}$ for $\theta \in \mathbb{R}$), rotating vectors by a phase factor without altering their norms.

Exercise 9.16 Provide an example of a 2×2 normal matrix, in its matrix representation, that is neither unitary, Hermitian, nor skew-Hermitian.

Exercise 9.17 Specify the condition that a matrix must satisfy to be both unitary and Hermitian.

2 Trace and Determinant of Normal Matrices

The spectral representation of normal matrices is particularly useful when studying functions of matrices. While a comprehensive discussion on this topic is reserved for later in § 11.3.1, below are trace and determinant as examples:

Theorem 9.22 If A is a normal matrix that can be diagonalized into the diagonal matrix $\Lambda = \text{diag}(\lambda_1, \lambda_2, \ldots, \lambda_n)$, then

$$\text{tr}\, A = \sum_{i=1}^{n} \lambda_i. \tag{9.22}$$

Theorem 9.23 If A is a normal matrix that can be diagonalized into the diagonal matrix $\Lambda = \text{diag}(\lambda_1, \lambda_2, \ldots, \lambda_n)$, then

$$\det A = \prod_{i=1}^{n} \lambda_i. \tag{9.23}$$

These theorems hold because trace and determinant are invariant under similarity transformations (Theorem 9.8).

Theorem 9.24 If A is a normal matrix that can be diagonalized into the diagonal matrix $\Lambda = \text{diag}(\lambda_1, \lambda_2, \ldots, \lambda_n)$, then

$$\text{tr}(A^\dagger A) = \sum_{i=1}^{n} |\lambda_i|^2. \tag{9.24}$$

The quantity $\|A\| \equiv \text{tr}(A^\dagger A)$, being real and non-negative, is known as the Frobenius norm of A.

Proof. We observe that:

$$A^\dagger A = \left(\sum_{i}^{n} \lambda_i^* \, |\phi_i\rangle \, \langle\phi_i| \right) \left(\sum_{j}^{n} \lambda_j \, |\phi_j\rangle \, \langle\phi_j| \right)$$

$$= \sum_{i}^{n} \sum_{j}^{n} \lambda_i^* \lambda_j \, |\phi_i\rangle \, \langle\phi_i| \, |\phi_j\rangle \, \langle\phi_j|$$

$$= \sum_{i}^{n} \sum_{j}^{n} \lambda_i^* \lambda_j \, |\phi_i\rangle \, \delta_{ij} \, \langle\phi_j|$$

$$= \sum_{i}^{n} |\lambda_i|^2 \, |\phi_i\rangle \, \langle\phi_i| .$$

This means $A^\dagger A$ has a spectral decomposition with eigenvalues $|\lambda_i|^2$. Using Theorem 9.22, we arrive at Eq. 9.24.   □

In these theorems, each occurrence in a degenerate multiplicity of eigenvalues is counted individually.

3   Projection Operators Revisited

The outer product $|\phi_i\rangle \langle\phi_i|$ in Theorem 9.19 is a projection operator onto the eigenspace associated with the eigenvalue λ_i.

An incomplete orthonormal set of vectors $S_P = \{|\phi_i\rangle\}$ for $i = 1, 2, \ldots, r$ within $\mathbb{C}^n$, where $r < n$, represents a subset of a complete orthonormal basis. The following summation of these vectors forms a projection operator that projects any vector onto the subspace spanned by S_P:

$$P = \sum_{i=1}^{r} |\phi_i\rangle \langle\phi_i| .$$ (9.25)

When P is applied to any vector $|\psi\rangle$, the resulting vector is:

$$P |\psi\rangle = \sum_{i=1}^{r} \langle\phi_i|\psi\rangle |\phi_i\rangle ,$$ (9.26)

ensuring $|\psi\rangle$ is confined within the span of S_P.

Exercise 9.18 Verify that $P^\dagger P = P$, establishing that P is indeed a projection operator.

The non-projected component $|\psi\rangle - P |\psi\rangle$ lies within the orthogonal complement $S^\perp$, reaffirming our earlier discussions in §§ 8.1.5.4 and 8.2.3.2.

Exercise 9.19 Demonstrate that $|\psi\rangle - P |\psi\rangle$ resides in $S_P^\perp$ by:

(a) identifying an orthonormal basis for $S_P^\perp$ and confirming that $|\psi\rangle - P |\psi\rangle$ can be expressed as a linear combination of these basis vectors,

(b) proving that $I - P$ acts as a projection operator onto $S_P^\perp$.

4   Rank-r Approximation of Matrices

The spectral decomposition allows an $n \times n$ matrix A to be decomposed into n rank-one matrices, as each component $\lambda_i |\phi_i\rangle\langle\phi_i|$ is of rank one. If A is a full-rank matrix (of rank n), it is possible to approximate A by a lower-rank matrix by retaining the components with the most significant eigenvalues.

Assuming without loss of generality that the eigenvalues are sorted in descending order by their magnitudes: $|\lambda_1| \geq |\lambda_2| \geq |\lambda_3| \geq \cdots \geq |\lambda_n| > 0$, a rank-$r$ approximation of A (where $r < n$) can be achieved by summing only the first r terms:

$$A \approx \sum_{i=1}^{r} \lambda_i |\phi_i\rangle \langle\phi_i| .$$ (9.27)

This approximation method proves particularly useful when many eigenvalues are zero or near zero, allowing significant data compression with minimal loss of information by storing only the first r eigenvalues and eigenvectors. This approach is also applicable through Singular Value Decomposition and can be extended to matrices of any size.

9.3.4   ＊Proof of the Spectral Theorem

A more concise statement of the spectral decomposition theorem (Theorem 9.17) is:

Theorem 9.25 A matrix is unitarily diagonalizable if and only if it is a normal matrix.

Proof.

1. If a matrix is unitarily diagonalizable, then it is normal.

Suppose A is unitarily diagonalizable. This means there exists a unitary matrix U (where $U^\dagger U = UU^\dagger = I$) and a diagonal matrix Λ such that $A = U\Lambda U^\dagger$.

Taking the conjugate transpose of both sides, we have

$$A^\dagger = (U\Lambda U^\dagger)^\dagger = U\Lambda^\dagger U^\dagger.$$

Then

$$\begin{aligned}
AA^\dagger &= (U\Lambda U^\dagger)(U\Lambda^\dagger U^\dagger) \\
&= U\Lambda(U^\dagger U)\Lambda^\dagger U^\dagger \\
&= U\Lambda\Lambda^\dagger U^\dagger,
\end{aligned}$$

and

$$\begin{aligned}
A^\dagger A &= (U\Lambda^\dagger U^\dagger)(U\Lambda U^\dagger) \\
&= U\Lambda^\dagger(U^\dagger U)\Lambda U^\dagger \\
&= U\Lambda^\dagger \Lambda U^\dagger.
\end{aligned}$$

Since Λ is a diagonal matrix, $\Lambda\Lambda^\dagger = \Lambda^\dagger\Lambda$, which means $AA^\dagger = A^\dagger A$, and A is normal.

2. If a matrix is normal, then it is unitarily diagonalizable.

We will proceed by induction on the size of the matrix.

Base Case: For a 1×1 matrix (a scalar), the statement is trivially true.

Inductive Hypothesis: Assume the statement is true for all normal matrices of size $(n-1) \times (n-1)$.

Inductive Step: Consider a normal matrix A of size $n \times n$. Since A is a complex matrix, it has at least one eigenvalue λ and a corresponding eigenvector v (see Theorem 9.4). We can normalize v and extend it to an orthonormal basis $\{v, u_2, \ldots, u_n\}$ of $\mathbb{C}^n$ using the Gram-Schmidt process (see § 6.5.3).

Let U_1 be the unitary matrix whose columns are $v, u_2, \ldots, u_n$. Since $Av = \lambda v$,

$$AU_1 = A \begin{bmatrix} | & | & & | \\ v & u_2 & \cdots & u_n \\ | & | & & | \end{bmatrix} = \begin{bmatrix} | & | & & | \\ \lambda v & Au_2 & \cdots & Au_n \\ | & | & & | \end{bmatrix}.$$

Let $B = U_1^\dagger A U_1$. Then

$$\begin{aligned}
B = U_1^\dagger A U_1 &= U_1^\dagger \begin{bmatrix} | & | & & | \\ \lambda v & Au_2 & \cdots & Au_n \\ | & | & & | \end{bmatrix} \\
&= \begin{bmatrix} - & v^\dagger & - \\ - & u_2^\dagger & - \\ & \vdots & \\ - & u_n^\dagger & - \end{bmatrix} \begin{bmatrix} | & | & & | \\ \lambda v & Au_2 & \cdots & Au_n \\ | & | & & | \end{bmatrix}
\end{aligned}$$

$$
= \begin{bmatrix} \lambda v^\dagger v & * & \cdots & * \\ \lambda u_2^\dagger v & * & \cdots & * \\ \vdots & \vdots & & \vdots \\ \lambda u_n^\dagger v & * & \cdots & * \end{bmatrix} = \begin{bmatrix} \lambda & * & \cdots & * \\ 0 & * & \cdots & * \\ \vdots & \vdots & & \vdots \\ 0 & * & \cdots & * \end{bmatrix},
$$

where the entries marked $*$ are some complex numbers.

We can write B as a block matrix:

$$
B = \begin{bmatrix} \lambda & * \\ 0 & C \end{bmatrix},
$$

where C is an $(n-1) \times (n-1)$ matrix, and $*$ is some row vector.

We claim that B is normal. Indeed,

$$
\begin{aligned}
BB^\dagger &= (U_1^\dagger A U_1)(U_1^\dagger A U_1)^\dagger \\
&= U_1^\dagger A U_1 U_1^\dagger A^\dagger U_1 \\
&= U_1^\dagger A A^\dagger U_1 \\
&= U_1^\dagger A^\dagger A U_1 \quad (\because \ A^\dagger A = A A^\dagger) \\
&= U_1^\dagger A^\dagger U_1 U_1^\dagger A U_1 \\
&= (U_1^\dagger A U_1)^\dagger (U_1^\dagger A U_1) \\
&= B^\dagger B.
\end{aligned}
$$

Then

$$
\begin{bmatrix} \lambda & * \\ 0 & C \end{bmatrix} \begin{bmatrix} \lambda^* & 0 \\ * & C^\dagger \end{bmatrix} = \begin{bmatrix} \lambda^* & 0 \\ * & C^\dagger \end{bmatrix} \begin{bmatrix} \lambda & * \\ 0 & C \end{bmatrix},
$$

so

$$
\begin{bmatrix} |\lambda|^2 + * & * \\ * & CC^\dagger \end{bmatrix} = \begin{bmatrix} |\lambda|^2 + * & * \\ * & C^\dagger C \end{bmatrix}.
$$

Hence, $CC^\dagger = C^\dagger C$, which means C is normal.

By the inductive hypothesis, C is unitarily diagonalizable, so $C = U_2 \Lambda_2 U_2^\dagger$ for some unitary matrix U_2 and diagonal matrix Λ_2.

Let

$$
U = U_1 \begin{bmatrix} 1 & 0 \\ 0 & U_2 \end{bmatrix}.
$$

Then

$$
\begin{aligned}
D = U^\dagger A U &= \begin{bmatrix} 1 & 0 \\ 0 & U_2^\dagger \end{bmatrix} U_1^\dagger A U_1 \begin{bmatrix} 1 & 0 \\ 0 & U_2 \end{bmatrix} \\
&= \begin{bmatrix} 1 & 0 \\ 0 & U_2^\dagger \end{bmatrix} \begin{bmatrix} \lambda & s \\ 0 & C \end{bmatrix} \begin{bmatrix} 1 & 0 \\ 0 & U_2 \end{bmatrix} \\
&= \begin{bmatrix} 1 & 0 \\ 0 & U_2^\dagger \end{bmatrix} \begin{bmatrix} \lambda & s \\ 0 & CU_2 \end{bmatrix} \\
&= \begin{bmatrix} \lambda & s \\ 0 & U_2^\dagger CU_2 \end{bmatrix} \\
&= \begin{bmatrix} \lambda & s \\ 0 & U_2^\dagger U_2 \Lambda_2 U_2^\dagger U_2 \end{bmatrix}
\end{aligned}
$$

$$= \begin{bmatrix} \lambda & s \\ 0 & \Lambda_2 \end{bmatrix}.$$

Here, λ is a scalar, Λ_2 is diagonal, and s is a row vector.

In addition, we can show that D is normal, using logic similar to the one used to prove B is normal, i.e., $DD^\dagger = D^\dagger D$. This condition translates to:

$$\begin{bmatrix} \lambda\lambda^* + ss^\dagger & \lambda\Lambda_2^* \\ \Lambda_2 s^\dagger & 0 \end{bmatrix} = \begin{bmatrix} \lambda\lambda^* & \lambda^* s + \Lambda_2^* \Lambda_2 \\ s^\dagger \lambda & s^\dagger s \end{bmatrix}.$$

This equality forces $ss^\dagger = 0$, which implies $s = 0$. Therefore, D is diagonal, meaning A is unitarily diagonalizable. $\qquad\blacksquare$

9.4 Hermitian Matrices

Hermitian matrices are a special type of normal matrix that satisfy $A = A^\dagger$. A defining characteristic of Hermitian matrices is that all their eigenvalues are real. They are also guaranteed to possess an orthonormal eigenbasis and allow for spectral decomposition. These properties underscore their critical role in quantum mechanics and quantum computing, where they frequently represent observable quantities that yield real values upon measurement.

9.4.1 Basic Properties

1 Unitarily Diagonalizable with Real Eigenvalues

Hermitian matrices are normal because $A = A^\dagger$ leads to $AA^\dagger = A^\dagger A = A^2$. As a subset of normal matrices, Hermitian matrices allow for spectral decomposition and enjoy all the associated properties, as discussed in § 9.3.

In particular,

Theorem 9.26 All eigenvalues of a Hermitian matrix are real.

Proof. Let λ_i be an eigenvalue of A with a corresponding eigenvector $|v_i\rangle$. Then we have:

$$A |v_i\rangle = \lambda_i |v_i\rangle.$$

Since $A = A^\dagger$, the inner product $\langle v_i|A|v_i\rangle = \langle v_i|A^\dagger|v_i\rangle$ can be computed in two ways:

$$\langle v_i|A|v_i\rangle = \langle v_i|Av_i\rangle = \langle v_i|\lambda_i v_i\rangle = \lambda_i \langle v_i|v_i\rangle,$$
$$\langle v_i|A^\dagger|v_i\rangle = \langle Av_i|v_i\rangle = \langle \lambda_i v_i|v_i\rangle = \lambda_i^* \langle v_i|v_i\rangle.$$

Since $\lambda_i \langle v_i|v_i\rangle = \lambda_i^* \langle v_i|v_i\rangle$ and $\langle v_i|v_i\rangle \neq 0$ (as $|v_i\rangle$ is a nonzero eigenvector), it follows that $\lambda_i = \lambda_i^*$, meaning λ_i is real. $\qquad\blacksquare$

2 Orthonormal Eigenbasis

Due to the spectral theorem, a Hermitian matrix possesses an orthonormal eigenbasis (see § 8.4.2.1). When all eigenvalues are distinct, any set of normalized eigenvectors corresponding to these eigenvalues will naturally form an orthonormal eigenbasis.

However, in cases where eigenvalues are degenerate, constructing an orthonormal basis for the associated eigenspace becomes necessary. This can be achieved using the Gram-Schmidt process.

While proving the spectral theorem for normal matrices is quite involved (see § 9.3.4), here we provide a simpler proof of the following key property of Hermitian matrices, which is frequently used in quantum mechanics:

Theorem 9.27 For a Hermitian matrix, the eigenvectors corresponding to distinct eigenvalues are orthogonal.

Proof. Let A be a Hermitian matrix, meaning $A = A^\dagger$. Suppose $|v_1\rangle$ and $|v_2\rangle$ are eigenvectors of A corresponding to distinct eigenvalues λ_1 and λ_2, respectively:

$$A\,|v_1\rangle = \lambda_1\,|v_1\rangle, \quad A\,|v_2\rangle = \lambda_2\,|v_2\rangle.$$

We will prove that $\langle v_1|v_2\rangle = 0$.

First, consider the inner product $\langle v_2|Av_1\rangle$. Using the fact that $A\,|v_1\rangle = \lambda_1\,|v_1\rangle$, we have:

$$\langle v_2|Av_1\rangle = \lambda_1\,\langle v_2|v_1\rangle.$$

Next, consider the inner product $\langle Av_2|v_1\rangle$. Since A is Hermitian, we have $A^\dagger = A$, so:

$$\langle Av_2|v_1\rangle = \langle v_2|A^\dagger v_1\rangle = \langle v_2|Av_1\rangle.$$

But from $A\,|v_2\rangle = \lambda_2\,|v_2\rangle$, we get:

$$\langle Av_2|v_1\rangle = \lambda_2^*\,\langle v_2|v_1\rangle.$$

Since eigenvalues of Hermitian matrices are always real, we have $\lambda_2^* = \lambda_2$. Thus,

$$\langle Av_2|v_1\rangle = \lambda_2\,\langle v_2|v_1\rangle.$$

Now, equating the two expressions for $\langle v_2|Av_1\rangle$, we obtain:

$$\lambda_1\,\langle v_2|v_1\rangle = \lambda_2\,\langle v_2|v_1\rangle.$$

Since $\lambda_1 \neq \lambda_2$, this implies that:

$$\langle v_2|v_1\rangle = 0.$$

Therefore, the eigenvectors $|v_1\rangle$ and $|v_2\rangle$ are orthogonal.   □

9.4.2 Quadratic Forms

Another important perspective that aids in understanding Hermitian matrices is by examining the properties they impose on a complex quadratic form:

$$\langle x|A|x\rangle = \boldsymbol{x}^\dagger A\boldsymbol{x},$$

where $|x\rangle \in \mathbb{C}^n$ and $A \in \mathbb{C}^{n\times n}$. For a regular matrix, this form invariably yields a complex scalar.

i | In some contexts, $\langle x|A|x\rangle$ is equivalently expressed as $\langle \boldsymbol{x}, A\boldsymbol{x}\rangle$.

Consider an example where $|x\rangle = [x_1 \ x_2]^T$ and $A = \begin{bmatrix} a & b \\ c & d \end{bmatrix}$. Matrix multiplication reveals that:

$$\langle x|A|x\rangle = ax_1^*x_1 + bx_1^*x_2 + cx_1x_2^* + dx_2^*x_2.$$

If all components and coefficients are real, this expression simplifies to the real quadratic form $ax_1^2 + (b+c)x_1x_2 + dx_2^2$.

Exercise 9.20 Identify the conditions that the scalars a, b, c, and d must meet to ensure that the expression $ax_1^*x_1 + bx_1^*x_2 + cx_1x_2^* + dx_2^*x_2$ remains real for any complex numbers x_1 and x_2.

For a Hermitian matrix A, the quadratic form $\langle x|A|x\rangle$ is always real, regardless of $|x\rangle$ (Theorem 7.13). The converse is also true: if $\langle x|A|x\rangle$ is real for any $|x\rangle$, then A is necessarily Hermitian:

Theorem 9.28 A matrix $A \in \mathbb{C}^{n\times n}$ is Hermitian if and only if the quadratic form $\langle x|A|x\rangle$ is real for any $|x\rangle \in \mathbb{C}^n$.

In quantum mechanics, the statistical average of measuring an observable represented by a Hermitian matrix A on a quantum state $|\psi\rangle$ is $\langle \psi|A|\psi\rangle$. Theorem 9.28 ensures this average value is real, affirming its measurability.

Proof. ✳

If Direction

Assume A is a Hermitian matrix. Taking the complex conjugate of $\langle x|A|x\rangle$, we have:

$$(\langle x|A|x\rangle)^* = \langle x|A^\dagger|x\rangle = \langle x|A|x\rangle,$$

where we used the fact that $A = A^\dagger$. Hence, the quadratic form is real.

Only-if Direction

Suppose $\langle x|A|x\rangle$ is real for any $|x\rangle \in \mathbb{C}^n$. Let $|x\rangle = |v\rangle + |w\rangle$, where $|v\rangle, |w\rangle \in \mathbb{C}^n$. Then

$$\langle x|A|x\rangle = (\langle v| + \langle w|)A(|v\rangle + |w\rangle)$$
$$= \langle v|A|v\rangle + \langle v|A|w\rangle + \langle w|A|v\rangle + \langle w|A|w\rangle.$$

Since $\langle v|A|v\rangle$, $\langle w|A|w\rangle$, and $\langle x|A|x\rangle$ are real, $\langle v|A|w\rangle + \langle w|A|v\rangle$ must also be real. Therefore,

$$\langle v|A|w\rangle + \langle w|A|v\rangle = (\langle v|A|w\rangle + \langle w|A|v\rangle)^*$$
$$= \langle w|A^\dagger|v\rangle + \langle v|A^\dagger|w\rangle$$
$$= \langle v|A^\dagger|w\rangle^* + \langle v|A^\dagger|w\rangle. \tag{9.28}$$

Now let $|x\rangle = |v\rangle + i\,|w\rangle$. Then

$$\langle x|A|x\rangle = (\langle v| - i\,\langle w|)A(|v\rangle + i\,|w\rangle)$$

$$= \langle v|A|v \rangle + i \langle v|A|w \rangle - i \langle w|A|v \rangle + \langle w|A|w \rangle .$$

Since $\langle v|A|v \rangle$, $\langle w|A|w \rangle$, and $\langle x|A|x \rangle$ are real, $\langle v|A|w \rangle - \langle w|A|v \rangle$ is purely imaginary. Therefore,

$$\begin{aligned} \langle v|A|w \rangle - \langle w|A|v \rangle &= -(\langle v|A|w \rangle - \langle w|A|v \rangle)^* \\ &= - \langle w|A^\dagger|v \rangle + \langle v|A^\dagger|w \rangle \\ &= - \langle v|A^\dagger|w \rangle^* + \langle v|A^\dagger|w \rangle . \end{aligned} \tag{9.29}$$

Adding Eqs. 9.28 and 9.29, we get:

$$2 \langle v|A|w \rangle = 2 \langle v|A^\dagger|w \rangle .$$

Thus, $\langle v|A|w \rangle = \langle v|A^\dagger|w \rangle$ for all $|v \rangle , |w \rangle \in \mathbb{C}^n$. Therefore, $A = A^\dagger$, and thus A is Hermitian. $\square$

9.4.3   Definite Hermitian Matrices

An important concept associated with Hermitian matrices is their definiteness. This property classifies matrices based on the behavior of their corresponding quadratic forms under specific conditions.

Definition 9.9 — Definiteness of a Hermitian Matrix. A Hermitian matrix $A \in \mathbb{C}^{n \times n}$ is described as follows based on the quadratic form $\langle x|A|x \rangle$ for all nonzero vectors $|x \rangle \in \mathbb{C}^n$:

- Positive definite if $\langle x|A|x \rangle > 0$.

- Positive semidefinite if $\langle x|A|x \rangle \geq 0$.

- Negative definite if $\langle x|A|x \rangle < 0$.

- Negative semidefinite if $\langle x|A|x \rangle \leq 0$.

If A does not qualify as either positive or negative semidefinite, it is considered indefinite.

When considering quadratic forms of real matrices, there are no constraints on the type of matrix A (though definiteness typically refers to real symmetric matrices), as the quadratic form invariably yields a real number. However, for complex matrices, A must be Hermitian to ensure that $\langle x|A|x \rangle$ is always real, validating the definition in Def. 9.9.

A common practical example of definite matrices is that $A^\dagger A$ (and similarly $AA^\dagger$) are always positive semidefinite for any matrix A.

Theorem 9.29 For any matrix $A \in \mathbb{C}^{m \times n}$, $A^\dagger A$ is a positive semidefinite Hermitian matrix.

Proof.   The matrix $A^\dagger A$ is Hermitian because $(A^\dagger A)^\dagger = A^\dagger (A^\dagger)^\dagger = A^\dagger A$. To establish its semidefiniteness, consider any vector $|v \rangle \in \mathbb{C}^n$. The associated quadratic form is:

$$\langle v|A^\dagger A|v \rangle = \langle Av|Av \rangle = \| A |v \rangle \|^2 \geq 0.$$

Hence, by definition, $A^\dagger A$ is positive semidefinite. $\square$

Exercise 9.21 Demonstrate that if A is a full-rank square matrix, then both $A^\dagger A$ and $AA^\dagger$ are positive definite.

The definiteness of a Hermitian matrix is closely related to its eigenvalues. The following theorem explicates this relationship.

Theorem 9.30 A Hermitian matrix A is positive definite (or semidefinite) if and only if all its eigenvalues are positive (or non-negative). Conversely, A is negative definite (or semidefinite) if and only if all its eigenvalues are negative (or non-positive).

Proof. We will demonstrate that a matrix A is positive definite if and only if all its eigenvalues are positive. This is established by utilizing A's orthonormal eigenbasis $\{|v_i\rangle\}$, corresponding to eigenvalues λ_i for $i = 1, 2, \ldots, n$.

Consider an arbitrary vector $|v\rangle$ decomposed onto this eigenbasis:

$$|v\rangle = \sum_i c_i |v_i\rangle.$$

The quadratic form $\langle v|A|v\rangle$ is then expressed as:

$$\begin{aligned}
\langle v|A|v\rangle &= \sum_{i,j} c_i^* c_j \langle v_i|A|v_j\rangle \\
&= \sum_{i,j} c_i^* c_j \langle v_i|\lambda_j v_j\rangle \\
&= \sum_i \lambda_j c_i^* c_j \delta_{ij} \\
&= \sum_i \lambda_i |c_i|^2,
\end{aligned}$$

where the orthogonality of the eigenbasis ($\langle v_i|v_j\rangle = \delta_{ij}$) simplifies the expression.

If all $\lambda_i > 0$, then $\langle v|A|v\rangle > 0$ for any nonzero $|v\rangle$, as there must be at least one nonzero $|c_i|$, confirming that A is positive definite.

Conversely, suppose $\langle v|A|v\rangle > 0$ for all nonzero $|v\rangle$. If any $\lambda_i \le 0$, consider $|v\rangle = |v_i\rangle$, the corresponding eigenvector. The quadratic form becomes $\langle v_i|A|v_i\rangle = \lambda_i$, which would not be positive, contradicting our assumption. Hence, all λ_i must be positive.

Thus, A is positive definite if and only if all its eigenvalues are positive. A similar methodology applies to proving the conditions for positive semidefiniteness, negative definiteness, and negative semidefiniteness. □

Theorems 9.29 and 9.30 suggest that both $A^\dagger A$ and $AA^\dagger$ invariably have non-negative eigenvalues. Readers are encouraged to independently verify this property directly.

Exercise 9.22 Demonstrate that $A^\dagger A$ invariably possesses non-negative eigenvalues without referencing Theorem 9.29 or Theorem 9.30.

Figure 9.3 illustrates the hierarchical relationships between different types of matrices and their diagonalization properties.

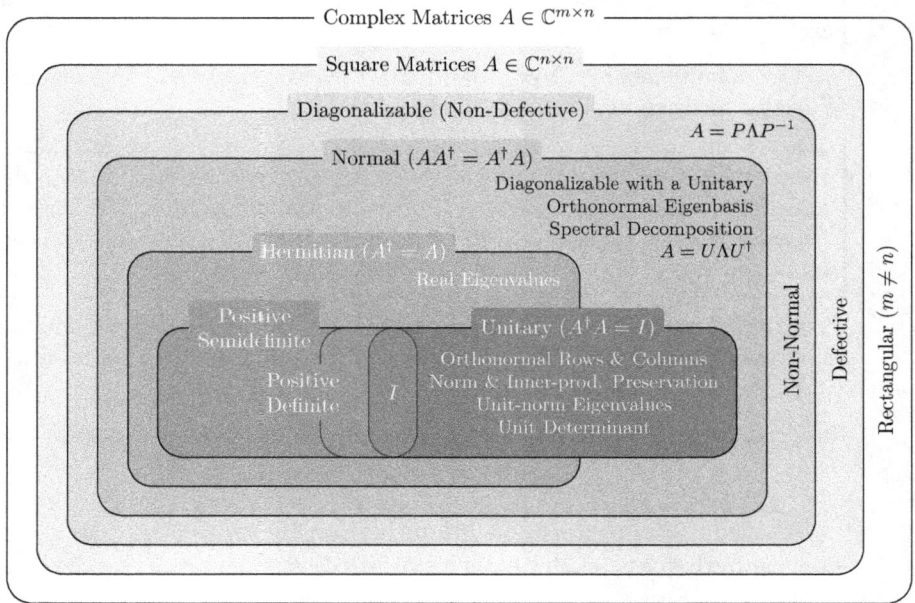

Figure 9.3: Relationships Among Matrix Types and Diagonalization

Exercise 9.23 Related to Fig. 9.3, show that the only matrix that is both unitary and positive-definite is the identity matrix I.

Exercise 9.24 Show that all invertible positive-semidefinite matrices are positive-definite matrices. How would you express this relationship in Fig. 9.3?

9.4.4 Symmetric Matrices

1 Real Symmetric Matrices

The term "symmetric matrix" typically refers to a *real* symmetric matrix within the field $\mathbb{R}$. In this realm, complex Hermitian matrices reduce to real symmetric matrices. Unsurprisingly, they share several fundamental properties, such as real eigenvalues, orthonormal eigenbases, and conditions for definiteness. Notably, real symmetric matrices are orthogonally diagonalizable, as opposed to unitarily diagonalizable like their complex counterparts.

Symmetric matrices frequently emerge as foundational representations or system characteristics across various fields:

- In statistics, they are utilized as correlation matrices, encapsulating the strength of linear relationships between variables.

- In network analysis, symmetric matrices are used as adjacency matrices, detailing symmetric connections within graphs.

- In physics, they serve as inertia tensors, describing the rotational properties of a rigid body.

While standard linear algebra textbooks primarily focus on real symmetric matrices and introduce complex Hermitian matrices as an extension, this book adopts the reverse approach due to its emphasis on quantum computing. Readers interested in a more systematic discussion of real symmetric matrices are encouraged to consult traditional linear algebra texts.

2 Complex Symmetric Matrices

Complex symmetric matrices are matrices with complex entries that satisfy $A = A^T$, where A^T is the transpose of A. These matrices differ fundamentally from complex Hermitian matrices, which satisfy $A = A^\dagger$ (the conjugate transpose of A). A complex symmetric matrix is Hermitian only when all its entries are real.

Unlike Hermitian matrices, complex symmetric matrices are generally neither orthogonally nor unitarily diagonalizable. Furthermore, their eigenvalues are not guaranteed to be real.

An important example of a complex symmetric matrix is the Discrete Fourier Transform (DFT) matrix, discussed in § 11.2.5.3.

9.4.5 ✳ Applications in Quantum Computing

Hermitian matrices are fundamental in quantum mechanics and quantum computing, underpinning several key concepts and operations. Below is an introduction to some of the most significant roles Hermitian matrices play. For a more detailed exploration of these applications, readers may refer to *Quantum Computing and Information: A Scaffolding Approach* [1].

1 Physical Observables

In quantum mechanics, physical quantities that are classically measurable are represented by Hermitian matrices. Consider an observable M with real eigenvalues λ_i and corresponding eigenvectors, referred to as **eigenstates**, denoted by $|\lambda_i\rangle$. The spectral decomposition of M is:

$$M = \sum_i \lambda_i |\lambda_i\rangle \langle\lambda_i| . \tag{9.30}$$

When a measurement is performed on an eigenstate $|\lambda_i\rangle$, the outcome is λ_i with certainty, i.e., with probability one. However, if the measurement is made on a general quantum state $|\psi\rangle$, the probability of observing each eigenvalue λ_i as the outcome is given by $|\langle\lambda_i|\psi\rangle|^2$. Since $\langle\lambda_i|\psi\rangle$ represents the magnitude of the projection of $|\psi\rangle$ onto $|\lambda_i\rangle$, this type of measurement is referred to as a **projective measurement**.

Using the orthonormality of the eigenstates and the completeness condition stated in Eq. 9.21, we can show that all probabilities sum to one:

$$\sum_i |\langle\lambda_i|\psi\rangle|^2 = \sum_i \langle\psi|\lambda_i\rangle \langle\lambda_i|\psi\rangle$$

$$= \langle\psi| \left(\sum_i |\lambda_i\rangle \langle\lambda_i| \right) |\psi\rangle$$

$$= \langle\psi| I |\psi\rangle$$

$$= \langle \psi | \psi \rangle = 1.$$

The statistical average or **expectation value** of M is computed as the weighted average of the eigenvalues:

$$\langle M \rangle = \sum_i |\langle \lambda_i | \psi \rangle|^2 \lambda_i$$

$$= \sum_i \lambda_i \langle \psi | \lambda_i \rangle \langle \lambda_i | \psi \rangle$$

$$= \langle \psi | \left(\sum_i \lambda_i | \lambda_i \rangle \langle \lambda_i | \right) | \psi \rangle$$

$$= \langle \psi | M | \psi \rangle.$$

Here, we have used the spectral decomposition of M.

2 POVM Operators

A Positive Operator-Valued Measure (POVM) is a more general measurement framework than projective measurements represented by a single Hermitian matrix M. A POVM consists of a set of Hermitian operators $\{E_i\}$ associated with specific measurement outcomes λ_i. Each operator E_i must satisfy two essential conditions:

(a) The operators sum to the identity: $\sum_i E_i = I$,

(b) Each operator E_i is positive semidefinite.

For any given quantum state vector $|\psi\rangle$, the probability of observing the outcome λ_i is given by:

$$p_i = \langle \psi | E_i | \psi \rangle. \tag{9.31}$$

The conditions imposed on E_i ensure that $p_i \geq 0$ and that the total probability sums to one, $\sum p_i = 1$.

Exercise 9.25 Verify that $\sum p_i = 1$ for the probabilities p_i defined in Eq. 9.31.

A POVM reduces to a projective measurement if $\{E_i\}$ are projectors onto the eigenstates of an observable M, i.e., $E_i = |\lambda_i\rangle \langle \lambda_i|$.

3 Density Matrices

As previously discussed in § 7.4.2.1, the general form of a density matrix representing mixed quantum states is given by:

$$\rho = \sum_i p_i |\phi_i\rangle \langle \phi_i|. \tag{9.32}$$

Here, $|\phi_i\rangle$ represents a set of normalized state vectors, though they are not necessarily orthogonal to each other. The probabilities p_i correspond to the likelihood of the system being in the state $|\phi_i\rangle$, where $p_i \geq 0$ and $\sum p_i = 1$. Thus, ρ represents a statistical mixture of the pure states $\{|\phi_i\rangle\}$.

Density matrices are Hermitian, meaning $\rho^\dagger = \rho$. To verify this, we note that:

$$\rho^\dagger = \left(\sum_i p_i |\phi_i\rangle \langle \phi_i| \right)^\dagger = \sum_i p_i \left(|\phi_i\rangle \langle \phi_i| \right)^\dagger = \sum_i p_i |\phi_i\rangle \langle \phi_i| = \rho.$$

Additionally, density matrices are positive semidefinite. This can be shown by considering the expectation value of ρ for any arbitrary vector $|\psi\rangle$:

$$\langle\psi|\rho|\psi\rangle = \sum_i p_i \langle\psi| \left(|\phi_i\rangle \langle\phi_i|\right) |\psi\rangle = \sum_i p_i |\langle\psi|\phi_i\rangle|^2 . \tag{9.33}$$

Since $p_i \geq 0$ and $|\langle\psi|\phi_i\rangle|^2 \geq 0$ for all i, it follows that $\langle\psi|\rho|\psi\rangle \geq 0$ for any vector $|\psi\rangle$. Therefore, ρ is positive semidefinite.

In Eq. 9.33, the term $|\langle\psi|\phi_i\rangle|^2$ represents the probability of the state $|\psi\rangle$ overlapping with the state $|\phi_i\rangle$. Therefore, $\langle\psi|\rho|\psi\rangle$ corresponds to the probability of observing the pure state $|\psi\rangle$ within the mixed state described by ρ.

4   Hamiltonian Operators

In quantum mechanics, the Hamiltonian of a system, which dictates the system's dynamics, is represented by a Hermitian matrix. A defining characteristic of a Hamiltonian is that its eigenvalues, which are inherently real, correspond to the energy levels of the system. This property stems directly from the nature of Hermitian matrices, with these energy levels constituting the **energy spectrum** of the system. The energy spectrum provides crucial physical insights into the potential states a system may occupy.

Furthermore, the time evolution of a system's quantum state is governed by the Schrödinger equation:

$$i\hbar\frac{\partial}{\partial t} |\psi(t)\rangle = H |\psi(t)\rangle , \tag{9.34}$$

where H is the Hamiltonian, t represents time, and $\hbar$ is the reduced Planck constant. For time-independent Hamiltonians, this relationship yields the general solution:

$$|\psi(t)\rangle = U(t) |\psi(0)\rangle , \quad \text{where} \quad U(t) = e^{-iHt/\hbar}. \tag{9.35}$$

Here, U is a unitary operator, ensuring conservation of probability and maintaining the norm of quantum states. This unitarity is a direct consequence of H being Hermitian. In this context, H functions as the "generator" of the unitary operator U, which explicitly describes how the system evolves over time. For a more detailed exploration of exponential functions of matrices, refer to § 11.3.3.

Problem Set 9

9.1 Determine the eigenvalues and corresponding eigenvectors for the following matrices by analyzing the transformation rules they represent:

(a) $P_\theta = \begin{bmatrix} \cos^2\theta & \cos\theta\sin\theta \\ \cos\theta\sin\theta & \sin^2\theta \end{bmatrix}$,   (b) $F_\theta = \begin{bmatrix} \cos 2\theta & \sin 2\theta \\ \sin 2\theta & -\cos 2\theta \end{bmatrix}$.

9.2 Find the eigenvalues and eigenvectors for the matrix $A = \begin{bmatrix} 1 & 1 & 0 \\ 1 & 1 & 0 \\ 0 & 0 & 1 \end{bmatrix}$.

9.3 Calculate the eigenvalues and corresponding eigenvectors for the matrix using its characteristic equation:

$$R_\theta = \begin{bmatrix} \cos\theta & -\sin\theta \\ \sin\theta & \cos\theta \end{bmatrix}.$$

9.4 Prove that the characteristic equation for any 2×2 matrix A is given by:

$$\lambda^2 - \text{tr}(A)\lambda + \det(A) = 0.$$

9.5 Demonstrate that:

(a) An eigenvalue λ of A is also an eigenvalue of A^T.

(b) An eigenvalue λ of A implies λ^* is an eigenvalue of $A^\dagger$.

Note the corresponding eigenvectors of A, A^T, and $A^\dagger$ do not share such simple relations. Hint: start with the characteristic equation for the proof.

9.6 Analyze the characteristics of the linear operator $A = \begin{bmatrix} a & b \\ 1-a & 1-b \end{bmatrix}$, where $0 < a, b < 1$. Compute $\lim_{k\to\infty} A^k x_0$, with $x_0 = [1\ 0]^T$.

9.7 ✳ Prove Theorem 9.5: The geometric multiplicity of any eigenvalue of a square matrix is always less than or equal to its algebraic multiplicity.

9.8 Demonstrate that any invertible matrix P acts as a basis transformation matrix between two bases of $\mathbb{C}^n$.

9.9 Address the following proofs regarding matrix similarity:

(a) Prove Theorem 9.10 which states that two diagonalizable matrices are similar if and only if they share the same characteristic polynomials.

(b) Prove that any diagonalizable matrix A is similar to its transpose A^T.

9.10 ✳ Theorem 9.15 suggests that commutative, diagonalizable matrices A and B share a common eigenbasis. Explore the relationship between the eigenvectors of diagonalizable matrices A and B when these matrices *anti-commute*.

9.11 Prove that skew-Hermitian matrices, which satisfy $A = -A^\dagger$, must have purely imaginary eigenvalues.

9.12 ✳ Prove that

(a) A matrix $A \in \mathbb{C}^{n\times n}$ is Hermitian if and only if the quadratic form $\langle x|A|x \rangle$ is real for any $|x\rangle \in \mathbb{C}$.

(b) A matrix $A \in \mathbb{C}^{n\times n}$ is Hermitian if and only if $\langle Ax, y \rangle = \langle x, Ay \rangle$ for any vectors $x, y \in \mathbb{C}$.

9.13 Prove that a matrix $A \in \mathbb{C}^{n\times n}$ is a normal matrix if and only if $\|Ax\| = \|A^\dagger x\|$ for any vector $|x\rangle \in \mathbb{C}^n$.

9.14 Prove that if P is a projection operator as defined in Eq. 9.25, then $2P - I$ is both a Hermitian and a unitary operator.

9.15 Using the completeness condition of orthonormal bases, prove Theorems 8.12 and 8.13.

9.16 Prove that if $A, B \in \mathbb{C}^{n \times n}$ are commutative normal matrices, then

    (a) A, B can be diagonalized by the same unitary matrix.

    (b) Both AB and $A + B$ are also normal matrices.

9.17 Prove that if A is a normal matrix, then there exists a unitary matrix U such that $A^\dagger = AU$.

9.18 Prove that an upper triangular matrix that is normal must also be diagonal.

9.19 Complete the following tasks for the matrix $A = \begin{bmatrix} 1 & i \\ i & 5 \end{bmatrix}$:

    (a) Determine if A is a normal matrix.

    (b) Find the eigenvalues λ_1, λ_2 and normalized eigenvectors $|\phi_1\rangle, |\phi_2\rangle$ for each eigenvalue.

    (c) Assess whether A possesses an orthonormal eigenbasis.

    (d) Verify if A adheres to the spectral decomposition formula
$$A = \lambda_1 |\phi_1\rangle \langle\phi_1| + \lambda_2 |\phi_2\rangle \langle\phi_2|.$$

9.20 Prove Theorem 9.21.

III

Matrix Methods for Quantum Computing

In Part II of this book, we conducted a systematic exploration of the fundamentals of linear algebra, tailored specifically for quantum computing applications. With Part III, our focus shifts to more advanced aspects of linear algebra that are essential for quantum computing. This part begins with the tensor product of vector spaces (Chapter 10), continues with functions of vectors and matrices (Chapter 11), examines Pauli matrices, Pauli strings, and Pauli groups (Chapter 12), and concludes with advanced matrix decompositions (Chapter 13).

The division between Part II and Part III is intentional. While Part II provides a foundational introduction to linear algebra, adapted to the framework of Dirac notation, Part III extends beyond the scope of conventional linear algebra textbooks, introducing concepts and tools critical for deeper studies in quantum computing.

Part III also adopts a different structure from Part II. In addition to introducing new concepts, it revisits and summarizes key topics from earlier chapters. This design enables readers already familiar with standard linear algebra to skip Part II and directly engage with Part III, allowing for a more efficient immersion into the theory of quantum computing.

10. Tensor Products of Vector Spaces

Contents

Quantum computing operates using qubits instead of classical bits to store and process information. The quantum states of a system with n qubits reside in a space of dimension $\mathbb{C}^{2^n}$. For instance, when $n = 100$, this dimension reaches an astonishing 1.27×10^{30}. This ability to *simultaneously* manipulate vast amounts of information within such high-dimensional spaces is a key reason why quantum computers have the potential to outperform classical computers.

To advance quantum computing theory and develop practical algorithms, it is crucial to understand how quantum states and operations are represented in these high-dimensional spaces. In this chapter, we introduce tensor products using Dirac notation, providing foundational tools for this purpose.

In previous chapters, we covered the basics of single-qubit systems, represented by vectors in $\mathbb{C}^2$ and matrices in $\mathbb{C}^{2\times 2}$. We now shift our focus to multi-qubit systems, extending our study from two-qubit and three-qubit systems to general n-qubit systems. This extension lays the groundwork for a deeper exploration of multi-qubit quantum computation.

10.1   Tensor Products of Vectors and Vector Spaces

10.1.1   Introduction

The concept of tensor products arises from the need to describe the *joint state* of a multi-qubit system in terms of the state vectors of individual qubits. This subsection provides an intuition, starting with two-qubit systems, and builds up to the general case for multi-qubit systems and vector spaces.

1   Joint States of Two Qubits

Consider two qubits, A and B, each of which can be in one of two basis states, $|0\rangle$ or $|1\rangle$. Ignoring superpositions for now, the possible joint states of qubits A and B are the four combinations:

$$\{|0\rangle_A \otimes |0\rangle_B, \quad |0\rangle_A \otimes |1\rangle_B, \quad |1\rangle_A \otimes |0\rangle_B, \quad |1\rangle_A \otimes |1\rangle_B\}.$$

Here, $|0\rangle_A$ denotes that qubit A is in state $|0\rangle$, and similarly for the other states. This example demonstrates that each qubit's two possible basis states combine to yield four possible joint basis states for the two-qubit system.

2   Generalization to Multi-Qubit Systems

Adding additional qubits extends this pattern. For instance, including a third qubit, C, with its own two basis states, brings the total number of joint basis states to eight, calculated as 2^3. In general, an n-qubit system has 2^n possible basis states, with each qubit being either in $|0\rangle$ or $|1\rangle$. This result follows from basic combinatorial principles, where each qubit independently contributes two possible states to the total.

3   Tensor Products of Vectors

Expanding on this concept, consider vectors $|u\rangle \in \mathbb{C}^m$ and $|v\rangle \in \mathbb{C}^n$. The joint states of these two vectors span a space with mn possible combinations, reflecting the product of m and n. The purpose of the tensor product operation is to create a single state vector that represents this joint space, providing a mapping:

$$\mathbb{C}^m \otimes \mathbb{C}^n \to \mathbb{C}^{mn}.$$

For an m-qubit system in $\mathbb{C}^{2^m}$ and an n-qubit system in $\mathbb{C}^{2^n}$, their combined system will have a state vector in $\mathbb{C}^{2^{m+n}}$, matching the dimensionality expected for an $(m+n)$-qubit system.

4   Superposition States and Bilinearity

Consider the case where qubit A is in a superposition state, $\frac{1}{\sqrt{2}}|0\rangle + \frac{1}{\sqrt{2}}|1\rangle$, which indicates a 50% probability of being in $|0\rangle$ and a 50% probability of being in $|1\rangle$. Here, the coefficients of the state vector are *probability amplitudes*, and their magnitudes squared yield the corresponding probabilities. Thus, the 50% probability arises from the square of the coefficient $\frac{1}{\sqrt{2}}$.

Suppose qubit B is in the basis state $|0\rangle$. The joint state of qubits A and B then represents a 50% probability of being in $|0\rangle \otimes |0\rangle$ and a 50% probability of being in $|1\rangle \otimes |0\rangle$. This joint state can be expressed as:

$$\left(\frac{1}{\sqrt{2}} |0\rangle + \frac{1}{\sqrt{2}} |1\rangle \right) \otimes |0\rangle = \frac{1}{\sqrt{2}} |0\rangle \otimes |0\rangle + \frac{1}{\sqrt{2}} |1\rangle \otimes |0\rangle .$$

This demonstrates that the tensor product is linear in its first argument. Similarly, linearity holds in the second argument as well, which implies that the tensor product operation is bilinear.

5   Tensor Product of Vector Spaces

Therefore, the tensor product can be understood as a bilinear map from $\mathbb{C}^m \otimes \mathbb{C}^n$ to $\mathbb{C}^{mn}$. This bilinearity allows us to define the operation by specifying the tensor products between basis vectors of each space, which enables us to compute tensor products for any pair of vectors in these spaces.

6   Overview of Next Steps

The following sections will formalize the definition of the tensor product between basis vectors, introduce computational rules for working with tensor products in matrix form, and extend these concepts to general vector spaces.

10.1.2   Tensor Products of Vectors

1   Tensor Products of Two Vectors

To understand tensor products, it is instructive to first look at its definition for two column vectors in matrix representation.

> **Definition 10.1 — Tensor Product of Column Vectors.** Given two column vectors $|u\rangle \in \mathbb{C}^m$ and $|v\rangle \in \mathbb{C}^n$, represented by:
>
> $$|u\rangle = \begin{bmatrix} u_1 \\ u_2 \\ \vdots \\ u_m \end{bmatrix} , \quad |v\rangle = \begin{bmatrix} v_1 \\ v_2 \\ \vdots \\ v_n \end{bmatrix} ,$$

the tensor product $|u\rangle \otimes |v\rangle$ is expressed in matrix form as:

$$|u\rangle \otimes |v\rangle = \begin{bmatrix} u_1 |v\rangle \\ u_2 |v\rangle \\ \vdots \\ u_m |v\rangle \end{bmatrix} = \begin{bmatrix} u_1 v_1 \\ u_1 v_2 \\ \vdots \\ u_2 v_1 \\ u_2 v_2 \\ \vdots \\ \vdots \\ u_m v_1 \\ u_m v_2 \\ \vdots \end{bmatrix}. \qquad (10.1)$$

Notation: The tensor product of two column vectors $|u\rangle \otimes |v\rangle$ is often abbreviated as $|u \otimes v\rangle$ or $|uv\rangle$. For instance, we write $|0\rangle \otimes |1\rangle$ simply as $|01\rangle$.

■ **Example 10.1** Compute the matrix representations of $|00\rangle, |01\rangle, |10\rangle, |11\rangle$.

Solution. Applying Def. 10.1, we find:

$$|00\rangle = \begin{bmatrix} 1 \\ 0 \end{bmatrix} \otimes \begin{bmatrix} 1 \\ 0 \end{bmatrix} = \begin{bmatrix} 1 \\ 0 \\ 0 \\ 0 \end{bmatrix}.$$

Similarly, we derive:

$$|01\rangle = \begin{bmatrix} 1 \\ 0 \end{bmatrix} \otimes \begin{bmatrix} 0 \\ 1 \end{bmatrix} = \begin{bmatrix} 0 \\ 1 \\ 0 \\ 0 \end{bmatrix}, \quad |10\rangle = \begin{bmatrix} 0 \\ 1 \end{bmatrix} \otimes \begin{bmatrix} 1 \\ 0 \end{bmatrix} = \begin{bmatrix} 0 \\ 0 \\ 1 \\ 0 \end{bmatrix}, \quad |11\rangle = \begin{bmatrix} 0 \\ 1 \end{bmatrix} \otimes \begin{bmatrix} 0 \\ 1 \end{bmatrix} = \begin{bmatrix} 0 \\ 0 \\ 0 \\ 1 \end{bmatrix}.$$

As shown, the set $\{|00\rangle, |01\rangle, |10\rangle, |11\rangle\}$ forms the standard basis for $\mathbb{C}^4$ in matrix representation.   ■

■ **Example 10.2** Prove that if $|u\rangle, |v\rangle \in \mathbb{C}^2$ are unit vectors, then their tensor product $|uv\rangle$ is also a unit vector.

Proof. Write $|u\rangle, |v\rangle$ in matrix form:

$$|u\rangle = \begin{bmatrix} \alpha \\ \beta \end{bmatrix}, \quad |v\rangle = \begin{bmatrix} \gamma \\ \delta \end{bmatrix},$$

where $\alpha, \beta, \gamma, \delta \in \mathbb{C}$ satisfy $|\alpha|^2 + |\beta|^2 = 1$ and $|\gamma|^2 + |\delta|^2 = 1$, since $|u\rangle, |v\rangle$ are normalized. Applying the tensor product, we get:

$$|u\rangle \otimes |v\rangle = \begin{bmatrix} \alpha\gamma \\ \alpha\delta \\ \beta\gamma \\ \beta\delta \end{bmatrix}.$$

The squared norm of this vector is:

$$|\alpha\gamma|^2 + |\alpha\delta|^2 + |\beta\gamma|^2 + |\beta\delta|^2 = |\alpha|^2(|\gamma|^2 + |\delta|^2) + |\beta|^2(|\gamma|^2 + |\delta|^2)$$
$$= |\alpha|^2 + |\beta|^2 = 1.$$

Therefore, $|uv\rangle$ remains a unit vector. ∎

Exercise 10.1 Compute the matrix representations of $|++\rangle, |-+\rangle, |+-\rangle, |--\rangle$ and verify that they form an orthonormal basis of $\mathbb{C}^4$. The states $|+\rangle$ and $|-\rangle$ are defined as:

$$|+\rangle = \frac{1}{\sqrt{2}}(|0\rangle + |1\rangle), \quad |-\rangle = \frac{1}{\sqrt{2}}(|0\rangle - |1\rangle).$$

2 Tensor Products of Multiple Vectors

The tensor product of multiple vectors can be expressed using a generalized product notation:

$$|v_1\rangle \otimes |v_2\rangle \otimes \cdots \otimes |v_n\rangle = \bigotimes_{i=1}^{n} |v_i\rangle. \tag{10.2}$$

For n identical vectors, this expression simplifies to an exponent notation:

$$|v\rangle \otimes |v\rangle \otimes \cdots \otimes |v\rangle = |v\rangle^{\otimes n} \quad (n \text{ times}). \tag{10.3}$$

(i) As dimensionality increases, matrix representations of tensor products become increasingly impractical. For instance, explicitly representing a basis vector in $\mathbb{C}^{16}$ for a 4-qubit system in matrix form is cumbersome. Using Dirac notation, such as $|0000\rangle$, provides greater clarity, conciseness, and simplicity for computation.

These notations are foundational for exploring the basis vectors of an n-qubit system. A standard basis vector of $\mathbb{C}^{2^n}$ can be represented as a tensor product of n vectors from the computational basis of $\mathbb{C}^2$:

$$|x_1\rangle \otimes |x_2\rangle \otimes \cdots |x_n\rangle \equiv |x_1 x_2 \cdots x_n\rangle, \text{ where } x_i \in \{0, 1\}.$$

Each $|x_i\rangle$ can be either $|0\rangle$ or $|1\rangle$, leading to a total of 2^n basis vectors. This setup is referred to as the **binary representation** of basis vectors for an n-qubit system, with the sequence $(x_1 x_2 \cdots x_n)$ representing a binary number.

For instance, the basis vector $|101\rangle$ corresponds to the column vector:

$$\begin{bmatrix} 0 & 0 & 0 & 0 & 0 & 1 & 0 & 0 \end{bmatrix}^T,$$

where the only non-zero element, the "1," is at index 5 (starting from 0), which matches the binary number $(101)_2 = 5$. The general case's proof is left as an exercise (Problem 10.4).

Exercise 10.2 Express the vector $|010\rangle$ in its matrix representation for a 3-qubit system.

Exercise 10.3 Express the vector $|1101\rangle$ in its matrix representation for a 4-qubit system.

10.1.3 Tensor Products of Vector Spaces

We can extend the concept of the tensor product from vectors to vector spaces. Specifically, given two vector spaces V and W, their tensor product $V \otimes W$ is a new vector space consisting of all possible linear combinations of tensor products $|v\rangle \otimes |w\rangle$, where $|v\rangle \in V$ and $|w\rangle \in W$.

The basis vectors of $V \otimes W$ are formed by all possible tensor products between the basis vectors of V and those of W. As a result, the dimension of $V \otimes W$ is the product of the dimensions of V and W: if $\dim(V) = m$ and $\dim(W) = n$, then $\dim(V \otimes W) = m \times n$.

In particular, the tensor product $\mathbb{C}^m \otimes \mathbb{C}^n$ is defined as follows:

Definition 10.2 — Tensor Product of Vector Spaces. Let $\{|u_i\rangle\}$ be a basis for $\mathbb{C}^m$ and $\{|v_j\rangle\}$ a basis for $\mathbb{C}^n$, where $i = 1, 2, \ldots, m$ and $j = 1, 2, \ldots, n$.

The tensor product of $\mathbb{C}^m$ and $\mathbb{C}^n$, denoted by $\mathbb{C}^m \otimes \mathbb{C}^n$, is the vector space with the bilinear map

$$\otimes : \mathbb{C}^m \times \mathbb{C}^n \to \mathbb{C}^m \otimes \mathbb{C}^n,$$

such that the set $\{|u_i\rangle \otimes |v_j\rangle\}$ forms a basis of $\mathbb{C}^m \otimes \mathbb{C}^n$. That is,

$$\mathbb{C}^m \otimes \mathbb{C}^n = \text{span}\{|u_i\rangle \otimes |v_j\rangle \mid 1 \leq i \leq m, 1 \leq j \leq n\}. \tag{10.4}$$

Multiple Possibilities for Tensor Products between Vectors

Knowing that the tensor product is a bilinear map from $\mathbb{C}^m \otimes \mathbb{C}^n$ to $\mathbb{C}^{mn}$ is sufficient to define it. Thus, Def. 10.2 also encompasses the rules for defining the tensor product between vectors. However, within this general framework, multiple conventions exist for defining the tensor products of vectors, with Def. 10.1 being the most commonly adopted one.

In fact, any reordering of the elements in Eq. 10.1 would also constitute a valid computational rule for vector tensor products. It can be shown that all such variations are isomorphic, meaning that, despite different explicit computational rules, they are fundamentally equivalent.

Thus, Def. 10.2 provides a general definition of tensor products for two vector spaces, without explicitly specifying a computational rule for the tensor products of vectors.

Definition 10.2 can be extended to multiple vector spaces. For example, the n-qubit Hilbert space is $\bigotimes^n \mathbb{C}^2$, often denoted as $\mathbb{C}^{2^n}$ for simplicity, even though the two are technically distinct. Specifically, the standard (computational) basis for $\mathbb{C}^2$, $\{|0\rangle, |1\rangle\}$, expands to $\{|x\rangle\}$ with $x \in \{0, 1\}^n$ for $\mathbb{C}^{2^n}$. Here, $x \in \{0, 1\}^n$ indicates that x is an n-bit string, where each bit is either 0 or 1 (see Example 4.9).

10.1.4 General Vectors in Joint Vector Spaces

1 General n-Qubit State Vectors

Using the binary representation of basis vectors, a general n-qubit state can be expressed as:

$$|\psi\rangle = \sum_{x_1, x_2, \ldots, x_n \in \{0, 1\}} c_{x_1 x_2 \cdots x_n} |x_1 x_2 \cdots x_n\rangle, \tag{10.5}$$

or more succinctly as:

$$|\psi\rangle = \sum_{x \in \{0,1\}^n} c_x |x\rangle,\tag{10.6}$$

where the sum is taken over all 2^n possible binary strings of length n. For the vector $|\psi\rangle$ to be normalized, it is required that $\sum |c_x|^2 = 1$.

2 Bipartite State Vectors

In quantum computing, it is common to divide an n-qubit system into two sub-systems—a k-qubit subsystem and an l-qubit subsystem, where $n = k + l$. This configuration is referred to as a bipartite system.

In such settings, we often express a general vector in the joint system using basis vectors that are tensor products of basis vectors from each subsystem:

$$|\psi\rangle = \sum_i \sum_j c_{ij} |u_i\rangle \otimes |v_j\rangle, \quad \text{where } |\psi\rangle \in \mathbb{C}^k \otimes \mathbb{C}^l,\tag{10.7}$$

where $\{|u_i\rangle\}$ is an orthonormal basis of $\mathbb{C}^k$, and $\{|v_j\rangle\}$ is an orthonormal basis of $\mathbb{C}^l$.

Mathematically, Eq. 10.7 allows us to represent a vector in $\mathbb{C}^k \otimes \mathbb{C}^l$ as a $k \times l$ matrix $[c_{ij}]_{k \times l}$ by rearranging its kl components. For instance, the state:

$$|\psi\rangle = |00\rangle + |01\rangle - |10\rangle + |11\rangle,$$

in $\mathbb{C}^2 \otimes \mathbb{C}^2$, is represented as $\begin{bmatrix} 1 & 1 & -1 & 1 \end{bmatrix}^T$. Using Eq. 10.7, $|\psi\rangle$ can be expressed as:

$$|\psi\rangle = \sum_{i=0}^{1} \sum_{j=0}^{1} c_{ij} |ij\rangle,$$

which can be rearranged into a 2×2 matrix:

$$C = \sum_{i=0}^{1} \sum_{j=0}^{1} c_{ij} |i\rangle \langle j| = \begin{bmatrix} 1 & 1 \\ -1 & 1 \end{bmatrix}.$$

Equation 10.7 will be used in the following subsection to distinguish between product and non-product states for pure quantum states (recall that a pure quantum state can be represented by a unit vector in Hilbert space), and in § 13.3 to introduce the Schmidt decomposition.

10.1.5 Product States and Non-Product States

While every basis vector in $\mathbb{C}^m \otimes \mathbb{C}^n$ is a tensor product between a basis vector in $\mathbb{C}^m$ and a basis vector in $\mathbb{C}^n$, not all vectors in $\mathbb{C}^m \otimes \mathbb{C}^n$ can be expressed as tensor products of vectors from $\mathbb{C}^m$ and $\mathbb{C}^n$. This leads to the distinction between product states and non-product states. Here, "states" refers to quantum state vectors.

A product state is one that can be decomposed into a tensor product of individual states. For example, the state $|00\rangle$ is expressible as $|0\rangle \otimes |0\rangle$, making it a typical example of a product state. Conversely, the state $\frac{1}{\sqrt{2}}(|00\rangle + |11\rangle)$, which cannot be decomposed into a tensor product of two individual qubits, is a non-product state.

Definition 10.3 — Product State. A state vector $|\psi\rangle \in \mathbb{C}^m \otimes \mathbb{C}^n$ is a product state if it can be expressed as:
$$|\psi\rangle = |v_1\rangle \otimes |v_2\rangle, \tag{10.8}$$
where $|v_1\rangle \in \mathbb{C}^m$ and $|v_2\rangle \in \mathbb{C}^n$. If no such decomposition exists, $|\psi\rangle$ is a non-product state.

To distinguish between product states and non-product states, we can rearrange the components of a state vector $|\psi\rangle \in \mathbb{C}^{mn}$ into an $m \times n$ matrix $C = [c_{ij}]$ using Eq. 10.7 and check the rank of C. Specifically, $|\psi\rangle$ is a product (or separable) state if and only if C is of rank one.

To understand why this is the case, consider the vector representation of $|u\rangle \otimes |v\rangle$ in Eq. 10.1. Rearranging all components into an $m \times n$ matrix yields:

$$\begin{bmatrix} u_1 v_1 & u_1 v_2 & \cdots & u_1 v_n \\ u_2 v_1 & u_2 v_2 & \cdots & u_2 v_n \\ \vdots & \vdots & \ddots & \vdots \\ u_m v_1 & u_m v_2 & \cdots & u_m v_n \end{bmatrix} = \begin{bmatrix} u_1 \\ u_2 \\ \vdots \\ u_m \end{bmatrix} \begin{bmatrix} v_1 & v_2 & \cdots & v_n \end{bmatrix} \equiv |u\rangle \langle v|^*.$$

This is a rank-one matrix since all rows (or columns) are linearly dependent.

■ **Example 10.3** Show that $|\psi\rangle = |00\rangle + |01\rangle \in \mathbb{C}^2 \otimes \mathbb{C}^2$ is a product state.

Proof. Rearrange the vector $|\psi\rangle = \begin{bmatrix} 1 & 1 & 0 & 0 \end{bmatrix}^T$ into a 2×2 matrix:

$$\begin{bmatrix} 1 & 1 \\ 0 & 0 \end{bmatrix},$$

which is a rank-one matrix. Therefore, $|\psi\rangle$ can be expressed as a product state:

$$|\psi\rangle = |0\rangle \otimes (|0\rangle + |1\rangle).$$

■

■ **Example 10.4** Show that $|\psi\rangle = \frac{1}{\sqrt{2}}(|01\rangle + |10\rangle) \in \mathbb{C}^2 \otimes \mathbb{C}^2$ is not a product state.

Proof. The vector representation of $\frac{1}{\sqrt{2}}(|01\rangle + |10\rangle) = \frac{1}{\sqrt{2}} \begin{bmatrix} 0 & 1 & 1 & 0 \end{bmatrix}^T$ rearranges into:

$$\begin{bmatrix} 0 & \frac{1}{\sqrt{2}} \\ \frac{1}{\sqrt{2}} & 0 \end{bmatrix},$$

which has a nonzero determinant, indicating a full-rank matrix. Therefore, $|\psi\rangle$ is a non-product state.

■

For a more formal approach to determine if a pure state $|\psi\rangle$ is a product state, we can perform singular value decomposition (SVD) on the matrix $[c_{ij}]_{m \times n}$ formed from the components of $|\psi\rangle$. This involves analyzing the number of nonzero singular values, referred to as the Schmidt rank. A state $|\psi\rangle$ is a product state if the rearranged matrix has only one nonzero singular value. Further details will be discussed in § 13.3.

Exercise 10.4 Determine whether each of the following pure states is a product state or not. For any product states, decompose the vector into its tensor product components.

> (a) $\frac{1}{2}(|00\rangle + |01\rangle + |10\rangle + |11\rangle)$,
>
> (b) $\frac{1}{2}(|00\rangle - |01\rangle + |10\rangle + |11\rangle)$,
>
> (c) $\frac{1}{2}(|00\rangle - |01\rangle - |10\rangle + |11\rangle)$.

Non-product states exist in a tensor-product joint space because the joint space, as a Hilbert space, must be complete. The components of the state vector described in Eq. 10.6 are independent and can be arbitrary complex numbers, rather than being restricted to combinations derived from product states.

> (i) In quantum mechanics, a product state is also known as a **separable state**, while a non-product state is referred to as an **entangled state** [1].

The above definitions apply to pure states. For mixed states, the terms product states and separable states are not synonymous, nor are non-product states and entangled states equivalent. The precise definition of entangled states for mixed states will be elaborated in § 10.3.5.

Cartesian Product vs Tensor Product

The Cartesian product (Def. 4.10) forms pairs of elements, while the tensor product creates a new space that encapsulates bilinear relationships between elements of the original spaces. Although they may seem similar, the Cartesian and tensor products serve distinct purposes and exhibit different properties.

For example, if $V = W = \mathbb{C}^2$, the Cartesian product $V \times W$ is a set:

$$\mathbb{C}^2 \times \mathbb{C}^2 = \{(|x\rangle, |y\rangle) \mid |x\rangle, |y\rangle \in \mathbb{C}^2\}.$$

This consists of pairs of vectors in $\mathbb{C}^2$. In contrast, the tensor product $V \otimes W$ is an inner product space:

$$\mathbb{C}^2 \otimes \mathbb{C}^2 = \text{span}\{|x\rangle \otimes |y\rangle \mid |x\rangle, |y\rangle \in \mathbb{C}^2\}.$$

Now consider the case where $V = W = \mathbb{R}$. The Cartesian product $V \times W$ is:

$$\mathbb{R} \times \mathbb{R} = \{(x, y) \mid x, y \in \mathbb{R}\}.$$

This represents the Euclidean plane $\mathbb{R}^2$, which is two-dimensional. On the other hand, the tensor product $V \otimes W$ is:

$$\mathbb{R} \otimes \mathbb{R} = \text{span}\{x \otimes y \mid x, y \in \mathbb{R}\}.$$

Since the basis of $\mathbb{R}$ is $\{1\}$, elements of $\mathbb{R} \otimes \mathbb{R}$, such as $x \otimes y$, effectively correspond to the scalar product xy. Thus, the tensor product space is one-dimensional and isomorphic to $\mathbb{R}$.

10.1.6 Fundamental Properties

Tensor products exhibit many elegant algebraic properties when combined with other algebraic operations involving vectors or matrices. This subsection introduces some fundamental properties of tensor products between two column vectors, while subsequent sections will explore properties involving matrix products.

1 Non-commutativity

Tensor products in quantum computing do not generally adhere to commutative rules, which can be expressed as:

$$|u\rangle \otimes |v\rangle \neq |v\rangle \otimes |u\rangle.$$

For instance, consider a simple two-qubit system: $|01\rangle \neq |10\rangle$. In the state $|01\rangle$, qubit A is in state $|0\rangle$ and qubit B in state $|1\rangle$. Conversely, in the state $|10\rangle$, qubit B is in state $|0\rangle$ and qubit A in state $|1\rangle$. These configurations represent distinctly different system states, underscoring the non-commutative nature of tensor products.

2 Associativity

The tensor product of column vectors satisfies the associative property. Specifically, for any vectors $|u\rangle$, $|v\rangle$, and $|w\rangle$, we have:

$$(|u\rangle \otimes |v\rangle) \otimes |w\rangle = |u\rangle \otimes (|v\rangle \otimes |w\rangle) = |u\rangle \otimes |v\rangle \otimes |w\rangle. \qquad (10.9)$$

This associativity allows us to use the concise notation $|uvw\rangle$ for the tensor product of multiple ket vectors without needing to specify the order of operations.

Exercise 10.5 Using Def. 10.1, prove that the tensor product is associative. That is, show that:

$$(|u\rangle \otimes |v\rangle) \otimes |w\rangle = |u\rangle \otimes (|v\rangle \otimes |w\rangle).$$

3 Construction of the Computational Basis

The computational basis (as defined in Def. 6.5) for the tensor product of two vector spaces can be constructed by taking the tensor products of the computational bases of each constituent space:

> **Theorem 10.1 — Tensor Product and Computational Basis.** Let $\{|i\rangle\}$ for $i = 0, 1, \ldots, m - 1$ represent the computational basis of $\mathbb{C}^m$, and $\{|j\rangle\}$ for $j = 0, 1, \ldots, n - 1$ represent the computational basis of $\mathbb{C}^n$. Then, the set $\{|i\rangle \otimes |j\rangle\}$ forms the computational basis of $\mathbb{C}^m \otimes \mathbb{C}^n$.

In fact, we have already used this property in § 10.1.2.2 and § 10.1.4 without proving it. The proof is left as an exercise to the reader (Problem 10.6), based on the product rule as defined in Def. 10.1.

4 Inner Product and Outer Product of Tensor Products

> **Theorem 10.2 — Inner Product of Tensor Products.** Given vectors $|\phi_1\rangle, |\psi_1\rangle \in \mathbb{C}^m$ and $|\phi_2\rangle, |\psi_2\rangle \in \mathbb{C}^n$, the following identity holds:
>
> $$\langle \phi_1 \otimes \phi_2 | \psi_1 \otimes \psi_2 \rangle = \langle \phi_1 | \psi_1 \rangle \langle \phi_2 | \psi_2 \rangle. \qquad (10.10)$$

This property states that the inner product of two tensor products of vectors is equal to the product of the inner products of the individual vectors.

> **Theorem 10.3 — Outer Product of Tensor Products.** Given vectors $|i\rangle, |k\rangle \in \mathbb{C}^m$
> and $|j\rangle, |l\rangle \in \mathbb{C}^n$, the following identity holds:
>
> $$(|i\rangle \otimes |j\rangle)(\langle k| \otimes \langle l|) = (|i\rangle \langle k|) \otimes (|j\rangle \langle l|). \qquad (10.11)$$

This property posits that the outer product of two tensor products of vectors is equal to the tensor product of the outer products of the individual vectors.

(i) | **Notation:** The outer product $(|i\rangle \otimes |j\rangle)(\langle k| \otimes \langle l|)$ is often abbreviated as $|ij\rangle\langle kl|$.

Theorems 10.2 and 10.3 follow as corollaries of Theorem 10.7. They are also included as part of Theorem 10.8.

| **Exercise 10.6** Prove that if $|\psi_1\rangle$ is a unit vector in $\mathbb{C}^m$ and $|\psi_2\rangle$ is a unit vector in $\mathbb{C}^n$, then their tensor product $|\psi_1\rangle \otimes |\psi_2\rangle$ is also a unit vector in $\mathbb{C}^m \otimes \mathbb{C}^n$.

10.2 Tensor Products of Matrices

In quantum computing, tensor products of matrices are essential for representing operations on multi-qubit systems. When applying quantum gates or performing measurements in multi-qubit or bipartite systems, these operations typically target specific subsystems. To represent an operator acting on the entire system, we use the tensor product of operators for each subsystem, denoted as $(U_A \otimes U_B)$. Here, U_A acts on subsystem A, while U_B acts on subsystem B.

Furthermore, quantum states can be represented not only by state vectors but also by density matrices, particularly in cases involving mixed (non-pure) states. To describe composite systems using density matrices, it is necessary to understand tensor products of matrices, denoted as $\rho_A \otimes \rho_B$, analogous to the tensor products $|u\rangle \otimes |v\rangle$ used for vectors.

This section will explore the fundamentals of tensor products of matrices and discuss the key algebraic properties that are widely used in both the theoretical foundations and practical applications of quantum computing.

10.2.1 Matrix Partition by Blocks

Before discussing tensor products between matrices, it's useful to review the concept of matrix partitioning. Any matrix can be partitioned into smaller submatrices, referred to as *blocks*, provided that these submatrices form a rectangular array:

$$A_{m \times n} = \begin{bmatrix} A_{11} & A_{12} \\ A_{21} & A_{22} \end{bmatrix}. \qquad (10.12)$$

Here, each A_{ij} represents a submatrix of A. The sizes of these submatrices must adhere to specific rules for consistency: if A_{11} is $p \times q$, then A_{12} must be $p \times (n - q)$, A_{21} must be $(m - p) \times q$, and A_{22} must be $(m - p) \times (n - q)$.

For instance, consider the 4×4 identity matrix I_4, which can be partitioned into:

$$I_4 = \begin{bmatrix} 1 & 0 & 0 & 0 \\ 0 & 1 & 0 & 0 \\ 0 & 0 & 1 & 0 \\ 0 & 0 & 0 & 1 \end{bmatrix} = \begin{bmatrix} I_2 & O \\ O & I_2 \end{bmatrix},$$

where each block is a 2×2 submatrix. In partitioned matrix notation, the letter "O" is typically used to denote zero matrices.

Matrix products can also be computed in partitioned form when two matrices are partitioned consistently. For example, consider a general 4×4 matrix A partitioned into four 2×2 blocks:

$$A = \begin{bmatrix} A_{11} & A_{12} \\ A_{21} & A_{22} \end{bmatrix}.$$

The product of I_4 and A in this partitioned form can be computed as:

$$I_4 A = \begin{bmatrix} I_2 & O \\ O & I_2 \end{bmatrix} \begin{bmatrix} A_{11} & A_{12} \\ A_{21} & A_{22} \end{bmatrix}$$

$$= \begin{bmatrix} I_2 A_{11} + O A_{21} & I_2 A_{12} + O A_{22} \\ O A_{11} + I_2 A_{21} & O A_{12} + I_2 A_{22} \end{bmatrix} = \begin{bmatrix} A_{11} & A_{12} \\ A_{21} & A_{22} \end{bmatrix} = A.$$

As demonstrated, the row-column rule remains applicable when computing the product of matrices in partitioned form. This method ensures that the interactions between blocks of matrices follow the same principles as those observed in the multiplication of non-partitioned matrices.

Exercise 10.7 Demonstrate that if a matrix A can be partitioned as:

$$A = \begin{bmatrix} A_{11} & A_{12} \\ A_{21} & A_{22} \end{bmatrix}, \tag{10.13}$$

then the transpose of A, denoted A^T, can be partitioned as:

$$A^T = \begin{bmatrix} A_{11}^T & A_{21}^T \\ A_{12}^T & A_{22}^T \end{bmatrix}. \tag{10.14}$$

Exercise 10.8 Consider a matrix A partitioned as:

$$A = \begin{bmatrix} A_{11} & A_{12} \\ O & A_{22} \end{bmatrix}, \tag{10.15}$$

where all blocks are square matrices of the same size. Prove that A is invertible if both A_{11} and A_{22} are invertible, and the inverse of A is given by:

$$A^{-1} = \begin{bmatrix} A_{11}^{-1} & -A_{11}^{-1} A_{12} A_{22}^{-1} \\ O & A_{22}^{-1} \end{bmatrix}. \tag{10.16}$$

Generally, a matrix can be partitioned into mn blocks arranged in an $m \times n$ array:

$$A = \begin{bmatrix} A_{11} & A_{12} & \cdots & A_{1n} \\ A_{21} & A_{22} & \cdots & A_{2n} \\ \vdots & \vdots & \ddots & \vdots \\ A_{m1} & A_{m2} & \cdots & A_{mn} \end{bmatrix}, \tag{10.17}$$

where each A_{ij} represents a block. This partitioning will be used in the subsequent Def. 10.4.

It is noteworthy that for a matrix partitioned as described in Eq. 10.17, its conjugate transpose is given by:

$$
A^\dagger = \begin{bmatrix} A_{11}^\dagger & A_{21}^\dagger & \cdots & A_{m1}^\dagger \\ A_{12}^\dagger & A_{22}^\dagger & \cdots & A_{m2}^\dagger \\ \vdots & \vdots & \ddots & \vdots \\ A_{1n}^\dagger & A_{2n}^\dagger & \cdots & A_{mn}^\dagger \end{bmatrix},
\tag{10.18}
$$

where each $A_{ij}^\dagger$ represents the conjugate transpose of the corresponding block. Readers are encouraged to verify this pattern by proving it in Problem 10.8.

10.2.2 Defining Matrix Tensor Products

1 Tensor Products in Matrix Representation

An $m \times n$ matrix in $\mathbb{C}^{m \times n}$ can be viewed as a generalized vector in $\mathbb{C}^{mn}$. This conceptualization allows us to apply the general definition of tensor products between two vector spaces (Def. 10.2) to matrices. Specifically, the rules for tensor products between matrices can be defined analogously to those described for column vectors in Def. 10.1.

Definition 10.4 — Tensor Product of Matrices. Consider matrices $A = [a_{ij}] \in \mathbb{C}^{m \times n}$ with entries a_{ij} for $i = 1, 2, \ldots, m$ and $j = 1, 2, \ldots, n$, and $B = [b_{kl}] \in \mathbb{C}^{p \times q}$ with entries b_{kl} for $k = 1, 2, \ldots, p$ and $l = 1, 2, \ldots, q$. The tensor product of the matrix spaces $\mathbb{C}^{m \times n} \otimes \mathbb{C}^{p \times q} = \mathbb{C}^{mp \times nq}$ is defined by:

$$
A \otimes B = \begin{bmatrix} a_{11}B & a_{12}B & \cdots & a_{1n}B \\ a_{21}B & a_{22}B & \cdots & a_{2n}B \\ \vdots & \vdots & \ddots & \vdots \\ a_{m1}B & a_{m2}B & \cdots & a_{mn}B \end{bmatrix},
\tag{10.19}
$$

where each $a_{ij}B$ is a submatrix of size $p \times q$ and the resulting matrix $A \otimes B$ is of size $mp \times nq$.

---

i | The product rule for matrices as defined in Def. 10.4 is called the **Kronecker product**. The Kronecker product provides a natural way to represent the tensor product between matrix spaces. In the context of quantum computing, it is largely synonymous with tensor products, as it is commonly used to model composite quantum states and operators.

■ **Example 10.5** Given matrices $A = I_2$ and $B = \begin{bmatrix} 0 & 1 \\ 1 & 0 \end{bmatrix}$, compute $A \otimes B$ and $B \otimes A$ according to Def. 10.4.

Solution.

$$
A \otimes B = \begin{bmatrix} B & O \\ O & B \end{bmatrix} = \begin{bmatrix} 0 & 1 & 0 & 0 \\ 1 & 0 & 0 & 0 \\ 0 & 0 & 0 & 1 \\ 0 & 0 & 1 & 0 \end{bmatrix},
$$

$$B \otimes A = \begin{bmatrix} O & I_2 \\ I_2 & O \end{bmatrix} = \begin{bmatrix} 0 & 0 & 1 & 0 \\ 0 & 0 & 0 & 1 \\ 1 & 0 & 0 & 0 \\ 0 & 1 & 0 & 0 \end{bmatrix}.$$

Note that $A \otimes B \neq B \otimes A$, illustrating the non-commutativity of tensor products. ∎

Exercise 10.9 Compute the tensor product $Z \otimes Z$, where Z is defined as the Pauli-Z matrix:

$$Z = \begin{bmatrix} 1 & 0 \\ 0 & -1 \end{bmatrix}. \tag{10.20}$$

2   Adjoint Property

Theorem 10.4 The conjugate transpose of the tensor product of two matrices is equal to the tensor product of their conjugate transposes:

$$(A \otimes B)^\dagger = A^\dagger \otimes B^\dagger. \tag{10.21}$$

ⓘ Note that the positions of A and B are not swapped in $(A \otimes B)^\dagger = A^\dagger \otimes B^\dagger$, unlike in $(AB)^\dagger = B^\dagger A^\dagger$.

Proof.   Let $A = [a_{ij}] \in \mathbb{C}^{m \times n}$ and $B \in \mathbb{C}^{p \times q}$ as per the notation in Def. 10.4. Using Eq. 10.18, $(A \otimes B)^\dagger$ in its block partition can be expressed as:

$$(A \otimes B)^\dagger = \begin{bmatrix} (a_{11}B)^\dagger & (a_{21}B)^\dagger & \cdots & (a_{m1}B)^\dagger \\ (a_{12}B)^\dagger & (a_{22}B)^\dagger & \cdots & (a_{m2}B)^\dagger \\ \vdots & \vdots & \ddots & \vdots \\ (a_{1n}B)^\dagger & (a_{2n}B)^\dagger & \cdots & (a_{mn}B)^\dagger \end{bmatrix}$$

$$= \begin{bmatrix} a_{11}^* B^\dagger & a_{21}^* B^\dagger & \cdots & a_{m1}^* B^\dagger \\ a_{12}^* B^\dagger & a_{22}^* B^\dagger & \cdots & a_{m2}^* B^\dagger \\ \vdots & \vdots & \ddots & \vdots \\ a_{1n}^* B^\dagger & a_{2n}^* B^\dagger & \cdots & a_{mn}^* B^\dagger \end{bmatrix}$$

$$= A^\dagger \otimes B^\dagger,$$

since $A^\dagger = \begin{bmatrix} a_{11}^* & a_{21}^* & \cdots & a_{m1}^* \\ a_{12}^* & a_{22}^* & \cdots & a_{m2}^* \\ \vdots & \vdots & \ddots & \vdots \\ a_{1n}^* & a_{2n}^* & \cdots & a_{mn}^* \end{bmatrix}.$ □

Arising directly from Theorem 10.4, for any vectors $|u\rangle \in \mathbb{C}^m$ and $|v\rangle \in \mathbb{C}^n$, the following identity holds:

$$\langle u| \otimes \langle v| = (|u\rangle \otimes |v\rangle)^\dagger. \tag{10.22}$$

Consistent with this, the adjoint $(|uv\rangle)^\dagger$ is abbreviated as $\langle uv|$ rather than $\langle vu|$.

Exercise 10.10 Prove that the tensor product $\Lambda_A \otimes \Lambda_B$ results in a diagonal matrix when both Λ_A and Λ_B are diagonal matrices.

3   Tensor Products of Matrices in Dirac Notation

In quantum computing, tensor products between matrices are often represented using Dirac notation, which utilizes outer products and tensor products between vector spaces. This notation significantly simplifies derivations and computations in theoretical analysis. For readers unfamiliar with Dirac notation, it is recommended to review the basics in § 6.1, § 6.3.3, and § 7.2.6.

Here, we present an equivalent definition of tensor products of matrices using Dirac notation, based on Def. 10.4 and matrix basis decomposition as given in Eq. 7.34.

Definition 10.5 — Tensor Product of Matrices in Dirac Notation. Consider matrices

$$A = \sum_{i,j} a_{ij} |i\rangle\langle j| \ \in \mathbb{C}^{m \times n}, \quad B = \sum_{k,l} b_{kl} |k\rangle\langle l| \ \in \mathbb{C}^{p \times q},$$

where

- $|i\rangle$ for $i = 0, 1, \ldots, m-1$ is the computational basis of $\mathbb{C}^m$,

- $|j\rangle$ for $j = 0, 1, \ldots, n-1$ is the computational basis of $\mathbb{C}^n$,

- $|k\rangle$ for $k = 0, 1, \ldots, p-1$ is the computational basis of $\mathbb{C}^p$, and

- $|l\rangle$ for $l = 0, 1, \ldots, q-1$ is the computational basis of $\mathbb{C}^q$.

The tensor product $A \otimes B$ is defined as:

$$A \otimes B = \sum_{i,j,k,l} a_{ij} b_{kl} (|i\rangle \otimes |k\rangle)(\langle j| \otimes \langle l|) \equiv \sum_{i,j,k,l} a_{ij} b_{kl} |ik\rangle\langle jl|, \qquad (10.23)$$

where $\{|ik\rangle\}$ for all i, k forms the computational basis of $\mathbb{C}^{mp}$ and $\{|jl\rangle\}$ for all j, l forms the computational basis of $\mathbb{C}^{nq}$.

■ **Example 10.6** Show that $I_m \otimes I_n = I_{mn}$ is consistent with Def. 10.5.

Proof. Decompose the identity matrices I_m and I_n in their respective computational bases:

$$I_m = \sum_{i=0}^{m-1} |i\rangle\langle i|, \quad I_n = \sum_{j=0}^{n-1} |j\rangle\langle j|.$$

According to Def. 10.5, the tensor product $I_m \otimes I_n$ is given by:

$$I_m \otimes I_n = \sum_{i,j} |ij\rangle \langle ij|.$$

Since $\{|ij\rangle\}$ forms the computational basis of $\mathbb{C}^{mn}$, the above expression is precisely I_{mn}. Therefore, $I_m \otimes I_n = I_{mn}$.

Using Theorem 10.3, we can also arrive at the same conclusion by computing the sums as follows:

$$I_m \otimes I_n = \sum_{i,j} (|i\rangle\langle i|) \otimes (|j\rangle\langle j|)$$

$$= \sum_{i,j} (|i\rangle \otimes |j\rangle)(\langle i| \otimes \langle j|)$$

$$= \sum_{i,j} |ij\rangle \langle ij| \, .$$

This confirms the consistency of Def. 10.5.   ∎

Exercise 10.11 Given matrices $A = \begin{bmatrix} 1 & 2 \\ 3 & 4 \end{bmatrix}$, and $B = \begin{bmatrix} 4 & 3 \\ 2 & 1 \end{bmatrix}$, demonstrate that Def. 10.5 is consistent with Def. 10.4 by computing $A \otimes B$.

Exercise 10.12 Redo the tensor product computation $Z \otimes Z$ from Exercise 10.9 using Dirac notation. Subsequently, show that your results are consistent with the matrix representation obtained from Exercise 10.9.

10.2.3 Properties of Matrix Tensor Products

1 Basic Properties

The basic properties of tensor products between matrices can be derived from Def. 10.4 or Def. 10.5. However, the following theorem is more general due to the bilinear nature of tensor products and is not limited to matrix tensor products.

Theorem 10.5 For the tensor product between matrices:

$$A \otimes (B + C) = A \otimes B + A \otimes C \qquad \text{(Left Distributive Law)} \qquad (10.24\text{a})$$
$$(A + B) \otimes C = A \otimes C + B \otimes C \qquad \text{(Right Distributive Law)} \qquad (10.24\text{b})$$
$$(kA) \otimes B = k(A \otimes B) \qquad \text{(Left Homogeneity Law)} \qquad (10.24\text{c})$$
$$A \otimes (kB) = k(A \otimes B) \qquad \text{(Right Homogeneity Law)} \qquad (10.24\text{d})$$

The proof of this theorem, based on Def. 10.4 or Def. 10.5, is straightforward and is left as an exercise for readers to familiarize themselves with tensor product properties.

2 Non-commutativity and Associativity

Like tensor products of column vectors, tensor products of matrices are generally non-commutative:

$$A \otimes B \neq B \otimes A, \qquad (10.25)$$

as can be verified using the product rule defined in Def. 10.4.

Despite their non-commutativity, tensor products of matrices do exhibit associativity, a property crucial in applications where multiple tensor products are involved.

Theorem 10.6 The Kronecker product is associative:

$$(A \otimes B) \otimes C = A \otimes (B \otimes C), \qquad (10.26)$$

for any matrices A, B, and C.

The proof of Theorem 10.6 is a useful exercise to reinforce understanding of Def. 10.5. Readers are encouraged to complete this proof as an exercise (Problem 10.11).

Exercise 10.13 Demonstrate that the tensor product, as defined in Def. 10.4, is associative by using an example.

3 Properties Involving Matrix Multiplication

The interaction between matrix multiplication and tensor products is a fundamental aspect of their utility in mathematics and quantum computing. A key property of this interaction is the compatibility between matrix multiplication and tensor products, as formalized in the following theorem:

Theorem 10.7 Let A, B, C, and D be matrices such that all products involved are well-defined. Then the following identity holds:

$$(A \otimes B)(C \otimes D) = (AC) \otimes (BD). \tag{10.27}$$

Proof. This proof can be derived using either Def. 10.4 or Def. 10.5. Here, we provide a proof using Def. 10.4.

Let $A \in \mathbb{C}^{m \times r}$, $B \in \mathbb{C}^{p \times s}$, $C \in \mathbb{C}^{r \times n}$, and $D \in \mathbb{C}^{s \times q}$, such that AC and BD are valid matrix products. Using Def. 10.4, the partitioned forms of $A \otimes B$ and $C \otimes D$ can be written as:

$$A \otimes B = \begin{bmatrix} a_{11}B & a_{12}B & \cdots & a_{1r}B \\ a_{21}B & a_{22}B & \cdots & a_{2r}B \\ \vdots & \vdots & \ddots & \vdots \\ a_{m1}B & a_{m2}B & \cdots & a_{mr}B \end{bmatrix},$$

$$C \otimes D = \begin{bmatrix} c_{11}D & c_{12}D & \cdots & c_{1n}D \\ c_{21}D & c_{22}D & \cdots & c_{2n}D \\ \vdots & \vdots & \ddots & \vdots \\ c_{r1}D & c_{r2}D & \cdots & c_{rn}D \end{bmatrix}.$$

Next, we compute the matrix product $P = (A \otimes B)(C \otimes D)$ using the row-column rule for block-partitioned matrices. Partition P into an array of $m \times n$ blocks, with each block being a $p \times q$ matrix:

$$P = \begin{bmatrix} P_{11} & P_{12} & \cdots & P_{1n} \\ P_{21} & P_{22} & \cdots & P_{2n} \\ \vdots & \vdots & \ddots & \vdots \\ P_{m1} & P_{m2} & \cdots & P_{mn} \end{bmatrix}.$$

For $i = 1, 2, \ldots, m$ and $j = 1, 2, \ldots, n$, we compute each block P_{ij}:

$$P_{ij} = (a_{i1}B)(c_{1j}D) + (a_{i2}B)(c_{2j}D) + \cdots + (a_{ir}B)(c_{rj}D)$$

$$= \left(\sum_{k=1}^{r} a_{ik}c_{kj} \right) BD$$

$$= (AC)_{ij}BD.$$

Thus, the partitioned form of P is:

$$\begin{bmatrix} (AC)_{11}BD & (AC)_{12}BD & \cdots & (AC)_{1n}BD \\ (AC)_{21}BD & (AC)_{22}BD & \cdots & (AC)_{2n}BD \\ \vdots & \vdots & \ddots & \vdots \\ (AC)_{m1}BD & (AC)_{m2}BD & \cdots & (AC)_{mn}BD \end{bmatrix}.$$

By Def. 10.4, this is equal to $(AC) \otimes (BD)$, completing the proof.   □

Theorem 10.7 holds for matrices of any size, including column vectors (kets) and row vectors (bras). The following theorem presents several commonly used formulas that involve composite operations of tensor products and matrix multiplication.

Theorem 10.8 The following identities hold, assuming all matrix products are valid:

$$\langle \phi_1 \otimes \phi_2 | \psi_1 \otimes \psi_2 \rangle \equiv \langle \phi_1 \phi_2 | \psi_1 \psi_2 \rangle = \langle \phi_1 | \psi_1 \rangle \langle \phi_2 | \psi_2 \rangle \tag{10.28a}$$

$$(|i\rangle \otimes |k\rangle)(\langle j| \otimes \langle l|) \equiv |ik\rangle\langle jl| = (|i\rangle \langle j|) \otimes (|k\rangle \langle l|) \tag{10.28b}$$

$$(A \otimes B)(|u\rangle \otimes |v\rangle) = |Au\rangle \otimes |Bv\rangle \tag{10.28c}$$

$$A \otimes B = (A \otimes I)(I \otimes B). \tag{10.28d}$$

It is instructive to observe that, due to property Eq. 10.28b, the tensor product definition in Dirac notation (Eq. 10.23) emerges naturally as an algebraic rearrangement of terms:

$$A \otimes B = \left(\sum_{i,j} a_{ij} |i\rangle\langle j| \right) \otimes \left(\sum_{k,l} b_{kl} |k\rangle\langle l| \right)$$

$$= \sum_{i,j,k,l} a_{ij} b_{kl} (|i\rangle\langle j|) \otimes (|k\rangle\langle l|)$$

$$= \sum_{i,j,k,l} a_{ij} b_{kl} |ik\rangle\langle jl| . \tag{10.29}$$

Exercise 10.14 Derive each of the identities in Theorem 10.8 using Theorem 10.7.

4 Eigenvalue-related Properties

Theorem 10.9 For square matrices A and B, if $\{\lambda_A, |a\rangle\}$ is an eigenvalue-eigenvector pair of A, and $\{\lambda_B, |b\rangle\}$ is an eigenvalue-eigenvector pair of B, then $\{\lambda_A \lambda_B, |a\rangle \otimes |b\rangle\}$ constitutes an eigenvalue-eigenvector pair of $A \otimes B$.

Proof. Since λ_A is the eigenvalue associated with $|a\rangle$ for A, we have:

$$A |a\rangle = \lambda_A |a\rangle .$$

Similarly, for B, we have:

$$B |b\rangle = \lambda_B |b\rangle .$$

Applying Theorem 10.8 for the tensor product of A and B, we find:

$$(A \otimes B)(|a\rangle \otimes |b\rangle) = (A |a\rangle) \otimes (B |b\rangle)$$

$$= (\lambda_A |a\rangle) \otimes (\lambda_B |b\rangle)$$

$$= \lambda_A \lambda_B (|a\rangle \otimes |b\rangle),$$

confirming that $\lambda_A \lambda_B$ is an eigenvalue of $A \otimes B$ associated with the eigenvector $|a\rangle \otimes |b\rangle$, or $|ab\rangle$.   □

The converse of Theorem 10.9 also holds for any square matrices A and B, as stated in the theorem below.

> **Theorem 10.10** For square matrices A and B, all eigenvalues of $A \otimes B$ are formed by the products $\lambda_A \lambda_B$, where λ_A is an eigenvalue of A and λ_B is an eigenvalue of B.

Specifically, let λ_i for $i = 1, 2, \ldots, m$ be the eigenvalues of $A \in \mathbb{C}^{m \times m}$, and let μ_j for $j = 1, 2, \ldots, n$ be the eigenvalues of $B \in \mathbb{C}^{n \times n}$, each repeated according to its algebraic multiplicity. The eigenvalues of $A \otimes B$ are then all possible combinations of $\lambda_i \mu_j$, where the algebraic multiplicity of each $\lambda_i \mu_j$ equals the number of times it appears among these combinations.

Moreover, the corresponding eigenspace for each eigenvalue $\lambda_i \mu_j$ is the tensor product of the eigenspaces corresponding to the eigenvalues λ_i of A and μ_j of B.

The proof of Theorem 10.10 for diagonalizable matrices A and B is left as an exercise for the reader (Problem 10.13). The proof for defective matrices, which involves the Jordan canonical form, is beyond the scope of this chapter.

5 Trace- and Determinant-related Properties

There are two important formulas that can be derived from Theorem 10.9 that relate to the trace and determinant of tensor products.

> **Theorem 10.11** For square matrices A and B, the following identity holds true:
> $$\operatorname{tr}(A \otimes B) = \operatorname{tr}(A) \operatorname{tr}(B). \tag{10.30}$$

Proof. Utilizing the properties of the trace and the tensor product eigenvalue relationships given in Eq. 9.13b, let:

$$\operatorname{tr}(A) = \sum_{i=1}^{m} a_i, \quad \operatorname{tr}(B) = \sum_{j=1}^{n} b_j,$$

where $\{a_i\}$ for $i = 1, 2, \ldots, m$ are the eigenvalues of $A \in \mathbb{C}^{m \times m}$ and $\{b_j\}$ for $j = 1, 2, \ldots, n$ are the eigenvalues of $B \in \mathbb{C}^{n \times n}$. The trace of the tensor product $A \otimes B$ can be calculated as follows:

$$
\begin{aligned}
\operatorname{tr}(A) \operatorname{tr}(B) &= \left(\sum_i a_i \right) \left(\sum_j b_j \right) \\
&= \sum_{i,j} a_i b_j \\
&= \operatorname{tr}(A \otimes B),
\end{aligned}
$$

since the eigenvalues $a_i b_j$ for all i, j constitute all eigenvalues of $A \otimes B$, as established by Theorem 10.10. This confirms the trace identity for tensor products.　□

Exercise 10.15 Prove Theorem 10.11 by directly applying Def. 10.4.

> **Theorem 10.12** For square matrices $A \in \mathbb{C}^{m \times m}$ and $B \in \mathbb{C}^{n \times n}$, the following identity holds true:
> $$\det(A \otimes B) = (\det A)^n (\det B)^m. \tag{10.31}$$

The proof of Theorem 10.12 utilizes the fact that the determinant of a square matrix is equal to the product of all its eigenvalues. Readers may complete the proof in Problem 10.16.

6   Inverse-related Properties

Theorem 10.13 Given two invertible matrices A and B, the tensor product $A \otimes B$ is also invertible, with its inverse given by:

$$(A \otimes B)^{-1} = A^{-1} \otimes B^{-1}. \tag{10.32}$$

Exercise 10.16 Prove Theorem 10.13 by demonstrating that multiplying $A \otimes B$ by $A^{-1} \otimes B^{-1}$ results in the identity matrix.

10.3   Tensor Products in Quantum Computing

In this section, we discuss properties of tensor products commonly utilized in quantum computing, focusing particularly on tensor products involving unitary matrices, Hermitian matrices, and unit vectors.

10.3.1   Tensor Products of Unit Vectors

Since quantum states are represented by unit vectors within Hilbert spaces, the tensor product of multiple state vectors representing joint states of independent quantum systems must also be a unit vector. This is confirmed by the following theorem:

Theorem 10.14 Given unit vectors $|\psi_1\rangle, |\psi_2\rangle, \ldots, |\psi_n\rangle$, their tensor product is also a unit vector:

$$\langle \psi_1 \psi_2 \cdots \psi_n | \psi_1 \psi_2 \cdots \psi_n \rangle = 1. \tag{10.33}$$

Proof.   Following Theorem 10.2, the inner product of the tensor product of the vectors is calculated as:

$$\langle \psi_1 \psi_2 \cdots \psi_n | \psi_1 \psi_2 \cdots \psi_n \rangle = \langle \psi_1 | \psi_1 \rangle \cdot \langle \psi_2 | \psi_2 \rangle \cdots \cdot \langle \psi_n | \psi_n \rangle = 1, \tag{10.34}$$

given that each $|\psi_i\rangle$ is a unit vector, so $\langle \psi_i | \psi_i \rangle = 1$ for all i.   □

10.3.2   Tensor Products of Unitary Matrices

1   Preservation of Unitarity

For a unitary matrix U_A associated with subsystem A, and a unitary matrix U_B for subsystem B, their composite operation on the joint system can be represented by the tensor product $U_A \otimes U_B$. As expected, the tensor product of these unitary matrices is itself unitary, as confirmed by the following theorem.

Theorem 10.15 Unitarity is preserved under tensor products. Specifically, $U_A \otimes U_B$ is a unitary matrix in $\mathbb{C}^{mn \times mn}$ for any unitary matrices $U_A \in \mathbb{C}^{m \times m}$ and $U_B \in \mathbb{C}^{n \times n}$.

Proof. Using the properties of tensor products, consider the composition of the adjoint and the matrix itself:

$$(U_A \otimes U_B)^\dagger (U_A \otimes U_B) = (U_A^\dagger \otimes U_B^\dagger)(U_A \otimes U_B)$$
$$= (U_A^\dagger U_A) \otimes (U_B^\dagger U_B)$$
$$= I_m \otimes I_n$$
$$= I_{mn}.$$

Therefore, by the definition of unitary matrices, $U_A \otimes U_B$ is confirmed to be a unitary matrix in $\mathbb{C}^{mn \times mn}$. $\qquad\qquad\qquad\qquad\qquad\qquad\qquad\qquad\qquad\qquad\qquad\qquad$ □

The properties outlined in Theorem 10.15 generalize to joint systems comprising more than two subsystems:

$$U_1 \otimes U_2 \otimes \cdots \otimes U_n \text{ is unitary, for unitary matrices } U_1, U_2, \ldots, U_n. \qquad (10.35)$$

2 Unitary Transformation Across Multiple Subsystems

When unitary operators are applied to a product state separable into n subsystems, the following identity holds:

$$(U_1 \otimes U_2 \otimes \cdots \otimes U_n) |\psi_1 \psi_2 \cdots \psi_n\rangle = (U_1 |\psi_1\rangle) \otimes (U_2 |\psi_2\rangle) \cdots \otimes (U_n |\psi_n\rangle). \quad (10.36)$$

Thus, applying a unitary matrix U_i to each subsystem in the state $|\psi_i\rangle$ for $i = 1, 2, \ldots, n$ is equivalent to applying the composite unitary matrix $\bigotimes_{i=1}^n U_i$ to the entire product state $\bigotimes_{i=1}^n |\psi_i\rangle$.

Equation 10.27 demonstrates the commutativity between matrix multiplication and tensor products. This principle can be generalized for multiple subsystems as follows:

$$(A_1 \otimes A_2 \otimes \cdots \otimes A_n)(B_1 \otimes B_2 \otimes \cdots \otimes B_n) = (A_1 B_1) \otimes (A_2 B_2) \otimes \cdots \otimes (A_n B_n), \tag{10.37a}$$

$$(A_1 \otimes B_1)(A_2 \otimes B_2) \cdots (A_n \otimes B_n) = (A_1 A_2 \cdots A_n) \otimes (B_1 B_2 \cdots B_n). \tag{10.37b}$$

For instance, consider A_i and B_i as unitary matrices. In Eq. 10.37a, each of the n subsystems undergoes operations B_i followed by A_i sequentially for each subsystem i. The collective operation on the entire system can be expressed equivalently by either side of the equation.

Conversely, Eq. 10.37b illustrates how two subsystems, A and B, undergoing n sequential operations, can have their combined operations on the entire system equivalently expressed by either side of the equation.

10.3.3 Tensor Products of Hermitian Matrices

1 Preservation of Hermiticity

Tensor products preserve Hermiticity when the matrices involved are Hermitian. This property is essential for describing the measurement processes in multi-qubit systems and the behavior of composite mixed states represented by density matrices.

> **Theorem 10.16** Given two Hermitian matrices H_A and H_B, their tensor product $H_A \otimes H_B$ is also Hermitian.

Proof. Applying Theorem 10.4, we observe that:

$$(H_A \otimes H_B)^\dagger = H_A^\dagger \otimes H_B^\dagger = H_A \otimes H_B,$$

because $H_A^\dagger = H_A$ and $H_B^\dagger = H_B$ by their Hermitian nature. Thus, $H_A \otimes H_B$ is confirmed to be Hermitian.   □

2 Statistical Average of Measurement

Measurements in quantum mechanics are represented using the bra-ket notation, where the statistical average of a measurement operator (i.e., matrix) M on a state $|\psi\rangle$ is calculated by the quadratic form $\langle\psi|M|\psi\rangle$. This formulation, a specific case of matrix multiplication also demonstrates commutativity with tensor products, as shown in the following theorem.

> **Theorem 10.17** Assuming all matrix multiplications are valid, the following identity holds true:
>
> $$\langle\psi_A\psi_B|M_A \otimes M_B|\psi_A\psi_B\rangle = \langle\psi_A|M_A|\psi_A\rangle \langle\psi_B|M_B|\psi_B\rangle. \qquad (10.38)$$

This formula is useful for computing the statistical average of joint measurements on a multi-qubit system, whether in a product state or an entangled state.

■ **Example 10.7** Consider a measurement of the operator $Z \otimes Z$ on the state $|01\rangle$, where $Z = \begin{bmatrix} 1 & 0 \\ 0 & -1 \end{bmatrix}$. What is the statistical average of this measurement?

Solution. Using the decomposition $Z = |0\rangle\langle 0| - |1\rangle\langle 1|$ and applying Eq. 10.38, we calculate:

$$\langle 01|Z \otimes Z|01\rangle = \langle 0|Z|0\rangle \langle 1|Z|1\rangle = 1 \times (-1) = -1.$$

Thus, the statistical average of the measurement is -1.   ■

> **Exercise 10.17** Using the same approach, determine the statistical average of performing a measurement using the operator $Y \otimes Y$ on the states
>
> $$\text{(a) } |01\rangle. \text{ (b) } \frac{1}{\sqrt{2}}(|00\rangle + |11\rangle), \text{ (c) } \frac{1}{\sqrt{2}}(|00\rangle - |11\rangle),$$
>
> where $Y = \begin{bmatrix} 0 & -i \\ i & 0 \end{bmatrix}$.

10.3.4 ✳ Spectral Properties of Tensor Products

1 Preservation of Normality

The properties of unitarity and Hermiticity preserved through tensor products can be extended to include the preservation of normality for all normal matrices.

> **Theorem 10.18** If A and B are normal matrices, then their tensor product $A \otimes B$ is also a normal matrix.

Proof. To demonstrate that $A \otimes B$ is a normal matrix, it is necessary to establish that $(A \otimes B)^{\dagger}(A \otimes B) = (A \otimes B)(A \otimes B)^{\dagger}$. This equivalence can be shown through the following steps:

$$\begin{aligned}
(A \otimes B)^{\dagger}(A \otimes B) &= (A^{\dagger} \otimes B^{\dagger})(A \otimes B) \\
&= (A^{\dagger}A) \otimes (B^{\dagger}B) \\
&= (AA^{\dagger}) \otimes (BB^{\dagger}) \\
&= (A \otimes B)(A^{\dagger} \otimes B^{\dagger}) \\
&= (A \otimes B)(A \otimes B)^{\dagger},
\end{aligned}$$

where algebraic properties of tensor products are used alongside the definition of normal matrices. This completes the proof that $A \otimes B$ is normal, leveraging the properties of tensor products and the inherent normality of A and B. $\qquad\square$

2 Spectral Decomposition - Dirac Notation

Normal matrices are unitarily diagonalizable, which permits their representation through spectral decomposition. Consider two normal matrices $A \in \mathbb{C}^{m \times m}$ and $B \in \mathbb{C}^{n \times n}$, expressed as:

$$A = \sum_{i=1}^{m} \lambda_i \, |a_i\rangle \, \langle a_i| \,, \quad B = \sum_{j=1}^{n} \mu_j \, |b_j\rangle \, \langle b_j| \,, \tag{10.39}$$

where λ_i are the eigenvalues of A, associated with the eigenvectors $|a_i\rangle$ for $i = 1, 2, \ldots, m$, and μ_j are the eigenvalues of B, associated with the eigenvectors $|b_j\rangle$ for $j = 1, 2, \ldots, n$.

Computing the tensor product $A \otimes B$ by taking their spectral representations in Eq. 10.39 gives the spectral decomposition of $A \otimes B$:

$$A \otimes B = \sum_{i,j} \lambda_i \mu_j (|a_i\rangle \otimes |b_j\rangle)(\langle a_i| \otimes \langle b_j|) = \sum_{i,j} \lambda_i \mu_j \, |a_i b_j\rangle \, \langle a_i b_j| \,, \tag{10.40}$$

because $\lambda_i \mu_j$ are the eigenvalues of $A \otimes B$, each corresponding to the eigenvector $|a_i\rangle \otimes |b_j\rangle$ for all pairs i, j according to § 10.2.3.4. This formula also verifies the theorem that $A \otimes B$ is a normal matrix by showcasing its spectral decomposition.

3 Spectral Decomposition - Matrix Representation

The spectral decomposition of normal matrices A and B can be expressed equivalently in matrix form:

$$A = U_A \Lambda_A U_A^{\dagger}, \quad B = U_B \Lambda_B U_B^{\dagger},$$

where U_A and U_B are unitary matrices, and Λ_A and Λ_B are diagonal matrices representing the eigenvalues of A and B, respectively.

When taking the tensor product of A and B, their decompositions can be utilized to form the following expression:

$$A \otimes B = (U_A \otimes U_B)(\Lambda_A \otimes \Lambda_B)(U_A \otimes U_B)^{\dagger}. \tag{10.41}$$

In this decomposition, $U_A \otimes U_B$ is still unitary, and $\Lambda_A \otimes \Lambda_B$ is still diagonal. Therefore, Eq. 10.41 reaffirms that $A \otimes B$ is unitarily diagonalizable.

Exercise 10.18 Determine the eigenvalues of the tensor product $A \otimes B$ by examining the diagonal matrix $\Lambda_A \otimes \Lambda_B$.

10.3.5 ✳ Tensor Products of Density Matrices

1 Introduction

As introduced in § 7.4.2.1, a pure quantum state in finite dimensions is represented by a unit vector in $\mathbb{C}^n$, while a mixed state is a statistical ensemble of such pure states. It is represented by a density matrix defined as:

$$\rho = \sum_i p_i |\psi_i\rangle\langle\psi_i|,$$

where $p_i \geq 0$ and $\sum_i p_i = 1$. Here, $|\psi_i\rangle$ are pure states, and when there are multiple terms in the sum, the state is mixed, with p_i being the probability of observing the pure state $|\psi_i\rangle$.

Density matrices have two key properties: unit trace $\operatorname{tr}\rho = 1$ and positive semidefiniteness. The former was discussed in § 7.4.2.1 and the latter in § 9.4.5.3.

In this subsection, we will investigate additional properties of density matrices in relation to tensor products.

2 Tensor Products of Density Matrices

Theorem 10.19 Let ρ_1 and ρ_2 be two valid density matrices. Then $\rho_1 \otimes \rho_2$ is also a valid density matrix.

Proof.

Trace one: Using Theorem 10.11, we obtain

$$\operatorname{tr}(\rho_1 \otimes \rho_2) = \operatorname{tr}(\rho_1)\operatorname{tr}(\rho_2) = 1.$$

Preservation of Hermiticity: see Theorem 10.16.

Preservation of positivity: as shown in § 10.3.4.2, if the eigenvalues of ρ_1 are $\{\lambda_{1i}\}$, and the eigenvalues of ρ_2 are $\{\lambda_{2j}\}$, then those for $\rho_1 \otimes \rho_2$ are given by $\{\lambda_{1i}\lambda_{2j}\}$. If $\lambda_{1i} \geq 0$ for all i as required by the semi-positivity of ρ_1, and similarly for $\lambda_{2j} \geq 0$ for all j, then $\lambda_{1i}\lambda_{2j} \geq 0$ for all combinations of i and j. This shows $\rho_1 \otimes \rho_2$ is positive semidefinite. □

3 Product, Separable, and Entangled States

In § 10.1.5, we discussed product states (separable states) and non-product states (entangled states) for pure states. However, when dealing with mixed states, the terms "product states" and "separable states" are no longer synonymous, and similarly, "non-product states" and "entangled states" diverge in meaning. To better understand the concept of entanglement for mixed states, it is necessary to define "separable states" in the context of mixed states.

In quantum mechanics, product states, separable states, and entangled states are distinguished based on their physical properties [1]:

- A product state involves subsystems that are independent of one another, so that it is possible to manipulate or measure one subsystem without affecting the others.

- A separable state has subsystems that are correlated in a classical way. The state represents a classical probability distribution over uncorrelated (product) states of the subsystems.

- An entangled state has subsystems that are correlated in a non-classical manner, typically beyond the realm of local realism.

In terms of density matrices, product states, separable states, and entangled states are distinguished as follows:

Definition 10.6 — Product Mixed State. If a density matrix ρ can be expressed as the tensor product of two density matrices, each representing the state of a subsystem:

$$\rho = \rho_A \otimes \rho_B, \tag{10.42}$$

then the quantum state represented by ρ, whether pure or mixed, is said to be a product state.

Definition 10.7 — Separable Mixed State. A density matrix ρ represents a separable state if it can be expressed as:

$$\rho = \sum_i p_i \rho_i^A \otimes \rho_i^B, \tag{10.43}$$

where $p_i \geq 0$ and $\sum_i p_i = 1$, and ρ_i^A, ρ_i^B are density matrices of subsystems A and B, which may be either pure or mixed states.

A product state is a special case of separable states. If ρ cannot be expressed as a separable state, it is considered an **entangled state**.

Determining whether a state is a product state is relatively straightforward using methods like the Schmidt decomposition (§ 13.3). However, distinguishing separable states from entangled states for mixed states is generally more challenging, and this remains an area of active research beyond the scope of this discussion. The following exercises aim to solidify these concepts for two-qubit systems.

Exercise 10.19 ✳ Consider the following density matrices for two-qubit states:

$$\rho_A = \frac{1}{2}(|00\rangle\langle 00| + |11\rangle\langle 11|),$$

$$\rho_B = \frac{1}{2}(|00\rangle\langle 00| + |01\rangle\langle 01|),$$

$$\rho_C = \frac{1}{2}(|00\rangle + |11\rangle)(\langle 00| + \langle 11|).$$

(a) Identify whether each state is a pure state or a mixed state.

(b) Determine whether each state is a product state, a separable state, or an entangled state.

Problem Set 10

10.1 Express the state vector $|+\rangle^{\otimes 3}$ using the binary representation format described in Eq. 10.6.

10.2 Using Eq. 10.7, express $|\psi\rangle = |001\rangle + |110\rangle$ in matrix form by considering it under two different bipartite splits: $\mathbb{C}^2 \otimes \mathbb{C}^4$ and $\mathbb{C}^4 \otimes \mathbb{C}^2$.

10.3 Prove that Def. 10.1 is a valid computational rule that complies with Def. 10.2.

10.4 Demonstrate that the vector $|x_1 x_2 \cdots x_n\rangle$, where $x_i \in \{0, 1\}$, can be represented by a column vector:

$$\begin{bmatrix} 0 & 0 & 0 & \cdots & 1 & \cdots & 0 \end{bmatrix}^T,$$

with the sole '1' at the index corresponding to the binary number $(x_1 x_2 \cdots x_n)_2$. Indexing starts from 0 and ends at $2^n - 1$.

10.5 $*$ Provide an example of a pure state in $\mathbb{C}^8$ which is a product state within the bipartite system $\mathbb{C}^2 \otimes \mathbb{C}^4$ but not within the bipartite system $\mathbb{C}^4 \otimes \mathbb{C}^2$.

10.6 Prove Theorem 10.1.

10.7 Prove that if $\{|u_i\rangle\}$ is an orthonormal basis of $\mathbb{C}^m$ and $\{|v_j\rangle\}$ is an orthonormal basis of $\mathbb{C}^n$, then $\{|u_i\rangle \otimes |v_j\rangle\}$ is an orthonormal basis of $\mathbb{C}^m \otimes \mathbb{C}^n$.

10.8 Prove that if a matrix A is partitioned into mn submatrices:

$$A = \begin{bmatrix} A_{11} & A_{12} & \cdots & A_{1n} \\ A_{21} & A_{22} & \cdots & A_{2n} \\ \vdots & \vdots & \ddots & \vdots \\ A_{m1} & A_{m2} & \cdots & A_{mn} \end{bmatrix},$$

then its conjugate transpose is given by:

$$A^\dagger = \begin{bmatrix} A_{11}^\dagger & A_{21}^\dagger & \cdots & A_{m1}^\dagger \\ A_{12}^\dagger & A_{22}^\dagger & \cdots & A_{m2}^\dagger \\ \vdots & \vdots & \ddots & \vdots \\ A_{1n}^\dagger & A_{2n}^\dagger & \cdots & A_{mn}^\dagger \end{bmatrix}.$$

10.9 $*$ Show that if a state vector $|\psi\rangle \in \mathbb{C}^m \otimes \mathbb{C}^n$ is a product state, expressed as $|\psi\rangle = |u\rangle \otimes |v\rangle$, where $|u\rangle \in \mathbb{C}^m$ and $|v\rangle \in \mathbb{C}^n$, then the density matrix ρ_ψ of the state $|\psi\rangle$ (as defined in § 7.4.2.1) is equal to the tensor product of the density matrices of $|u\rangle$ and $|v\rangle$:

$$\rho_\psi = \rho_u \otimes \rho_v.$$

10.10 Prove Theorem 10.5 using Def. 10.4 or Def. 10.5.

10.11 $*$ Prove Theorem 10.6 using Def. 10.5.

10.12 ✳ Prove Theorem 10.7 using Def. 10.5.

10.13 ✳ Prove the converse of Theorem 10.9 for diagonalizable matrices. Specifically, let λ_i for $i = 1, 2, \ldots, m$ be the eigenvalues of $A \in \mathbb{C}^{m \times m}$, and μ_j for $j = 1, 2, \ldots, n$ be the eigenvalues of $B \in \mathbb{C}^{n \times n}$, with each eigenvalue repeated according to its algebraic multiplicity. Assume both A and B are diagonalizable.

   (a) Prove that the eigenvalues of $A \otimes B$ are all possible combinations $\lambda_i \mu_j$ for $i = 1, 2, \ldots, m$ and $j = 1, 2, \ldots, n$, with the algebraic multiplicity of each eigenvalue $\lambda_i \mu_j$ equal to the number of times it appears among these combinations.

   (b) Demonstrate that the corresponding eigenspace is the direct sum of eigenspaces formed by the tensor product of the corresponding eigenvectors from A and B.

10.14 Verify the conclusions of Problem 10.13 concerning the eigenvalues and eigenvectors of tensor products by computing these for the tensor products of the Pauli matrices:

$$\text{(a) } X \otimes X, \text{ (b) } Y \otimes Y, \text{ (c) } Z \otimes Z, \text{ (d) } X \otimes Z, \text{ (e) } I_2 \otimes Z,$$

where $X = \begin{bmatrix} 0 & 1 \\ 1 & 0 \end{bmatrix}$, $Y = \begin{bmatrix} 0 & -i \\ i & 0 \end{bmatrix}$, and $Z = \begin{bmatrix} 1 & 0 \\ 0 & -1 \end{bmatrix}$ are Pauli matrices.

10.15 If λ_A is an eigenvalue of $A \in \mathbb{C}^{m \times m}$ associated with an eigenvector $|a\rangle$, and λ_B is an eigenvalue of $B \in \mathbb{C}^{n \times n}$ associated with an eigenvector $|b\rangle$, demonstrate that $\lambda_A + \lambda_B$ is an eigenvalue of $A \otimes I + I \otimes B$ associated with the eigenvector $|a\rangle \otimes |b\rangle$.

10.16 Prove Theorem 10.12.

10.17 ✳ The Controlled-U gate is mathematically represented as:

$$\text{CU} = |0\rangle\langle 0| \otimes I + |1\rangle\langle 1| \otimes U,$$

where I is the 2×2 identity matrix and U is a generic 2×2 unitary matrix representing a specific quantum gate.

   (a) Show that the CU gate applies U to the target qubit only when the control qubit is $|1\rangle$:

$$\text{CU} |0\psi\rangle = |0\psi\rangle, \quad \text{CU} |1\psi\rangle = |1\rangle |U\psi\rangle.$$

   (b) Show that when $U^n = I_2$, $(\text{CU})^n = I_4$ for any positive integer n.

10.18 The four Bell states are defined as:

$$|\beta_{00}\rangle = \frac{1}{\sqrt{2}}(|00\rangle + |11\rangle), \quad |\beta_{01}\rangle = \frac{1}{\sqrt{2}}(|01\rangle + |10\rangle),$$

$$|\beta_{10}\rangle = \frac{1}{\sqrt{2}}(|00\rangle - |11\rangle), \quad |\beta_{11}\rangle = \frac{1}{\sqrt{2}}(|01\rangle - |10\rangle).$$

Show that the following relationships hold:

   (a) $|\beta_{00}\rangle \langle\beta_{00}| = \frac{1}{2} (I \otimes I + Z \otimes Z)$,

   (b) $|\beta_{01}\rangle \langle\beta_{01}| = \frac{1}{2} (X \otimes X + Y \otimes Y)$,

(c) $|\beta_{10}\rangle \langle\beta_{10}| = \frac{1}{2}(I \otimes I - Z \otimes Z)$,

(d) $|\beta_{11}\rangle \langle\beta_{11}| = \frac{1}{2}(X \otimes X - Y \otimes Y)$.

for Pauli matrices X, Y, Z given in Problem 10.14.

11. Functions of Vectors and Matrices

Contents

A *function* f from a set A to a set B, denoted by $f : A \to B$, assigns exactly one element of B to each element of A. This relationship is expressed as $f(a) = b$, where b is the element of B assigned to an element a of A. (Refer to § 4.3.1 for more details.)

For example, consider the function $f(x) = x^2$, where $f : \mathbb{R} \to \mathbb{R}$. Similarly, the function defined by $r = \sqrt{x^2 + y^2}$ maps pairs of real numbers to real numbers and can be denoted as $f : \mathbb{R} \times \mathbb{R} \to \mathbb{R}$.

In this chapter, we expand the concept of functions to include those with domains and codomains that consist of vectors and matrices, extending beyond just sets of real or complex numbers. Functional analysis with vectors and matrices is widely used in quantum computing, from analyzing quantum algorithms to designing error-correcting codes, as well as in various fields of mathematics and engineering. This exploration will also strengthen your understanding and skills with vectors and matrices, which are essential in many areas of advanced study.

> *(i)* In this chapter, we will briefly revisit concepts introduced in earlier chapters, focusing on their connections to functions. When doing so, definitions and theorems will be restated directly in the narrative rather than presented in formal blocks.

11.1   Scalar-Valued Functions

We begin by examining functions that yield scalar values but accept vector or matrix arguments. You are familiar with several examples of such functions:

- Inner product of vectors: $\langle x|y \rangle : \mathbb{C}^n \times \mathbb{C}^n \to \mathbb{C}$

- Norm of a vector: $\|x\| : \mathbb{C}^n \to \mathbb{R}$

- Trace of a matrix: $\operatorname{tr} A : \mathbb{C}^{n \times n} \to \mathbb{C}$

- Determinant of a matrix: $\det A : \mathbb{C}^{n \times n} \to \mathbb{C}$

In this section, we will revisit these familiar concepts and explore them from the perspective of functional analysis.

> *(i)* **Dirac Notation:** Consistent with general conventions in quantum mechanics literature, we employ the Dirac (bra-ket) notation in the following manner:
>
> - $|x\rangle$ represents a column vector, often expressed as $\boldsymbol{x}$ in other contexts.
>
> - $\langle x|$ denotes the adjoint (conjugate transpose) of $|x\rangle$.
>
> - $\langle u|v \rangle \equiv \langle u||v \rangle$ specifies the inner product of $|u\rangle$ with $|v\rangle$.
>
> - The expression $|\alpha y + \beta z\rangle$ is shorthand for $\alpha |y\rangle + \beta |z\rangle$, where α and β are scalars.
>
> - A function f applied to $|x\rangle$ is written as $f(x)$, which is equivalent to $f(|x\rangle)$.
>
> - The norm of $|x\rangle$ is expressed as $\|x\|$.

11.1.1   Inner (Dot) Products and Norms of Vectors

The inner product, also known as the dot product, is a fundamental operation in linear spaces (see Chapter 6). Defined as a function from $\mathbb{C}^n \times \mathbb{C}^n \to \mathbb{C}$, the inner product of two vectors exhibits the following essential properties:

- Conjugate Linearity in the First Argument: For any $\alpha, \beta \in \mathbb{C}$,

$$\langle \alpha x_1 + \beta x_2 | y \rangle = \alpha^* \langle x_1 | y \rangle + \beta^* \langle x_2 | y \rangle. \tag{11.1}$$

- Linearity in the Second Argument: For any $\alpha, \beta \in \mathbb{C}$,

$$\langle x | \alpha y_1 + \beta y_2 \rangle = \alpha \langle x | y_1 \rangle + \beta \langle x | y_2 \rangle. \tag{11.2}$$

- Conjugate Symmetry:

$$\langle x | y \rangle = \langle y | x \rangle^*. \tag{11.3}$$

- Positive Definiteness:

  ○ $\langle x|x \rangle$ is real and non-negative,

  ○ $\langle x|x \rangle = 0$ if and only if $|x \rangle = 0$.

The inner product function is termed *sesquilinear* because it is linear in one argument and conjugate-linear (or anti-linear) in the other. The prefix "sesqui-" comes from the Latin, meaning "one and a half."

The vector norm (or amplitude) is directly derived from the inner product:

$$\|x\| = \sqrt{\langle x|x \rangle}. \tag{11.4}$$

Exercise 11.1 Given $\lambda \in \mathbb{C}$ and $|x \rangle, |y \rangle \in \mathbb{C}^n$, complete the following tasks:

(a) Prove these identities:

$$\langle \lambda x|y \rangle = \lambda^* \langle x|y \rangle \tag{11.5a}$$
$$\langle x|y \rangle \langle y|x \rangle = |\langle x|y \rangle|^2 \tag{11.5b}$$
$$\langle x|y \rangle + \langle y|x \rangle = 2\,\mathrm{Re}(\langle x|y \rangle) \tag{11.5c}$$
$$\langle x|y \rangle - \langle y|x \rangle = 2i\,\mathrm{Im}(\langle x|y \rangle) \tag{11.5d}$$
$$\langle y - \lambda x|y - \lambda x \rangle = \langle y|y \rangle - \lambda^* \langle x|y \rangle - \lambda \langle y|x \rangle + |\lambda|^2 \langle x|x \rangle \tag{11.5e}$$

(b) Expand $\langle x + \lambda y|x - \lambda y \rangle$.

11.1.2 Inner Products, Norms, and Positivity of Matrices

1 Inner Product

The inner product, or dot product, can be extended to matrices, as discussed in § 7.4.2.3. For matrices $A, B \in \mathbb{C}^{m \times n}$, the inner product is defined as a function mapping $\mathbb{C}^{m \times n} \times \mathbb{C}^{m \times n} \to \mathbb{C}$:

$$\langle A, B \rangle \equiv A \cdot B = \mathrm{tr}(A^\dagger B) = \sum_{i=1}^{m} \sum_{j=1}^{n} a_{ij}^* b_{ij}. \tag{11.6}$$

This definition of matrix inner product is known as the Hilbert-Schmidt inner product. It retains the same properties as the vector inner product outlined in § 11.1.1.

Exercise 11.2 Prove the following properties for the Hilbert-Schmidt inner product $\langle A, B \rangle$:

- Conjugate Linearity in the First Argument: For any $\alpha, \beta \in \mathbb{C}$,

$$\langle \alpha A_1 + \beta A_2|B \rangle = \alpha^* \langle A_1|B \rangle + \beta^* \langle A_2|B \rangle. \tag{11.7}$$

- Linearity in the Second Argument: For any $\alpha, \beta \in \mathbb{C}$,

$$\langle A|\alpha B_1 + \beta B_2 \rangle = \alpha \langle A|B_1 \rangle + \beta \langle A|B_2 \rangle. \tag{11.8}$$

- Conjugate Symmetry:

$$\langle A|B \rangle = \langle B|A \rangle^*. \tag{11.9}$$

- Positive Definiteness:
    - $\langle A|A \rangle$ is real and non-negative,
    - $\langle A|A \rangle = 0$ if and only if $A = 0$.

2   Frobenius Norm

The matrix norm associated with the above inner product definition is termed the Frobenius norm:

$$\|A\|_F = \sqrt{\langle A, A \rangle} = \sqrt{\mathrm{tr}(A^\dagger A)} = \sqrt{\sum_{i=1}^{m} \sum_{j=1}^{n} |a_{ij}|^2}. \tag{11.10}$$

Theorem 11.1 $\|A\|_F^2$ equals the sum of the eigenvalues of $A^\dagger A$. (Note that $A^\dagger A$ is Hermitian for any A, and therefore has real non-negative eigenvalues.)

Proof.

$$\begin{aligned}
\|A\|_F^2 &= \mathrm{tr}(A^\dagger A) \\
&= \mathrm{tr}(U\Lambda U^\dagger) && \text{(eigendecomposition of } A^\dagger A) \\
&= \mathrm{tr}(\Lambda) && \text{(cyclic property of the trace)} \\
&= \sum_i \lambda_i.
\end{aligned}$$

$\square$

3   Spectral Norm

A more frequently used form of matrix norm is the spectral norm. It is an operator norm defined as:

$$\|A\|_2 = \sup \left\{ \sqrt{\frac{\langle Av|Av \rangle}{\langle v|v \rangle}} : v \neq 0 \right\} = \sup \left\{ \sqrt{\langle Av|Av \rangle} : \langle v|v \rangle = 1 \right\}, \tag{11.11}$$

where sup is the supremum function, defined as the least upper bound of all elements of a set.

The subscript 2 in the notation $\|A\|_2$ signifies that the norm measures the maximum stretching factor of A on the Euclidean (2-norm) length of a vector. Sometimes, the subscript is omitted, and the spectral norm is simply written as $\|A\|$.

For a normal matrix, the spectral norm is the maximum absolute values of its eigenvalues. It can be interpreted as a measure of the maximum amount a matrix can stretch a vector.

Theorem 11.2 If A is a normal matrix, then $\|A\|_2 = \sup\{|\lambda_i|\}$, where $\{\lambda_i\}$ are the eigenvalues of A.

Proof.   Since A is a normal matrix, there exists a unitary matrix U such that $A = U\Lambda U^\dagger$, where Λ is a diagonal matrix containing the eigenvalues of A. Thus, for any unit vector $|v\rangle$, we have:

$$\langle Av|Av\rangle = \langle U\Lambda U^\dagger v|U\Lambda U^\dagger v\rangle$$
$$= \langle v|U^\dagger \Lambda^\dagger U U^\dagger \Lambda U|v\rangle$$
$$= \langle v|\Lambda^\dagger \Lambda|v\rangle$$
$$= \sum_i |\lambda_i|^2 |v_i|^2$$
$$\leq (\sup\{|\lambda_i|\})^2 \sum_i |v_i|^2$$
$$= (\sup\{|\lambda_i|\})^2 \langle v|v\rangle,$$

where v_i represents the coordinates of $|v\rangle$ in the eigenbasis of A. The inequality follows because $|\lambda_i|^2 \leq (\sup\{|\lambda_i|\})^2$ for all i. Since $\langle v|v\rangle = 1$ for normalized vectors,

$$\|A\|_2 = \sqrt{(\sup\{|\lambda_i|\})^2} = \sup\{|\lambda_i|\}.$$

$\blacksquare$

4   Positivity

An important concept associated with Hermitian matrices is their definiteness, as introduced in § 9.4.3. In particular, a Hermitian matrix A is said to be positive definite if for any non-zero vector $|x\rangle$, $\langle x|A|x\rangle > 0$. A Hermitian matrix A is said to be positive semidefinite (or Gramian) if for any vector $|x\rangle$, $\langle x|A|x\rangle \geq 0$.

Let $\{\lambda_i\}$ be the eigenvalues of A. Then, by similar arguments as in the proof of Theorem 11.2, positive definiteness implies $\lambda_i > 0$ for all i, while positive semidefiniteness implies $\lambda_i \geq 0$ for all i.

If A is positive definite, then $\det(A) > 0$, which means that A is invertible and A^{-1} is also positive definite.

If A and B are positive definite, then $A + B$ is also positive definite.

Exercise 11.3 Show that for any square matrix A, $AA^\dagger$ is positive semidefinite.

11.1.3   Cauchy-Schwarz Inequality and Associated Relations

For real vectors in $\mathbb{R}^2$, the cosine of the angle θ between them is given by

$$\cos\theta = \frac{\langle x|y\rangle}{\|x\|\|y\|}.$$

Since $|\cos\theta| \leq 1$, it follows that

$$|\langle x|y\rangle| \leq \|x\|\|y\|.$$

This relationship generalizes to $\mathbb{R}^n$, $\mathbb{C}^n$, and any inner product space. It is commonly known as the Cauchy-Schwarz inequality.

This inequality states that the absolute value (or modulus) of the inner product between two vectors is bounded by the product of their norms. It is a fundamental result in linear algebra and functional analysis, particularly significant in quantum mechanics as it underpins the uncertainty principle.

Theorem 11.3 — Cauchy-Schwarz Inequality. For any $|x\rangle, |y\rangle$ in an inner product space such as $\mathbb{C}^n$,

$$|\langle x|y\rangle| \leq \|x\|\|y\|, \tag{11.12}$$

or, equivalently,

$$\langle x|y\rangle \langle y|x\rangle \leq \langle x|x\rangle \langle y|y\rangle, \tag{11.13}$$

with equality if and only if $|x\rangle$ and $|y\rangle$ are linearly dependent. Specifically,

$$|y\rangle = \frac{\langle x|y\rangle}{\langle x|x\rangle}|x\rangle, \quad \text{if } |x\rangle \neq 0. \tag{11.14}$$

Proof.   Assume $|x\rangle \neq 0$ and $|y\rangle \neq 0$; otherwise, the inequality holds trivially. Consider the vector $|y - t\lambda x\rangle$, where $t \in \mathbb{R}$ and $\lambda \in \mathbb{C}$. By the positive-definiteness of the inner product,

$$p(t) \equiv \langle y - t\lambda x|y - t\lambda x\rangle \geq 0.$$

Choose λ such that $\lambda \langle x|y\rangle = |\langle x|y\rangle|$; then $\lambda^* \langle y|x\rangle = |\langle x|y\rangle|$ as well, and $|\lambda| = 1$. With this choice,

$$p(t) = \langle x|x\rangle t^2 - 2|\langle x|y\rangle|t + \langle y|y\rangle \geq 0, \tag{11.15}$$

a quadratic function in t that must be non-negative for all t. Thus, its discriminant must be non-positive:

$$\Delta = 4\left(|\langle x|y\rangle|^2 - \langle x|x\rangle \langle y|y\rangle\right) \leq 0,$$

which immediately yields the Cauchy-Schwarz inequality.

For the equality case $\langle x|y\rangle \langle y|x\rangle = \langle x|x\rangle \langle y|y\rangle$, we can cancel the common factor $\langle y|$ and obtain $\langle x|y\rangle |x\rangle = \langle x|x\rangle |y\rangle$, which leads to the condition in Eq. 11.14.

Equations 11.12 and 11.13 are equivalent because, for any complex number z, $|z|^2 = z^*z = zz^*$.   □

(i) This proof relies solely on the three defining properties of inner products, not on any specific form. Consequently, the Cauchy-Schwarz inequality holds for all inner product spaces, including spaces of matrices.

The Cauchy-Schwarz inequality leads directly to the following:

Theorem 11.4 — Triangle Inequality. For any $|x\rangle, |y\rangle$ in an inner product space,

$$\|x\| + \|y\| \geq \|x + y\|. \tag{11.16}$$

Proof. Note that

$$\|x + y\|^2 = \langle x + y|x + y\rangle = \langle x|x\rangle + \langle x|y\rangle + \langle y|x\rangle + \langle y|y\rangle.$$

Since $\langle y|x\rangle = \langle x|y\rangle^*$, we have

$$\langle x|y\rangle + \langle y|x\rangle = 2\,\text{Re}(\langle x|y\rangle).$$

Thus,

$$\|x + y\|^2 = \|x\|^2 + 2\,\text{Re}(\langle x|y\rangle) + \|y\|^2 \leq \|x\|^2 + 2|\langle x|y\rangle| + \|y\|^2,$$

since $\text{Re}(z) \leq |z|$ for any complex number z.

By the Cauchy–Schwarz inequality, $|\langle x|y \rangle| \leq \|x\|\|y\|$. Therefore,

$$\|x + y\|^2 \leq \|x\|^2 + 2\|x\|\|y\| + \|y\|^2 = (\|x\| + \|y\|)^2.$$

Taking square roots yields the triangle inequality.   □

Setting $t = 1$ in Eq. 11.15 yields another useful inequality:

Theorem 11.5 For any $|x\rangle, |y\rangle$ in an inner product space,

$$|\langle x|y\rangle| \leq \frac{1}{2}\left(\|x\|^2 + \|y\|^2\right). \tag{11.17}$$

This is a weaker form of the Cauchy-Schwarz inequality, as it follows from the inequality $2ab \leq a^2 + b^2$ for any real numbers a and b.

Exercise 11.4 Prove Theorem 11.5 by combining the Cauchy-Schwarz inequality with the fact that $\left(\|x\| - \|y\|\right)^2 \geq 0$.

A related identity is:

Theorem 11.6 — Parallelogram Law. For any $|x\rangle, |y\rangle$ in an inner product space,

$$2\|x\|^2 + 2\|y\|^2 = \|x + y\|^2 + \|x - y\|^2. \tag{11.18}$$

Proof.   Adding the equations

$$\langle x + y|x + y \rangle = \langle x|x\rangle + \langle y|y\rangle + \langle x|y\rangle + \langle y|x\rangle,$$
$$\langle x - y|x - y \rangle = \langle x|x\rangle + \langle y|y\rangle - \langle x|y\rangle - \langle y|x\rangle,$$

we obtain

$$\langle x + y|x + y\rangle + \langle x - y|x - y\rangle = 2\langle x|x\rangle + 2\langle y|y\rangle.$$

This yields the parallelogram law.   □

Exercise 11.5 Let $a, b, c, x, y, z \in \mathbb{R}$. Find the maximum of $ax + by + cz$ given that $x^2 + y^2 + z^2 = 1$. Hint: Consider

$$(ax + by + cz)^2 \leq (a^2 + b^2 + c^2)(x^2 + y^2 + z^2).$$

Theorem 11.7 — Cauchy-Schwarz Inequality for Traces. For square matrices A and B,

$$|\operatorname{tr}(A^\dagger B)|^2 \leq \operatorname{tr}(A^\dagger A)\operatorname{tr}(B^\dagger B). \tag{11.19}$$

If A and B are Hermitian matrices, this simplifies to:

$$|\operatorname{tr}(AB)|^2 \leq \operatorname{tr}(A^2)\operatorname{tr}(B^2). \tag{11.20}$$

Proof.   Consider the Hilbert-Schmidt inner product between two matrices A and B (see § 11.1.2), defined as:

$$\langle A, B \rangle = \operatorname{tr}(A^\dagger B).$$

By applying the Cauchy-Schwarz inequality for inner products (Eq. 11.13), we directly obtain Eq. 11.19.   □

The Heisenberg Uncertainty Principle

The famous Heisenberg Uncertainty Principle (HUP) in quantum mechanics can be expressed in terms of the standard deviations of measurements for two Hermitian operators A and B:

$$\Delta A \cdot \Delta B \geq \frac{1}{2} |\langle [A, B] \rangle|, \tag{11.21}$$

where $\Delta A = \sqrt{\langle A^2 \rangle - \langle A \rangle^2}$ and similarly for ΔB. Here, $[A, B] = AB - BA$ represents the commutator of A and B.

This inequality stems from the Cauchy-Schwarz inequality. Below is a proof outline:

Consider a quantum state vector $|\psi\rangle$. Define $|\alpha\rangle$ and $|\beta\rangle$ as follows:

$$|\alpha\rangle = (A - \langle A \rangle) |\psi\rangle, \quad |\beta\rangle = (B - \langle B \rangle) |\psi\rangle,$$

where $\langle A \rangle = \langle \psi | A | \psi \rangle$ and $\langle B \rangle = \langle \psi | B | \psi \rangle$ are the expected values of A and B.

The Cauchy-Schwarz inequality states that for any vectors $|\alpha\rangle$ and $|\beta\rangle$,

$$|\langle \alpha | \beta \rangle|^2 \leq \langle \alpha | \alpha \rangle \langle \beta | \beta \rangle.$$

First, calculate $\langle \alpha | \alpha \rangle$ and $\langle \beta | \beta \rangle$:

$$\langle \alpha | \alpha \rangle = \langle (A - \langle A \rangle) \psi | (A - \langle A \rangle) \psi \rangle = \langle A^2 \rangle - \langle A \rangle^2 = (\Delta A)^2,$$

$$\langle \beta | \beta \rangle = \langle (B - \langle B \rangle) \psi | (B - \langle B \rangle) \psi \rangle = \langle B^2 \rangle - \langle B \rangle^2 = (\Delta B)^2.$$

Next, note that the imaginary part of $\langle \alpha | \beta \rangle = \langle \psi | (A - \langle A \rangle)(B - \langle B \rangle) | \psi \rangle$ is related to the commutator (see Eq. 11.5d):

$$\langle [A, B] \rangle = 2i \operatorname{Im}(\langle \alpha | \beta \rangle).$$

Thus,

$$|\langle [A, B] \rangle| = 2|\operatorname{Im}(\langle \alpha | \beta \rangle)| \leq 2|\langle \alpha | \beta \rangle|.$$

Combining this with the Cauchy-Schwarz inequality, we obtain the Uncertainty inequality, Eq. 11.21.

11.1.4 Quadratic Forms

While the inner product is a sesquilinear function, functions of higher degrees, particularly quadratic functions, are also significant in linear algebra and quantum computing.

A quadratic form is a vector function of the form (see also § 9.4.2)

$$g(x) = \langle x | A | x \rangle, \tag{11.22}$$

where A is a square matrix, assumed to be constant. For example, consider

$$A = \begin{bmatrix} 1 & 2 \\ 3 & 4 \end{bmatrix},$$

then the quadratic form $g(x)$ is given by

$$g(x) = x_1^* x_1 + 2x_1^* x_2 + 3x_1 x_2^* + 4x_2^* x_2.$$

This quadratic form arises from the more general sesquilinear form when both inputs are the same vector:

$$L(x, y) = \langle x|A|y \rangle, \tag{11.23}$$

where evaluating this function at $L(x, x)$ gives us the quadratic form $g(x)$.

Exercise 11.6 Express $Q = 5x_1^2 - 10x_1x_2 + x_2^2$ to a quadratic form using a symmetric matrix.

If A is Hermitian, then $g(x)$ is real. Many problems, such as those encountered in the variational quantum eigensolver (VQE), relate to finding the minimum of $g(x)$ under the constraint $\langle x|x \rangle = 1$. Equivalently, we aim to solve for $|x\rangle$ such that

$$\tilde{g}(x) = \frac{\langle x|A|x \rangle}{\langle x|x \rangle} \tag{11.24}$$

is minimized, given that A is Hermitian.

Theorem 11.8 Given a Hermitian matrix A, the minimum of $\langle x|A|x \rangle$ under the constraint $\langle x|x \rangle = 1$ corresponds to the smallest eigenvalue of A, and $|x\rangle$ is the corresponding eigenvector.

Proof. Because A is Hermitian, it can be diagonalized through eigendecomposition $A = K^\dagger \Lambda K$, where K is unitary and Λ is a diagonal matrix with eigenvalues arranged in non-decreasing order. We can then express $g(x)$ as follows:

$$
\begin{aligned}
g(x) &= \langle x|A|x \rangle \\
&= \langle x|K^\dagger \Lambda K|x \rangle \\
&= \langle y|\Lambda|y \rangle \quad \text{where } |y\rangle = K|x\rangle \\
&= \sum_i |y_i|^2 \lambda_i \\
&\geq \sum_i |y_i|^2 \lambda_1 \\
&= \lambda_1 \quad (\because \sum_i |y_i|^2 = \langle y|y \rangle = 1).
\end{aligned}
$$

Here, y_i represents the coordinates of $|y\rangle$ in the eigenbasis of A. The constraint $\langle x|x \rangle = 1$ ensures $\langle y|y \rangle = 1$. The minimum is attained when $|x\rangle$ aligns with the eigenvector $|k_1\rangle$ corresponding to λ_1, yielding $g(k_1) = \lambda_1$. $\qquad \square$

Variational Quantum Eigensolver (VQE)

The Variational Quantum Eigensolver (VQE) is a hybrid quantum-classical algorithm that finds the ground state energy (the smallest eigenvalue) of a quantum system. This algorithm is particularly useful in quantum chemistry and material science.

The goal of VQE is to minimize the expectation value of the Hamiltonian of the system, analogous to the quadratic form discussed earlier:

$$E(\theta) = \langle \psi(\theta)|H|\psi(\theta) \rangle, \tag{11.25}$$

where H is the Hamiltonian matrix of the system, and $|\psi(\theta)\rangle$ is the quantum state parameterized by θ. A quantum state vector is always normalized.

VQE employs an iterative process where a quantum computer evaluates $E(\theta)$, and a classical computer optimizes the parameter θ to minimize $E(\theta)$. This process leverages the strengths of both quantum and classical computing paradigms to efficiently solve problems that are otherwise too challenging for classical computers alone.

The minimum $E(\theta)$ is the ground state energy, and the corresponding eigenvector is the ground state.

Exercise 11.7 Given $|x\rangle = \begin{bmatrix} \cos\theta \\ \sin\theta \end{bmatrix}$ and $X = \begin{bmatrix} 0 & 1 \\ 1 & 0 \end{bmatrix}$, where θ is a real number, find the minimum and maximum of $\langle x|X|x\rangle$.

While this simple problem has an analytic solution, the Variational Quantum Eigensolver (VQE) aims to solve for a multidimensional θ with much more complicated $|x\rangle$ and X.

11.1.5 ＊Regression Functions

A regression function is a vector function of the form

$$h(x) = \langle Ax - y|Ax - y\rangle = \|Ax - y\|^2, \tag{11.26}$$

where $|y\rangle \in \mathbb{C}^m$, $|x\rangle \in \mathbb{C}^n$, and $A \in \mathbb{C}^{m \times n}$. Usually, the number of predictors n is smaller than the number of observations m, $n \le m$. Both A and $|y\rangle$ are assumed to be constant.

The minimum of $h(x)$ plays a key role in linear regression analysis and many other applications, including in quantum computation. In linear regression, A is a matrix of predictors (or the design matrix), $|x\rangle$ is the vector of coefficients to be estimated, and $|y\rangle$ is the vector of observations. The function $h(x)$ represents the squared norm of the residual vector $|Ax - y\rangle$, and minimizing this function is the core objective of linear regression.

In the classic least-squares fit problem (see § 8.5.5.2), where we aim to find the optimal coefficients k and b such that $|v\rangle = k|u\rangle + b$ fits best for m data points $(u_1, v_1), (u_2, v_2), \ldots, (u_m, v_m)$, the above model (Eq. 11.26 with $n = 2$) translates to:

$$|x\rangle = \begin{bmatrix} b \\ k \end{bmatrix}, \quad |y\rangle = \begin{bmatrix} v_1 \\ v_2 \\ \vdots \\ v_m \end{bmatrix}, \quad A = \begin{bmatrix} 1 & u_1 \\ 1 & u_2 \\ \vdots & \\ 1 & u_m \end{bmatrix}.$$

Theorem 11.9 Assuming $A^\dagger A$ is invertible, the minimum of the regression function defined in Eq. 11.26 is $\langle y|y\rangle - \langle y|A(A^\dagger A)^{-1}A^\dagger|y\rangle$, and is attained for $|x\rangle = (A^\dagger A)^{-1}A^\dagger|y\rangle$.

The invertibility of $A^\dagger A$ generally requires A to have full column rank, meaning that the columns of A are linearly independent.

Proof. We can rewrite $h(x)$ as follows:

$$h(x) = \langle Ax - y|Ax - y\rangle$$

$$= \langle x|A^\dagger A|x\rangle - \langle x|A^\dagger|y\rangle - \langle y|A|x\rangle + \langle y|y\rangle$$
$$= \langle x|A^\dagger A|x\rangle - \langle x|A^\dagger|y\rangle - \langle y|A|x\rangle + \langle y|A(A^\dagger A)^{-1}A^\dagger|y\rangle + C,$$
$$\text{where } C = \langle y|y\rangle - \langle y|A(A^\dagger A)^{-1}A^\dagger|y\rangle,$$
$$= \|Ax - A(A^\dagger A)^{-1}A^\dagger y\|^2 + C.$$

Since C is independent of $|x\rangle$ and $\|\ldots\|^2 \geq 0$, it is apparent that the minimum of $h(x)$ is C, achieved when $\|\ldots\|^2 = 0$, which leads to $|x\rangle = (A^\dagger A)^{-1}A^\dagger|y\rangle$.

The last step uses the following matrix properties (with $Q \equiv A(A^\dagger A)^{-1}A^\dagger$):

$$\begin{array}{ll}
(U^\dagger V)^\dagger = V^\dagger U, & \text{for any compatible matrices } U, V, \\
(S^{-1})^\dagger = (S^\dagger)^{-1}, & S \text{ is an invertible square matrix}, \\
(A^\dagger A)^\dagger = A^\dagger A, & A^\dagger A \text{ is Hermitian}, \\
Q^\dagger = Q, & \text{due to the above properties}, \\
A^\dagger Q = A^\dagger, & \\
Q^\dagger A = A, & \\
Q^\dagger Q = Q. &
\end{array}$$

□

11.2 Vector-Valued Functions

A familiar example of a vector-valued function is $r(t) = (x(t), y(t), z(t))$ representing a parametric curve in space, where $t \in \mathbb{R}$. While such functions with scalar arguments are common, our focus here shifts to vector-valued functions involving vectors or matrices as arguments. These are crucial in fields like quantum computing, where managing vector and matrix operations is fundamental. Here are some typical examples of these functions:

1. Vector Normalization: Normalizing a vector $|x\rangle$ involves scaling it to unit length, achieved by dividing the vector by its norm:

$$|u\rangle = \frac{|x\rangle}{\|x\|}. \tag{11.27}$$

Here, $|u\rangle$ is the normalized version of $|x\rangle$, ensuring that $\|u\| = 1$.

2. Linear Transformation: A linear transformation is given by $|y\rangle = A|x\rangle$, where A is a matrix representing the function $f : \mathbb{C}^n \to \mathbb{C}^m$ and $A \in \mathbb{C}^{m \times n}$.

3. Affine Transformation: An affine transformation is given by $|y\rangle = A|x\rangle + |b\rangle$, where $|b\rangle \in \mathbb{C}^m$, adding a translation component to the linear transformation.

4. Eigenvalue Problem: Calculating the eigenvector $|v\rangle$ corresponding to a specific eigenvalue of a Hermitian matrix H, notably the ground state in quantum simulation, is a fundamental vector-valued function.

5. **Tensor Product of Two Vectors:** Given $|u\rangle \in \mathbb{C}^n$ and $|v\rangle \in \mathbb{C}^m$, the tensor product $|u\rangle \otimes |v\rangle$ results in a vector in $\mathbb{C}^{mn}$. For example:

$$|uv\rangle = \begin{bmatrix} u_1 \\ u_2 \end{bmatrix} \otimes \begin{bmatrix} v_1 \\ v_2 \end{bmatrix} = \begin{bmatrix} u_1 v_1 \\ u_1 v_2 \\ u_2 v_1 \\ u_2 v_2 \end{bmatrix}.$$

6. **The vec Function:** By $\text{vec}(A)$, we mean the vector containing all columns of A, stacked one below another. For example:

$$A = \begin{bmatrix} 1 & 2 \\ 3 & 4 \end{bmatrix}, \text{ thus } \text{vec}(A) = \begin{bmatrix} 1 \\ 3 \\ 2 \\ 4 \end{bmatrix}, \text{ and } \text{vec}(A^T) = \begin{bmatrix} 1 \\ 2 \\ 3 \\ 4 \end{bmatrix}.$$

The following result allows us to express the matrix product ABC in terms of $\text{vec}(B)$:

$$\text{vec}(ABC) = (C^T \otimes A)\,\text{vec}(B). \tag{11.28}$$

In the rest of this section, we will delve deeper into several cases that are fundamental in both linear algebra and quantum computing: linear transformations, unitary transformations, change of basis, projections, and the discrete Fourier transform (DFT). Although some of these topics have been introduced in preceding chapters, we revisit them from the angle of functional mappings and emphasize their interconnections, also serving as a comprehensive review.

11.2.1   Linear Transformations

Linear transformations are a cornerstone of linear algebra. We detailed these as matrix transformations in § 8.1. In general, linear transformations can be defined through their actions on vectors.

1   Definition

A linear transformation from a vector space V over a field F into a vector space W is a function $T : V \to W$ that satisfies the following two properties for all vectors $|u\rangle, |v\rangle \in V$ and all scalars $c \in F$:

1. Additivity: $T(|u\rangle + |v\rangle) = T(|u\rangle) + T(|v\rangle)$,

2. Homogeneity: $T(c\,|u\rangle) = cT(|u\rangle)$.

In quantum theories, W and V are generally Hilbert spaces, which can be infinite in dimension. However, for simplicity and without sacrificing generality, we will focus our discussion on linear transformations where W and V are $\mathbb{C}^n$ or its subspaces, and F is $\mathbb{C}$. In this case, T is represented by a matrix in $\mathbb{C}^{n \times n}$. It should be noted that the principles outlined here are equally applicable to the broader context of infinite-dimensional Hilbert spaces.

2   Geometric Interpretation

Geometrically, linear transformations can be interpreted as transformations like rotation, scaling (stretching), projection, and shearing. For two-dimensional real

Euclidean space, typical transformations are listed in Table 8.1 along with their transformation matrices.

> **Exercise 11.8** Find the transformation matrix in 2D Euclidean space for:
>
> (a) Reflection through the line $y = kx + b$,
>
> (b) Projection onto the line $y = kx + b$,
>
> (c) Projection onto the line at angle θ from the x axis.

3   Relationship with Eigenvalues and Eigenvectors

Eigenvectors and eigenvalues fundamentally characterize a linear transformation. For real-valued eigenvalues and eigenvectors, eigenvectors are vectors that undergo pure scaling (stretched or compressed) without any rotational or shearing transformation. The corresponding eigenvalue represents the scale factor, and a negative eigenvalue indicates a reversal in direction.

To illustrate this in two dimensions, consider $|v\rangle = A\,|u\rangle$ where $A = \begin{bmatrix} 1 & -1 \\ -0.25 & 1 \end{bmatrix}$. The eigenvalues of A are $\lambda_1 = 1.5$ and $\lambda_2 = 0.5$ with corresponding eigenvectors $|\lambda_1\rangle \approx \begin{bmatrix} 0.894 \\ -0.447 \end{bmatrix}$ and $|\lambda_2\rangle \approx \begin{bmatrix} 0.894 \\ 0.447 \end{bmatrix}$.

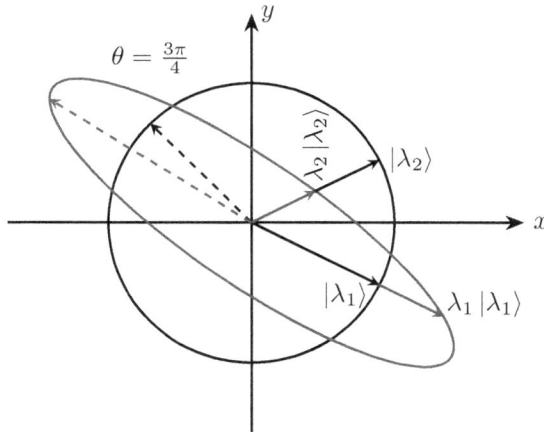

Figure 11.1: Relationship Between Linear Transformation and Eigenvectors

As shown in Fig. 11.1, when $|u\rangle$ is a unit vector, it traces a unit circle defined by $x = \cos\theta$, $y = \sin\theta$. The image $|v\rangle$ under transformation by A traces an ellipse defined by $x = \cos\theta - \sin\theta$, $y = \sin\theta - 0.25\cos\theta$.

The eigenvectors are depicted by arrows labeled $|\lambda_1\rangle$ and $|\lambda_2\rangle$, corresponding to directions where the transformation involves pure scaling. Points on the circle and ellipse for $\theta = \frac{3\pi}{4}$ (as example) demonstrate that the vectors at this angle are not collinear, indicating rotation as well as scaling.

Note that the directions of the eigenvectors do not necessarily correspond to the maximum and minimum scaling ratios, which relate to the singular values and singular vectors of A (see § 13.2.2) and are found by solving $\arg\max_{u,\|u\|=1} \|Au\|$ and $\arg\min_{u,\|u\|=1} \|Au\|$, as opposed to the eigenvalue equation $A\,|u\rangle = \lambda\,|u\rangle$.

Furthermore, when a matrix has degenerate eigenvalues (eigenvalues with more than one linearly independent eigenvector), it signifies subspaces in which every vector is uniformly scaled by the same factor. These subspaces are essential for understanding the geometric impact of transformations in multidimensional spaces. For more details on this topic, see § 9.1.3.

When dealing with complex eigenvalues, such as for a matrix A representing a rotation in the 2D plane: $A = \begin{bmatrix} \cos\theta & -\sin\theta \\ \sin\theta & \cos\theta \end{bmatrix}$, the eigenvalues are $e^{i\theta}$ and $e^{-i\theta}$, signifying rotations without any real stretching or compression. However, it is still often useful to employ the intuition from real vector spaces, where eigenvalues are thought of as stretching factors along the directions of the eigenvectors.

4 Order Dependence and Commutativity

The product of two linear transformations, represented as matrices A and B, is another linear transformation, but the order of multiplication is crucial. Generally, $AB \neq BA$. This non-commutativity must be considered when transformations are applied sequentially. The final outcome can significantly depend on the order in which transformations are applied.

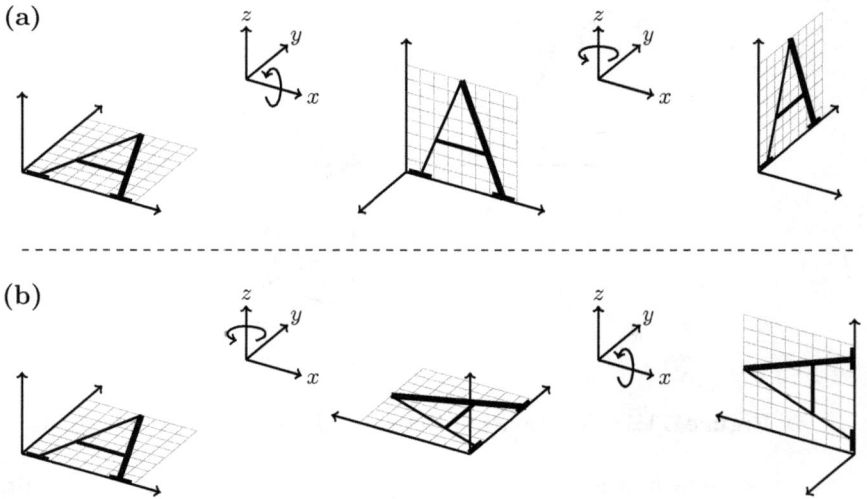

Figure 11.2: Demonstration of the Non-commutative Nature of 3D Rotations

For example, as depicted in Fig. 11.2, rotating around the x-axis by 90° followed by the z-axis (part a) does not result in the same orientation as when the rotation sequence is reversed (part b). This can be mathematically expressed as $R_z R_x \neq R_x R_z$, or equivalently, R_z and R_x do not commute: $[R_z, R_x] \neq 0$. Here,

$$R_z(\theta) = \begin{bmatrix} \cos\theta & -\sin\theta & 0 \\ \sin\theta & \cos\theta & 0 \\ 0 & 0 & 1 \end{bmatrix}, \quad R_x(\theta) = \begin{bmatrix} 1 & 0 & 0 \\ 0 & \cos\theta & -\sin\theta \\ 0 & \sin\theta & \cos\theta \end{bmatrix}. \tag{11.29}$$

Exercise 11.9 Verify that $[R_z, R_x] \neq 0$.

11.2.2 Unitary Transformations and Symmetry Groups

1 Unitary Transformations

Unitary transformations are a special class of linear transformations characterized by their ability to preserve the inner products (and norms) of vectors. Mathematically, a transformation $U \in \mathbb{C}^{n \times n}$ is unitary if $U^\dagger U = UU^\dagger = I$, where I denotes the identity matrix (see § 8.3). These transformations are fundamental in quantum mechanics as they govern the evolution of states in closed quantum systems.

Unitary transformations possess the following essential properties:

1. Preservation of Inner Products: The overlap of two vectors, $|\phi\rangle$ and $|\psi\rangle$, remains invariant under the same unitary evolution. Explicitly,

$$\langle U\phi | U\psi \rangle = \langle \phi | U^\dagger U | \psi \rangle = \langle \phi | \psi \rangle. \tag{11.30}$$

This property ensures that an orthonormal basis remains orthonormal under unitary evolution, preserving the probabilistic interpretation of quantum mechanics.

2. Reversibility: Unitary transformations can be reversed by applying their inverse, given by $U^{-1} = U^\dagger$.

3. Linearity: Being linear transformations, unitary transformations allow for the superposition of quantum states.

4. Orthonormal Row and Column Vectors: The column vectors of a unitary matrix form an orthonormal set, as do the transposed row vectors.

5. Unit Eigenvalues: The eigenvalues of a unitary matrix are complex numbers with an absolute value of 1, enabling them to be interpreted as pure rotations and changes of basis.

Exercise 11.10 If U and V are unitary, are $U + V$, UV, $U \otimes V$, and $U \oplus V$ also unitary?

Here, $U \oplus V$ represents the direct sum of U and V, defined as the block diagonal matrix:

$$U \oplus V \equiv \begin{bmatrix} U & 0 \\ 0 & V \end{bmatrix}.$$

Exercise 11.11 Show that the eigenvectors of a unitary transformation belonging to distinct eigenvalues are orthogonal.

2 Orthogonal Transformations

Orthogonal transformations involve matrices $Q \in \mathbb{R}^{n \times n}$ that are orthogonal, i.e., $Q^T Q = QQ^T = I$, where Q^T is the transpose of Q. These matrices are the real-number counterparts of unitary matrices and represent rotations in $\mathbb{R}^n$, possessing all the properties of unitary matrices. Despite being real, their eigenvalues and eigenvectors may be complex, particularly in the context of rotations in higher dimensions. Examples of orthogonal transformations include three-dimensional rotations, represented by matrices R_z and R_x in Eq. 11.29.

Both orthogonal and unitary matrices are *normal*, meaning they commute with their conjugate transpose: $U^\dagger U = UU^\dagger$. This property ensures that such matrices can be diagonalized by an orthonormal basis of eigenvectors, a crucial result known as the spectral theorem.

3 ✷ Symmetry Groups

Symmetry groups are a foundational concept in both mathematics and physics, representing the set of all symmetries that a geometric object, physical system, or mathematical set can have. These symmetries include transformations like rotations, reflections, translations, and glide reflections that preserve some property of the object or system. The study of symmetry groups is a part of group theory.

A symmetry group is formally defined as a group G whose elements are transformations that preserve some structure of a set S. The operation of the group is the composition of transformations, which satisfies the group axioms (refer to § 4.2): closure, associativity, the existence of an identity element, and the existence of inverse elements.

SO(N)

The special orthogonal group in N dimensions, SO(N), includes all $N \times N$ orthogonal matrices with determinant 1, representing rotations in N-dimensional space that preserve distances and orientation. For example, SO(2) governs the symmetries of a circle, while SO(3) relates to the symmetries of a sphere.

The determinant of an orthogonal matrix can only be ± 1, because $Q^T Q = I$ implies $(\det Q)^2 = 1$. The determinant being $+1$ specifically ensures that the transformation preserves orientation, while a determinant of -1 would imply a reflection that reverses orientation.

SU(2)

The special unitary group of degree 2, SU(2), is the set of 2×2 complex unitary matrices with determinant 1. Without the determinant constraint, the group is called U(2).

In quantum computing, qubit states are represented by two-dimensional complex vectors. Single-qubit unitary gates, which perform rotations on these qubits, are elements of SU(2). If a unitary operator U representing a gate has determinant -1, the operator $-U$ (with determinant 1) can be used instead without affecting the underlying physics, as global phase factors are physically irrelevant.

SU(2) as a Double Cover of SO(3)

The group SU(2) is closely related to SO(3), acting as a double cover of SO(3). This means that each rotation in SO(3) corresponds to two distinct elements in SU(2).

To understand this relationship, consider a sphere where each diameter represents a possible axis of rotation. In SO(3), a 360° rotation around any axis returns the system to its original orientation. In contrast, in SU(2), achieving the same physical orientation requires a 720° rotation. This double cover nature reflects the topology of SU(2), which can be visualized as a 3-sphere (S^3) rather than the 2-sphere (S^2) associated with SO(3). A 360° rotation in SO(3) corresponds to a 360° traversal in the interior of SU(2), emerging on the opposite side, and then another 360° traversal along the exterior to return to the original element.

This property is fundamental to understanding quantum systems where SU(2) describes spinor representations. We will demonstrate this relationship explicitly using spin rotation operators in § 12.2.5.3.

SU(N)

SU(N) encompasses the special unitary group of degree N, extending the concepts of SU(2) to higher dimensions. These groups are pivotal in the study of quantum computing where N-level quantum systems (quNits) are considered. The eigenvalues of unitary matrices, having absolute values of 1, imply that their product, the determinant, also has an absolute value of 1. Requiring the determinant to be exactly 1 ensures that transformations within SU(N) preserve not just lengths and angles but also the overall orientation and volume of objects in complex spaces.

11.2.3 Change of Basis

A change of basis allows the same vector space to be described using different sets of basis vectors, which can facilitate easier computations and offer deeper insights into the structure of a problem—especially in contexts like quantum mechanics where different bases highlight different properties.

In linear algebra, the concepts of change of basis and linear transformations are tightly interconnected (see § 8.4). A basis change can also be viewed as a vector-valued function, where vectors in the original basis are mapped to their representations in the new basis. This mapping is defined by the basis transformation operator, which enables seamless transitions between different perspectives within the same vector space.

1 Basis

A basis provides a method for describing vector spaces similar to Cartesian coordinate systems. A basis consists of a set of vectors $\{|b_i\rangle\}$, where each $|b_i\rangle \in \mathbb{C}^n$ and $i = 1, 2, \ldots, n$, analogous to unit vectors in traditional coordinate systems.

A basis set is termed *orthonormal* if all the basis vectors are unit vectors and mutually orthogonal:

$$\langle b_i | b_j \rangle = \delta_{ij}, \tag{11.31}$$

where δ_{ij} is the Kronecker delta function, defined as $\delta_{ij} = 1$ if $i = j$ and $\delta_{ij} = 0$ if $i \neq j$. Orthogonality ensures that the basis vectors do not overlap in any dimension, providing a clear, distinct direction for each component.

A basis set is considered *complete* if it spans the entire vector space:

$$\sum_{i=1}^{n} |b_i\rangle\langle b_i| = I, \tag{11.32}$$

where I is the identity matrix. This completeness condition guarantees that any vector in the space can be represented as a linear combination of the basis vectors.

A general vector $|\psi\rangle$ in this space can be expressed in the basis $\{|b_i\rangle\}$ as:

$$|\psi\rangle = \sum_{i=1}^{n} c_i |b_i\rangle, \tag{11.33}$$

where the coefficients $\{c_i\}$, or the vector components of $|\psi\rangle$, are given by:

$$c_i = \langle b_i | \psi \rangle. \tag{11.34}$$

These coefficients are analogous to coordinates in a Cartesian system, allowing every point or vector in the space to be uniquely determined by its components along the basis vectors.

2 Basis-Dependent Representation of Vectors

Different bases in a linear space can represent the same physical objects such as quantum states, which are invariant across these bases, although their vector or matrix representations differ. Consider two complete and orthonormal bases $\{|b_i\rangle\}$ and $\{|b'_i\rangle\}$, characterized by:

$$\langle b_i | b_j \rangle = \langle b'_i | b'_j \rangle = \delta_{ij}, \quad \sum_i |b_i\rangle\langle b_i| = \sum_i |b'_i\rangle\langle b'_i| = I. \tag{11.35}$$

A vector $|\psi\rangle$ can be expressed in either basis:

$$|\psi\rangle = \sum_i c_i |b_i\rangle = \sum_i c'_i |b'_i\rangle. \tag{11.36}$$

The coefficients or vector components are:

$$c_i = \langle b_i | \psi \rangle, \quad c'_i = \langle b'_i | \psi \rangle. \tag{11.37}$$

In Eq. 11.36, $|\psi\rangle$ is considered a physical vector (such as a quantum state) which is basis independent, while (c_i) as a column vector $|c\rangle$ and (c'_i) as $|c'\rangle$ are its representations in two different bases. Change of basis is concerned with the relationship between $|c\rangle$ and $|c'\rangle$.

3 The Basis Transformation Operator

To change from the basis $\{|b_i\rangle\}$ to $\{|b'_i\rangle\}$, we define a unitary operator (referred to as the basis transformation operator) U such that

$$|b'_i\rangle = U |b_i\rangle. \tag{11.38}$$

This relationship is analogous to a coordinate rotation, as illustrated in Fig. 8.12.

Specifically, U can be constructed as:

$$U = \sum_i |b'_i\rangle \langle b_i| = \sum_{i,j} \langle b_j | b'_i \rangle |b_j\rangle \langle b_i|. \tag{11.39}$$

The matrix form of U, in the basis of $\{|b_i\rangle\}$, is given by:

$$U = \begin{bmatrix} \langle b_1 | b'_1 \rangle & \langle b_1 | b'_2 \rangle & \cdots \\ \langle b_2 | b'_1 \rangle & \langle b_2 | b'_2 \rangle & \\ \vdots & & \ddots \end{bmatrix}. \tag{11.40}$$

Given the change-of-basis operator U, the relationship between the vector representations $|c\rangle$ and $|c'\rangle$ is given by:

$$|c'\rangle = U^\dagger |c\rangle. \tag{11.41}$$

Exercise 11.12 Derive the formulas of U in Eq. 11.39 from its definition $|b_i'\rangle = U |b_i\rangle$. Consider the effect of U on each basis vector $|b_i\rangle$ and verify that it aligns with $|b_i'\rangle$.

4   Invariance in Basis Change

Since the basis transformation operator U is unitary, a change of basis conserves inner products and norms:

$$\langle u'|v'\rangle = \langle u|v\rangle , \tag{11.42}$$

given that $|u'\rangle = U^\dagger |u\rangle$ and $|v'\rangle = U^\dagger |v\rangle$ according to Eq. 11.41.

For a square matrix A, $\langle u|A|u\rangle$ represents a scalar quantity. It therefore should remain constant across different bases. This requires the matrix form of A to transform as follows:

$$A' = U^\dagger AU. \tag{11.43}$$

With this unitary similarity transformation, the quadratic form $\langle u|A|u\rangle$ is indeed invariant:

$$
\begin{aligned}
\langle u'|A'|u'\rangle &= \langle U^\dagger u|U^\dagger AU|U^\dagger u\rangle \\
&= \langle u|UU^\dagger AUU^\dagger|u\rangle \\
&= \langle u|A|u\rangle .
\end{aligned}
$$

The trace of a matrix is a scalar, and should also remain constant across different bases. Indeed,

$$\operatorname{tr} A' = \operatorname{tr}(U^\dagger AU) = \operatorname{tr}(UU^\dagger A) = \operatorname{tr} A. \tag{11.44}$$

The inner product between two matrices is also invariant under basis change:

$$\langle U^\dagger AU, U^\dagger BU\rangle = \langle A, B\rangle . \tag{11.45}$$

Proof.

$$
\begin{aligned}
\langle U^\dagger AU, U^\dagger BU\rangle &= \operatorname{tr}\left((U^\dagger AU)^\dagger U^\dagger BU\right) \\
&= \operatorname{tr}\left(U^\dagger A^\dagger UU^\dagger BU\right) \\
&= \operatorname{tr}\left(U^\dagger A^\dagger BU\right) \quad \text{(since } UU^\dagger = U^\dagger U = I\text{)} \\
&= \operatorname{tr}\left(UU^\dagger A^\dagger B\right) \quad \text{(using the cyclic property of the trace)} \\
&= \operatorname{tr}(A^\dagger B) = \langle A, B\rangle .
\end{aligned}
$$

□

Exercise 11.13 Show that the spectral norm of a normal matrix defined in § 11.1.2 is invariant under change of basis.

11.2.4   Projection and Reflection

We previously discussed orthogonal projections in § 6.4.3 as constructs within inner product spaces. Here, we revisit the topic in the context of vector-valued functions.

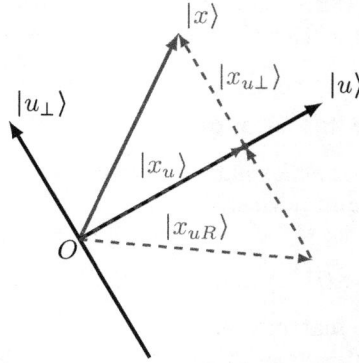

Figure 11.3: Illustration of Projection and Reflection

1   Projection onto and Reflection Across a Vector

Let $|u\rangle$ be a unit vector, i.e., $\langle u|u\rangle = 1$. The projection of $|x\rangle$ in the direction of $|u\rangle$ is given by:

$$|x_u\rangle = \Pi\,|x\rangle = |u\rangle\langle u|\,|x\rangle = \langle u|x\rangle\,|u\rangle. \tag{11.46}$$

Here,

$$\Pi = |u\rangle\langle u| \tag{11.47}$$

is termed a projection operator. If $|u\rangle$ is not normalized, we can construct Π as:

$$\Pi = \frac{|u\rangle\langle u|}{\langle u|u\rangle}. \tag{11.48}$$

The component of $|x\rangle$ that is orthogonal to $|u\rangle$ (see Fig. 11.3) is given by:

$$|x_{u\perp}\rangle = |x\rangle - |x_u\rangle = |x\rangle - |u\rangle\langle u|\,|x\rangle = (I - |u\rangle\langle u|)\,|x\rangle. \tag{11.49}$$

To show that $|x_{u\perp}\rangle$ is orthogonal to $|u\rangle$, we calculate:

$$\langle u|x_{u\perp}\rangle = \langle u|(I - |u\rangle\langle u|)|x\rangle = \langle u|x\rangle - \langle u|x\rangle = 0.$$

The reflection of $|x\rangle$ across $|u\rangle$ is given by:

$$|x_{uR}\rangle = |x_u\rangle - |x_{u\perp}\rangle = (2\,|u\rangle\langle u| - I)\,|x\rangle. \tag{11.50}$$

■ **Example 11.1** Consider the projection onto a unit vector $|u\rangle = \begin{bmatrix} \cos\theta \\ \sin\theta \end{bmatrix}$ in two-dimensional real Euclidean space. Here, $|u\rangle$ is at an angle θ from the x-axis, as shown in Fig. 11.4. The projection operator is given by:

$$\Pi = |u\rangle\langle u| = \begin{bmatrix} \cos^2\theta & \sin\theta\cos\theta \\ \sin\theta\cos\theta & \sin^2\theta \end{bmatrix}. \tag{11.51}$$

Let $|\psi\rangle$ be a vector at an angle ϕ from the x-axis: $|\psi\rangle = \begin{bmatrix} r\cos\phi \\ r\sin\phi \end{bmatrix}$. Then the projection is given by:

$$|\psi_u\rangle = \Pi\,|\psi\rangle = \langle u|\psi\rangle\,|u\rangle = r\cos(\theta - \phi) \begin{bmatrix} \cos\theta \\ \sin\theta \end{bmatrix}. \tag{11.52}$$

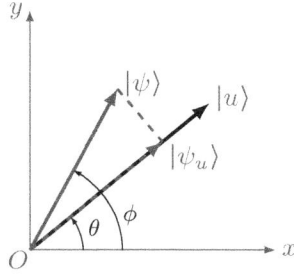

Figure 11.4: Two-Dimensional Projection Example

We can also calculate $|\psi_u\rangle$ using matrix multiplication:

$$|\psi_u\rangle = \begin{bmatrix} \cos^2\theta & \sin\theta\cos\theta \\ \sin\theta\cos\theta & \sin^2\theta \end{bmatrix} \begin{bmatrix} r\cos\phi \\ r\sin\phi \end{bmatrix}$$

$$= r\cos(\theta - \phi) \begin{bmatrix} \cos\theta \\ \sin\theta \end{bmatrix}.$$

The projection has a length $r\cos(\theta - \phi)$, as expected.   ∎

Exercise 11.14 Using the unit vector $|u\rangle$ and the vector $|x\rangle$ from Example 11.1, perform the following tasks:

(a) Calculate $|x_{u\perp}\rangle$, the component of $|x\rangle$ that is orthogonal to $|u\rangle$, and verify that it satisfies $\langle u|x_{u\perp}\rangle = 0$.

(b) Calculate $|x_{uR}\rangle$, the reflection of $|x\rangle$ across $|u\rangle$, and confirm that $|x_{uR}\rangle$ aligns with geometric intuition.

(c) Verify that the decomposition $|x_u\rangle = |x_{uR}\rangle + |x_{u\perp}\rangle$ holds.

The projection operator is idempotent, meaning $\Pi^2 = \Pi$. This property can be demonstrated as follows:

$$\Pi^2 = |u\rangle\langle u|\,|u\rangle\langle u| = |u\rangle\,(\langle u|u\rangle)\,\langle u| = |u\rangle\langle u| = \Pi.$$

2   Projection onto and Reflection Across a Subspace

Projection onto a subspace generalizes the concept of projecting a vector onto another vector. A subspace in linear algebra is defined as a set of vectors that forms a smaller vector space under vector addition and scalar multiplication (see § 5.3.3). For any vectors $|u\rangle$ and $|v\rangle$ in the subspace, and any scalar c, both $|u\rangle + |v\rangle$ and $c|u\rangle$ must also reside within the subspace. An intuitive example of a subspace is a plane in 3D space, shown in Fig. 11.5.

Consider an n-dimensional vector space $\mathbb{C}^n$ represented by the basis set $\{|b_1\rangle, |b_2\rangle, ..., |b_k\rangle, ..., |b_n\rangle\}$. Let a subspace S be spanned by the first k basis vectors $\{|b_1\rangle, |b_2\rangle, ..., |b_k\rangle\}$. The projection operator onto this subspace is:

$$\Pi = \sum_{i=1}^{k} |b_i\rangle\langle b_i|. \tag{11.53}$$

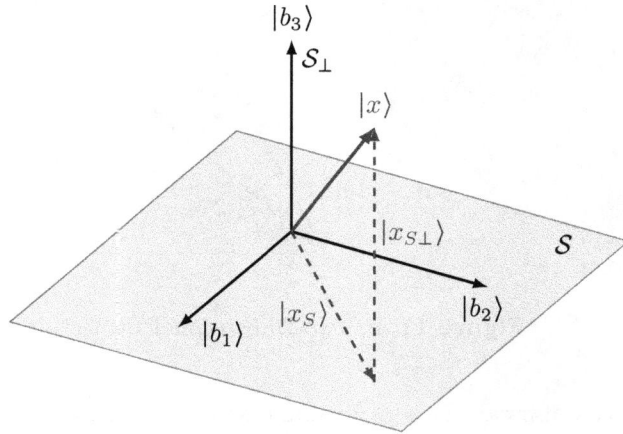

Figure 11.5: Illustration of Projection onto a 3D Plane

The projection of any vector $|x\rangle$ onto $\mathcal{S}$ is:

$$|x_S\rangle = \Pi |x\rangle. \tag{11.54}$$

The component of $|x\rangle$ orthogonal to $\mathcal{S}$ is:

$$|x_{S\perp}\rangle = (I - \Pi) |x\rangle. \tag{11.55}$$

To verify this orthogonality, consider:

$$\langle b_j | x_{S\perp}\rangle = \langle b_j | I - \Pi | x\rangle = \langle b_j | x\rangle - \langle b_j | x\rangle = 0, \quad \text{for } j = 1, 2, \ldots, k.$$

The reflection of $|x\rangle$ across $\mathcal{S}$ is:

$$|x_{SR}\rangle = (2\Pi - I) |x\rangle. \tag{11.56}$$

Here, $\Xi = 2\Pi - I$ acts as the reflection operator.

3 Properties of Projection Operators

Projection operators have the following salient properties:

1. The image of Π is spanned by $\{|b_1\rangle, |b_2\rangle, \ldots, |b_k\rangle\}$. Therefore, the dimension of the image of Π is k, and the rank of Π is k.

2. The subspace $\mathcal{S}$ is the column space of Π, while $\mathcal{S}_\perp$ is the null space of Π.

3. Π is Hermitian and idempotent:

$$\Pi^\dagger = \Pi, \quad \Pi^2 = \Pi. \tag{11.57}$$

4. The eigenvalues of Π are 0 and 1. Any vector $|v\rangle$ in $\mathcal{S}$ is an eigenvector of Π with eigenvalue 1:

$$\Pi |v\rangle = |v\rangle.$$

Vectors orthogonal to $\mathcal{S}$ are eigenvectors of Π with eigenvalue 0:

$$\Pi |p\rangle = 0.$$

5. Π is non-invertible unless $\Pi = I$, because if Π^{-1} exists, then $\Pi^{-1}\Pi^2 = \Pi^{-1}\Pi = I$, which implies $\Pi = I$.

Exercise 11.15 Verify that the reflection operator $\Xi = 2\Pi - I$ has eigenvalues ± 1. Determine the eigenvectors corresponding to each eigenvalue, illustrating how Ξ affects vectors within and outside $\mathcal{S}$.

4 Projection onto the Column Space of a Matrix

A subspace can often be defined by the columns of a matrix A, as in the regression problems discussed in § 11.1.5. Here, A need not be square, but the number of columns must not exceed the number of rows to allow for the possibility of linear independence among the columns. The linearly independent columns of A form a basis for the subspace, with its dimensionality determined by the number of such columns.

Theorem 11.10 Assume $A^\dagger A$ is invertible, indicating that A has full rank and its columns are linearly independent. The projection of a vector $|x\rangle$ onto the column space of A is given by:

$$|x_A\rangle = \Pi |x\rangle, \quad \text{where } \Pi = A(A^\dagger A)^{-1}A^\dagger. \quad (11.58)$$

Proof. ✳

The Orthogonality Principle: The vector $|x_A\rangle$, representing the projection of $|x\rangle$ onto the subspace $\mathcal{S}$ (column space of A), must satisfy the condition that the error vector $|e\rangle = |x\rangle - |x_A\rangle$ is orthogonal to every vector in $\mathcal{S}$, hence every column of A. This condition can be expressed in matrix form as:

$$A^\dagger(|x\rangle - |x_A\rangle) = 0.$$

Expressing $|x_A\rangle$ in Terms of A: Since $|x_A\rangle$ lies in the column space of A, it can be written as $|x_A\rangle = A|v\rangle$ for some vector $|v\rangle$. Substituting into the orthogonality condition gives:

$$A^\dagger(|x\rangle - A|v\rangle) = 0.$$

Solving for $|v\rangle$ and $|x_A\rangle$: Rearranging terms and solving for $|v\rangle$ assuming $A^\dagger A$ is invertible:

$$|v\rangle = (A^\dagger A)^{-1}A^\dagger |x\rangle.$$

Multiplying both sides by A and substituting $|x_A\rangle = A|v\rangle$, we find:

$$|x_A\rangle = A(A^\dagger A)^{-1}A^\dagger |x\rangle.$$

This confirms the formula given in Eq. 11.58, completing the proof. □

If $A \in \mathbb{C}^{m\times n}$ with $n \le m$, then $\Pi \in \mathbb{C}^{m\times m}$ with a rank of n.

Exercise 11.16 Prove that Π in Eq. 11.58 is idempotent: $\Pi^2 = \Pi$.

Regression Problem Revisited

You may have noticed that the same projection operator Π defined in Eq. 11.58 also appeared in the regression solution detailed in Theorems 8.39 and 11.9. This recurrence is not coincidental. In regression, the function $h(x) = \langle Av - x | Av - x \rangle$

reaches its minimum when the error vector $|x\rangle - A|v\rangle$ is orthogonal to the column space of A.

This orthogonality condition implies that $A^\dagger(|x\rangle - A|v\rangle) = 0$, leading to $A^\dagger A|v\rangle = A^\dagger|x\rangle$. Solving this equation yields $|v\rangle = (A^\dagger A)^{-1}A^\dagger|x\rangle$, which means $A|v\rangle$ is the projection of $|x\rangle$ onto the column subspace of A, thus demonstrating a deep connection between projection in linear algebra and regression analysis.

11.2.5   ∗ Discrete Fourier Transform (DFT)

The Discrete Fourier Transform (DFT) is a central tool in signal processing and linear algebra. It also serves as the foundation for the quantum Fourier transform and quantum phase estimation, two essential quantum algorithms. The DFT can be understood as a function mapping between two vector spaces or as a change of basis within a vector space.

> ⓘ Although this topic is not essential for studying the second book, *Quantum Computing and Information*, it will become important in the third book, *Quantum Algorithms and Applications*. Readers may defer studying it until encountering the relevant quantum algorithms.

1   Background: Fourier Transform

The Fourier transform is a powerful mathematical tool traditionally used to analyze the frequency components of a function or signal, commonly known as spectral analysis. By converting a function from its original domain (often time or space) to the frequency domain, the Fourier transform allows us to identify and study periodic features. In essence, it decomposes the function into a sum of sinusoidal components, each with a specific frequency, amplitude, and phase. This decomposition reveals the underlying periodic structure, making it easier to understand and analyze the behavior of the function in terms of its frequency content.

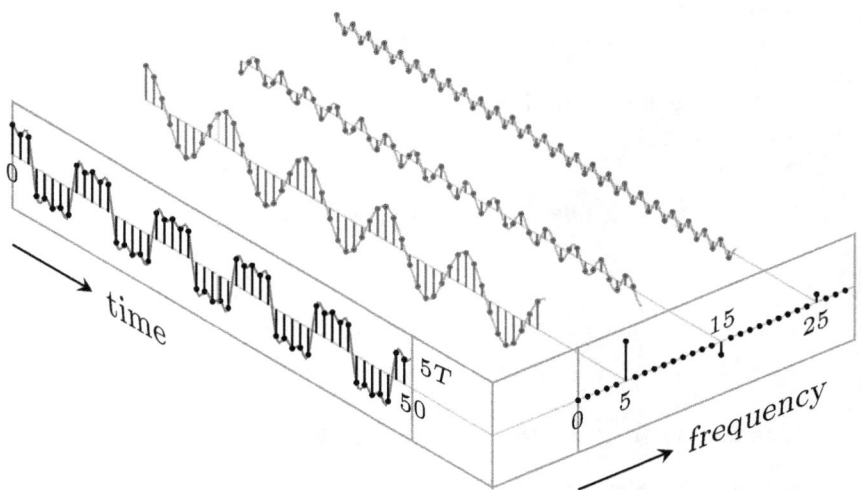

Figure 11.6: Example of Discrete Fourier Transform (DFT)

To illustrate, consider Fig. 11.6. The time-domain function shown is given by:

$$r(t) = \cos\left(\frac{2\pi}{N}5t\right) - \frac{1}{3}\cos\left(\frac{2\pi}{N}15t\right) + \frac{1}{5}\cos\left(\frac{2\pi}{N}25t\right),$$

which represents the real part of a complex function whose Fourier transform we will compute:

$$x(t) = e^{\frac{2\pi i}{N}5t} - \frac{1}{3}e^{\frac{2\pi i}{N}15t} + \frac{1}{5}e^{\frac{2\pi i}{N}25t}.$$

Here, $N = 50$ represents the maximum value of the domain and corresponds to the number of samples in the discrete Fourier transform (DFT), as explained below.

In Fig. 11.6, $x(t)$ consists of three harmonic components with periods $T = \frac{1}{5}$, $T = \frac{1}{15}$, and $T = \frac{1}{25}$ (in units of N), corresponding to frequencies $f = 5$, $f = 15$, and $f = 25$ (in units of $\frac{1}{N}$), respectively. If the number of harmonic components is extended indefinitely, the combined $x(t)$ approximates a square wave.

The Fourier transform of $x(t)$, denoted $\tilde{x}(f)$, represents the frequency domain. It exhibits non-zero values at three distinct frequencies, namely $f = 5$, $f = 15$, and $f = 25$, corresponding to the three harmonic components of $x(t)$.

2 Discrete Fourier Transform

The preceding discussion outlined the basic idea behind the *continuous* Fourier transform. In the *discrete* Fourier transform (DFT), we sample the function $x(t)$ as a sequence of N complex numbers:

$$x_k = x(k), \quad k = 0, 1, \ldots, N-1.$$

The DFT of this sequence x_k is defined by

$$\tilde{x}_n = \frac{1}{\sqrt{N}} \sum_{k=0}^{N-1} x_k\, e^{-\frac{2\pi i}{N} kn}, \quad n = 0, 1, \ldots, N-1. \tag{11.59}$$

(i) | It is often helpful to regard k as the "time" index and n as the "frequency" index.

To evaluate $\tilde{x}_n$, we resort to an identity developed in § 3.3.4:

$$\sum_{k=0}^{N-1} \omega^k = \begin{cases} 0, & \omega \neq 1, \\ N, & \omega = 1, \end{cases} \tag{11.60}$$

where ω is any Nth root of unity.

Applying this identity, one finds:

$$\sum_{k=0}^{N-1} e^{-\frac{2\pi i}{N} kn}\, e^{\frac{2\pi i}{N} kn'} = \sum_{k=0}^{N-1} e^{-\frac{2\pi i}{N} k(n-n')} = N\,\delta_{nn'}, \tag{11.61}$$

where $\delta_{nn'}$ is the Kronecker delta function, and $n, n' = 0, 1, \ldots, N-1$.

If we allow n and n' to be any integers in Eq. 11.61, then we obtain a more general DFT identity (see Example 3.5):

$$\sum_{k=0}^{N-1} e^{-\frac{2\pi i}{N} k(n-n')} = N\,\delta_{n-n' \bmod N}, \quad n, n' \in \mathbb{Z}, \tag{11.62}$$

where $\delta_{n-n' \bmod N} = 1$ if and only if $n \equiv n' \pmod{N}$, and 0 otherwise.

Returning to the example in Fig. 11.6, where

$$x_k = e^{\frac{2\pi i}{N} 5k} - \frac{1}{3} e^{\frac{2\pi i}{N} 15k} + \frac{1}{5} e^{\frac{2\pi i}{N} 25k},$$

we use Eq. 11.61 to obtain:

$$\tilde{x}_n = \sqrt{N}\left(\delta_{n,5} - \tfrac{1}{3}\delta_{n,15} + \tfrac{1}{5}\delta_{n,25}\right),$$

reflecting the three harmonics present in x_k. (The factor $\sqrt{N}$ arises because the vectors x_k are not normalized.)

> **Exercise 11.17** Calculate the DFT of the sequence $x_k = e^{\frac{2\pi i}{N}nk}$, where the real part of x_k is the sinusoidal function $r(t) = \cos\left(\frac{2\pi}{N}nt\right)$, for $n = 0$, 1, and 2. Explain the meaning of the results.

Before delving into the matrix presentation of the DFT, we briefly describe two key aspects related to the practical application of the DFT.

Nyquist Limit

In Fig. 11.6, we observe that the harmonic component $e^{\frac{2\pi i}{N}25k}$ has only two samples in each period, because 25 is exactly $\frac{N}{2}$. This represents the limit where the digitized sequence x_k can accurately represent the continuous function $x(k)$. If the frequency of the harmonic component is higher than $f_N = \frac{N}{2}$, the digitization process will irreversibly distort the original function.

This concept is referred to as the Nyquist sampling limit. The Nyquist frequency f_N is defined as $\frac{N}{2}$. A signal can be accurately represented in the DFT provided the highest harmonic frequency is below the Nyquist frequency.

Cyclic Property

The DFT assumes that the input sequence is periodic, with period N. This periodicity implies that the values of the sequence repeat every N samples:

$$x_k = x_{k+N},$$

for any integer k.

This cyclic nature implies that the DFT coefficients $\tilde{x}_k$ are also periodic with period N, if k is extended to include all integers.

3   Matrix Representation

The DFT can be represented using the DFT matrix F:

$$F = \frac{1}{\sqrt{N}}\begin{bmatrix} 1 & 1 & 1 & \cdots & 1 \\ 1 & \omega & \omega^2 & \cdots & \omega^{N-1} \\ 1 & \omega^2 & \omega^4 & \cdots & \omega^{2(N-1)} \\ \vdots & \vdots & \vdots & \ddots & \vdots \\ 1 & \omega^{N-1} & \omega^{2(N-1)} & \cdots & \omega^{(N-1)^2} \end{bmatrix}, \tag{11.63}$$

where

$$\omega = e^{-\frac{2\pi i}{N}}. \tag{11.64}$$

In Dirac notation, F is given by:

$$F = \frac{1}{\sqrt{N}} \sum_{n=0}^{N-1} \sum_{k=0}^{N-1} \omega^{kn} |n\rangle\langle k| . \tag{11.65}$$

Treating x_k and $\tilde{x}_k$ as vectors, the DFT is then given by:

$$|\tilde{x}\rangle = F |x\rangle . \tag{11.66}$$

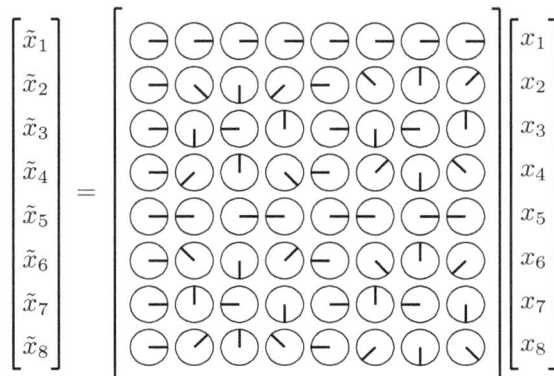

$$
\begin{bmatrix} \tilde{x}_1 \\ \tilde{x}_2 \\ \tilde{x}_3 \\ \tilde{x}_4 \\ \tilde{x}_5 \\ \tilde{x}_6 \\ \tilde{x}_7 \\ \tilde{x}_8 \end{bmatrix}
=
\begin{bmatrix} & & & & & & & \end{bmatrix}
\begin{bmatrix} x_1 \\ x_2 \\ x_3 \\ x_4 \\ x_5 \\ x_6 \\ x_7 \\ x_8 \end{bmatrix}
$$

Figure 11.7: DFT Matrix as Phase Angles

The DFT matrix can be visualized in terms of phase angles, as illustrated in Fig. 11.7 for $N = 8$. Each element of the DFT matrix is represented by a unit circle with a line indicating the phase angle of the complex exponential $e^{-\frac{2\pi i}{N} kn}$. The phase angle in each column (or row) increases linearly. The matrix transforms a vector from the time domain to the frequency domain by adjusting each element of the input vector according to these phase angles.

4 Properties of the DFT Matrix

The DFT matrix has several noteworthy and significant properties.

1. Symmetry: The DFT matrix F is symmetric (not Hermitian). Specifically,

$$F_{jk} = F_{kj} = \frac{1}{\sqrt{N}} \omega^{jk}.$$

2. Orthogonality: The columns (and rows) of F are orthonormal vectors. Namely, for any two distinct columns (or rows) j and k:

$$F_j \cdot F_k = \frac{1}{N} \sum_{n=0}^{N-1} (\omega^{nj})^* \omega^{nk} = \delta_{jk}.$$

This property follows directly from Eq. 11.61.

3. Unitarity: The orthogonality property of the DFT matrix leads to its unitarity, meaning $F^{-1} = F^{\dagger}$, or $FF^{\dagger} = F^{\dagger}F = I$.

Note that the matrix $F^{\dagger}$ is simply F with ω replaced by ω^*.

Consequently, the inverse discrete Fourier transform (IDFT) is given by:

$$|x\rangle = F^\dagger |\tilde{x}\rangle, \tag{11.67}$$

or, equivalently,

$$x_k = \frac{1}{\sqrt{N}} \sum_{n=0}^{N-1} \tilde{x}_n e^{\frac{2\pi i}{N} kn}, \quad k = 0, 1, \ldots, N - 1. \tag{11.68}$$

4. Coordinate Inversion: The matrix $J \equiv F^2$, representing applying the DFT twice, results in a coordinate inversion, or reflection. Namely, it maps x_k to $x_{-k \mod N}$. This follows from Eq. 11.62:

$$J_{jk} = \frac{1}{N} \sum_{n=0}^{N-1} e^{-\frac{2\pi i}{N}(j+k)n} = \delta_{j+k \mod N}.$$

   As a result, $J_{jk} = 0$ unless $j \equiv N - k \mod N$.

5. Eigenvalues: The eigenvalues of the DFT matrix F are $\{1, -1, i, -i\}$. This is because applying coordinate inversion twice restores the coordinates, and consequently, $J^2 = F^4 = I$.

5 DFT as a Change of Basis

The DFT matrix F serves as a change of basis matrix from the standard basis to the Fourier basis.

Recall that a basis of a vector space is a set of vectors that are linearly independent and span the space. The standard basis for $\mathbb{C}^N$, denoted as $\{|k\rangle\}$, is the set of vectors where each vector has a 1 in one coordinate and 0's elsewhere. They are the columns of the identity matrix I.

The Fourier basis for $\mathbb{C}^N$ consists of the columns of the DFT matrix F. Each column of F is a vector in $\mathbb{C}^N$ formed by evaluating the complex exponential function at different points. Specifically, the $(k + 1)$th column of F is given by:

$$|f_k\rangle = \frac{1}{\sqrt{N}} \begin{bmatrix} 1 \\ \omega^k \\ \omega^{2k} \\ \vdots \\ \omega^{(N-1)k} \end{bmatrix},$$

where $\omega = e^{-\frac{2\pi i}{N}}$.

Let $|x\rangle$ be a vector in the standard basis, then the transformed vector $|\tilde{x}\rangle = F |x\rangle$ represents the same vector in the Fourier basis. Conversely, to transform a vector $|\tilde{x}\rangle$ from the Fourier basis back to the standard basis, we use the inverse DFT: $|x\rangle = F^\dagger |\tilde{x}\rangle$.

This change of basis facilitates operations like spectral analysis, filtering, compression, and convolution by leveraging the properties of the Fourier basis.

1. Spectral Analysis: In spectral analysis, the Fourier basis vectors correspond to different frequency components of the signal. Thus, the DFT change of basis

is equivalent to representing the same signal in both the time and frequency domains. Specifically, each component of $|\tilde{x}\rangle$ corresponds to the amplitude and phase of a specific frequency component in the original signal.

2. Filtering: Applying a filter to a signal is straightforward in the Fourier basis. By transforming the signal to the frequency domain, multiplying by the filter's frequency response, and transforming back, filtering operations become efficient and intuitive.

3. Compression: Many signals have sparse representations in the Fourier basis, meaning most of the energy is concentrated in a few coefficients. This property is exploited in compression algorithms to reduce the amount of data needed to represent a signal without significant loss of information.

4. Convolution: Convolution in the time domain corresponds to pointwise multiplication in the frequency domain. This property simplifies the convolution of signals, which is a key operation in signal processing, by allowing it to be performed as simple multiplications in the Fourier basis.

In addition, since the DFT is unitary, it preserves the inner products and norms of vectors. In signal processing, this corresponds to the energy conservation of the signal. In other words, the sum of the squares of the signal's amplitudes in the time domain is equal to that in the frequency domain.

6   Hadamard Transform

The DFT can be generalized to other types of transforms that serve similar purposes in different contexts. One such transform is the Hadamard transform, which is useful in quantum computing for creating superposition states and analyzing error correction codes.

The Hadamard transform is a unitary and symmetric matrix that transforms a vector from the standard basis to the Hadamard basis. For an N-dimensional vector, where $N = 2^n$ for some integer n, the Hadamard matrix H is defined recursively as:

$$H_1 = \begin{bmatrix} 1 \end{bmatrix}, \quad H_{2^n} = \frac{1}{\sqrt{2}} \begin{bmatrix} H_{2^{n-1}} & H_{2^{n-1}} \\ H_{2^{n-1}} & -H_{2^{n-1}} \end{bmatrix}. \tag{11.69}$$

For example, the Hadamard matrices H_2 and H_4 are:

$$H_2 = \frac{1}{\sqrt{2}} \begin{bmatrix} 1 & 1 \\ 1 & -1 \end{bmatrix}, \tag{11.70}$$

$$H_4 = \frac{1}{2} \begin{bmatrix} 1 & 1 & 1 & 1 \\ 1 & -1 & 1 & -1 \\ 1 & 1 & -1 & -1 \\ 1 & -1 & -1 & 1 \end{bmatrix}. \tag{11.71}$$

The Hadamard transform is particularly useful due to its simplicity (involving only $+1$ and -1) and the fact that it operates in the real domain.

11.3   Matrix-Valued Functions

This section explores matrix-valued functions with vectors or matrices as arguments. Below are some common examples:

1. Matrix Operations: Includes transpose, adjoint, inverse, and multiplication.

2. Commutator: Defined as $[A, B] = AB - BA$.

3. Anti-Commutator: Defined as $\{A, B\} = AB + BA$.

4. Outer Product of Two Vectors: $|u\rangle\langle v|$, which yields a matrix.

5. Diagonal Matrix Creation: $\text{diag}(v)$ forms a diagonal matrix from vector v.

6. Direct Sum: $A \oplus B$ constructs the block diagonal matrix from A and B.

7. Tensor Product of Two Matrices: For matrices $A \in \mathbb{C}^{m \times m}$ and $B \in \mathbb{C}^{n \times n}$, the tensor product $A \otimes B$ results in a matrix in $\mathbb{C}^{mn \times mn}$. This operation involves creating a block matrix where each block (i, j) is the product of A_{ij} (the (i, j)th element of A) and the matrix B (see § 10.2.2). Key properties of the tensor product include:

   - Non-commutativity: Generally, $A \otimes B \neq B \otimes A$.

   - Adjoint: $(A \otimes B)^\dagger = A^\dagger \otimes B^\dagger$.

   - Inverse: If both A and B are invertible, then $(A \otimes B)^{-1} = A^{-1} \otimes B^{-1}$.

   - Multiplication: $(A \otimes B)(C \otimes D) = AC \otimes BD$.

We will next examine a specific category of these functions, namely, analytic functions that accept normal matrices as inputs. These functions are particularly useful in quantum computing applications. We start with power functions which serves as the foundation of analytic functions.

11.3.1 Power Functions

1 Zero and Negative Exponents

The power function of matrices extends the power function of regular numbers. Let's start with A^k with k being a positive integer. For $k = 0$, we define $A^0 = I$, where I is the identity matrix. If A is invertible, we define $A^{-k} = (A^{-1})^k$, where A^{-1} is the inverse of A. Thus, A^k is defined for all integers k.

2 Spectral Representation

Assume A is a normal matrix, i.e., $A^\dagger A = AA^\dagger$. Then A is unitarily diagonalizable. In this case, the computation of A^k can be efficiently handled using the spectral representation of A (see § 9.3).

Theorem 11.11 If $A = U\Lambda U^\dagger$ is the spectral decomposition of A, where Λ is a diagonal matrix containing the eigenvalues of A and U is a unitary matrix, then for any integer k:

$$A^k = U\Lambda^k U^\dagger. \tag{11.72}$$

Here, Λ^k can be computed efficiently by raising each diagonal element of Λ to the power k:

$$\Lambda = \begin{bmatrix} \lambda_1 & & & \\ & \lambda_2 & & \\ & & \ddots & \\ & & & \lambda_n \end{bmatrix}, \quad \Lambda^k = \begin{bmatrix} \lambda_1^k & & & \\ & \lambda_2^k & & \\ & & \ddots & \\ & & & \lambda_n^k \end{bmatrix}. \tag{11.73}$$

Proof. Because of $U^\dagger U = I$, the proof is straightforward for $k > 0$:

$$A^k = U\Lambda U^\dagger U\Lambda U^\dagger \cdots U\Lambda U^\dagger = U\Lambda^k U^\dagger.$$

For $k = 0$, we have $A^0 = I$ by definition, consistent with $A^0 = U\Lambda^0 U^\dagger = UIU^\dagger = UU^\dagger = I$.

For $k < 0$, assuming A is invertible, then all the eigenvalues of A are non-zero, and

$$A^{-1} = (U\Lambda U^\dagger)^{-1} = (U^\dagger)^{-1}\Lambda^{-1}U^{-1} = U\Lambda^{-1}U^\dagger.$$

From this, denoting $m = -k$, we can deduce $A^{-m} = (A^{-1})^m = U\Lambda^{-m}U^\dagger.$ □

In Dirac notation, the spectral representation of A is given by:

$$A = \sum_{i=1}^{n} \lambda_i \, |\phi_i\rangle\langle\phi_i|, \tag{11.74}$$

where $\{|\phi_i\rangle\}$ is the eigenbasis of A, formed by the columns of U, and $\{\lambda_i\}$ are the corresponding eigenvalues. As a consequence, Theorem 11.11 can be reformulated in a more general form:

Theorem 11.12 Let $\{|\phi_i\rangle\}$ be a complete orthonormal basis set and $\lambda_i \in \mathbb{C}$. Then for any integer k:

$$A^k = \left(\sum_{i=1}^{n} \lambda_i \, |\phi_i\rangle\langle\phi_i| \right)^k = \sum_{i=1}^{n} \lambda_i^k \, |\phi_i\rangle\langle\phi_i|. \tag{11.75}$$

Proof.

$$A^2 = \left(\sum_{i=1}^{n} \lambda_i \, |\phi_i\rangle\langle\phi_i| \right) \left(\sum_{j=1}^{n} \lambda_j \, |\phi_j\rangle\langle\phi_j| \right)$$

$$= \sum_{i=1}^{n}\sum_{j=1}^{n} \lambda_i \, |\phi_i\rangle\langle\phi_i| \, \lambda_j \, |\phi_j\rangle\langle\phi_j|$$

$$= \sum_{i=1}^{n} \lambda_i \, |\phi_i\rangle \sum_{j=1}^{n} \lambda_j \, \langle\phi_i|\phi_j\rangle \, \langle\phi_j|$$

$$= \sum_{i=1}^{n} \lambda_i \, |\phi_i\rangle \sum_{j=1}^{n} \lambda_j \delta_{ij} \, \langle\phi_j|$$

$$= \sum_{i=1}^{n} \lambda_i \, |\phi_i\rangle \lambda_i \, \langle\phi_i| = \sum_{i=1}^{n} \lambda_i^2 \, |\phi_i\rangle\langle\phi_i|.$$

Similarly, we can compute A^3 as A^2A, and so on, all the way to A^k, assuming k is positive.

We leave the proof for the $k = 0$ and $k < 0$ cases to the reader.   □

3 Square Roots of Matrices

The square root of a square matrix A, denoted by $\sqrt{A}$ or $A^{\frac{1}{2}}$, is a matrix such that:

$$(\sqrt{A})^2 = A. \tag{11.76}$$

In the complex domain, the square root of a matrix always exists but may not be unique. Furthermore, $-\sqrt{A} \equiv (-1)\sqrt{A}$ is also a square root of A.

Square Root of Normal Matrices

If A is a normal matrix, the square root of A can be defined in terms of its spectral decomposition. Recall that a normal matrix A can be diagonalized as:

$$A = \sum_{i=1}^{n} \lambda_i |\phi_i\rangle \langle\phi_i| ,$$

where λ_i are the eigenvalues of A, and $\{|\phi_i\rangle\}$ are the corresponding orthonormal eigenvectors. The square root of A is then given by:

$$\sqrt{A} = \sum_{i=1}^{n} \sqrt{\lambda_i} |\phi_i\rangle \langle\phi_i| , \tag{11.77}$$

where $\sqrt{\lambda_i}$ denotes the square root of each eigenvalue λ_i.

In general, λ_i can be complex, and $\sqrt{\lambda_i}$ can have multiple branches. However, if A is a positive semidefinite matrix, its eigenvalues λ_i are real and non-negative. In this case, $\sqrt{\lambda_i}$ is a well-defined real number.

> **Exercise 11.18** Show that the definition of $\sqrt{A}$ in Eq. 11.77 satisfies $(\sqrt{A})^2 = A$ and $\sqrt{A^2} = A$, as expected.

Inverse Square Root

If A is invertible, all its eigenvalues are non-zero. As a result, the square root of A is also invertible, and the inverse square root is given by:

$$\sqrt{A^{-1}} = A^{-\frac{1}{2}} = \sum_{i=1}^{n} \frac{1}{\sqrt{\lambda_i}} |\phi_i\rangle \langle\phi_i| . \tag{11.78}$$

Non-Uniqueness of the Square Root

The square root of a matrix is not necessarily unique. For example, the Pauli matrices X, Y, Z (defined in § 12.1.1), along with the identity matrix I, are all square roots of I, since they satisfy:

$$X^2 = Y^2 = Z^2 = I^2 = I.$$

In addition, $-X, -Y, -Z, -I$ are also square roots of I. Moreover, any linear combination of the form:

$$\alpha X + \beta Y + \gamma Z,$$

where α, β, γ are real numbers satisfying $\alpha^2 + \beta^2 + \gamma^2 = 1$, is also a square root of I.

Exercise 11.19 Given the matrices:

$$X = \begin{bmatrix} 0 & 1 \\ 1 & 0 \end{bmatrix} \quad \text{and} \quad Z = \begin{bmatrix} 1 & 0 \\ 0 & -1 \end{bmatrix},$$

find $\sqrt{X}$ and $\sqrt{Z}$. Verify your results by showing that $(\sqrt{X})^2 = X$ and $(\sqrt{Z})^2 = Z$.

4   Real Exponents

Using Eq. 11.75, we can extend the matrix power function to real exponents. Assume that a normal matrix A has a spectral expansion as given by Eq. 11.74. For any real x, A^x is defined as:

$$A^x = \sum_{i=1}^{n} \lambda_i^x |\phi_i\rangle \langle\phi_i|. \tag{11.79}$$

Here, if A is positive semidefinite, the eigenvalues λ_i are real and non-negative, so λ_i^x is well-defined in the real domain. If A is a general normal matrix, the eigenvalues λ_i may be complex, and λ_i^x is understood in the complex domain.

5   The Cayley-Hamilton Theorem

The Cayley-Hamilton theorem has significant applications in various fields of mathematics and engineering. It allows us to express higher powers of matrices in terms of lower powers, which simplifies certain computations.

Theorem 11.13 — Cayley-Hamilton Theorem. Any square matrix A satisfies its own characteristic equation.

Let the characteristic equation of A be:

$$f(\lambda) = \det(A - \lambda I) = \lambda^n + c_1 \lambda^{n-1} + c_2 \lambda^{n-2} + \cdots + c_{n-1}\lambda + c_n = 0, \tag{11.80}$$

where λ is a scalar, I is the identity matrix, and $c_1, c_2, \ldots, c_n$ are constants. The solutions of this polynomial equation are the eigenvalues of A.

The matrix A satisfies this equation when λ is replaced by A itself:

$$f(A) = A^n + c_1 A^{n-1} + c_2 A^{n-2} + \cdots + c_{n-1}A + c_n I = 0. \tag{11.81}$$

A rigorous proof of the Cayley-Hamilton theorem is complex and beyond the scope of this text. However, the proof is straightforward if A has a spectral decomposition. With the decomposition $A = V\Lambda V^{-1}$, where Λ is a diagonal matrix of eigenvalues and V is the matrix of eigenvectors, substituting A into its characteristic polynomial simplifies to:

$$f(A) = V f(\Lambda)V^{-1}.$$

Here, $f(\Lambda)$ is a diagonal matrix with entries $f(\lambda_i)$, the values of the characteristic polynomial evaluated at each eigenvalue λ_i. Since $f(\lambda_i) = 0$ by definition of the characteristic polynomial, it follows that $f(\Lambda) = 0$, and thus:

$$f(A) = V 0 V^{-1} = 0.$$

■ **Example 11.2** Consider the matrix A:

$$A = \begin{bmatrix} -2 & 4 \\ -1 & 3 \end{bmatrix}.$$

The characteristic equation for this matrix is:

$$\lambda^2 - \lambda - 2 = 0,$$

with eigenvalues:

$$\lambda_1 = -1, \quad \lambda_2 = 2.$$

Using the Cayley-Hamilton theorem, we can substitute A into the characteristic equation:

$$A^2 - A - 2I = 0.$$

Calculating powers of A:

$$A^2 = 2I + A,$$

$$A^3 = 2A + A^2 = 2A + (2I + A) = 2A + 2I + A = 2I + 3A.$$

Similarly:

$$A^4 = 6I + 5A,$$

$$A^5 = 10I + 11A,$$

$$A^6 = 22I + 21A.$$

The recursive relation for the powers of A is:

$$A^n = q_n A + r_n I \quad \Rightarrow \quad A^{n+1} = (q_n + r_n)A + 2q_n I,$$

with initial conditions:

$$q_2 = 1, \quad r_2 = 2.$$

The pattern shows that any power of a 2×2 matrix A can be written as a linear combination of A and the identity matrix I.

Now consider the matrix exponential (which we will delve into soon):

$$e^{At} = I + At + \frac{A^2 t^2}{2!} + \frac{A^3 t^3}{3!} + \cdots.$$

Using the expressions for A^2, A^3, and so on that we have derived above, we can substitute and simplify:

$$e^{At} = I + At + \frac{(2I + A)t^2}{2!} + \frac{(2I + 3A)t^3}{3!} + \cdots.$$

By grouping terms with A and I, we get a linear combination of A and I, where the coefficients are functions of t:

$$e^{At} = h(t)I + g(t)A,$$

where $h(t)$ and $g(t)$ are scalar functions that can be determined by collecting terms in the expansion. In some cases, the functions $h(t)$ and $g(t)$ can be expressed in closed form, giving us a simple and elegant formula for the matrix exponential. ∎

This example demonstrates that the Cayley-Hamilton theorem allows us to avoid calculating higher powers of A directly, which can be computationally expensive for large matrices. Instead, we only need to compute a few lower powers. The theorem can also be used to simplify matrix exponentials and other analytic functions.

11.3.2 Analytic Functions

Certain functions can be generalized to accept matrices as arguments. We will start with an illustrative example.

1 An Example

Suppose we have a normal matrix $A \in \mathbb{C}^{n \times n}$, and we want to compute the series $I + A + A^2 + A^3 + \dots$. It turns out that

$$\sum_{i=0}^{\infty} A^i = (I - A)^{-1}, \tag{11.82}$$

provided $\|A\| < 1$. Here, $\|A\|$ is the spectral norm of A, defined in § 11.1.2.3, which corresponds to the largest magnitude of the eigenvalues of A.

Consider the matrix $A = \begin{bmatrix} 0.4 & 0.1 \\ 0.1 & 0.4 \end{bmatrix}$. It has eigenvalues of 0.5 and 0.3, thus a norm of 0.5.

$$(I - A)^{-1} = \begin{bmatrix} 0.6 & -0.1 \\ -0.1 & 0.6 \end{bmatrix}^{-1} = \begin{bmatrix} \frac{12}{7} & \frac{2}{7} \\ \frac{2}{7} & \frac{12}{7} \end{bmatrix}.$$

This result can be confirmed numerically by summing the series $\sum_{i=0}^{\infty} A^i$ and observing that it converges to $(I - A)^{-1}$.

Now consider $B = 2A = \begin{bmatrix} 0.8 & 0.2 \\ 0.2 & 0.8 \end{bmatrix}$. It is straightforward to verify that $B^2 = B$, which implies that $B^i = B$ for all $i \geq 1$. Consequently, the series $\sum_{i=0}^{\infty} B^i$ does not converge, as each term from $i = 1$ onwards simply adds another B without approaching a limit.

This result is consistent with the fact that the eigenvalues of B are 1 and 0.6, giving a matrix norm $\|B\| = 1$. Since $\|B\| \not< 1$, the condition necessary for the convergence of $\sum_{i=0}^{\infty} B^i$ to $(I - B)^{-1}$ is not met.

In function notation, we may write $(I - A)^{-1}$ as $\frac{1}{I-A}$. This example is reminiscent of the geometric series

$$\frac{1}{1 - x} = \sum_{k=0}^{\infty} x^k = 1 + x + x^2 + x^3 + \cdots \quad \text{(for } |x| < 1\text{)}. \tag{11.83}$$

Indeed, that's the connection.

2 Analytic Functions of Matrices

Analytic functions are a class of functions that can be represented by convergent power series (Taylor expansions). Here are more examples ($x \in \mathbb{C}$):

$$(1 + x)^\alpha = \sum_{k=0}^{\infty} \frac{(\alpha)_k}{k!} x^k$$

$$= 1 + \alpha x + \alpha(\alpha - 1)x^2 + \alpha(\alpha - 1)(\alpha - 2)x^3 + \cdots \quad \text{(for } |x| < 1\text{)}, \tag{11.84}$$

$$\ln(1 + x) = \sum_{k=1}^{\infty} \frac{(-1)^{k+1}}{k} x^k = x - \frac{x^2}{2} + \frac{x^3}{3} - \cdots \quad \text{(for } |x| < 1\text{)}, \tag{11.85}$$

$$\ln(x) = \sum_{k=1}^{\infty} \frac{(-1)^{k+1}}{k} (x - 1)^k = x - 1 - \frac{(x - 1)^2}{2} + \cdots \quad \text{(for } |x - 1| < 1\text{)}, \tag{11.86}$$

$$e^x = \sum_{k=0}^{\infty} \frac{x^k}{k!} = 1 + x + \frac{x^2}{2!} + \frac{x^3}{3!} + \cdots \quad \text{(for } |x| < \infty\text{)}, \tag{11.87}$$

$$\sin x = \sum_{k=0}^{\infty} \frac{(-1)^k}{(2k + 1)!} x^{2k+1} = x - \frac{x^3}{3!} + \frac{x^5}{5!} - \cdots \quad \text{(for } |x| < \infty\text{)}, \tag{11.88}$$

$$\cos x = \sum_{k=0}^{\infty} \frac{(-1)^k}{(2k)!} x^{2k} = 1 - \frac{x^2}{2!} + \frac{x^4}{4!} - \cdots \quad \text{(for } |x| < \infty\text{)}, \tag{11.89}$$

An analytic function of a matrix can be expressed through its Taylor expansion, analogous to the way analytic functions are defined for scalars.

Definition — Analytic Functions of Matrices. Let $f(x)$ be an analytic function expressed as a power series:

$$f(x) = \sum_{k=0}^{\infty} c_k x^k. \tag{11.90}$$

For a normal matrix A, the function $f(A)$ is defined as

$$f(A) = \sum_{k=0}^{\infty} c_k A^k, \tag{11.91}$$

provided the series converges. (See Theorem 11.16 for the convergence condition.)

With the above definition, familiar scalar relations derived from Taylor expansions, such as $\sqrt{x^2} = x$, $\ln e^x = x$, and $\cos^2 x + \sin^2 x = 1$, extend naturally to matrix functions: $\sqrt{A^2} = A$, $\ln e^A = A$, and $\cos^2 A + \sin^2 A = I$. However, a key distinction in matrix algebra is that if A and B do not commute, then $AB \neq BA$, which can lead to non-intuitive results such as $e^{A+B} \neq e^A e^B$.

3 Spectral Representation

The spectral representation of A enables us to relate $f(A)$ with $f(\lambda_i)$, where λ_i are the eigenvalues of A.

Theorem 11.14 Let $\{|\phi_i\rangle\}$ be a complete orthonormal basis and $\lambda_i \in \mathbb{C}$. Suppose A has the spectral decomposition:

$$A = \sum_{i=1}^{n} \lambda_i |\phi_i\rangle\langle\phi_i|.$$

Then, for any analytic function $f(x)$, the matrix function $f(A)$ is given by:

$$f(A) = \sum_{i=1}^{n} f(\lambda_i) |\phi_i\rangle\langle\phi_i| . \tag{11.92}$$

Proof. Using the power series expansion of $f(x)$, we write:

$$f(x) = \sum_{k=0}^{\infty} c_k x^k \quad \text{so that} \quad f(A) = \sum_{k=0}^{\infty} c_k A^k.$$

Substituting the spectral decomposition of A:

$$A = \sum_{i=1}^{n} \lambda_i |\phi_i\rangle\langle\phi_i| ,$$

we have:

$$\begin{aligned}
f(A) &= \sum_{k=0}^{\infty} c_k \left(\sum_{i=1}^{n} \lambda_i |\phi_i\rangle\langle\phi_i| \right)^k \\
&= \sum_{k=0}^{\infty} c_k \sum_{i=1}^{n} \lambda_i^k |\phi_i\rangle\langle\phi_i| \quad \text{(by Theorem 11.12)} \\
&= \sum_{i=1}^{n} |\phi_i\rangle\langle\phi_i| \sum_{k=0}^{\infty} c_k \lambda_i^k \\
&= \sum_{i=1}^{n} |\phi_i\rangle\langle\phi_i| f(\lambda_i).
\end{aligned}$$

Thus, the result follows.   ☐

A direct corollary of the above theorem is as follows:

Theorem 11.15 Let A be a normal matrix, and let $f(x)$ be an analytic function such that $f(A)$ is well-defined and converges. Then:

(a) $f(A)$ is also a normal matrix.

(b) The eigenvalues of $f(A)$ are $f(\lambda_i)$, where λ_i are the eigenvalues of A.

(c) The eigenvectors of $f(A)$ are the same as those of A.

---

4   Convergence Condition

From Theorem 11.14, it follows that the convergence of every $f(\lambda_i)$ ensures the convergence of $f(A)$. Therefore, if $f(x)$ converges for $|x| < R$, and every $|\lambda_i| < R$, then $f(A)$ converges. Since $\|A\| = \sup\{|\lambda_i|\}$, we have the following theorem:

Theorem 11.16 If an analytic function $f(x)$ converges for $|x| < R$, then $f(A)$ converges for $\|A\| < R$, where A is a normal matrix with spectral norm $\|A\|$.

The principles delineated in this subsection enable the application of many familiar analytic functions to matrices, providing a crucial tool in various fields, including quantum computing and systems theory. Two particularly essential functions are exponential and logarithmic functions, which we will delve into next.

11.3.3 The Exponential Function

1 Definition and Properties

The exponential function of any square matrix A is defined as:

$$\exp(A) \equiv e^A \equiv \sum_{k=0}^{\infty} \frac{1}{k!} A^k, \tag{11.93}$$

where $A^0 = I$ is the identity matrix.

> **Exercise 11.20** Determine the power series representations of 2^A and 2^{-A}. Compare their forms to the definition of e^A.

> **Exercise 11.21** Let $A = \begin{bmatrix} 0 & 1 \\ 0 & 1 \end{bmatrix}$ and $B = \begin{bmatrix} 1 & 0 \\ 1 & 0 \end{bmatrix}$. Compute the following matrix exponentials:
> $$e^A, \quad e^B, \quad e^{iA}, \quad \text{and} \quad e^{iB}.$$

If A is a normal matrix with the diagonal form

$$A = \sum_{i=1}^{n} \lambda_i \, |\phi_i\rangle\langle\phi_i|,$$

where λ_i are the eigenvalues of A and $|\phi_i\rangle$ are the corresponding orthonormal eigenvectors, then the matrix exponential e^A is given by:

$$e^A = \sum_{i=1}^{n} e^{\lambda_i} \, |\phi_i\rangle\langle\phi_i|. \tag{11.94}$$

The function e^A has the following basic properties:

- e^A converges for any square matrix A.
- If A is a normal matrix and λ_i are the eigenvalues of A, then e^{λ_i} are the eigenvalues of e^A, with the same eigenvectors as A.
- If U is unitary, $e^{U^\dagger A U} = U^\dagger e^A U$.
- If A and B are similar matrices, i.e., there exists an invertible matrix P such that $B = P^{-1}AP$, then $e^B = P^{-1}e^A P$.
- If A and B commute (i.e., $AB = BA$), then e^A and e^B also commute: $e^A e^B = e^B e^A$.
- $e^{nA} = \left(e^A\right)^n$ for $n \in \mathbb{Z}$. In particular, $e^0 = I$ and $e^{-A} = (e^A)^{-1}$.
- $\det(e^A) = e^{\mathrm{tr}(A)}$.
- $\frac{d}{dt} e^{tA} = A e^{tA} = e^{tA} A$.
- $\|e^A\| \le e^{\|A\|}$.
- If A is Hermitian, then e^{iA} is unitary, where i is the imaginary unit.
- In general, $e^{A \otimes B} \ne e^A \otimes e^B$, but $e^{A \otimes I + I \otimes B} = e^A \otimes e^B$.
- In general, $e^{A+B} \ne e^A e^B$. See § 11.3.3.3 for more details.

Exercise 11.22 Prove property 11.3.3.1, i.e., $e^{P^{-1}AP} = P^{-1}e^A P$.

Next, we will delve into several additional formulas that find extensive applications in quantum computing. Their proofs involve essential skills related to matrix exponentials, making them valuable topics to explore.

2 Generalized Euler Formula

Theorem 11.17 Let A be an involutory matrix such that $A^2 = I$, and let θ be a real number. Then the following relationship holds:

$$e^{i\theta A} = \cos\theta I + i\sin\theta A. \qquad (11.95)$$

Equation 11.95 is analogous to the Euler formula, $e^{i\theta} = \cos\theta + i\sin\theta$.

Proof. Given the definition of the exponential of a matrix,

$$e^{i\theta A} = \sum_{k=0}^{\infty} \frac{1}{k!}(i\theta A)^k,$$

we separate the series into even and odd terms:

$$e^{i\theta A} = \sum_{k=0}^{\infty} \frac{1}{(2k)!}(i\theta A)^{2k} + \sum_{k=0}^{\infty} \frac{1}{(2k+1)!}(i\theta A)^{2k+1}.$$

Given $A^2 = I$ and $A^{2k} = I^k = I$, we can simplify the terms.

For even powers, $(i\theta A)^{2k} = (i^2)^k \theta^{2k} A^{2k} = (-1)^k \theta^{2k} I$.

For odd powers, $(i\theta A)^{2k+1} = (i\theta A)(-1)^k \theta^{2k} A^{2k} = i(-1)^k \theta^{2k+1} A$.

Thus, the series can be rewritten as:

$$e^{i\theta A} = \sum_{k=0}^{\infty} \frac{(-1)^k \theta^{2k}}{(2k)!} I + \sum_{k=0}^{\infty} \frac{i(-1)^k \theta^{2k+1}}{(2k+1)!} A$$

$$= \cos\theta I + i\sin\theta A.$$

In the final step, we recognize the Taylor series expansions of $\cos\theta$ and $\sin\theta$. Therefore, we have proved that $e^{i\theta A} = \cos\theta I + i\sin\theta A$ when $A^2 = I$. □

Exercise 11.23 Consider the Pauli matrices $X = \begin{bmatrix} 0 & 1 \\ 1 & 0 \end{bmatrix}$ and $Z = \begin{bmatrix} 1 & 0 \\ 0 & -1 \end{bmatrix}$. For a real parameter θ, compute the matrices for $e^{i\theta X}$ and $e^{i\theta Z}$.

3 ✳ BCH, Trotter, and Trotter-Suzuki Formulas

You might expect the familiar property $e^{A+B} = e^A e^B$ to hold. However, this relation is valid only when A and B commute, i.e., $[A, B] = 0$. The non-commutativity of A and B introduces complexities that merit a dedicated discussion.

The Baker-Campbell-Hausdorff (BCH) formula, the Trotter formula, and the Trotter-Suzuki formula provide methods for approximating e^{A+B} when $[A, B] \neq 0$. These formulas are powerful tools in the study of matrix exponentials, with applications in quantum mechanics, statistical mechanics, and other fields.

The BCH Formula

The BCH formula provides a way to express e^{A+B} in terms of e^A and e^B, along with additional correction terms involving their commutator $[A, B]$ and higher-order commutators. Specifically, the BCH formula states:

$$e^A e^B = e^{A+B+C},$$ (11.96)

where C is given by an infinite series involving commutators of A and B:

$$C = \frac{1}{2}[A, B] + \frac{1}{12}([A, [A, B]] + [B, [B, A]]) + \cdots.$$ (11.97)

Here, we assume that successive higher-order commutators become progressively smaller:

$$\|A\|, \|B\| \gg \|[A, B]\| \gg \|[B, [A, B]]\|, \|[A, [A, B]]\| \gg \ldots,$$

which is often the case in scenarios where:

$$\|A\|, \|B\| \propto \frac{1}{n}, \quad \|[A, B]\| \propto \frac{1}{n^2}, \quad \|[B, [A, B]]\|, \|[A, [A, B]]\| \propto \frac{1}{n^3}, \quad \ldots,$$

and n is a large integer.

From the BCH formula, we see that if $[A, B] = 0$, $e^{A+B} = e^A e^B$.

Furthermore, if $[A, B]$ commutes with both A and B, the BCH formula simplifies to:

$$e^A e^B = e^{A+B+\frac{1}{2}[A,B]}, \quad \text{or,} \quad e^{A+B} = e^A e^B e^{-\frac{1}{2}[A,B]}.$$ (11.98)

This formula can be generalized to n matrices $A_1, A_2, \ldots, A_n$:

$$e^{A_1+A_2+\cdots+A_n} \approx e^{A_1} e^{A_2} \cdots e^{A_n} e^{-\frac{1}{2}\sum_{1\leq i<j\leq n}[A_i,A_j]}.$$ (11.99)

If second-order commutators are included, the truncated BCH formula becomes:

$$e^{A+B} \approx e^A e^B e^{-\frac{1}{2}[A,B]} e^{-\frac{1}{12}([A,[A,B]]+[B,[A,B]])}.$$ (11.100)

Proof. $*$ Proving the full BCH formula is quite involved and typically involves techniques from Lie theory and differential equations. Here we provide a simplified proof of the BCH formula up the first order commutator.

Consider the following error function, expanded up to the order of $\frac{1}{n^2}$:

$$e^{A+B+C} - e^A e^B = I + (A + B + C) + \frac{1}{2}(A + B + C)^2 + \ldots$$

$$- \left(I + A + \frac{1}{2}A^2 + \ldots\right)\left(I + B + \frac{1}{2}B^2 + \ldots\right)$$

$$= I + (A + B + C) + \frac{1}{2}(A^2 + AB + BA + B^2) + \ldots$$

$$- \left(I + A + B + \frac{1}{2}A^2 + \frac{1}{2}B^2 + AB + \ldots\right)$$

$$\approx C + \frac{1}{2}(BA - AB) = C - \frac{1}{2}[A, B].$$

Here, we have also omitted higher-order terms C^2, CA, BC, etc., because $C \propto \frac{1}{n^2}$. Therefore, to the order of $\frac{1}{n^2}$, $C = \frac{1}{2}[A, B]$. $\qquad \square$

The Trotter Formula

The Trotter formula (or Trotter product formula) is used to approximate the exponential of the sum of two non-commuting matrices. For operators A and B, it states:

$$e^{A+B} = \lim_{n\to\infty} \left(e^{\frac{A}{n}} e^{\frac{B}{n}} \right)^n. \tag{11.101}$$

The basic idea of the Trotter formula is to divide the exponentiation process into n micro steps. For sufficiently small steps, the order of exponentiation becomes less significant. In other words, $e^{\frac{A+B}{n}} \approx e^{\frac{A}{n}} e^{\frac{B}{n}}$ holds approximately for each small step. By taking the limit as the number of steps n approaches infinity, the approximation improves, and the product of exponentials converges to the exponential of the sum:

$$\left(e^{\frac{A}{n}} e^{\frac{B}{n}} \right)^n \to e^{A+B} \quad \text{as} \quad n \to \infty.$$

Proof. ✳ Using the BCH formula, we can approximate the product $e^{\frac{A}{n}} e^{\frac{B}{n}}$ as:

$$e^{\frac{A}{n}} e^{\frac{B}{n}} \approx e^{\frac{A+B}{n} + \frac{1}{2n^2}[A,B]}.$$

To approximate e^{A+B}, we raise this expression to the power n:

$$\left(e^{\frac{A}{n}} e^{\frac{B}{n}} \right)^n \approx \left(e^{\frac{A+B}{n} + \frac{1}{2n^2}[A,B]} \right)^n.$$

Using the property $(e^X)^n = e^{nX}$, we get:

$$\left(e^{\frac{A}{n}} e^{\frac{B}{n}} \right)^n \approx e^{A+B+\frac{1}{2n}[A,B]}.$$

We now calculate the error function:

$$e^{A+B} - \left(e^{\frac{A}{n}} e^{\frac{B}{n}} \right)^n \approx e^{A+B} - e^{A+B+\frac{1}{2n}[A,B]}.$$

Expanding $e^{A+B+\frac{1}{2n}[A,B]}$ and e^{A+B} as Taylor series, we have:

$$e^{A+B} - e^{A+B+\frac{1}{2n}[A,B]} \approx -\frac{1}{2n}[A,B].$$

Therefore, the leading-order error term is:

$$e^{A+B} - \left(e^{\frac{A}{n}} e^{\frac{B}{n}} \right)^n \approx -\frac{1}{2n}[A,B],$$

which diminishes as $\frac{1}{n}$ when $n \to \infty$. ☐

This result demonstrates that the Trotter formula approximates e^{A+B} with an error that is of order $O\left(\frac{1}{n}\right)$, which decreases as n increases. The leading error term is proportional to the commutator $[A,B]$, reflecting the fact that the Trotter approximation is exact when A and B commute.

For example, in quantum mechanics, A and B could represent different parts of a Hamiltonian, the operator that describes the total energy of the system. The Trotter formula helps in simulating the time evolution of quantum systems by breaking down the evolution operator $e^{i(A+B)\tau}$ over a time period τ into micro time intervals of duration $\frac{\tau}{n}$. Even though A and B don't commute, so $e^{i(A+B)\tau} \neq e^{iA\tau}e^{iB\tau}$, yet in each micro interval,

$$e^{\frac{i(A+B)\tau}{n}} \approx e^{\frac{iA\tau}{n}}e^{\frac{iB\tau}{n}}.$$

As a result,

$$\left(e^{\frac{iA\tau}{n}}e^{\frac{iB\tau}{n}}\right)^n \approx e^{i(A+B)\tau}$$

to the first order of $\frac{1}{n}$.

The Trotter-Suzuki Formula

The Trotter-Suzuki formula, an extension of the Trotter product formula, improves the accuracy of the approximation by reducing the error terms associated with the simple Trotter product formula. This is achieved by symmetrically composing the exponential operators:

$$e^{A+B} = \lim_{n\to\infty}\left(e^{\frac{A}{2n}}e^{\frac{B}{n}}e^{\frac{A}{2n}}\right)^n. \tag{11.102}$$

Proof. ✳ Using the BCH formula $e^X e^Y \approx e^{X+Y+\frac{1}{2}[X,Y]}$ with $X = \frac{A}{2n}$ and $Y = \frac{B}{n}$, we obtain:

$$e^{\frac{A}{2n}}e^{\frac{B}{n}} \approx e^{\frac{A}{2n}+\frac{B}{n}+\frac{1}{4n^2}[A,B]}.$$

Applying the BCH formula again with $X = \frac{A}{2n} + \frac{B}{n} + \frac{1}{4n^2}[A,B]$ and $Y = \frac{A}{2n}$, we obtain:

$$e^{\frac{A}{2n}}e^{\frac{B}{n}}e^{\frac{A}{2n}} \approx e^{\frac{A+B}{n}+\frac{1}{4n^2}[A,B]+\frac{1}{4n^2}[B,A]+\frac{1}{16n^3}[[A,B],A]}$$

$$= e^{\frac{A+B}{n}+\frac{1}{16n^3}[[A,B],A]} \quad (\because [B,A] = -[A,B]).$$

To approximate e^{A+B}, we raise the above expression to the power n:

$$\left(e^{\frac{A}{2n}}e^{\frac{B}{n}}e^{\frac{A}{2n}}\right)^n \approx e^{A+B+\frac{1}{16n^2}[[A,B],A]}.$$

We now compute the difference:

$$e^{A+B} - \left(e^{\frac{A}{2n}}e^{\frac{B}{n}}e^{\frac{A}{2n}}\right)^n \approx e^{A+B} - e^{A+B+\frac{1}{16n^2}[[A,B],A]}.$$

Expanding both exponentials as Taylor series, we obtain the leading error term:

$$e^{A+B} - \left(e^{\frac{A}{2n}}e^{\frac{B}{n}}e^{\frac{A}{2n}}\right)^n \approx -\frac{1}{16n^2}[[A,B],A],$$

which diminishes as $\frac{1}{n^2}$ as $n \to \infty$. ☐

   This proof demonstrates that the Trotter-Suzuki formula achieves higher accuracy than the basic Trotter formula by reducing the leading-order error term to $O\left(\frac{1}{n^2}\right)$. The symmetrized form effectively cancels out the first-order commutator $[A,B]$, leaving a higher-order commutator $[[A,B],A]$ as the leading error term.

Exercise 11.24 Consider the Pauli matrices $X = \begin{bmatrix} 0 & 1 \\ 1 & 0 \end{bmatrix}$ and $Z = \begin{bmatrix} 1 & 0 \\ 0 & -1 \end{bmatrix}$. The following approximation methods have wide applications in quantum computation.

(a) Calculate the commutator $[X, Z]$. Then, find the second-order commutators $[X, [X, Z]]$ and $[Z, [X, Z]]$.

(b) Using the BCH formula (Eq. 11.100), approximate $e^{i\theta(X+Z)}$ as a product of matrix exponentials for a small real parameter θ.

(c) Using the Trotter product formula (Eq. 11.101), approximate $e^{i\tau(X+Z)}$ as a product of matrix exponentials for a real parameter τ (not necessarily small).

11.3.4 ∗The Logarithmic Function

1 Definition and Properties

For a normal matrix A close to the identity matrix, $\ln(A)$ (or $\log(A)$) can be expressed as a power series:

$$\ln(A) = \sum_{k=1}^{\infty} \frac{(-1)^{k+1}}{k} (A - I)^k. \tag{11.103}$$

This series converges if $\|A - I\| < 1$, which mirrors Eq. 11.86.

If A is positive definite, it can be expressed as $A = U\Lambda U^\dagger$, where U is a unitary matrix and Λ is a diagonal matrix containing the eigenvalues λ_i of A. The matrix logarithm $\ln(A)$ is then given by:

$$\ln(A) = U \ln(\Lambda) U^\dagger, \tag{11.104}$$

where $\ln(\Lambda)$ is a diagonal matrix with entries $\ln(\lambda_i)$. Equivalently,

$$\ln(A) = \sum_{i=1}^{n} \ln(\lambda_i) |\phi_i\rangle\langle\phi_i|, \tag{11.105}$$

where $\{|\phi_i\rangle\}$ is the orthonormal eigenbasis comprising the columns of U.

Exercise 11.25 Formulate the power series and eigenbasis representations of $\log_2(A)$.

The concept of $\ln(A)$ can even be generalized to matrices with complex (but non-zero) eigenvalues, even though less relavent to quantum computing. In this scenario, $\ln(A)$ exists if A is invertible. For a normal matrix, $\ln(A)$ is unique if we restrict the logarithm function to its principal branch, typically chosen such that the imaginary part of $\ln(\lambda_i)$ lies in $(-\pi, \pi]$.

The function $\ln(A)$ has the following basic properties. (If $\ln(A)$ is extended to matrices with non-zero complex eigenvalues, then ln is taken on the principal branch.)

1. If A is a normal matrix, $\ln(A)$ is also a normal matrix.

2. If A is a normal matrix, $e^{\ln(A)} = A$ and $\ln(e^A) = A$.

3. If A is a normal matrix and $\{\lambda_i\}$ are the eigenvalues of A, then $\{\ln(\lambda_i)\}$ are the eigenvalues of $\ln(A)$.

4. If A and B are normal matrices that commute (i.e., $AB = BA$), then $\ln(AB) = \ln(B) + \ln(A)$.

2   von Neumann Entropy

A significant application of the matrix logarithm is the von Neumann entropy in quantum information theory, which serves as a measure of uncertainty in quantum systems. The von Neumann entropy is defined as:

$$S(\rho) = -\operatorname{tr}(\rho \log \rho), \tag{11.106}$$

where ρ is a matrix representing a quantum state, known as a density matrix (see §§ 7.4.2.1 and 9.4.5.3). Here, log usually represents the base-2 logarithm.

Any density matrix ρ is Hermitian and positive semidefinite by definition, meaning $\langle \phi | \rho | \phi \rangle \geq 0$ for any vector $|\phi\rangle$, which ensures that ρ can be diagonalized and that all its eigenvalues are non-negative.

Equation (11.106) can be simplified using the spectral decomposition of ρ. In the eigenbasis of ρ, where ρ is diagonal:

$$S(\rho) = -\sum_i \lambda_i \log \lambda_i, \tag{11.107}$$

However, λ_i can be zero for certain ρ. To ensure the validity of the above equation, we extend $\lambda_i \log \lambda_i$ to $\lambda_i = 0$ by simply defining $0 \log 0 = 0$, consistent with $\lim_{\lambda_i \to 0} \lambda_i \log \lambda_i = 0$.

This result is invariant under basis changes, thanks to the basis-independence of the trace operation (see § 11.2.3.4).

> ### Exercise 11.26
>
> (a) Show that $S(\rho) = \log(d)$ if $\rho = \frac{1}{d} I$.
>
> (b) Find $S(\rho)$ for $\rho = \frac{1}{4} \begin{bmatrix} 2 & 1 \\ 1 & 2 \end{bmatrix}$.

3   BCH and Trotter Formulas

If A and B do not commute, normally $\ln(AB) \neq \ln(A) + \ln(B)$, and $\ln(e^A e^B) \neq A + B$. In this scenario, the BCH and Trotter formulas provide ways to approximate these product logarithms.

The BCH formula (Eq. 11.100) can be written in terms of the logarithm function:

$$\ln(e^A e^B) = A + B + \frac{1}{2}[A, B] + \frac{1}{12}([A, [A, B]] + [B, [B, A]]) + \cdots, \tag{11.108}$$

or

$$\ln(AB) = \ln(A) + \ln(B) + \frac{1}{2}[\ln(A), \ln(B)] + \frac{1}{12}(\dots) + \cdots. \tag{11.109}$$

The Trotter formula (Eq. 11.101) becomes:

$$\lim_{n \to \infty} n \ln \left(e^{\frac{A}{n}} e^{\frac{B}{n}} \right) = A + B, \tag{11.110}$$

or

$$\lim_{n \to \infty} n \ln \left(A^{\frac{1}{n}} B^{\frac{1}{n}} \right) = \ln(A) + \ln(B). \tag{11.111}$$

Problem Set 11

11.1 Given $a_1, a_2, \ldots, a_n \in \mathbb{R}$ such that $a_1 + a_2 + \cdots + a_n = 1$, prove that:

$$a_1^2 + a_2^2 + \cdots + a_n^2 \geq \frac{1}{n},$$

using the Cauchy-Schwarz inequality.

11.2 Given $x, y, z \in \mathbb{R}$, find the minimum value of the expression $x^2 + 4y^2 + 9z^2$ subject to the constraint $x + 2y + 3z = 1$.

11.3 Show that if two Hermitian matrices commute in the original basis ($[A, B] = 0$), they will also commute in the transformed basis ($[A', B'] = 0$), where $A' = U^\dagger A U$ and $B' = U^\dagger B U$, and U is the basis transformation operator.

11.4 Given

$$A = \begin{bmatrix} 1 & 1 & -2 \\ -1 & 2 & 1 \\ 0 & 1 & -1 \end{bmatrix},$$

find A^{10} using the spectral representation of A^k and the Cayley-Hamilton theorem, respectively.

11.5 Given

$$A = \begin{bmatrix} 7 & 3 \\ -3 & 1 \end{bmatrix},$$

find A^m as a single 2×2 matrix in closed form, where m is a positive integer.

11.6 The SWAP operator is defined as:

$$\text{SWAP} = |00\rangle\langle 00| + |01\rangle\langle 10| + |10\rangle\langle 01| + |11\rangle\langle 11| = \begin{bmatrix} 1 & 0 & 0 & 0 \\ 0 & 0 & 1 & 0 \\ 0 & 1 & 0 & 0 \\ 0 & 0 & 0 & 1 \end{bmatrix}.$$

Find an operator U such that $U^2 = \text{SWAP}$ (i.e., find $\sqrt{\text{SWAP}}$).

11.7 Given $A^2 = -I$ and $\theta \in \mathbb{R}$, derive a formula for $e^{i\theta A}$ similar to Eq. 11.95.

11.8 Given $A^2 = I$ and $\theta \in \mathbb{R}$, show that:

$$e^{\theta A} = \cosh(\theta)I + \sinh(\theta)A,$$

where $\cosh(\theta) = \frac{1}{2}(e^\theta + e^{-\theta})$ and $\sinh(\theta) = \frac{1}{2}(e^\theta - e^{-\theta})$.

11.9 Show that if U is unitary and A is normal, then:

$$e^{U^\dagger A U} = U^\dagger e^A U.$$

11.10 Prove that $\det(e^A) = e^{\text{tr}(A)}$, where A is a normal matrix.

11.11 Show that if A and B are normal matrices, then:

$$e^{A \otimes I + I \otimes B} = e^A \otimes e^B.$$

11.12 Prove that:

$$\|e^A\|_F \leq e^{\|A\|_F},$$

where A is a normal matrix, and $\|A\|_F$ represents the Frobenius norm of A.

11.13 Below, i is the imaginary unit, and $\theta \in \mathbb{R}$.

(a) Show that if A is Hermitian, then $e^{i\theta A}$ is unitary.

(b) Show that if $e^{i\theta A}$ is unitary, then A is Hermitian.

(c) If A is Hermitian, is $e^{\theta A}$ also Hermitian? Justify your conclusion.

11.14 Show that under any unitary transformation U, the equality:

$$U^{\dagger} f(A) U = f(U^{\dagger} A U),$$

holds, where $f(A)$ is an analytic function of A.

11.15 The two-dimensional projection operator can be represented by a 2×2 matrix (see § 8.1.5.4):

$$P = \begin{bmatrix} \cos^2 \theta & \sin \theta \cos \theta \\ \sin \theta \cos \theta & \sin^2 \theta \end{bmatrix}.$$

Find e^P.

11.16 In $\mathbb{R}^2$, consider the projection onto the straight line $y = 2x$.

(a) What is the projection matrix P?

(b) Calculate $\cos \left(\frac{\pi}{2} P \right)$.

(c) Does $\ln(P)$ exist? If yes, find the matrix; if not, explain why.

11.17 For each of the following cases of matrix A, calculate $E = e^{i\theta A}$, and confirm that $\ln E = A$. Here, $X = \begin{bmatrix} 0 & 1 \\ 1 & 0 \end{bmatrix}$ and $Z = \begin{bmatrix} 1 & 0 \\ 0 & -1 \end{bmatrix}$ are the Pauli matrices, $\theta \in \mathbb{R}$, and $i = \sqrt{-1}$:

(a) $A = iZX$.

(b) $A = Z \otimes X$.

(c) $A = I \oplus X$, which represents the block diagonal matrix formed by I and X.

11.18 Given $X = \begin{bmatrix} 0 & 1 \\ 1 & 0 \end{bmatrix}$ and n is a positive integer, find $X^{1/n}$.

11.19 $*$ The Frobenius norm of a matrix A is defined as:

$$\|A\|_F = \sqrt{\operatorname{tr}(A^T A)}.$$

(a) Prove that $\|A\|_F^2$ is the sum of the squares of all the entries of A.

(b) Show that the Frobenius norm is invariant under orthogonal transformations, i.e., $\|QA\|_F = \|AQ\|_F = \|A\|_F$ for any orthogonal matrix Q.

12. Pauli Matrices, Strings, and Groups

Contents

Pauli matrices are a set of three 2×2 complex matrices named after the physicist Wolfgang Pauli.

The Pauli matrices themselves are Hermitian and unitary, making them useful in studying properties of such matrices. Their eigenvalues and eigenvectors are often employed in various matrix decomposition techniques and in solving matrix equations.

Pauli matrices, along with the identity matrix, form a basis for the space of 2×2 complex matrices. Any 2×2 matrix can be expressed as a linear combination of these matrices. This is useful in various mathematical formulations and quantum mechanical applications.

In quantum mechanics, Pauli matrices represent spins by acting as the generators of rotations in quantum mechanical spin space. Specifically, they describe the spin angular momentum operators for many types of particles with spin-$\frac{1}{2}$, such as electrons, protons, neutrons and neutrinos. These matrices facilitate the calculation of spin dynamics and interactions in a straightforward algebraic framework.

In quantum computing, Pauli matrices represent fundamental quantum gates. The Pauli-X, Y, and Z gates correspond to the Pauli matrices X, Y, and Z, respectively. These gates are essential in the manipulation and measurement of qubits, forming the building blocks of quantum circuits.

Pauli strings are tensor products of Pauli matrices and the identity matrix, which are used extensively in quantum computing and quantum information theory. They play a critical role in various aspects of quantum algorithms, error correction, and the representation of quantum states and operators.

12.1 Pauli Matrices

12.1.1 Definitions and Properties

1 Definitions and Notations

> **Definition 12.1 — Pauli Matrices.** Pauli matrices are a set of three 2×2 complex matrices defined as:
>
> $$\sigma_1 = \begin{bmatrix} 0 & 1 \\ 1 & 0 \end{bmatrix}, \quad \sigma_2 = \begin{bmatrix} 0 & -i \\ i & 0 \end{bmatrix}, \quad \sigma_3 = \begin{bmatrix} 1 & 0 \\ 0 & -1 \end{bmatrix}. \quad (12.1)$$

Additionally, the identity matrix is often included in discussions involving Pauli matrices:

$$\sigma_0 = I = \begin{bmatrix} 1 & 0 \\ 0 & 1 \end{bmatrix}.$$

Alternative symbols are commonly used for these matrices:

$$X \equiv \sigma_x \equiv \sigma_1, \quad Y \equiv \sigma_y \equiv \sigma_2, \quad Z \equiv \sigma_z \equiv \sigma_3.$$

In the standard basis, where $|0\rangle = \begin{bmatrix} 1 \\ 0 \end{bmatrix}$ and $|1\rangle = \begin{bmatrix} 0 \\ 1 \end{bmatrix}$, the Pauli matrices can be expressed as:

$$\sigma_1 = |0\rangle\langle 1| + |1\rangle\langle 0|,$$
$$\sigma_2 = -i\,|0\rangle\langle 1| + i\,|1\rangle\langle 0|,$$
$$\sigma_3 = |0\rangle\langle 0| - |1\rangle\langle 1|. \tag{12.2}$$

Exercise 12.1 Confirm the following relations:

$$\omega_1 \equiv \frac{I+Z}{2} = \begin{bmatrix} 1 & 0 \\ 0 & 0 \end{bmatrix}, \quad \omega_4 \equiv \frac{I-Z}{2} = \begin{bmatrix} 0 & 0 \\ 0 & 1 \end{bmatrix},$$
$$\omega_2 \equiv \frac{X+iY}{2} = \begin{bmatrix} 0 & 1 \\ 0 & 0 \end{bmatrix}, \quad \omega_3 \equiv \frac{X-iY}{2} = \begin{bmatrix} 0 & 0 \\ 1 & 0 \end{bmatrix}. \tag{12.3}$$

2   Basic Properties

Theorem 12.1 Pauli matrices possess the following properties (for $i, j, k \in \{1, 2, 3\}$):

- Hermitian: $\sigma_i^\dagger = \sigma_i$.

- Unitary: $\sigma_i^\dagger \sigma_i = I$.

- Involutory: $\sigma_i^2 = I$.

- Trace: $\operatorname{tr}\sigma_i = 0$.

- Determinant: $\det \sigma_i = -1$.

- Cyclic Relations:

$$\sigma_1\sigma_2 = -\sigma_2\sigma_1 = i\sigma_3, \quad \sigma_2\sigma_3 = -\sigma_3\sigma_2 = i\sigma_1, \quad \sigma_3\sigma_1 = -\sigma_1\sigma_3 = i\sigma_2. \tag{12.4}$$

- Commutation Relations:

$$[\sigma_1, \sigma_2] = 2i\sigma_3, \quad [\sigma_2, \sigma_3] = 2i\sigma_1, \quad [\sigma_3, \sigma_1] = 2i\sigma_2. \tag{12.5}$$

- Anticommutation Relations:

$$\{\sigma_i, \sigma_j\} = \sigma_i\sigma_j + \sigma_j\sigma_i = 2\delta_{ij}\sigma_0, \tag{12.6}$$

where δ_{ij} is the Kronecker delta, defined as $\delta_{ij} = 1$ if $i = j$ and $\delta_{ij} = 0$ if $i \neq j$. That is,

$$\{\sigma_1, \sigma_2\} = \{\sigma_2, \sigma_3\} = \{\sigma_3, \sigma_1\} = 0,$$
$$\{\sigma_1, \sigma_1\} = \{\sigma_2, \sigma_2\} = \{\sigma_3, \sigma_3\} = 2\sigma_0.$$

Combining the commutation and anticommutation properties, we observe that two Pauli matrices commute only if they are the same and anticommute if they are different.

Exercise 12.2 Find four distinct Hermitian and unitary matrices A such that $A^2 = I$ where I is the 2×2 identity matrix.

Exercise 12.3 Prove the following relations:

$$\begin{aligned}
XXX &= X, & XYX &= -Y, & XZX &= -Z, \\
YXY &= -X, & YYY &= Y, & YZY &= -Z, \\
ZXZ &= -X, & ZYZ &= -Y, & ZZZ &= Z.
\end{aligned}$$

Here, $X \equiv \sigma_1$, $Y \equiv \sigma_2$, and $Z \equiv \sigma_3$.

12.1.2 Basis for 2×2 Matrices

The 2×2 complex matrices in $\mathbb{C}^{2\times2}$ form an inner product space under the inner product of matrices (refer to § 11.1.2), defined as:

$$\langle A, B \rangle = A \cdot B = \operatorname{tr}(A^\dagger B) = a_{11}^* b_{11} + a_{12}^* b_{12} + a_{21}^* b_{21} + a_{22}^* b_{22}. \tag{12.7}$$

Theorem 12.2 The three Pauli matrices, along with the identity matrix, $\{\sigma_0, \sigma_1, \sigma_2, \sigma_3\}$, form a complete, orthogonal basis for $\mathbb{C}^{2\times2}$:

$$\langle \sigma_i, \sigma_j \rangle = \operatorname{tr}(\sigma_i^\dagger \sigma_j) = 2\delta_{ij}, \quad \text{where } i,j = 0,1,2,3. \tag{12.8}$$

Proof.

Orthogonality: Equation 12.8 holds because:

- For $i = j$, we use the property $\sigma_i^2 = I$. Thus:

$$\operatorname{tr}(\sigma_i \sigma_i) = \operatorname{tr}(I) = 2.$$

- For $i \neq j$, the Pauli matrices anticommute: $\sigma_i \sigma_j + \sigma_j \sigma_i = 0$. Hence:

$$\operatorname{tr}(\sigma_i \sigma_j) = \frac{1}{2} \operatorname{tr}(\sigma_i \sigma_j + \sigma_j \sigma_i) = \frac{1}{2} \operatorname{tr}(0) = 0.$$

Completeness: The matrices $\{\sigma_0, \sigma_1, \sigma_2, \sigma_3\}$ span the entire space of $\mathbb{C}^{2\times2}$ because they are linearly independent. The linear dimension of $\mathbb{C}^{2\times2}$ is four, as four independent complex numbers are required to define a 2×2 complex matrix. ◻

(i) Note that the Pauli basis $\{\sigma_i\}$ is not normalized because $\langle \sigma_i, \sigma_i \rangle = 2$, for $i = 0, 1, 2, 3$.

Theorem 12.2 implies that any matrix $C \in \mathbb{C}^{2\times2}$ can be expressed as a linear combination of $\sigma_0, \sigma_1, \sigma_2, \sigma_3$:

$$C = \sum_{i=0}^{3} \gamma_i \sigma_i, \tag{12.9}$$

where each coefficient γ_i is a complex number. Thus, any matrix C can be viewed as a vector in a four-dimensional complex vector space with the Pauli matrices as a basis.

The Pauli matrices are not the only possible basis for 2×2 matrices. For example, the matrices $\{w_i\}$ defined in Eq. 12.3 form an alternative orthonormal basis for the space of 2×2 matrices. Any matrix C can also be expressed as:

$$C = c_{11}w_1 + c_{12}w_2 + c_{21}w_3 + c_{22}w_4,$$

where c_{ij} is the (i, j)th component of C.

1 Expansion Coefficients

By taking the inner product between C and σ_i, we obtain the coefficients γ_i:

$$\gamma_i = \frac{1}{2} \langle \sigma_i, C \rangle = \frac{1}{2} \operatorname{tr}(\sigma_i^\dagger C) = \frac{1}{2} \operatorname{tr}(\sigma_i C). \tag{12.10}$$

Here, the factor $\frac{1}{2}$ ensures the coefficients γ_i are correctly normalized, given that $\langle \sigma_i, \sigma_i \rangle = 2$.

Thus, the matrix C can be written as:

$$C = \frac{1}{2} \left[(c_{11} + c_{22})\sigma_0 + (c_{12} + c_{21})\sigma_1 + i(c_{12} - c_{21})\sigma_2 + (c_{11} - c_{22})\sigma_3 \right]. \tag{12.11}$$

Exercise 12.4 Express $\begin{bmatrix} 1 & 2 \\ 3 & 4 \end{bmatrix}$ as a linear combination of the identity and Pauli matrices.

2 Hermitian Condition

Since each Pauli matrix σ_i is Hermitian, C will be Hermitian if and only if each coefficient γ_i is real:

$$\gamma_i \in \mathbb{R} \quad \text{for} \quad i = 0, 1, 2, 3.$$

This implies that Hermitian matrices in the Pauli basis correspond to real vectors in the four-dimensional vector space of 2×2 complex matrices.

3 Unitary Condition

For C to be unitary, it must satisfy $CC^\dagger = I$. This condition implies $\operatorname{tr}(CC^\dagger) = 2$, which is equivalent to:

$$\sum_{i=0}^{3} |\gamma_i|^2 = 1. \tag{12.12}$$

■ **Example 12.1** Let's decompose the unitary matrix $U = \frac{1}{\sqrt{2}} \begin{bmatrix} 1 & 1 \\ -1 & 1 \end{bmatrix}$ in the Pauli basis:

$$\gamma_0 = \frac{1}{2} \operatorname{tr}(U\sigma_0) = \frac{1}{2} \operatorname{tr}(U) = \frac{\sqrt{2}}{2},$$

$$\gamma_1 = \frac{1}{2} \operatorname{tr}(U\sigma_1) = \frac{1}{2} \operatorname{tr}\left(\frac{1}{\sqrt{2}} \begin{bmatrix} 1 & 1 \\ -1 & 1 \end{bmatrix} \begin{bmatrix} 0 & 1 \\ 1 & 0 \end{bmatrix} \right) = 0,$$

$$\gamma_2 = \frac{1}{2}\operatorname{tr}(U\sigma_2) = \frac{1}{2}\operatorname{tr}\left(\frac{1}{\sqrt{2}}\begin{bmatrix}1 & 1 \\ -1 & 1\end{bmatrix}\begin{bmatrix}0 & -i \\ i & 0\end{bmatrix}\right) = \frac{i\sqrt{2}}{2},$$

$$\gamma_3 = \frac{1}{2}\operatorname{tr}(U\sigma_3) = \frac{1}{2}\operatorname{tr}\left(\frac{1}{\sqrt{2}}\begin{bmatrix}1 & 1 \\ -1 & 1\end{bmatrix}\begin{bmatrix}1 & 0 \\ 0 & -1\end{bmatrix}\right) = 0.$$

Thus, we have:

$$U = \frac{\sqrt{2}}{2}\sigma_0 + \frac{i\sqrt{2}}{2}\sigma_2.$$

This expansion satisfies Eq. 12.12 because:

$$|\gamma_0|^2 + |\gamma_1|^2 + |\gamma_2|^2 + |\gamma_3|^2 = \left(\frac{\sqrt{2}}{2}\right)^2 + 0 + \left(\frac{\sqrt{2}}{2}\right)^2 + 0 = 1.$$

∎

However, Eq. 12.12 alone is insufficient to ensure the unitarity of C, as illustrated by the following example.

■ **Example 12.2** Consider

$$C = \sqrt{\frac{1}{2}}\,\sigma_0 + \sqrt{\frac{1}{2}}\,\sigma_1 = \sqrt{\frac{1}{2}}\begin{bmatrix}1 & 1 \\ 1 & 1\end{bmatrix}.$$

Clearly, $\left(\sqrt{\frac{1}{2}}\right)^2 + \left(\sqrt{\frac{1}{2}}\right)^2 = 1$. However,

$$CC^\dagger = \frac{1}{2}\begin{bmatrix}1 & 1 \\ 1 & 1\end{bmatrix}\begin{bmatrix}1 & 1 \\ 1 & 1\end{bmatrix} = \begin{bmatrix}1 & 1 \\ 1 & 1\end{bmatrix} \neq I.$$

∎

To derive the sufficient condition for unitarity, we expand $CC^\dagger$ in the Pauli matrix basis using Eq. 12.10:

$$CC^\dagger = \sum_{i=0}^{3}\sum_{j=0}^{3}\gamma_i\gamma_j^*\sigma_i\sigma_j$$

$$= \left(\sum_{i=0}^{3}|\gamma_i|^2\right)\sigma_0$$
$$+ \left(\gamma_0\gamma_1^* + \gamma_0^*\gamma_1 + i\gamma_2\gamma_3^* - i\gamma_2^*\gamma_3\right)\sigma_1$$
$$+ \left(\gamma_0\gamma_2^* + \gamma_0^*\gamma_2 + i\gamma_3\gamma_1^* - i\gamma_3^*\gamma_1\right)\sigma_2$$
$$+ \left(\gamma_0\gamma_3^* + \gamma_0^*\gamma_3 + i\gamma_1\gamma_2^* - i\gamma_1^*\gamma_2\right)\sigma_3. \tag{12.13}$$

To satisfy $CC^\dagger = I$, we require:

(a) The coefficient of $\sigma_0 \equiv I$ equals 1, which is ensured by Eq. 12.12.

(b) The coefficients of σ_1, σ_2, and σ_3 all vanish:

$$\gamma_0\gamma_1^* + \gamma_0^*\gamma_1 + i\gamma_2\gamma_3^* - i\gamma_2^*\gamma_3 = 0,$$
$$\gamma_0\gamma_2^* + \gamma_0^*\gamma_2 + i\gamma_3\gamma_1^* - i\gamma_3^*\gamma_1 = 0,$$
$$\gamma_0\gamma_3^* + \gamma_0^*\gamma_3 + i\gamma_1\gamma_2^* - i\gamma_1^*\gamma_2 = 0. \tag{12.14}$$

Thus, 2×2 unitary matrices form a subset of unit vectors in the complex vector space spanned by the Pauli matrices, constrained by Eq. 12.14.

12.1.3   Eigenvalues, Eigenvectors, and Diagonalization

Theorem 12.3 The eigenvalues of each Pauli matrix are -1 and 1. The corresponding eigenvectors, $|0\rangle$, $|1\rangle$, $|+\rangle$, $|-\rangle$, $|+_i\rangle$, and $|-_i\rangle$, are listed in Table 12.1.

The Pauli Z matrix is already diagonal. The Hadamard matrix H diagonalizes X to Z, and R_x diagonalizes Y to Z, where:

$$H = \frac{1}{\sqrt{2}} \begin{bmatrix} 1 & 1 \\ 1 & -1 \end{bmatrix}, \quad R_x = \frac{1}{\sqrt{2}} \begin{bmatrix} 1 & i \\ i & 1 \end{bmatrix}.$$

Pauli Matrix	Eigenvalue	Eigenvector	Diagonalization
$\sigma_3 \equiv Z$	1	$\|0\rangle = \begin{bmatrix} 1 & 0 \end{bmatrix}^T$	$IZI = Z$
	-1	$\|1\rangle = \begin{bmatrix} 0 & 1 \end{bmatrix}^T$	
$\sigma_1 \equiv X$	1	$\|+\rangle = \frac{1}{\sqrt{2}} (\|0\rangle + \|1\rangle)$	$HXH = Z$
	-1	$\|-\rangle = \frac{1}{\sqrt{2}} (\|0\rangle - \|1\rangle)$	
$\sigma_2 \equiv Y$	1	$\|+_i\rangle = \frac{1}{\sqrt{2}} (\|0\rangle + i\|1\rangle)$	$R_x^\dagger Y R_x = Z$
	-1	$\|-_i\rangle = \frac{1}{\sqrt{2}} (\|0\rangle - i\|1\rangle)$	

Table 12.1: Eigenvectors and Diagonalization of Pauli Matrices

(i) R_x can also be defined as $\frac{1}{\sqrt{2}} \begin{bmatrix} 1 & 1 \\ i & -i \end{bmatrix}$ because each normalized eigenvector is unique up to a global phase factor. This definition of R_x differs from the previous one only by a phase factor of i for the second eigenvector.

Exercise 12.5 Show that the Pauli Y matrix can be diagonalized as

$$Y = SHZHS^\dagger, \quad \text{or equivalently,} \quad Z = HS^\dagger Y SH,$$

where $S = \sqrt{Z}$ and H is the Hadamard matrix given above.

12.2   Representation of Spin-$\frac{1}{2}$ Particles

In quantum mechanics, Pauli matrices are used to represent the spin of a spin-$\frac{1}{2}$ particle, such as an electron, proton, or neutron. The spin of such a particle can be described as angular momentum along any direction in three-dimensional space. To understand how the Pauli matrices represent spin in any direction, we need to examine how they relate to the spin operators and spin states.

The transformation properties of spin states differ fundamentally from those of ordinary vectors under rotations. Because Pauli matrices are used to describe quantum states of particles with half-integer spin, they are often referred to as spin operators or "spinors" to distinguish them from regular vectors.

12.2.1   The Spin Operator in Any Direction

The spin operators for a spin-$\frac{1}{2}$ particle along the x, y, and z directions are represented by the Pauli matrices $\sigma_x \equiv \sigma_1$, $\sigma_y \equiv \sigma_2$, and $\sigma_z \equiv \sigma_3$, respectively. To represent spin in an arbitrary direction, we use the unit vector $\hat{n}$ to specify the direction. This vector can be expressed in terms of its components along the x, y, and z axes:

$$\hat{n} = (n_x, n_y, n_z), \quad \text{where } n_x^2 + n_y^2 + n_z^2 = 1. \tag{12.15}$$

In the spherical coordinate system, with polar angle θ and azimuthal angle ϕ,

$$\hat{n} = (\sin\theta\cos\phi, \ \sin\theta\sin\phi, \ \cos\theta). \tag{12.16}$$

The operator for the spin in the direction $\hat{n}$ is a linear combination of σ_x, σ_y, and σ_z, given by:

$$\sigma_{\hat{n}} = \hat{n} \cdot \boldsymbol{\sigma} = n_x\sigma_x + n_y\sigma_y + n_z\sigma_z. \tag{12.17}$$

Upon substitution, we obtain:

$$\begin{aligned}\sigma_{\hat{n}} &= \sin\theta\cos\phi\,\sigma_x + \sin\theta\sin\phi\,\sigma_y + \cos\theta\,\sigma_z \\ &= \begin{bmatrix} \cos\theta & \sin\theta e^{-i\phi} \\ \sin\theta e^{i\phi} & -\cos\theta \end{bmatrix}.\end{aligned} \tag{12.18}$$

Exercise 12.6 Verify that $\sigma_{\hat{n}}$ is Hermitian and unitary.

12.2.2   Spin States in Any Direction

A key discovery by Stern and Gerlach [1] is that a spin-$\frac{1}{2}$ particle oriented in any direction has two distinct states, usually referred to as the spin-up and spin-down states. These states correspond to the eigenvectors of $\sigma_{\hat{n}}$, associated with the eigenvalues ± 1.

The eigenvectors of $\sigma_{\hat{n}}$ can be found by solving the eigenvalue equation:

$$\sigma_{\hat{n}} \ket{\psi} = \lambda \ket{\psi}. \tag{12.19}$$

The eigenvector corresponding to spin-up ($\lambda = 1$) along $\hat{n}$ is:

$$\ket{\psi_+} = \cos\frac{\theta}{2}\ket{0} + e^{i\phi}\sin\frac{\theta}{2}\ket{1} = \begin{bmatrix} \cos\frac{\theta}{2} \\ \sin\frac{\theta}{2}e^{i\phi} \end{bmatrix}, \tag{12.20}$$

and the eigenvector corresponding to spin-down ($\lambda = -1$) along $\hat{n}$ is:

$$\ket{\psi_-} = \sin\frac{\theta}{2}\ket{0} - e^{i\phi}\cos\frac{\theta}{2}\ket{1} = \begin{bmatrix} \sin\frac{\theta}{2} \\ -\cos\frac{\theta}{2}e^{i\phi} \end{bmatrix}. \tag{12.21}$$

(i) The appearance of $\frac{\theta}{2}$ in Eq. 12.20 arises mathematically from solving for the eigenvectors of $\sigma_{\hat{n}}$ in Eq. 12.18. Physically, the $\frac{1}{2}$ factor is due to the fact that a spin-$\frac{1}{2}$ particle represents angular momentum with half-integer quantization.

Exercise 12.7 Verify that $\ket{\psi_+}$ and $\ket{\psi_-}$ correspond to the eigenvectors of σ_x, σ_y, and σ_z given in Table 12.1 at their corresponding polar and azimuthal angles.

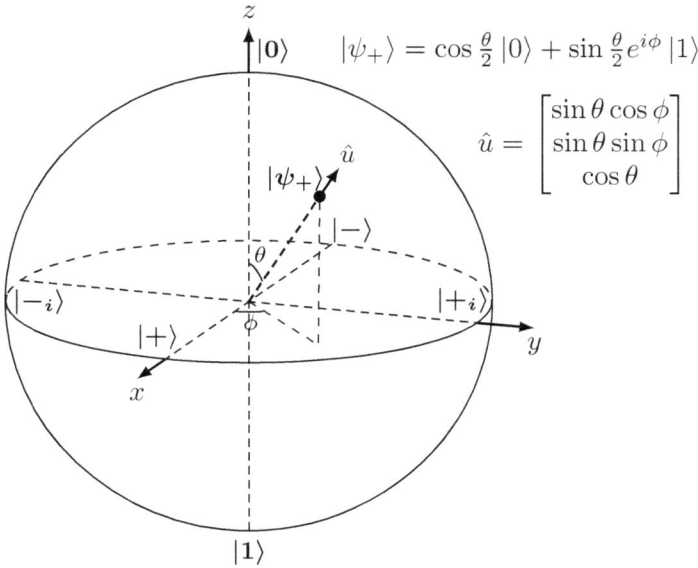

$$|\psi_+\rangle = \cos\frac{\theta}{2}|0\rangle + \sin\frac{\theta}{2}e^{i\phi}|1\rangle$$

$$\hat{u} = \begin{bmatrix} \sin\theta\cos\phi \\ \sin\theta\sin\phi \\ \cos\theta \end{bmatrix}$$

Figure 12.1: Representation of Spin States on the Bloch Sphere

12.2.3 The Bloch Sphere

The spin state $|\psi_+\rangle$ given in Eq. 12.20 represents a generic state specified by the polar angle θ and azimuthal angle ϕ. It is natural to map each point on the surface of a unit sphere, specified by these angles θ and ϕ, to a corresponding spin state $|\psi_+\rangle$, as shown in Fig. 12.1. Conversely, each possible spin state can be represented as a point on this sphere. This sphere, known as the Bloch Sphere, serves as an intuitive geometric representation for any spin state.

It is worth noting that orthogonal states are positioned at antipodal points on the sphere, separated by an angle of 180°, rather than the 90° separation one might intuitively expect. For instance, the orthogonal complement of a state $|\psi_+\rangle$ represented by (θ, ϕ), namely $|\psi_-\rangle$, is located at $(\pi - \theta, \pi + \phi)$ or equivalently at $(\theta + \pi, \phi)$. This arrangement reflects the unique property of quantum spin, where a full rotation requires a 720° turn to return to the original state, unlike the classical 360° rotation.

12.2.4 ✳ Expected Values

When we measure the particle's spin in the state $|\psi_+\rangle$ along another direction (e.g., along the x, y, or z-axis), we obtain the eigenvalues 1 or -1 randomly each time, indicating the quantized nature of spin. The average of many such measurements is given by the expected value of the form $\langle\psi|\sigma|\psi\rangle$. We can also consider a spin state aligned with the x, y, or z-axis while measuring along the direction $\hat{n}$.

The computed expected values for these scenarios are listed in Table 12.2. In each case, the spin behaves like a 3D vector: the measurement average corresponds to the projection of the spin vector in the measurement direction. Yet, the spin is always quantized into two distinct states. This ensures that in the macroscopic limit, quantum mechanics aligns with classical mechanics. This alignment arises naturally due to the mathematical properties of the Pauli matrices.

Spin Direction	Spin State	Measurement Direction	Expected Value
$\hat{n}$	$|\psi_+\rangle$	x-axis	$\langle\psi_+|\sigma_x|\psi_+\rangle = \sin\theta\cos\phi$
$\hat{n}$	$|\psi_+\rangle$	y-axis	$\langle\psi_+|\sigma_y|\psi_+\rangle = \sin\theta\sin\phi$
$\hat{n}$	$|\psi_+\rangle$	z-axis	$\langle\psi_+|\sigma_z|\psi_+\rangle = \cos\theta$
x-axis	$|+\rangle$	$\hat{n}$	$\langle+|\sigma_{\hat{n}}|+\rangle = \sin\theta\cos\phi$
y-axis	$|+_i\rangle$	$\hat{n}$	$\langle+_i|\sigma_{\hat{n}}|+_i\rangle = \sin\theta\sin\phi$
z-axis	$|0\rangle$	$\hat{n}$	$\langle0|\sigma_{\hat{n}}|0\rangle = \cos\theta$

Table 12.2: Expected Values of Spin Measurements

12.2.5 $*$ Exponentiation and Rotation Operators

1 Spin Rotation Operators

When a spin-$\frac{1}{2}$ particle undergoes a rotation, its spin state transforms according to an exponential of the Pauli matrices, referred to as a spin rotation operator.

For any real γ, we define the rotation operators about the x, y, and z-axis as

$$R_j(\gamma) = e^{-i\frac{\gamma}{2}\sigma_j}, \quad \text{where } j = 1, 2, 3. \tag{12.22}$$

Since $\sigma_j^2 = I$, according to the generalized Euler formula in Eq. 11.95,

$$R_j(\gamma) = \sigma_0 \cos\frac{\gamma}{2} - i\sigma_j \sin\frac{\gamma}{2}. \tag{12.23}$$

Explicitly,

$$R_x(\gamma) \equiv R_1(\gamma) = \begin{bmatrix} \cos\frac{\gamma}{2} & -i\sin\frac{\gamma}{2} \\ -i\sin\frac{\gamma}{2} & \cos\frac{\gamma}{2} \end{bmatrix}, \tag{12.24a}$$

$$R_y(\gamma) \equiv R_2(\gamma) = \begin{bmatrix} \cos\frac{\gamma}{2} & -\sin\frac{\gamma}{2} \\ \sin\frac{\gamma}{2} & \cos\frac{\gamma}{2} \end{bmatrix}, \tag{12.24b}$$

$$R_z(\gamma) \equiv R_3(\gamma) = \begin{bmatrix} e^{-i\frac{\gamma}{2}} & 0 \\ 0 & e^{i\frac{\gamma}{2}} \end{bmatrix}. \tag{12.24c}$$

Similarly, we can define the rotation operator about the axis $\hat{n}$ specified by the polar angle θ and azimuthal angle ϕ:

$$R_{\hat{n}}(\gamma) = e^{-i\frac{\gamma}{2}\sigma_{\hat{n}}} = \sigma_0 \cos\frac{\gamma}{2} - i\sigma_{\hat{n}} \sin\frac{\gamma}{2}. \tag{12.25}$$

2 Rotation from z-axis to $\hat{n}$

The spin-up state $|\psi_+\rangle$ along $\hat{n}$ at spherical angles (θ, ϕ) is given by Eq. 12.20. The spin-up state along the z-axis, corresponding to $\theta = \phi = 0$, is $|0\rangle$.

Geometrically, we expect the rotations from $|0\rangle$ to $|\psi_+\rangle$, and vice versa, to be governed by

$$|\psi_+\rangle = R_z(\phi)R_y(\theta)|0\rangle, \tag{12.26a}$$

$$|0\rangle = R_y(-\theta)R_z(-\phi)\,|\psi_+\rangle\,. \qquad (12.26b)$$

You can verify (as an exercise) mathematically that the above equations indeed hold.

> **Exercise 12.8** Formulate the rotation from $|0\rangle$ to $|\psi_-\rangle$ (spin-down along $\hat{n}$), and the rotation from $|+\rangle$ (spin-up along the x-axis) to $|0\rangle$, respectively.

3 SU(2) and SO(3) Revisited

The symmetry groups SU(2) and SO(3) were introduced in § 11.2.2 in the context of linear transformations. Here, we demonstrate why SU(2) acts as a double cover of SO(3).

In the definition of spin rotation operators in Eq. 12.22, the rotation angle is halved to $\frac{\gamma}{2}$. This adjustment ensures that the mathematics accurately reflects the geometric behavior of spin. (See the remark in § 12.2.2.)

As a result of this $\frac{1}{2}$ factor, when γ varies from $0°$ to $360°$, the spin state does not return to its initial configuration; instead, a full $720°$ rotation is required. This behavior implies that each rotation in SO(3) corresponds to two distinct elements in SU(2), denoted as U and $-U$, which represent the same physical rotation.

This $720°$ property is mathematically embedded in SU(2) through the use of half-angles, which trace a path that only closes after an additional $360°$. An analogy is to imagine walking on a Möbius strip: after traveling $360°$, you end up on the "opposite side" and need to complete another $360°$ to return to your starting point. Here, U and $-U$ are analogous to these two opposite points, corresponding to the same rotation in SO(3).

Thus, SU(2) is a "double cover" of SO(3), meaning that each physical rotation in three-dimensional space (SO(3)) has two representations in SU(2). This distinction is fundamental for spin-$\frac{1}{2}$ particles, whose states only return to their original orientation after a $720°$ rotation, unlike classical rotations, which reset after $360°$.

12.2.6 Summary

The key results for the spin operator (Pauli matrix) along an arbitrary direction are summarized in the following theorem. These properties enable the Pauli matrices to represent spin-$\frac{1}{2}$ particles.

> **Theorem 12.4** The spin operator along the direction
> $$\hat{n} = (n_x, n_y, n_z) = (\sin\theta\cos\phi,\ \sin\theta\sin\phi,\ \cos\theta)$$
> is given by:
> $$\sigma_{\hat{n}} = \hat{n}\cdot\boldsymbol{\sigma} = \sin\theta\cos\phi\,\sigma_x + \sin\theta\sin\phi\,\sigma_y + \cos\theta\,\sigma_z.$$
> Its eigenvectors for eigenvalues $+1$ and -1, respectively, are:
> $$|\psi_+\rangle = \cos\frac{\theta}{2}\,|0\rangle + e^{i\phi}\sin\frac{\theta}{2}\,|1\rangle,$$
> $$|\psi_-\rangle = \sin\frac{\theta}{2}\,|0\rangle - e^{i\phi}\cos\frac{\theta}{2}\,|1\rangle.$$

The expected values of $\sigma_{\hat{n}}$ coincide with the components of $\hat{n}$:

$$\langle\sigma_{\hat{n}}\rangle_x = \langle+|\sigma_{\hat{n}}|+\rangle = \sin\theta\cos\phi,$$
$$\langle\sigma_{\hat{n}}\rangle_y = \langle+_i|\sigma_{\hat{n}}|+_i\rangle = \sin\theta\sin\phi,$$
$$\langle\sigma_{\hat{n}}\rangle_z = \langle0|\sigma_{\hat{n}}|0\rangle = \cos\theta.$$

The rotation operator about the axis $\hat{n}$ is:

$$R_{\hat{n}}(\gamma) = e^{-i\frac{\gamma}{2}\sigma_{\hat{n}}} = \sigma_0\cos\frac{\gamma}{2} - i\sigma_{\hat{n}}\sin\frac{\gamma}{2}.$$

12.3 Pauli String Operators

Pauli string operators (or simply Pauli strings), along with the Pauli string basis and the Pauli group, are extensions of the Pauli matrices and form essential concepts in quantum information theory and quantum computing. While Pauli matrices operate on a two-dimensional complex vector space, $\mathbb{C}^2$, Pauli strings act on a 2^n-dimensional complex space, $\mathbb{C}^{2^n}$. In quantum computing, a two-dimensional system is referred to as a qubit. Thus, Pauli matrices act on single qubits, while Pauli strings are used for multi-qubit systems.

12.3.1 Definitions and Notations

A Pauli string operator, or simply Pauli string, is a tensor product (refer to § 10.2) of Pauli matrices, used to describe multi-qubit quantum states and operations.

Definition 12.2 — Pauli String Operator. A Pauli string operator for n qubits is defined as:

$$\boldsymbol{P} = \bigotimes_{j=1}^{n} P_j = P_1 \otimes P_2 \otimes \cdots \otimes P_n, \tag{12.27}$$

where each P_j is one of the Pauli matrices or the identity matrix I, i.e., $P_j \in \{I, X, Y, Z\}$. These Pauli strings represent $2^n \times 2^n$ complex matrices.

For brevity, we often drop the $\otimes$ symbol and write

$$\boldsymbol{P} = P_1 P_2 \cdots P_n, \tag{12.28}$$

implicitly indicating that P_j acts on qubit j.

Additionally, we use the notation $\boldsymbol{P} \in \{I, X, Y, Z\}^{\otimes n}$.

The definition and properties of the Pauli matrices are provided in § 12.1.1 and replicated here for convenience:

$$I = \begin{bmatrix} 1 & 0 \\ 0 & 1 \end{bmatrix}, \quad X = \begin{bmatrix} 0 & 1 \\ 1 & 0 \end{bmatrix}, \quad Y = \begin{bmatrix} 0 & -i \\ i & 0 \end{bmatrix}, \quad Z = \begin{bmatrix} 1 & 0 \\ 0 & -1 \end{bmatrix}.$$

(i) Bold capital symbols, such as $\boldsymbol{P}$, are typically used to represent Pauli strings to distinguish them from their components, the Pauli matrices.

In the context of Pauli strings, ZY (for example) represents the tensor product $Z \otimes Y$ rather than the matrix product of Z and Y.

For example, for two qubits, a Pauli string might be written as $X \otimes I$ or $Z \otimes Y$, or simply XI or ZY. Specifically,

$$X \otimes Y = \begin{bmatrix} 0 & Y \\ Y & 0 \end{bmatrix} = \begin{bmatrix} 0 & 0 & 0 & -i \\ 0 & 0 & i & 0 \\ 0 & -i & 0 & 0 \\ i & 0 & 0 & 0 \end{bmatrix}.$$

12.3.2 Basic Properties

Theorem 12.5 Pauli strings exhibit several key properties inherited from Pauli matrices:

1. Hermiticity: $\boldsymbol{P}^\dagger = \boldsymbol{P}$.

2. Unitarity: $\boldsymbol{P}^\dagger \boldsymbol{P} = I$.

3. Involutiveness: $\boldsymbol{P}^2 = I$.

4. Zero Trace: $\operatorname{tr} \boldsymbol{P} = 0$ for non-identity Pauli strings.

5. Eigenvalues: ± 1.

Proof.

1. Hermiticity: Each Pauli matrix is Hermitian, and $(A \otimes B)^\dagger = A^\dagger \otimes B^\dagger$.

2. Unitarity: Each Pauli matrix is unitary, and the tensor product of unitary matrices is also unitary:

$$(U \otimes V)^\dagger (U \otimes V) = (U^\dagger U) \otimes (V^\dagger V) = I \otimes I = I.$$

3. Involutiveness: $\boldsymbol{P}^2 = I$, as $\boldsymbol{P}$ is Hermitian and unitary.

4. Zero Trace: For non-identity Pauli strings, $\operatorname{tr}(A \otimes B) = \operatorname{tr}(A)\operatorname{tr}(B)$, and $\operatorname{tr}(P_j) = 0$ for $P_j \in \{X, Y, Z\}$.

5. Eigenvalues: The eigenvalues of Pauli matrices are ± 1, and tensor products of such operators retain this property.

$\square$

12.3.3 Products of Pauli Strings

Consider two Pauli strings of length n, each representing a $2^n \times 2^n$ matrix:

$$\boldsymbol{P} = P_1 P_2 \cdots P_n \equiv P_1 \otimes P_2 \otimes \cdots \otimes P_n,$$
$$\boldsymbol{Q} = Q_1 Q_2 \cdots Q_n \equiv Q_1 \otimes Q_2 \otimes \cdots \otimes Q_n.$$

Their product is given by:

$$\boldsymbol{PQ} = (P_1 P_2 \cdots P_n)(Q_1 Q_2 \cdots Q_n)$$
$$\equiv (P_1 \otimes P_2 \otimes \cdots \otimes P_n)(Q_1 \otimes Q_2 \otimes \cdots \otimes Q_n). \tag{12.29}$$

Because P_i and Q_i both act on qubit i only, this product can be expressed as:

$$PQ = (P_1Q_1) \otimes (P_2Q_2) \otimes \cdots \otimes (P_nQ_n), \qquad (12.30)$$

where P_jQ_j represents a regular matrix product instead of a tensor product. This in turn can be expressed succinctly as:

$$PQ = \bigotimes_{j=1}^{n} P_jQ_j. \qquad (12.31)$$

Pauli string products can be further simplified using the properties (see § 12.1.1.2):

$$X^2 = Y^2 = Z^2 = I, \qquad (12.32a)$$

$$XY = -YX = iZ, \quad YZ = -ZY = iX, \quad ZX = -XZ = iY. \qquad (12.32b)$$

■ Example 12.3

$$(X \otimes Z \otimes I \otimes Y)(X \otimes Y \otimes Z \otimes I) = (XX) \otimes (ZY) \otimes (IZ) \otimes (YI)$$
$$= I \otimes (-iX) \otimes Z \otimes Y$$
$$= -iI \otimes X \otimes Z \otimes Y.$$

This is often shorthanded as:

$$(XZIY)(XYZI) = -i\,IXZY.$$

If we reverse the order of the product, we find:

$$(XYZI)(XZIY) = i\,IXZY.$$

Similarly,

$$(XZIYY)(XYZIX) = -IXZYZ,$$
$$(XYZIX)(XZIYY) = -IXZYZ.$$

■

From this example, we observe a general pattern for Pauli string products:

Lemma 12.6 For any pair of corresponding Pauli matrices P_j and Q_j:

1. If $P_j = Q_j$, their product is I, with a phase factor of $+1$.

2. If either P_j or Q_j is I, their product is Q_j or P_j, with a phase factor of $+1$.

3. If P_j and Q_j are distinct and non-identity, their product is the third Pauli matrix with a phase factor of $\pm i$. For instance, $XY = iZ$ and $YX = -iZ$.

Noticing that in the first two cases P_j and Q_j commute, while in the last case, P_j and Q_j anti-commute, we arrive at the following rules:

Theorem 12.7 If the number of anti-commuting pairs of distinct elements in $\{X, Y, Z\}$ in PQ is α:

> 1. If α is odd, the phase factor of PQ is $\pm i$; if α is even, it is ± 1. The sign depends on the count as well as the order inside these pairs.
>
> 2. PQ and QP differ only by an overall sign. If α is odd, one product has a phase of $+i$ and the other $-i$; if α is even, both have the same phase.

Exercise 12.9 Calculate $(ZZIYXIZ)(XYZIYXZ)$.

12.3.4 Commutation Relations

Understanding the commutation relations of Pauli strings is fundamental for analyzing quantum circuits, designing error correction codes, and formulating Hamiltonians for quantum simulations.

By definition, two Pauli strings P and Q commute if $PQ = QP$, and anti-commute if $PQ = -QP$. The commutation property of Pauli strings can be stated as follows:

Theorem 12.8 Two Pauli strings either commute or anti-commute. This follows from the multiplication rule developed earlier: PQ and QP differ only by an overall sign, which depends on the number of anti-commuting pairs. Specifically:

- P and Q commute if the number of positions j where P_j and Q_j anti-commute is even, as PQ and QP then have the same sign.

- P and Q anti-commute if the number of positions j where P_j and Q_j anti-commute is odd, as PQ and QP then have opposite signs.

■ **Example 12.4** Consider two Pauli strings $P = X \otimes Z \otimes I$ and $Q = Z \otimes X \otimes Y$. There are two anti-commuting pairs:

- X and Z anti-commute: $XZ = -ZX$.

- Z and X anti-commute: $ZX = -XZ$.

- I and Y commute.

Since the count of anti-commuting pairs is even, P and Q commute:

$$(X \otimes Z \otimes I)(Z \otimes X \otimes Y) = (Z \otimes X \otimes Y)(X \otimes Z \otimes I).$$

■

Exercise 12.10 Determine whether $ZZIYXIZ$ and $XYZIYXZ$ commute or anti-commute.

Exercise 12.11 Provide an example where Pauli strings P_1 and Q_1 anti-commute, and P_2 and Q_2 also anti-commute, yet the products P_1Q_1 and P_2Q_2 commute.

12.3.5 Pauli Exponentials

Pauli exponentials refer to exponentials of Pauli strings. Due to the involutory property of Pauli strings, $P^2 = I$, these exponentials simplify according to the generalized Euler formula in § 11.3.3.2:

$$e^{i\alpha P} = \cos \alpha\, I + i \sin \alpha\, P, \tag{12.33}$$

where α is a real scalar.

Exercise 12.12 Show that if P and Q commute, then $e^{i\alpha P}Q = Qe^{i\alpha P}$.

Exercise 12.13 Show that if P and Q anti-commute, then $e^{i\alpha P}Q = Qe^{-i\alpha P}$.

If P and Q commute, the Baker-Campbell-Hausdorff (BCH) formula (Eq. 11.98) allows the following simplification:

$$e^{i(\alpha P + \beta Q)} = e^{i\alpha P}e^{i\beta Q}. \tag{12.34}$$

For anti-commuting Pauli strings and small α and β, the exponential of their sum can be approximated using the first-order BCH formula:

$$e^{i(\alpha P + \beta Q)} \approx e^{i\alpha P}e^{i\beta Q}e^{-\frac{1}{2}[i\alpha P, i\beta Q]}. \tag{12.35}$$

Since $[P, Q] = 2PQ$, this becomes:

$$e^{i(\alpha P + \beta Q)} \approx e^{i\alpha P}e^{i\beta Q}e^{\alpha\beta PQ}. \tag{12.36}$$

12.3.6 Diagonalization

As discussed in § 12.1.3, the Pauli X matrix can be diagonalized using $H = \frac{1}{\sqrt{2}}\begin{bmatrix} 1 & 1 \\ 1 & -1 \end{bmatrix}$, and the Pauli Y matrix using $R_x = \frac{1}{\sqrt{2}}\begin{bmatrix} 1 & i \\ i & 1 \end{bmatrix}$:

$$H^\dagger X H = Z, \tag{12.37a}$$

$$R_x^\dagger Y R_x = Z. \tag{12.37b}$$

The identity matrix I and the Pauli Z matrix are already diagonal. Therefore:

Theorem 12.9 A Pauli string P can be diagonalized using a unitary matrix U:

$$U^\dagger P U = D, \tag{12.38}$$

where D is a tensor product of I's and Z's, obtained from P by replacing X and Y components with Z, and U is constructed by replacing X with H, Y with R_x, and Z with I in P.

■ **Example 12.5** Consider $P = ZXXIY$. It can be diagonalized with $U = IHHIR_x$. The resulting diagonal matrix is $D = ZZZIZ$.   ■

Since matrix exponentials satisfy $e^{U^\dagger A U} = U^\dagger e^A U$, the same U diagonalizes e^P and $e^{i\alpha P}$:

$$U^\dagger e^{i\alpha P} U = e^{i\alpha D}. \tag{12.39}$$

Exercise 12.14 Diagonalize $e^{i\alpha P}$ where $P = ZXIYX$.

12.3.7 ✳ Matrix Representations

For an n-qubit system, a Pauli string $P = P_1 \otimes P_2 \otimes \cdots \otimes P_n$ can be represented as a $2^n \times 2^n$ sparse matrix. The matrix representation of Pauli strings reveals distinct patterns of non-zero elements, each associated with a specific type of Pauli string. These patterns are essential for understanding the behavior of multi-qubit systems, quantum error correction, and the implementation of quantum algorithms.

Background Information: Permutation Matrices

A permutation matrix P is a square binary matrix (containing only 0s and 1s) that has exactly one entry of 1 in each row and each column, with all other entries being 0. Its key properties include:

1. Permutation matrices are orthogonal: $P^T = P^{-1}$.

2. The determinant of a permutation matrix is either $+1$ or -1, depending on whether the permutation is even or odd.

3. When a permutation matrix multiplies another matrix or vector, it permutes its rows or columns.

For example, consider the permutation matrix

$$P = \begin{bmatrix} 0 & 1 & 0 \\ 1 & 0 & 0 \\ 0 & 0 & 1 \end{bmatrix}.$$

For a vector v, multiplying by P gives:

$$Pv = \begin{bmatrix} 0 & 1 & 0 \\ 1 & 0 & 0 \\ 0 & 0 & 1 \end{bmatrix} \begin{bmatrix} v_1 \\ v_2 \\ v_3 \end{bmatrix} = \begin{bmatrix} v_2 \\ v_1 \\ v_3 \end{bmatrix}.$$

This operation swaps v_1 and v_2, leaving v_3 unchanged.

The matrix representation of P has the following properties:

1. Monomial: The matrix of P has exactly one non-zero element per row and per column, with values in $\{\pm 1, \pm i\}$. This type of matrix is known as a monomial matrix, which generalizes permutation matrices by allowing non-zero elements to be any non-zero value, not just 1. This property arises because $P_i \otimes P_j$ has non-zero elements only at positions determined by the non-zero elements of P_i and P_j.

2. Patterns of Non-Zero Element Positions: Interchanging $I \leftrightarrow Z$ and $X \leftrightarrow Y$ in a Pauli string retains the positions of the non-zero elements while altering only their values. This holds because I and Z share the same pattern of non-zero elements, as do X and Y.

■ **Example 12.6** Consider two-qubit Pauli strings:

$$P = I \otimes X = \begin{bmatrix} 1 & 0 \\ 0 & 1 \end{bmatrix} \otimes \begin{bmatrix} 0 & 1 \\ 1 & 0 \end{bmatrix} = \begin{bmatrix} 0 & 1 & 0 & 0 \\ 1 & 0 & 0 & 0 \\ 0 & 0 & 0 & 1 \\ 0 & 0 & 1 & 0 \end{bmatrix}$$

$$Q = Z \otimes Y = \begin{bmatrix} 1 & 0 \\ 0 & -1 \end{bmatrix} \otimes \begin{bmatrix} 0 & -i \\ i & 0 \end{bmatrix} = \begin{bmatrix} 0 & -i & 0 & 0 \\ i & 0 & 0 & 0 \\ 0 & 0 & 0 & i \\ 0 & 0 & -i & 0 \end{bmatrix}$$

Notice that the non-zero elements of $I \otimes X$ and $Z \otimes Y$ share the same structure, differing only in values. This demonstrates that interchanging $I \leftrightarrow Z$ and $X \leftrightarrow Y$ in a Pauli string retains the positions of non-zero elements, altering only their values. ■

Sparsity Patterns in Pauli String Matrices

The non-zero patterns in the $2^n \times 2^n$ matrix of P are invariant under $I \leftrightarrow Z$ and $X \leftrightarrow Y$. This property allows us to define the matrix sparsity patterns of Pauli strings. Specifically, the positions of the non-zero elements are determined by the tensor product structure, with each pattern having a unique, non-intersecting configuration of non-zero elements in the matrix representation.

■ **Example 12.7** Consider $P = I \otimes X \otimes X \otimes I \otimes I$. The 0-based index (i.e., starting from 0) of the non-zero element in the *first column* is $(01100)_2$, which is 12 in decimal. (Work out the first column of the 32×32 matrix P to visualize this.)

The Pauli string $Z \otimes Y \otimes X \otimes Z \otimes I$ has the same non-zero pattern. Thus, they belong to the same type. ■

The first column of P defines its pattern. By examining the indices (in binary notation) of the non-zero elements in the first column of P, we can determine the pattern of the Pauli string.

Specifically, if the non-zero element of the first column of P is at position k (1-based index), then converting $k - 1$ to binary (with a length of n) reveals the pattern: the 1's in $k - 1$ correspond to the X's and Y's in the Pauli string, and the 0's correspond to the I's and Z's.

For an n-qubit system, each Pauli string corresponds to a $2^n \times 2^n$ matrix, and there are a total of 4^n unique matrices because there are four choices (I, X, Y, Z) for each of the n qubits. Each matrix P has 2^n non-zero elements. There are 2^n distinct Pauli string patterns, with 2^n Pauli strings sharing the same pattern.

By grouping Pauli strings into types based on their matrix sparsity patterns, we can compress the storage of their matrices and achieve more efficient computations.

The value of the non-zero element in the first column of P, denoted p_1, depends on the number of Y's in the Pauli string, y: $p_1 = i^y$. Note that the values of non-zero elements in other columns may differ and can change under $I \leftrightarrow Z$ and $X \leftrightarrow Y$.

Exercise 12.15 For a 16-qubit system, the non-zero element of the first column of P is at position 1022 (1-based decimal). Provide four examples of P such that the non-zero element is 1, -1, i, and $-i$, respectively.

Hint: The binary of 1021 is $(0000001111111101)_2$. One possible Pauli string is $ZZZZZZXXXXXXXIX$.

12.3.8 * Applications of Pauli Strings

Pauli strings are used extensively in quantum computing and quantum information theory. They play a critical role in various aspects of quantum algorithms, error correction, and the representation of quantum states and operators. Below are some key applications of Pauli strings, which we will explore in more detail in the following subsections.

Basis for Operators on Multiple Qubits: Pauli strings form a basis for the space of operators acting on multiple qubits. Any operator on a system of n qubits can be expressed as a linear combination of Pauli strings.

Representation of Quantum Operations: In quantum algorithms, Pauli strings provide a convenient way to represent quantum operations. They describe the action of quantum gates and circuits on quantum states. This includes describing transformations, rotations, and other operations on qubit states through sequences of Pauli operators.

Hamiltonian Representation: In quantum mechanics, the Hamiltonian of a system, which describes its total energy, can often be expressed as a sum of Pauli strings. This decomposition is especially useful in quantum simulations where breaking down the Hamiltonian into simpler parts enables more efficient computation.

Quantum Error Correction: In quantum error correction, Pauli strings are used to represent errors and the corrective operations. Stabilizer codes, for instance, employ Pauli strings to define the stabilizers of the code space, which helps detect and correct errors in a robust manner.

Measurement and Observables: In quantum computing, measurements are frequently performed in the basis of Pauli operators. Observables in quantum algorithms are often decomposed into sums of Pauli strings, enabling efficient measurement of their expected values with respect to a quantum state.

12.4 Pauli String Basis

(i) Although this topic is not essential for studying the second book, *Quantum Computing and Information*, it will become important in the third book, *Quantum Algorithms and Applications*. Readers may defer studying it until encountering the relevant quantum algorithms.

Just as the Pauli matrices $\{I, X, Y, Z\}$ form a complete orthogonal basis in $\mathbb{C}^{2\times2}$ (see § 12.1.2), we have the following result:

Theorem 12.10 The 4^n distinct Pauli strings for n qubits form a complete orthogonal basis for matrices in $\mathbb{C}^{2^n \times 2^n}$, denoted as:

$$\mathcal{B}_n = \{I, X, Y, Z\}^{\otimes n}. \tag{12.40}$$

Below is an example illustrating this theorem, with the proof provided in the next subsection.

■ **Example 12.8** For a 2-qubit system, the Pauli string basis is:

$$\mathcal{B}_2 = \{II, IX, IY, IZ, XI, XX, XY, XZ, YI, YX, YY, YZ, ZI, ZX, ZY, ZZ\}.$$

Note that each element in this basis set is a tensor product, e.g., $XY \equiv X \otimes Y$. A 4×4 matrix has 16 elements, equal to the number of elements in the set. ■

12.4.1 Orthogonality and Completeness

The complex matrices in $\mathbb{C}^{2^n \times 2^n}$ form an inner product space under the Hilbert-Schmidt inner product (refer to § 11.1.2):

$$\langle A, B \rangle \equiv A \cdot B = \text{tr}(A^\dagger B). \tag{12.41}$$

1 Orthogonality

The Pauli string basis $\mathcal{B}_n$ is orthogonal, meaning that for any $P_i, P_j \in \mathcal{B}_n$:

$$P_i \cdot P_j \equiv \text{tr}(P_i^\dagger P_j) = 2^n \delta_{ij}, \text{ where } i, j = 1, 2, \ldots, 4^n. \tag{12.42}$$

Proof. Consider two Pauli strings P_i and P_j:

$$P_i = P_{\alpha_1} \otimes P_{\alpha_2} \otimes \cdots \otimes P_{\alpha_n},$$

$$P_j = P_{\beta_1} \otimes P_{\beta_2} \otimes \cdots \otimes P_{\beta_n}.$$

The Hilbert-Schmidt inner product of P_i and P_j is:

$$\text{tr}(P_i^\dagger P_j) = \text{tr}\left((P_{\alpha_1} \otimes P_{\alpha_2} \otimes \cdots \otimes P_{\alpha_n})^\dagger (P_{\beta_1} \otimes P_{\beta_2} \otimes \cdots \otimes P_{\beta_n}) \right).$$

Since each Pauli matrix is Hermitian, $P_i^\dagger = P_i$. Therefore:

$$\text{tr}(P_i P_j) = \text{tr}\left((P_{\alpha_1} \otimes P_{\alpha_2} \otimes \cdots \otimes P_{\alpha_n})(P_{\beta_1} \otimes P_{\beta_2} \otimes \cdots \otimes P_{\beta_n}) \right).$$

Using the properties of the tensor product and the trace operation, this simplifies to:

$$\text{tr}(P_i P_j) = \text{tr}(P_{\alpha_1} P_{\beta_1}) \text{tr}(P_{\alpha_2} P_{\beta_2}) \cdots \text{tr}(P_{\alpha_n} P_{\beta_n}).$$

Each term $\text{tr}(P_{\alpha_k} P_{\beta_k})$ evaluates as follows:

$$\text{tr}(P_{\alpha_k} P_{\beta_k}) = \begin{cases} 2 & \text{if } \alpha_k = \beta_k, \\ 0 & \text{if } \alpha_k \neq \beta_k. \end{cases}$$

Consequently, the product is non-zero only if $\alpha_k = \beta_k$ for all k. If $P_i = P_j$, then:

$$\text{tr}(P_i P_j) = 2^n.$$

If $P_i \neq P_j$, then there exists at least one k such that $\alpha_k \neq \beta_k$, which implies:

$$\text{tr}(P_i P_j) = 0.$$

Thus, we have shown:

$$\text{tr}(P_i P_j) = 2^n \delta_{ij}.$$

$\square$

2 Completeness

For n qubits, the basis set $\mathcal{B}_n$ consists of 4^n unique matrices, one for each possible combination of Pauli matrices (I, X, Y, Z) acting on the n qubits. These matrices are linearly independent, meaning that no Pauli string in the basis can be expressed as a linear combination of the others. Since the dimension of the space of $2^n \times 2^n$ matrices is 4^n, and we have 4^n linearly independent matrices, the Pauli string basis is complete, spanning the entire vector space of $2^n \times 2^n$ matrices.

3   Normalization

The Pauli string basis $\mathcal{B}_n$ as defined above is not normalized, which aligns with conventions in quantum computing literature. We could normalize each basis element by dividing it by $\sqrt{2^n}$, but the un-normalized version has practical advantages. Notably, each Pauli string, being a tensor product of Pauli matrices, retains properties (such as unitarity) that are advantageous for computations.

12.4.2   Matrix Expansion

> **Theorem 12.11** Any $2^n \times 2^n$ complex matrix A can be expanded in terms of the Pauli string basis as:
>
> $$A = \sum_{i=1}^{4^n} c_i \boldsymbol{P}_i, \tag{12.43}$$
>
> where c_i are the complex coefficients given by:
>
> $$c_i = \frac{1}{2^n} \langle \boldsymbol{P}_i, A \rangle = \frac{1}{2^n} \operatorname{tr}(\boldsymbol{P}_i^\dagger A) = \frac{1}{2^n} \operatorname{tr}(\boldsymbol{P}_i A) = \frac{1}{2^n} \operatorname{tr}(A \boldsymbol{P}_i). \tag{12.44}$$

Proof.   To prove Eq. 12.44, we take the inner product $\langle \boldsymbol{P}_i, A \rangle$. On the right-hand side, only the term associated with $\boldsymbol{P}_i$ is non-zero, as all other terms vanish due to the orthogonality condition. Therefore,

$$\langle \boldsymbol{P}_i, A \rangle = \langle \boldsymbol{P}_i, c_i \boldsymbol{P}_i \rangle = c_i \langle \boldsymbol{P}_i, \boldsymbol{P}_i \rangle = 2^n c_i. \tag{12.45}$$

In Eq. 12.44, we have used the Hermitian property of $\boldsymbol{P}_i$.   □

Combining Eq. 12.43 and Eq. 12.44, we obtain the following identity:

$$A = \frac{1}{2^n} \sum_{i=1}^{4^n} \operatorname{tr}(A \boldsymbol{P}_i) \, \boldsymbol{P}_i. \tag{12.46}$$

We can also express the expansion equation Eq. 12.43 in terms of individual Pauli matrices in the Pauli string:

$$A = \sum_{\alpha_1, \alpha_2, \ldots, \alpha_n} c_{\alpha_1 \alpha_2 \cdots \alpha_n} \left(P_{\alpha_1} \otimes P_{\alpha_2} \otimes \cdots \otimes P_{\alpha_n} \right), \tag{12.47}$$

where $c_{\alpha_1 \alpha_2 \cdots \alpha_n}$ are the complex coefficients corresponding to each combination of $\alpha_1, \alpha_2, \ldots, \alpha_n$, and $\alpha_i \in \{0, 1, 2, 3\}$ corresponds to $P_{\alpha_i} \in \{I, X, Y, Z\}$.

The coefficients $c_{\alpha_1 \alpha_2 \cdots \alpha_n}$ can be calculated as:

$$c_{\alpha_1 \alpha_2 \cdots \alpha_n} = \frac{1}{2^n} \operatorname{tr} \left(A \left(P_{\alpha_1} \otimes P_{\alpha_2} \otimes \cdots \otimes P_{\alpha_n} \right) \right). \tag{12.48}$$

Exercise 12.16 Expand the following matrix in the Pauli string basis:

$$\text{CNOT} = \begin{bmatrix} I & 0 \\ 0 & X \end{bmatrix} = \begin{bmatrix} 1 & 0 & 0 & 0 \\ 0 & 1 & 0 & 0 \\ 0 & 0 & 0 & 1 \\ 0 & 0 & 1 & 0 \end{bmatrix}.$$

12.4.3   Expansion of Hermitian and Unitary Matrices

1   Hermitian Matrices

Theorem 12.12 In the expansion of matrix A in the Pauli string basis given by Eq. 12.43, if A is Hermitian, then all the coefficients $\{c_i\}$ are real.

Proof.   Take the conjugate transpose of both sides of the expansion:

$$A^\dagger = \left(\sum_i c_i P_i \right)^\dagger = \sum_i c_i^* P_i^\dagger.$$

Since Pauli strings are Hermitian ($P_i^\dagger = P_i$), we have:

$$A^\dagger = \sum_i c_i^* P_i.$$

Since A is Hermitian ($A = A^\dagger$), we get:

$$\sum_i c_i P_i = \sum_i c_i^* P_i.$$

By the linear independence of Pauli strings, the coefficients of corresponding Pauli strings on both sides must be equal:

$$c_i = c_i^* \quad \text{(for all } i\text{)}.$$

This implies that the imaginary part of each coefficient c_i is zero. Therefore, all coefficients c_i are real numbers.   □

2   Unitary Matrices

Theorem 12.13 In the expansion of matrix A in the Pauli string basis given by Eq. 12.43, if A is unitary, then:

$$\sum_i |c_i|^2 = 1.$$

Proof.   Substitute the expansion of A into the unitarity condition $A^\dagger A = I$:

$$\left(\sum_i c_i^* P_i^\dagger \right) \left(\sum_j c_j P_j \right) = I.$$

Since Pauli strings are Hermitian ($P_i^\dagger = P_i$), this simplifies to:

$$\sum_i \sum_j c_i^* c_j P_i P_j = I.$$

Due to the orthogonality and unitarity of Pauli strings ($P_i P_j = 0$ for $i \neq j$, and $P_i^2 = I$), we get:

$$\sum_i |c_i|^2 I = I.$$

Comparing coefficients, we find:

$$\sum_i |c_i|^2 = 1.$$

$\square$

Note that the converse does not hold: satisfying $\sum_i |c_i|^2 = 1$ alone does not guarantee that A is unitary. An example illustrating this is given in § 12.1.2.3.

Exercise 12.17 Expand $U = H \otimes R_x$ in the Pauli basis, where $H = \frac{1}{\sqrt{2}} \begin{bmatrix} 1 & 1 \\ 1 & -1 \end{bmatrix}$ is the Hadamard matrix, and $R_x = \frac{1}{\sqrt{2}} \begin{bmatrix} 1 & i \\ i & 1 \end{bmatrix}$ is the x-rotation matrix.

12.4.4 * Applications of Pauli String Basis

In quantum computing, expanding the Hamiltonian (a Hermitian matrix representing system energy) in the Pauli string basis provides several advantages. Pauli strings can be directly translated into quantum gates, the building blocks of quantum circuits. This makes it possible to implement the Hamiltonian on a quantum computer. In addition, many quantum algorithms are formulated using Pauli strings. Expressing the Hamiltonian in this basis is essential for applying these algorithms. Mastery of Pauli strings and the Pauli basis is crucial for understanding these advanced topics. Below are a few examples:

1. Molecular Hamiltonians: The electronic structure of molecules is described by quantum mechanical Hamiltonians that include terms for the kinetic energy of electrons, the attraction between electrons and nuclei, and the repulsion between electrons. These Hamiltonians can be complex, but they can be systematically expressed in the Pauli string basis using techniques like the Jordan-Wigner or Bravyi-Kitaev transformations. This representation is crucial for simulating molecular systems on quantum computers, as it allows us to map the Hamiltonian onto qubit operations.

2. Quantum Phase Estimation (QPE): QPE is a powerful quantum algorithm used to estimate the ground state energy of a Hamiltonian (such as a molecular Hamiltonian). Expressing the Hamiltonian in the Pauli string basis is essential for applying QPE, as it enables us to construct the necessary quantum circuits for the algorithm.

3. Spin Hamiltonians: Many materials exhibit interesting magnetic properties due to the interactions between electron spins. These interactions are described by spin Hamiltonians, which are naturally expressed in terms of Pauli matrices.

Various other problems, such as logistics optimization and financial modeling, can be mapped to spin Hamiltonian problems. The Pauli string basis facilitates their analysis and simulation on quantum computers.

4. Hubbard Model: The Hubbard model is a simplified model used to study strongly correlated electron systems. It can be written in terms of fermionic creation and annihilation operators, which can be mapped to Pauli strings using transformations like Jordan-Wigner. This allows for the simulation of the Hubbard model on quantum computers.

12.5 Pauli Groups

In addition to Pauli string operators and bases, Pauli groups are another central concept related to Pauli matrices. While Pauli strings focus on representing quantum states as linear combinations of tensor products of Pauli operators, Pauli groups deal with the algebraic structure formed by these operators under multiplication. Pauli groups are foundational in quantum error correction, where they characterize errors and help design codes to protect quantum information. They are also essential in the stabilizer formalism, a powerful framework for describing quantum states and operations.

Although this topic is not essential for studying the second book, Quantum Computing and Information, it will become important in the third book, Quantum Algorithms and Applications. Readers may defer studying it until encountering the relevant quantum algorithms.

12.5.1 Definition

The concept of groups has been introduced in § 4.2. Essentially, a group is a set of elements equipped with an operation that combines any two elements to form a third element within the set. This operation must obey specific rules: it must be associative (the order of combining elements does not matter), there must be an identity element (which does not change other elements when combined), and each element must have an inverse (which, when combined with the original element, yields the identity element).

The Pauli group on one qubit, $\mathcal{P}_1$, is defined as the set of matrices formed by the Pauli matrices multiplied by ± 1 and $\pm i$:

$$\mathcal{P}_1 = \{\pm I, \pm X, \pm Y, \pm Z, \pm iI, \pm iX, \pm iY, \pm iZ\}. \qquad (12.49)$$

The operation in the group is ordinary matrix multiplication.

The factors $\{\pm 1, \pm i\}$ are included because the product of two Pauli matrices may yield an additional factor of ± 1 or $\pm i$, depending on their commutation or anti-commutation properties. This ensures that the group is closed under multiplication.

Definition 12.3 — Pauli Group. The Pauli group on n qubits, denoted as $\mathcal{P}_n$, is the group formed by Pauli strings (the tensor products of Pauli matrices) with the multiplicative factors $\{\pm 1, \pm i\}$. The elements of the Pauli group are thus of the form:

$$\mathcal{P}_n = \{(\pm 1, \pm i)P_1 \otimes P_2 \otimes \cdots \otimes P_n \mid P_i \in \{I, X, Y, Z\}\}. \qquad (12.50)$$

Alternatively, this can be written as:

$$\mathcal{P}_n = \{\pm 1, \pm i\}\{I, X, Y, Z\}^{\otimes n}, \tag{12.51}$$

or, shorthanded as:

$$\mathcal{P}_n = \{I, X, Y, Z\}^{\otimes n}. \tag{12.52}$$

Each element of $\mathcal{P}_n$ is a $2^n \times 2^n$ matrix, and there are 4^{n+1} distinct elements in $\mathcal{P}_n$. This arises from 4^n distinct tensor products of Pauli matrices, combined with four multiplicative factors $\{\pm 1, \pm i\}$ for each.

Exercise 12.18 List all the elements in $\mathcal{P}_2$.

12.5.2 Group Properties

To confirm that $\mathcal{P}_n$ is a group, we need to verify that it satisfies the group axioms under matrix multiplication:

1. Closure: For any $P, Q \in \mathcal{P}_n$, their product PQ is also in $\mathcal{P}_n$. This is because the product of any two Pauli strings results in another Pauli string of the same length, possibly multiplied by a phase factor of ± 1 or $\pm i$. See § 12.3.3 for justification and a method to determine the phase factor $\{\pm 1, \pm i\}$ in the product.

2. Identity: The identity element is $I \equiv I^{\otimes n}$, the identity matrix on n qubits. For any $P \in \mathcal{P}_n$:
$$P \cdot I = P = I \cdot P.$$

3. Inverse: For every $P \in \mathcal{P}_n$, there exists an inverse $P^{-1} \in \mathcal{P}_n$ such that $P \cdot P^{-1} = I$. This holds because each Pauli matrix is invertible, and the tensor product property $(A \otimes B)^{-1} = A^{-1} \otimes B^{-1}$ applies.

4. Associativity: Matrix multiplication is associative. For any $P, Q, R \in \mathcal{P}_n$:
$$P \cdot (Q \cdot R) = (P \cdot Q) \cdot R.$$

5. Non-Abelian: The Pauli group is non-Abelian, meaning that the order of multiplication matters. For example, $XY = iZ$ and $YX = -iZ$, so $XY \neq YX$.

The inclusion of the phase factors ± 1 and $\pm i$ is crucial for the Pauli group to satisfy the group axioms, particularly closure and the existence of inverses. Thus, the Pauli group $\mathcal{P}_n$ is a superset of the Pauli string basis set $\mathcal{B}_n$ discussed in § 12.4.

Exercise 12.19 What is the inverse of $-iXYZIZXY$ in $\mathcal{P}_7$? Do you need to reverse the order of the Pauli matrices in the string when finding the inverse? Why is the inverse unique?

12.5.3 Subgroups

A subgroup H of a group G is a subset of G that is itself a group with the group operation inherited from G.

■ **Example 12.9** The set $\{\pm I, \pm iI\}$ forms a subgroup of the Pauli group $\mathcal{P}_n$. It qualifies as a subgroup because it satisfies closure, contains the identity, and contains inverses:

1. Closure: If we take any two elements from the set and multiply them (using the group operation, which is matrix multiplication), the result must also be in the set. For instance:

   - $(\pm I) \cdot (\pm I) = \pm I$
   - $(\pm I) \cdot (\pm iI) = \pm iI$
   - $(\pm iI) \cdot (\pm iI) = -I$

2. Identity: The identity element of the Pauli group is the identity matrix I, which is included in the set.

3. Inverses: For every element in the set, its inverse must also be in the set. For example:

   - The inverse of I is I.
   - The inverse of $-I$ is $-I$.
   - The inverse of iI is $-iI$.
   - The inverse of $-iI$ is iI.

This subgroup is termed the *center* of $\mathcal{P}_n$, denoted as $Z(\mathcal{P}_n)$. In general, the center of a group G, denoted $Z(G)$, is the set of all elements in G that commute with every element of G. ∎

12.5.4 Group Generators

Generators provide a concise way to describe a group. Instead of listing all the elements, it is sufficient to specify the generators and the rules for combining them.

> **Definition 12.4 — Generators of Pauli Groups.** A set of elements S is said to generate a Pauli group (or subgroup) G, denoted as $G = \langle S \rangle$, if every element of G can be expressed as a finite product of elements from S and their inverses, possibly multiplied by phase factors ± 1 and $\pm i$.
>
> The set S is referred to as a generator set for G. Note that generator sets are not necessarily unique. A *minimal generator set*, also called a set of *independent generators*, is one where no proper subset of S generates G.

In listing the generators of Pauli groups or subgroups, we typically ignore the scalar factors ± 1 and $\pm i$. In addition, the identity I does not need to be listed, since by definition the inverses of the generators are included in the generation process.

> The convention regarding phase factors, commonly adopted in quantum computing literature, differs from that in pure abstract algebra. In abstract algebra, iI would typically be included as a generator to produce the phases $\{\pm 1, \pm i\}$.
>
> Additionally, in this context, $\langle \cdot \rangle$ denotes the group generated by a set, rather than the inner product.

■ **Example 12.10** In the single qubit Pauli group $\mathcal{P}_1$, any two of X, Y, and Z, along with the phase factors ± 1 and $\pm i$, can serve as generators. The third Pauli operator can be expressed in terms of the other two: for example, $Y = iXZ$. We express this as:

$$\mathcal{P}_1 = \langle X, Y \rangle = \langle Y, Z \rangle = \langle Z, X \rangle.$$

Although $\{X, Y, Z\}$ is also a generating set of $\mathcal{P}_1$, it is not minimal.

The two-qubit Pauli group can be generated by the following set of elements:

$$\{X \otimes I, Y \otimes I, I \otimes X, I \otimes Y\}.$$

By combining these generators through matrix multiplication, along with the phase factors ± 1 and $\pm i$, we can generate all 64 elements of the Pauli group $\mathcal{P}_2$, denoted as:

$$\mathcal{P}_2 = \langle X \otimes I, Y \otimes I, I \otimes X, I \otimes Y \rangle \equiv \langle XI, YI, IX, IY \rangle.$$

Note that in the context of Pauli strings and groups, the above tensor products are often written in shorthand as XI, YI, ZI, and IX. Here, $XI \neq X$! ∎

The minimum number of generators of $\mathcal{P}_n$ is $2n$. For example, we can employ one X and one Z for each of the n qubits; Y does not need to be included because it can be generated through $Y = iXZ$. Alternatively, we can also use one X and one Y, or other combinations, for each of the n qubits.

> **Exercise 12.20** List the generators of the subgroup $Z(\mathcal{P}_n)$ discussed in Example 12.9.

> **Exercise 12.21** In Example 12.10, why are $Z \otimes I$ and $X \otimes Y$ not included in the minimal generator set? Come up with a minimal generator set including $Z \otimes I$ and $X \otimes Y$, respectively.

12.5.5 ✳ Stabilizer Groups, Normalizer Groups, and Clifford Groups

The Pauli group $\mathcal{P}_n$ has subgroups and related groups that play important roles in quantum information theory and quantum computing. Here we introduce a few of these groups.

> **Definition 12.5 — Stabilizer Group.** A stabilizer group $\mathcal{S}$ is an Abelian subgroup of the Pauli group $\mathcal{P}_n$ that satisfies the following properties:
>
> 1. Closure Under Multiplication: For any $S_1, S_2 \in \mathcal{S}$, $S_1 S_2 \in \mathcal{S}$.
>
> 2. Abelian (Commutativity): For any $S_1, S_2 \in \mathcal{S}$, $S_1 S_2 = S_2 S_1$.
>
> 3. Exclusion of $-I$: $-I \notin \mathcal{S}$.

The Abelian property ensures that all elements of the stabilizer group commute, allowing for the simultaneous diagonalization of the elements of $\mathcal{S}$ and the definition of a subspace of $\mathcal{H}$.

Excluding $-I$ from the stabilizer group is essential to ensure that the code space is well-defined and non-empty. It guarantees that the eigenvalues of the stabilizers can be consistently interpreted as ± 1, which is necessary for the framework of stabilizer codes to function correctly. If $-I$ were included in the stabilizer group, every state in the code space would need to be a $+1$ eigenstate of $-I$, which is not possible because $-I$ acts as -1 on all states.

Given a stabilizer group $\mathcal{S}$, the subspace of $\mathcal{H}$ stabilized by $\mathcal{S}$ is called the code space $\mathcal{C}$:

Definition 12.6 — Code Space. The code space $\mathcal{C}$ defined by a stabilizer group $\mathcal{S}$ is the subspace of the 2^n-dimensional Hilbert space $\mathcal{H}$ given by:

$$\mathcal{C} = \{|\psi\rangle \in \mathcal{H} \mid P|\psi\rangle = |\psi\rangle \text{ for all } P \in \mathcal{S}\}.$$

In other words, the code space consists of all vectors in $\mathcal{H}$ that are $+1$ eigenvectors of every element in the stabilizer group.

■ **Example 12.11** Consider the stabilizer group generated by $\{Z \otimes Z, X \otimes X\}$ for a two-qubit system, i.e., $\mathcal{S} = \langle Z \otimes Z, X \otimes X \rangle$.

The stabilizer group also includes $Y \otimes Y$ and $I \otimes I$, as they are products of the generators. Thus, $\mathcal{S} = \{I \otimes I, X \otimes X, Y \otimes Y, Z \otimes Z\}$, with phase factors $\{\pm 1, \pm i\}$ omitted. However, the group does not include $X \otimes Y$ or $I \otimes Z$.

The generators commute:

$$(Z \otimes Z)(X \otimes X) = (X \otimes X)(Z \otimes Z),$$

allowing for a well-defined code space stabilized by both generators.

The code space stabilized by this group is one-dimensional and is spanned by $|\psi_1\rangle = \frac{1}{\sqrt{2}}(|00\rangle + |11\rangle)$, because

$$(Z \otimes Z)|\psi_1\rangle = |\psi_1\rangle,$$
$$(X \otimes X)|\psi_1\rangle = |\psi_1\rangle.$$

However, the following states do not belong to the code space stabilized by this group:

- $|\psi_2\rangle = \frac{1}{\sqrt{2}}(|01\rangle + |10\rangle)$: a -1 eigenvector of $Z \otimes Z$ and $+1$ of $X \otimes X$,

- $|\psi_3\rangle = \frac{1}{\sqrt{2}}(|01\rangle - |10\rangle)$: a -1 eigenvector of $Z \otimes Z$ and -1 of $X \otimes X$,

- $|\psi_4\rangle = \frac{1}{\sqrt{2}}(|00\rangle - |11\rangle)$: a $+1$ eigenvector of $Z \otimes Z$ and -1 of $X \otimes X$.

■

Stabilizer Generators and Dimension of the Code Space

The Hilbert space of an n-qubit system is 2^n-dimensional. The stabilizer generators impose constraints that reduce the dimension of the subspace (the code space) stabilized by the group. Each independent generator halves the dimension of the space it stabilizes. Thus, k independent generators stabilize a subspace of dimension 2^{n-k}. Therefore, in an n-qubit system, the stabilizer group can have at most n independent generators.

Each generator can be thought of as a constraint or a condition that the code space must satisfy. These generators must commute since the subgroup is Abelian. If a stabilizer group is generated by a set of k independent commuting Pauli operators, the number of elements in $\mathcal{S}$ is 2^k (with phase factors $\{\pm 1, \pm i\}$ omitted). In Example 12.11, $k = 2$, and there are $k^2 = 4$ elements in $\mathcal{S}$.

■ **Example 12.12** Continuing from Example 12.11, we examine the stabilizer group S generated by $\{Z \otimes Z, X \otimes X\}$.

Each generator of S imposes a constraint on states in the code space, as follows:

(a) The generator $Z \otimes Z$ acts as follows on the computational basis states:

$$Z{\otimes}Z \, |00\rangle = |00\rangle, \; Z{\otimes}Z \, |11\rangle = |11\rangle, \; Z{\otimes}Z \, |01\rangle = - \, |01\rangle, \; Z{\otimes}Z \, |10\rangle = - \, |10\rangle \, .$$

Thus, the states that satisfy $(Z \otimes Z) \, |\psi\rangle = |\psi\rangle$ (i.e., the $+1$-eigenstates of $Z \otimes Z$) are:

$$|\psi\rangle = a \, |00\rangle + b \, |11\rangle \, ,$$

where a and b are complex coefficients. This defines a two-dimensional subspace. The constraint imposed by the generator $Z \otimes Z$ reduces the dimension of the space from 4 to 2.

(b) Similarly, the generator $X \otimes X$ acts as follows on the computational basis states:

$$X{\otimes}X \, |00\rangle = |11\rangle, \; X{\otimes}X \, |11\rangle = |00\rangle, \; X{\otimes}X \, |01\rangle = |10\rangle, \; X{\otimes}X \, |10\rangle = |01\rangle \, .$$

For $|\psi\rangle = a \, |00\rangle + b \, |11\rangle$ to be a $+1$-eigenstate of $X \otimes X$, we require:

$$X \otimes X \, |\psi\rangle = a \, |11\rangle + b \, |00\rangle = |\psi\rangle \, .$$

This implies that $a = b$, so:

$$|\psi\rangle = a(|00\rangle + |11\rangle).$$

The constraint imposed by the generator $X \otimes X$ further reduces the dimension of the subspace from 2 to 1.

Together, these constraints define the code space, which in this case is the one-dimensional space spanned by $|\psi_1\rangle = \frac{1}{\sqrt{2}}(|00\rangle + |11\rangle)$. ■

Error Detection

Stabilizers help detect errors by measuring the eigenvalues of the stabilizer generators. When an error occurs, the corrupted state will no longer be an eigenstate of the stabilizer operators with eigenvalue $+1$. Instead, it may become an eigenstate with eigenvalue -1. By checking the eigenvalues of the stabilizer generators, we can identify which stabilizers yield a -1 value, indicating the presence of an error.

This process, known as syndrome measurement, allows us to determine the error syndrome, which specifies which stabilizers have detected an error. The error syndrome is then used to infer the type of error that has occurred and to apply the appropriate correction to restore the state to the code space.

1 Normalizer Groups

A normalizer group is defined relative to a stabilizer group $S \subset P_n$. It consists of all elements in P_n that commute with every element of the stabilizer group S. Formally:

> **Definition 12.7 — Normalizer Group.** The normalizer group of a stabilizer group $S \subset P_n$ is defined as:
>
> $$\mathcal{N}(S) = \{P \in P_n \mid PQ = QP, \ \forall Q \in S\}. \tag{12.53}$$

The normalizer group contains the stabilizer group, i.e., $\mathcal{N}(S) \supseteq S$. This inclusion holds because every element in S commutes trivially with itself and with other elements in S. Consequently, the normalizer group $\mathcal{N}(S)$ typically has a larger dimension than the stabilizer group S.

Specifically, if S has k independent generators, $\mathcal{N}(S)$ generally has $n + k$ or more independent generators, depending on the structure of S.

■ **Example 12.13** Consider the stabilizer group $S = \langle Z \otimes Z \rangle$ generated by $Z \otimes Z$ in a two-qubit system.

(a) The stabilizer group is:
$$S = \{I \otimes I, Z \otimes Z\}.$$

(b) The normalizer group is:
$$\mathcal{N}(S) = \{I \otimes I, Z \otimes Z, Z \otimes I, I \otimes Z, X \otimes X, Y \otimes Y\}.$$

All elements of $\mathcal{N}(S)$ commute with $Z \otimes Z$, as each has an even number (0 or 2) of anti-commuting pairs with respect to $Z \otimes Z$.

(c) Since $X \otimes I$ anti-commutes with $Z \otimes Z$, it is not in $\mathcal{N}(S)$.

(d) The minimal set of independent generators for $\mathcal{N}(S)$ is:
$$\{Z \otimes I, I \otimes Z, X \otimes X\}.$$

(e) The cosets of S in $\mathcal{N}(S)$ partition $\mathcal{N}(S)$ into disjoint sets, each having the same size as S. To construct the cosets, we multiply each representative element from $\mathcal{N}(S)$ by all the elements in S. The cosets are:
$$\{I \otimes I, Z \otimes Z\}, \quad \{Z \otimes I, I \otimes Z\}, \quad \{X \otimes X, Y \otimes Y\}.$$

■

In quantum computing, logical operations on encoded qubits correspond to elements in $\mathcal{N}(S)$. These operations can be performed on the encoded data without leaving the code space. The quotient group $\mathcal{N}(S)/S$ characterizes the logical operators. Each coset of S in $\mathcal{N}(S)$ represents a distinct logical operation on the code space, with S itself corresponding to the logical identity operator.

Next, we illustrate these concepts using Shor's three-qubit bit-flip code, which encodes one logical qubit into three physical qubits. For more information on Shor's error correction codes, readers are referred to *Quantum Computing and Information: A Scaffolding Approach* [1].

■ **Example 12.14 — Shor's Three-Qubit Bit-Flip Code.**

The Stabilizer Group and Code Space

The stabilizer group $\mathcal{S}$ for the three-qubit bit-flip code is generated by:

$$\mathcal{S} = \langle Z \otimes Z \otimes I, I \otimes Z \otimes Z \rangle.$$

The code space consists of all states $|\psi\rangle$ that satisfy

$$(Z \otimes Z \otimes I) |\psi\rangle = |\psi\rangle \quad \text{and} \quad (I \otimes Z \otimes Z) |\psi\rangle = |\psi\rangle.$$

The two basis states (codewords) of the code space are:

$$|0_L\rangle = |000\rangle, \quad |1_L\rangle = |111\rangle,$$

where $|0_L\rangle$ and $|1_L\rangle$ represent the encoded logical $|0\rangle$ and $|1\rangle$ states, respectively.

The code space is any linear combination of $|0_L\rangle$ and $|1_L\rangle$:

$$|\psi\rangle = \alpha |0_L\rangle + \beta |1_L\rangle.$$

Logical X_L and Z_L Operators in the Normalizer Group

Logical operators, which act on the encoded qubit, are elements in the normalizer $\mathcal{N}(\mathcal{S})$. They preserve the code space by mapping valid code states to other valid code states. Specifically, if $L \in \mathcal{N}(\mathcal{S})$, then for any stabilizer $S \in \mathcal{S}$:

$$S(L |\psi\rangle) = L(S |\psi\rangle) = L |\psi\rangle.$$

This follows because $S |\psi\rangle = |\psi\rangle$ (as $|\psi\rangle$ is in the code space) and L commutes with all elements of $\mathcal{S}$ by definition of the normalizer.

In particular, the logical X and Z operators are as follows:

- The logical Z_L operator, acting as a Z operation on the encoded qubit, can be represented by:

$$Z_L = Z \otimes Z \otimes Z.$$

This operator commutes with all elements of $\mathcal{S}$ and belongs to $\mathcal{N}(\mathcal{S})$. However, it is not part of $\mathcal{S}$, as it changes $|1_L\rangle$ to $-|1_L\rangle$, failing to stabilize the code space.

$Z \otimes I \otimes I$, $I \otimes Z \otimes I$, and $I \otimes I \otimes Z$ can also serve as the logical Z_L operator.

- The logical X_L operator, acting as an X operation on the encoded qubit, can be represented by:

$$X_L = X \otimes X \otimes X.$$

This operator commutes with all elements of $\mathcal{S}$ and is in $\mathcal{N}(\mathcal{S})$, but it is not in $\mathcal{S}$ since it changes $|0_L\rangle$ to $|1_L\rangle$ and vice versa, rather than stabilizing the code space.

The logical Y_L operator, derived from the Pauli relation $Y = iXZ$, is given by:

$$Y_L = iX_L Z_L.$$

Cosets and Logical Operations

The quotient group $\mathcal{N}(\mathcal{S})/\mathcal{S}$ has distinct cosets, each representing a different logical operation on the code space:

- The coset containing $\mathcal{S}$ itself represents the logical identity operation on the encoded qubit.

- The coset containing $X_L = X \otimes X \otimes X$ represents the logical X operation on the encoded qubit.

- The coset containing $Z_L = Z \otimes Z \otimes Z$ represents the logical Z operation on the encoded qubit.

- The coset containing $Y_L = Y \otimes Y \otimes Y$ represents the logical Y operation on the encoded qubit.

Thus, each coset of $\mathcal{S}$ in $\mathcal{N}(\mathcal{S})$ corresponds to a distinct logical operation on the encoded qubit. Logical operations can be performed without affecting the stabilizer conditions and thus remain within the code space.   ∎

2 Clifford Groups

The n-qubit Clifford group $\mathcal{C}_n$ consists of all unitary operators that, when conjugated with an element of the Pauli group $\mathcal{P}_n$, produce another element of the Pauli group. The Clifford group itself is not a subgroup of $\mathcal{P}_n$. Formally:

Definition 12.8 — Clifford Group. The n-qubit Clifford group $\mathcal{C}_n$ is defined as:

$$\mathcal{C}_n = \{U \in \mathrm{U}(2^n) \mid U\mathcal{P}_n U^\dagger = \mathcal{P}_n\},$$

where $\mathrm{U}(2^n)$ represents the group of unitary operators on n qubits.

The Clifford group preserves the Pauli group structure under conjugation. For U to be a member of the Clifford group $\mathcal{C}_n$, the conjugation $UPU^\dagger$ must map P to another element of $\mathcal{P}_n$, possibly up to a phase factor.

■ **Example 12.15 — Single-qubit Clifford Group.** The Hadamard gate $H = \frac{1}{\sqrt{2}}\begin{bmatrix} 1 & 1 \\ 1 & -1 \end{bmatrix}$ is an element of the Clifford group $\mathcal{C}_1$. Under conjugation, H transforms each Pauli operator as follows:

$$HZH = X, \quad HXH = Z, \quad HYH = -Y.$$

Since H maps each Pauli operator in $\mathcal{P}_1$ to another Pauli operator (up to a phase), it satisfies the definition of being an element of $\mathcal{C}_1$.

Another element in $\mathcal{C}_1$ is the phase gate $S = \begin{bmatrix} 1 & 0 \\ 0 & i \end{bmatrix}$, which also maps Pauli operators to Pauli operators under conjugation:

$$SZS^\dagger = Z, \quad SXS^\dagger = Y, \quad SYS^\dagger = -X.$$

The entire single-qubit Clifford group, $\mathcal{C}_1$, contains 24 elements. These are generated by H and S, and include compositions of the two, as follows:

$$
\begin{array}{llll}
I = H^2 = S^4 & H & S & Z = S^2 \\
X = HZH & Y = SXS^\dagger & Y^\dagger = S^\dagger XS & S^\dagger = S^3 \\
HS & SH & SHS & HSH \\
HS^\dagger & S^\dagger H & S^\dagger HS & SHS^\dagger \\
ZH = S^2 H & HZ = HS^2 & ZHS = S^2 HS & SHZ = SHS^2 \\
S^\dagger HS^\dagger & HS^\dagger H & ZHS^\dagger = S^2 HS^\dagger & S^\dagger HZ = S^\dagger HS^2
\end{array}
$$

The T-gate, defined as:

$$
T = \sqrt{S} = \begin{bmatrix} 1 & 0 \\ 0 & e^{i\frac{\pi}{4}} \end{bmatrix},
$$

is not a member of $\mathcal{C}_1$. For instance, it fails the Clifford test for X:

$$
TXT^\dagger = \begin{bmatrix} 1 & 0 \\ 0 & e^{i\frac{\pi}{4}} \end{bmatrix} \begin{bmatrix} 0 & 1 \\ 1 & 0 \end{bmatrix} \begin{bmatrix} 1 & 0 \\ 0 & e^{-i\frac{\pi}{4}} \end{bmatrix} = \begin{bmatrix} 0 & e^{i\frac{\pi}{4}} \\ e^{-i\frac{\pi}{4}} & 0 \end{bmatrix},
$$

which is not a Pauli operator (even up to a phase). ∎

Exercise 12.22 Demonstrate that X and Y are elements of $\mathcal{C}_1$ using the Clifford test:

$$
XPX = cQ, \quad YPY = cQ,
$$

where $P, Q \in \{I, X, Y, Z\}$, and $c \in \mathbb{C}$ is a phase factor.

■ **Example 12.16 — Two-qubit Clifford Group Generators.** A minimal set of generators for the two-qubit Clifford group $\mathcal{C}_2$ includes:

$$
H \otimes I, \ I \otimes H,
$$
$$
S \otimes I, \ I \otimes S, \ S^\dagger \otimes I, \ I \otimes S^\dagger,
$$
$$
\text{CNOT}_{1 \to 2}, \ \text{CNOT}_{2 \to 1}.
$$

Here:

- H and S are single-qubit gates acting independently on the first or second qubit.

- $\text{CNOT}_{1 \to 2} \equiv |00\rangle\langle 00| + |01\rangle\langle 01| + |11\rangle\langle 10| + |10\rangle\langle 11|$ is the controlled-NOT gate with the first qubit as control and the second as target. $\text{CNOT}_{2 \to 1}$ is the CNOT gate with the second qubit as control and the first as target.

Products of these generators yield the complete set of 11,520 elements of $\mathcal{C}_2$. Examples include:

$$
S \otimes S = (S \otimes I) \otimes (I \otimes S), \ H \otimes H = (H \otimes I) \otimes (I \otimes H),
$$
$$
\text{SWAP} = \text{CNOT}_{1 \to 2} \, \text{CNOT}_{2 \to 1} \, \text{CNOT}_{1 \to 2},
$$
$$
(H \otimes I)\text{CNOT}_{1 \to 2}, \ (I \otimes H)\text{CNOT}_{2 \to 1},
$$
$$
(S \otimes I)\text{CNOT}_{1 \to 2}, \ (I \otimes S)\text{CNOT}_{2 \to 1},
$$
$$
(S^\dagger \otimes S)\text{CNOT}_{1 \to 2}, \ (H \otimes H)\text{CNOT}_{1 \to 2}(H \otimes H).
$$

The two-qubit Clifford group is essentially $\mathcal{C}_1$ applied independently to each qubit, combined with $\text{CNOT}_{1 \to 2}$ and $\text{CNOT}_{2 \to 1}$ applied to both qubits. ∎

The Clifford group encapsulates symmetries of the Pauli group under conjugation. These symmetries reflect how quantum states and errors transform under logical operations, enabling universal error detection and correction strategies.

In quantum cryptography, Clifford groups help in developing secure quantum communication protocols, leveraging its properties to ensure data security. Additionally, in quantum algorithms, Clifford groups form the basis of many quantum gates and operations, facilitating efficient algorithm design and implementation.

Quantum gates implementing elements in a Clifford group are referred to as Clifford gates. Clifford gates have a unique property: circuits composed entirely of Clifford gates acting on stabilizer states can be efficiently simulated on a classical computer. This efficiency arises because the conjugation of Pauli operators by Clifford gates maintains the simplicity of tracking stabilizers, as they remain Pauli operators. This simplifies calculations and reduces computational complexity.

Problem Set 12

Note: In the following exercises, $X \equiv \sigma_1$, $Y \equiv \sigma_2$, and $Z \equiv \sigma_3$ represent the Pauli matrices, and $I \equiv \sigma_0$ represents the identity matrix.

12.1 Prove the following relations:

(a) $HXH = Z$

(b) $HYH = -Y$

(c) $HZH = X$

where $H = \frac{1}{\sqrt{2}} \begin{bmatrix} 1 & 1 \\ 1 & -1 \end{bmatrix}$. **Hint:** Express H in terms of X, Y, and Z.

12.2 Prove the following relations:

(a) $SXS^\dagger = Y$

(b) $SYS^\dagger = -X$

(c) $SZS^\dagger = Z$

where $S = \begin{bmatrix} 1 & 0 \\ 0 & i \end{bmatrix}$. **Hint:** Express S in terms of X, Y, and Z.

12.3 Let $C = \sum_{i=0}^{3} \gamma_i \sigma_i$, where $\gamma_i \in \mathbb{C}$. Expand $CC^\dagger$, $C^\dagger C$, and $CC^\dagger + C^\dagger C$ in the basis of Pauli matrices. (See Eq. 12.13.)

12.4 Given $\sigma_u = \begin{bmatrix} \cos\theta & \sin\theta e^{-i\phi} \\ \sin\theta e^{i\phi} & -\cos\theta \end{bmatrix}$, and the following state vectors:

$$|0\rangle = \begin{bmatrix} 1 \\ 0 \end{bmatrix}, \qquad\qquad |1\rangle = \begin{bmatrix} 0 \\ 1 \end{bmatrix},$$

$$|+\rangle = \frac{1}{\sqrt{2}} (|0\rangle + |1\rangle), \qquad |+_i\rangle = \frac{1}{\sqrt{2}} (|0\rangle + i|1\rangle),$$

$$|\psi_1\rangle = -\frac{1}{2}\,|0\rangle - \frac{\sqrt{3}}{2}\,|1\rangle\,, \qquad\qquad |\psi_2\rangle = -\frac{1}{2}\,|0\rangle + \frac{\sqrt{3}}{2}\,|1\rangle\,,$$

compute the following values:

(a) $\langle 0|\sigma_u|0\rangle$

(b) $\langle +|\sigma_u|+\rangle$

(c) $\langle +_i|\sigma_u|+_i\rangle$

(d) $\langle 0|\sigma_u|\psi_1\rangle$

(e) $\langle \psi_1|\sigma_u|\psi_2\rangle$

(f) $\langle \psi_2|\sigma_u|0\rangle$

12.5 Find a general formula for the square root of the 2×2 identity matrix I.

Furthermore, investigate the cubic and fourth roots of the 2×2 identity matrix I, beyond diagonal matrices.

12.6 Find $\alpha \in \mathbb{R}$ such that $e^{i\alpha\sigma_3}\sigma_1 e^{-i\alpha\sigma_3} = \sigma_2$.

12.7 Consider the linear operator:

$$\sigma_{\boldsymbol{n}} = \frac{1}{2}\left(I + \sum_{j=1}^{3} n_j \sigma_j\right),$$

where $\boldsymbol{n} = (n_1, n_2, n_3)$ is a unit vector, i.e., $n_1^2 + n_2^2 + n_3^2 = 1$ with $n_j \in \mathbb{R}$.

(a) Find $\sigma_{\boldsymbol{n}}^{\dagger}$, $\mathrm{tr}(\sigma_{\boldsymbol{n}})$, and $\sigma_{\boldsymbol{n}}^2$.

(b) Compute $\sigma_{\boldsymbol{n}} \begin{bmatrix} e^{i\phi}\cos\theta \\ \sin\theta \end{bmatrix}$ for $\phi, \theta \in \mathbb{R}$.

12.8 Show that for a Pauli string $\boldsymbol{P}$ and real scalars α and β:

$$e^{i\alpha\boldsymbol{P}}e^{i\beta\boldsymbol{P}} = e^{i(\alpha+\beta)\boldsymbol{P}}.$$

12.9 List the minimum set of generators for the Pauli group $\mathcal{P}_3$ and $\mathcal{P}_4$, respectively, using only the operators I, X, and Z. Then, generalize and formulate a minimum set of generators for the Pauli group $\mathcal{P}_n$.

12.10 ✳ Within the n-qubit Pauli group $\mathcal{P}_n$, for each Pauli matrix (X, Y, or Z), consider the subsets generated by all tensor products of that matrix with the identity matrix. Demonstrate that these subsets are subgroups of $\mathcal{P}_n$, and that these subgroups are Abelian.

12.11 ✳ For the stabilizer group $\mathcal{S} = \langle Z \otimes Z, X \otimes X \rangle$ in a two-qubit system:

(a) List all elements in the normalizer group $\mathcal{N}(\mathcal{S})$.

(b) Determine whether $X \otimes I$ is in $\mathcal{N}(\mathcal{S})$.

(c) Find the number of independent generators in $\mathcal{N}(\mathcal{S})$.

(d) Find the cosets of $\mathcal{S}$ in $\mathcal{N}(\mathcal{S})$.

12.12 ✳ **Generalized Pauli Matrices.** For general dimension d, the generalized Pauli matrices form a basis for the space of $d \times d$ matrices. These are constructed as follows:

    i. Shift Operator X:

$$X = \sum_{j=0}^{d-1} |(j+1) \mod d\rangle\langle j|.$$

ii. Phase Operator Z:

$$Z = \sum_{j=0}^{d-1} \omega^j \, |j\rangle\langle j| \,, \quad \text{where } \omega = e^{2\pi i/d}.$$

iii. The generalized Pauli matrices are the set of products of these operators:

$$\{X^a Z^b \mid a, b \in \{0, 1, \ldots, d-1\}\}.$$

Demonstrate that these generalized Pauli matrices retain key properties of the original Pauli matrices:

(a) They are unitary (but not necessarily Hermitian).

(b) Commutation relations: $XZ = \omega ZX$, and
$(X^a Z^b)(X^c Z^k) = \omega^{bc} X^{a+c} Z^{b+k}$, where $\omega = e^{2\pi i/d}$.

(c) Orthogonality relation: $\text{tr}((X^a Z^b)^\dagger (X^c Z^k)) = d\delta_{ac}\delta_{bk}$.

(d) They form a complete basis for the space of $d \times d$ matrices.

13. Advanced Matrix Decompositions

Contents

In this chapter, we revisit spectral (eigenvalue) decomposition and introduce Jordan forms. Building on this foundation, we explore singular value decomposition (SVD) and Schmidt decomposition. These decompositions have numerous applications in quantum computing and quantum information theory, including quantum algorithms, quantum circuit optimization, quantum state tomography, tensor networks, and the study of quantum entanglement.

> ⓘ Although this topic is not essential for studying the second book, *Quantum Computing and Information*, it will become important in the third book, *Quantum Algorithms and Applications*. Readers may defer studying it until encountering the relevant quantum algorithms.

13.1 Spectral Decomposition and Jordan Forms

Spectral decomposition was introduced in Chapter 9. In this section, we revisit related concepts, emphasize their interrelationships, and generalize them using Jordan forms. This lays the foundation for our exploration of singular value decomposition and Schmidt decomposition.

13.1.1 Diagonalization and Eigenvalue Decomposition

1 Diagonalization

Diagonalization refers to the process of transforming a square matrix into a diagonal matrix through a similarity transformation:

$$A = V\Lambda V^{-1}, \quad \text{or, equivalently, } V^{-1}AV = \Lambda, \tag{13.1}$$

where Λ is a diagonal matrix. Diagonalization is often used to simplify matrix calculations, such as computing powers or exponentials.

2 Diagonalizability

An $n \times n$ matrix A is diagonalizable if and only if it has n linearly independent eigenvectors.

Furthermore, if A is diagonalizable, then Λ consists of the eigenvalues of A, and the columns of V are the corresponding eigenvectors. Thus, Eq. 13.1 is also referred to as eigenvalue decomposition, or eigendecomposition.

3 Diagonalization Procedure

To diagonalize a matrix A, we normally:

1. Calculate the eigenvalues of A by solving the characteristic equation:

$$\det(A - \lambda I) = 0. \tag{13.2}$$

2. For each eigenvalue λ, find the associated eigenvectors by solving the equation:

$$(A - \lambda I)v = 0. \tag{13.3}$$

3. Determine if the eigenvectors form a linearly independent set. If they do, construct matrix V using the eigenvectors as columns, and compute V^{-1}.

■ **Example 13.1** Diagonalizability hinges on having a full set of linearly independent eigenvectors. Not all square matrices are diagonalizable. An example is:

$$A = \begin{bmatrix} 2 & 1 \\ 0 & 2 \end{bmatrix}.$$

The characteristic polynomial of A is:

$$\det(A - \lambda I) = (2 - \lambda)^2 = 0.$$

This means the eigenvalue $\lambda = 2$ has algebraic multiplicity 2.

To find the eigenvectors, we solve:

$$(A - 2I)v = 0.$$

The solution to this system is any vector of the form:

$$v = \begin{bmatrix} x \\ 0 \end{bmatrix},$$

where x is a free variable. This means the eigenspace associated with $\lambda = 2$ is one-dimensional, spanned by the vector $\begin{bmatrix} 1 \\ 0 \end{bmatrix}$.

Since we only have one linearly independent eigenvector, we cannot form an invertible matrix V whose columns are eigenvectors. Therefore, matrix A cannot be diagonalized.

Note that A is not a normal matrix, i.e., $AA^\dagger \neq A^\dagger A$:

$$AA^\dagger = \begin{bmatrix} 5 & 2 \\ 2 & 4 \end{bmatrix}, \quad A^\dagger A = \begin{bmatrix} 4 & 2 \\ 2 & 5 \end{bmatrix}.$$

∎

13.1.2 Spectral Decomposition

Spectral decomposition is a specific case of eigenvalue diagonalization. It applies only to normal matrices (matrices that commute with their conjugate transpose, $AA^\dagger = A^\dagger A$).

1 Unitary Diagonalization

Normal matrices have a special property: their eigenvectors corresponding to distinct eigenvalues are orthogonal. Furthermore, normal matrices are unitarily diagonalizable, meaning V in Eq. 13.1 can be made into a unitary matrix U, such that $U^\dagger = U^{-1}$. Then Eq. 13.1 becomes:

$$A = U\Lambda U^\dagger, \quad \text{or, equivalently,} \quad U^\dagger A U = \Lambda. \tag{13.4}$$

This is referred to as the spectral decomposition of the normal matrix A.

2 Dirac Notation

The Dirac notation offers a compact and intuitive way to express the orthonormal properties of eigenvectors and the completeness relation.

First, the orthonormal property of the eigenvectors can be expressed in Dirac notation as:

$$\langle \phi_i | \phi_j \rangle = \delta_{ij}, \tag{13.5}$$

where $|\phi_i\rangle$ represents the eigenvectors, i.e., the columns of U, and δ_{ij} is the Kronecker delta function.

The completeness relation states that the set of eigenvectors $\{|\phi_i\rangle\}$ forms a complete basis for the vector space. This is a direct result of $UU^\dagger = I$ and can be written as:

$$\sum_{i=1}^{n} |\phi_i\rangle\langle \phi_i| = I. \tag{13.6}$$

In Dirac notation, the spectral decomposition can be expressed as:

$$A = \sum_{i=1}^{n} \lambda_i |\phi_i\rangle\langle \phi_i|. \tag{13.7}$$

This reveals the direct relationship between a normal matrix and its eigenvalues and eigenvectors.

3 Special Matrix Types

A normal matrix allows for spectral decomposition with a unitary matrix collecting its orthonormal eigenvectors, but its eigenvalues can be complex and may include zero. Specific types of normal matrices exhibit unique characteristics in their spectral decomposition:

- The eigenvalues of a Hermitian matrix $(A = A^\dagger)$ are real.

- The eigenvalues of a skew-Hermitian matrix $(A = -A^\dagger)$ are purely imaginary.

- The eigenvalues of a unitary matrix $(AA^\dagger = I)$ have a modulus of 1.

- A positive semidefinite matrix is a Hermitian matrix with non-negative eigenvalues (see § 11.1.2.4).

- If a positive semidefinite matrix P has all positive eigenvalues, it is called positive definite. In this case, P is invertible, and its inverse, given by $P^{-1} = U\Lambda^{-1}U^\dagger$, is also positive definite.

■ **Example 13.2** Consider the positive definite matrix:

$$P = \frac{1}{2}\begin{bmatrix} 5 & -3 \\ -3 & 5 \end{bmatrix}.$$

The eigenvalues and eigenvectors of P are given by:

$$\Lambda = \begin{bmatrix} 1 & 0 \\ 0 & 4 \end{bmatrix}, \quad U = \frac{1}{\sqrt{2}}\begin{bmatrix} 1 & 1 \\ 1 & -1 \end{bmatrix}.$$

We can verify the spectral decomposition of P:

$$P = U\Lambda U^\dagger = \frac{1}{\sqrt{2}}\begin{bmatrix} 1 & 1 \\ 1 & -1 \end{bmatrix}\begin{bmatrix} 1 & 0 \\ 0 & 4 \end{bmatrix}\frac{1}{\sqrt{2}}\begin{bmatrix} 1 & 1 \\ 1 & -1 \end{bmatrix}.$$

The square root of Λ is:

$$\Lambda^{1/2} = \begin{bmatrix} 1 & 0 \\ 0 & 2 \end{bmatrix}.$$

Using U and $\Lambda^{1/2}$, we can construct:

$$S = U\Lambda^{1/2} = \frac{1}{\sqrt{2}}\begin{bmatrix} 1 & 1 \\ 1 & -1 \end{bmatrix}\begin{bmatrix} 1 & 0 \\ 0 & 2 \end{bmatrix} = \frac{1}{\sqrt{2}}\begin{bmatrix} 1 & 2 \\ 1 & -2 \end{bmatrix},$$

and verify:

$$SS^\dagger = \frac{1}{\sqrt{2}}\begin{bmatrix} 1 & 2 \\ 1 & -2 \end{bmatrix}\frac{1}{\sqrt{2}}\begin{bmatrix} 1 & 1 \\ 2 & -2 \end{bmatrix} = P.$$

Furthermore, the inverse square root of P is:

$$P^{-1/2} = U\Lambda^{-1/2}U^\dagger,$$

where:

$$\Lambda^{-1/2} = \begin{bmatrix} 1 & 0 \\ 0 & \frac{1}{2} \end{bmatrix}.$$

■

Exercise 13.1 As a refresher, prove the following using Dirac notation:

(a) If U is unitary and Λ is diagonal, then $U\Lambda U^\dagger$ is a normal matrix.

(b) The eigenvalues of a skew-Hermitian matrix $(A = -A^\dagger)$ are purely imaginary.

(c) A positive semidefinite matrix can always be expressed as the product of a matrix and its adjoint, $AA^\dagger$.

13.1.3 Jordan Decomposition

While normal matrices are guaranteed to be diagonalizable, they are not the only matrices that can be diagonalized. However, not all matrices are diagonalizable; matrices that cannot be diagonalized are called defective matrices.

1 Jordan Forms

Even though defective matrices cannot be diagonalized, they can still be transformed into a nearly diagonal form through a process called Jordan decomposition.

Theorem 13.1 — Jordan Decomposition. Any square matrix A can be transformed into a Jordan form through a process known as the Jordan-Chevalley decomposition, or simply Jordan decomposition:

$$A = VJV^{-1}, \qquad (13.8)$$

where J is the Jordan form of A, and V is the matrix of generalized eigenvectors of A.

We defer the proof of this theorem as well as the definition of generalized eigenvectors to § 13.1.3.2, while focusing first on understanding Jordan forms.

Definition 13.1 — Jordan Form. A Jordan form J consists of Jordan blocks along its diagonal, where each Jordan block corresponds to an eigenvalue of A. Within each Jordan block, the eigenvalue is repeated along the diagonal, and 1 appears on the superdiagonal. All other entries are zero.

Below is an illustration of a Jordan form, with missing entries implicitly assumed to be zero, and each Jordan block framed in a box:

$$J = \begin{bmatrix} \boxed{\begin{matrix} \lambda_1 & 1 & \\ & \lambda_1 & 1 \\ & & \lambda_1 \end{matrix}} & & & & \\ & \boxed{\lambda_2} & & & \\ & & \ddots & & \\ & & & \boxed{\begin{matrix} \lambda_n & 1 \\ & \lambda_n \end{matrix}} \end{bmatrix}.$$

■ **Example 13.3** Consider the matrix

$$A = \begin{bmatrix} -2 & 1 & 4 \\ -5 & 2 & 5 \\ -1 & 1 & 3 \end{bmatrix}.$$

This matrix has eigenvalues -1, 2, and 2, but only two linearly independent eigenvectors:

$$\boldsymbol{v}_1 = \begin{bmatrix} -7 \\ -15 \\ 2 \end{bmatrix}, \qquad \text{for } \lambda_1 = -1,$$

$$\boldsymbol{v}_2 = \begin{bmatrix} 1 \\ 0 \\ 1 \end{bmatrix}, \qquad \text{for } \lambda_2 = 2.$$

Thus, A is not diagonalizable (it is defective).

The Jordan decomposition of A is VJV^{-1}, where:

$$V = \begin{bmatrix} -7 & 1 & 0 \\ -15 & 0 & 1 \\ 2 & 1 & 0 \end{bmatrix}, \quad J = \begin{bmatrix} -1 & 0 & 0 \\ 0 & 2 & 1 \\ 0 & 0 & 2 \end{bmatrix}, \quad \text{and } V^{-1} = \frac{1}{9}\begin{bmatrix} -1 & 0 & 1 \\ 2 & 0 & 7 \\ -15 & 9 & 15 \end{bmatrix}.$$

The Jordan form J contains two Jordan blocks.   ■

2 ✳ Process and Proof of Jordan Decomposition

Let us now explore the fundamental reasoning behind why Jordan decomposition is valid.

Jordan Chain Generation

Consider an eigenvalue λ of A, with an algebraic multiplicity m and a geometric multiplicity $g \leq m$. We begin with one of the g linearly independent eigenvectors, $\boldsymbol{v}_1$:

$$A\boldsymbol{v}_1 = \lambda\boldsymbol{v}_1, \quad \text{or equivalently,} \quad (A - \lambda I)\boldsymbol{v}_1 = 0. \tag{13.9}$$

Next, we solve for a non-zero $\boldsymbol{v}_2$ using the equation:

$$(A - \lambda I)\boldsymbol{v}_2 = \boldsymbol{v}_1.$$

This process continues iteratively as:

$$(A - \lambda I)\boldsymbol{v}_i = \boldsymbol{v}_{i-1}, \quad \text{for } 1 < i \leq k, \tag{13.10}$$

where k is the largest integer such that a non-zero solution $\boldsymbol{v}_k$ exists.

The sequence of vectors $\boldsymbol{v}_1, \boldsymbol{v}_2, \ldots, \boldsymbol{v}_k$ is called a Jordan chain. If the final vector in the chain is $\boldsymbol{v}_k$, the chain can be expressed compactly as:

$$\boldsymbol{v}_1 \equiv (A - \lambda I)^{k-1}\boldsymbol{v}, \ \boldsymbol{v}_2 \equiv (A - \lambda I)^{k-2}\boldsymbol{v}, \ \ldots, \ \boldsymbol{v}_{k-1} \equiv (A - \lambda I)\boldsymbol{v}, \ \boldsymbol{v}_k \equiv \boldsymbol{v}. \tag{13.11}$$

■ **Example 13.4** Consider the matrix:

$$A = \begin{bmatrix} 2 & 1 & 0 \\ 0 & 2 & 1 \\ 0 & 0 & 2 \end{bmatrix}.$$

- Eigenvalue $\lambda = 2$, with an algebraic multiplicity of 3 and a geometric multiplicity of 1 (only one eigenvector).

- Eigenvector: $v_1 = \begin{bmatrix} 1 \\ 0 \\ 0 \end{bmatrix}$.

- Generalized eigenvectors: Solve $(A - 2I)v_2 = v_1$ and $(A - 2I)v_3 = v_2$ to obtain:

$$v_2 = \begin{bmatrix} 0 \\ 1 \\ 0 \end{bmatrix}, \quad v_3 = \begin{bmatrix} 0 \\ 0 \\ 1 \end{bmatrix}.$$

Here, A is already in Jordan form, with a single Jordan block. In the Jordan decomposition, $A = VJV^{-1}$, where $V = \begin{bmatrix} v_1 & v_2 & v_3 \end{bmatrix}$. In this case, V is the identity matrix, as expected. ∎

Generalized Eigenvectors

Due to the structure of the Jordan chain in Eq. 13.11 and the property $(A - \lambda I)v_1 = 0$, the vectors in the chain satisfy:

$$(A - \lambda I)^i v_i = 0, \quad \text{for } 1 \leq i \leq k. \tag{13.12}$$

This is a direct consequence of the iterative relation in Eq. 13.10. For this reason, the vectors $\{v_i\}$ are called generalized eigenvectors of A for the eigenvalue λ, except for v_1, which is a standard eigenvector.

Maximum Length of a Jordan Chain

The length of a Jordan chain, k, cannot exceed the algebraic multiplicity m of λ, because $(A - \lambda I)^m = 0$ by the Cayley-Hamilton theorem (see § 11.3.1.5). If $k > m$, the leading vector in the chain, $v_1 = (A - \lambda I)^{k-1}v$ from Eq. 13.11, would necessarily be zero, contradicting the assumption that v_1 is an eigenvector of A.

Jordan Chain and Jordan Block Correspondence

A Jordan block $J_k(\lambda)$ of size $k \times k$ corresponding to the eigenvalue λ is defined as:

$$J_k(\lambda) = \begin{bmatrix} \lambda & 1 & & \\ & \lambda & \ddots & \\ & & \ddots & 1 \\ & & & \lambda \end{bmatrix}_{k \times k}. \tag{13.13}$$

The iterative relation in Eq. 13.10 implies:

$$Av_i = \lambda v_i + v_{i-1}, \quad \text{for } 1 < i \leq k. \tag{13.14}$$

This corresponds to the action of the Jordan block $J_k(\lambda)$. Thus, each Jordan chain of length k gives rise to a Jordan block $J_k(\lambda)$ in the Jordan form of A. The vectors in the Jordan chain form the columns of the transformation matrix V that diagonalizes A into its Jordan form.

■ **Example 13.5** Consider Example 13.3, where we computed the eigenvalues -1 and 2 along with the eigenvectors v_1 and v_2. Since the algebraic multiplicity of $\lambda = 2$

is 2 but only one eigenvector v_2 was found, we compute a generalized eigenvector v_3:

$$(A - 2I)v_3 = v_2 \quad \Rightarrow \quad v_3 = \begin{bmatrix} 0 \\ 1 \\ 0 \end{bmatrix}.$$

Using v_1, v_2, and v_3 as columns, we construct the matrix V shown in Example 13.3.   ∎

Total Length of Jordan Blocks and Uniqueness of Jordan Decomposition

For the eigenvalue λ, the Jordan chain generation process can be repeated for each of the g independent eigenvectors, resulting in the construction of g Jordan blocks. Consequently, the number of Jordan blocks associated with λ is equal to its geometric multiplicity g, with each Jordan block contributing precisely one linearly independent eigenvector.

The total length of all Jordan chains associated with λ, which equals the sum of the sizes of the Jordan blocks, is always equal to the algebraic multiplicity m. Furthermore, the vectors within each Jordan chain are linearly independent. Lastly, the Jordan form of a matrix A is unique up to the order of its Jordan blocks. However, a formal proof of these results is beyond the scope of this text.

Exercise 13.2 Compute the Jordan decomposition of

$$A = \begin{bmatrix} 1 & -1 & -2 & 3 \\ 0 & 0 & -2 & 3 \\ 0 & 1 & 1 & -1 \\ 0 & 0 & -1 & 2 \end{bmatrix}.$$

3 Applications

The Jordan decomposition is a powerful tool for analyzing the structure of matrices that are not fully diagonalizable. It quantifies the extent of a matrix's deviation from diagonalizability by revealing the presence and size of Jordan blocks associated with its eigenvalues. This form has numerous applications in mathematics and quantum computing, as illustrated below:

- Analyzing Non-Diagonalizable Operators: In quantum computing, operations are typically represented by unitary or Hermitian matrices, which are diagonalizable. However, non-diagonalizable matrices can arise in scenarios such as open quantum systems or noisy processes. For instance, in quantum error correction, noise models often involve non-unitary operations that are not diagonalizable. The Jordan form enables the decomposition of such matrices into simpler components (Jordan blocks), offering valuable insights into their structure and behavior.

- Time Evolution in Open Quantum Systems: The dynamics of quantum systems interacting with open environments (e.g., thermal baths) are often governed by non-Hermitian Hamiltonians or Lindblad operators. These operators may not be diagonalizable, but their Jordan form allows the effective representation of the system's time evolution.

- Decomposition of Quantum Channels: Quantum channels, which describe the evolution of density matrices, are often represented by Kraus operators or matrix forms. In some cases, applying the Jordan decomposition simplifies the analysis of these matrices.

- Simplifying Solutions of Matrix Differential Equations: In quantum control and simulation, differential equations involving non-Hermitian matrices or dissipative systems frequently appear. These equations often require the computation of matrix exponentials, which is simplified by transforming the matrices into their Jordan form.

  Specifically, for a matrix A, the solution to $\frac{d}{dt}\boldsymbol{v}(t) = A\boldsymbol{v}(t)$ is given by $\boldsymbol{v}(t) = e^{At}\boldsymbol{v}(0)$. When A is non-diagonalizable, the Jordan form provides an efficient method for computing e^{At} through its Jordan blocks.

13.2 Singular Value Decomposition

While a square matrix with a complete set of linearly independent eigenvectors, particularly a normal matrix, admits an eigenvalue decomposition, any matrix, including non-square ones, can be decomposed through a more generalized method known as Singular Value Decomposition (SVD), also referred to as the Eckart-Young decomposition. Since SVD provides insights into the geometric and structural properties of the original matrix, it serves as a fundamental tool for understanding and analyzing matrices.

13.2.1 Fundamental Concepts

1 The SVD Theorem

The Singular Value Decomposition (SVD) is a cornerstone of linear algebra:

Theorem 13.2 — Singular Value Decomposition (SVD). Any $m \times n$ matrix A can be factored into three matrices:

$$A = U\Sigma V^{\dagger}, \tag{13.15}$$

where:

- Σ is a diagonal matrix with the same dimensions as A, containing non-negative real numbers on the diagonal, arranged in descending order as $\sigma_1, \sigma_2, \ldots$. These are known as the singular values of A.

- U is an $m \times m$ unitary matrix whose columns are called the left singular vectors of A.

- V is an $n \times n$ unitary matrix whose columns are called the right singular vectors of A.

Moreover, if the rank of A is r, which cannot exceed the smaller of m and n, then A has exactly r non-zero singular values.

(i) Unlike eigenvalues, which can be complex numbers, singular values are always non-negative real numbers by definition.

Next, we will explore the implications of SVD and develop an intuitive understanding of this powerful matrix decomposition before presenting its formal proof.

2 Dimensionality Relations

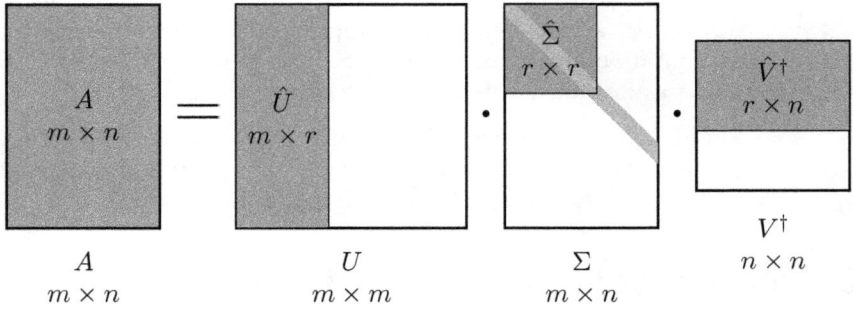

Figure 13.1: SVD Matrix Shapes and Dimensions

The component matrices of SVD, as well as the relationships between m, n, and r, are illustrated in Fig. 13.1. The figure shows the case where $m \geq n$, but the analysis is equally valid for $m < n$. Since only non-zero singular values contribute to the product $U\Sigma V^\dagger$, the submatrices corresponding to the non-zero singular values are highlighted and labeled as $\hat{U}$, $\hat{\Sigma}$, and $\hat{V}^\dagger$.

It follows that A can be expressed in terms of the reduced component matrices $\hat{U}$, $\hat{\Sigma}$, and $\hat{V}^\dagger$, which have dimensions $m \times r$, $r \times r$, and $r \times n$, respectively:

$$A = \hat{U}\hat{\Sigma}\hat{V}^\dagger. \tag{13.16}$$

The entries of U and V outside $\hat{U}$ and $\hat{V}$ do not contribute to A.

3 Alternative Representations

Since Σ is diagonal, Eq. 13.15 can be written as:

$$A = \sum_{i=1}^{r} \sigma_i \boldsymbol{u}_i \boldsymbol{v}_i^\dagger, \tag{13.17}$$

where $\boldsymbol{u}_i$ and $\boldsymbol{v}_i$ are the column vectors of U and V, respectively.

Using Dirac notation, this can be expressed as:

$$A = \sum_{i=1}^{r} \sigma_i \left| u_i \right\rangle \left\langle v_i \right|. \tag{13.18}$$

This form is often referred to as the *Rank-1 Decomposition* because each term is a column-row product, a rank-1 matrix.

Since only the first r singular values are non-zero, the summation limits in the above equations can be reduced from n to r.

Moreover, due to the unitarity of V, the SVD decomposition in Eq. 13.15 can be rewritten as:

$$AV = U\Sigma. \tag{13.19}$$

Equivalently,

$$Av_i = \sigma_i u_i, \quad \text{or,} \quad A\ket{v_i} = \sigma_i \ket{u_i}. \tag{13.20}$$

In matrix form, this becomes:

$$A \begin{bmatrix} \vdots & \vdots & \vdots & \vdots \\ v_1 & v_2 & \cdots & v_n \\ \vdots & \vdots & \vdots & \vdots \end{bmatrix} = \begin{bmatrix} \vdots & \vdots & \vdots & \vdots \\ \sigma_1 u_1 & \sigma_2 u_2 & \cdots & \sigma_n u_n \\ \vdots & \vdots & \vdots & \vdots \end{bmatrix}. \tag{13.21}$$

Exercise 13.3 Let Σ_A denote the singular value matrix of A. Show that $\Sigma_{A^\dagger} = (\Sigma_A)^T$.

Exercise 13.4 If A is a square matrix, show that $\det A$ is the product of the singular values of A.

i The term "singular value" has historical roots in linear algebra, where a matrix is termed singular if it is not invertible. While the smallest singular value does provide some indication of how close a matrix is to being singular, SVD offers a much broader understanding of a matrix's structure and properties. It identifies the dominant directions of a linear transformation, enables efficient low-rank approximations, and has numerous applications in fields ranging from machine learning to quantum computing.

13.2.2 Geometric Interpretations

The SVD reveals the essential geometry of a linear transformation by showing that every linear transformation can be expressed as a composition of three fundamental operations. Interpreting Eq. 13.15 from right to left:

- $V^\dagger$ represents a rotation or reflection of vectors in the n-dimensional domain.

- Σ represents a linear scaling (dilation or contraction) along the n coordinate axes. If $n \neq m$, this step also embeds the m-dimensional domain into the n-dimensional range or projects it onto an m-dimensional subspace.

- U represents a rotation or reflection of vectors in the m-dimensional range.

■ **Example 13.6** Consider a transformation from $\mathbb{R}^2$ to $\mathbb{R}^2$ to illustrate the concept. This example involves only 2D real vectors, making it easy to visualize (see Fig. 13.2).

Take the following 2×2 matrix:

$$A = \frac{1}{3} \begin{bmatrix} 3 & 0 \\ 4 & 5 \end{bmatrix}.$$

The U, Σ, and V matrices in the SVD decomposition of A are:

$$U = \frac{1}{\sqrt{10}} \begin{bmatrix} 1 & -3 \\ 3 & 1 \end{bmatrix}, \quad \Sigma = \frac{\sqrt{5}}{3} \begin{bmatrix} 3 & 0 \\ 0 & 1 \end{bmatrix}, \quad V = \frac{1}{\sqrt{2}} \begin{bmatrix} 1 & -1 \\ 1 & 1 \end{bmatrix}.$$

The matrices U, Σ, and V are typically constructed through the diagonalization of $AA^\dagger$ and $A^\dagger A$. While this process can be complex, it is now routinely performed by computer algorithms.

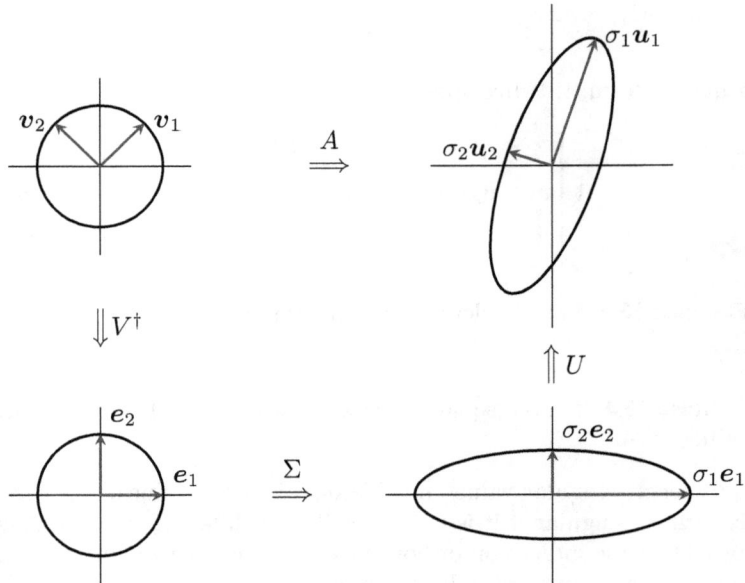

Figure 13.2: Illustration of SVD as $\mathbb{R}^2 \rightarrow \mathbb{R}^2$ (Example 13.6)

In this example, A acts on vectors in 2D space (domain) and maps them into another 2D space (range). The domain $\mathbb{R}^2$ is represented by the basis vectors:

$$v_1 = \frac{1}{\sqrt{2}} \begin{bmatrix} 1 \\ 1 \end{bmatrix}, \quad v_2 = \frac{1}{\sqrt{2}} \begin{bmatrix} -1 \\ 1 \end{bmatrix}.$$

$V^\dagger$ transforms these into the standard basis vectors (and V would perform the inverse transformation):

$$e_1 = \begin{bmatrix} 1 \\ 0 \end{bmatrix}, \quad e_2 = \begin{bmatrix} 0 \\ 1 \end{bmatrix}.$$

Σ scales e_1 and e_2 using the singular values $\sqrt{5}$ and $\sqrt{5}/3$, respectively:

$$\sigma_1 e_1 = \sqrt{5} \begin{bmatrix} 1 \\ 0 \end{bmatrix}, \quad \sigma_2 e_2 = \frac{\sqrt{5}}{3} \begin{bmatrix} 0 \\ 1 \end{bmatrix}.$$

Finally, U rotates these scaled vectors into the following:

$$\sigma_1 u_1 = \frac{1}{\sqrt{2}} \begin{bmatrix} 1 \\ 3 \end{bmatrix}, \quad \sigma_2 u_2 = \frac{1}{3\sqrt{2}} \begin{bmatrix} -3 \\ 1 \end{bmatrix}.$$

∎

Exercise 13.5 Describe the geometric interpretation of the SVD of $A^\dagger$, where A is given in Example 13.6.

■ **Example 13.7** This example illustrates a transformation from $\mathbb{R}^2$ to $\mathbb{R}^3$, as depicted in Fig. 13.3. It generalizes the geometric interpretation of SVD, particularly in the context of changing dimensionality through embedding or projection.

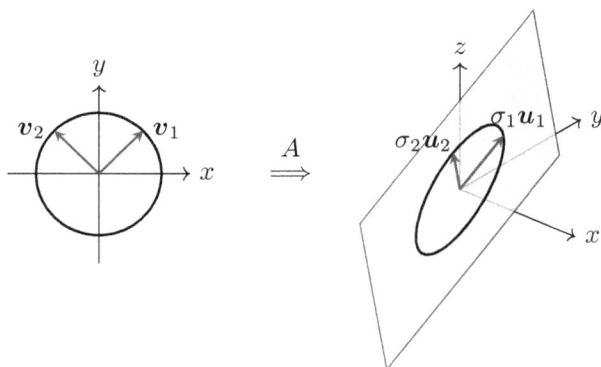

Figure 13.3: Illustration of SVD as $\mathbb{R}^2 \to \mathbb{R}^3$ (Example 13.7)

Consider the following 3×2 matrix:

$$A = \begin{bmatrix} 1 & 0 \\ 0 & 1 \\ 1 & 1 \end{bmatrix}.$$

The U, Σ, and V matrices in the SVD decomposition of A are:

$$U = \frac{1}{\sqrt{6}} \begin{bmatrix} 1 & -\sqrt{3} & \sqrt{2} \\ 1 & \sqrt{3} & \sqrt{2} \\ 2 & 0 & -\sqrt{2} \end{bmatrix}, \quad \Sigma = \begin{bmatrix} \sqrt{3} & 0 \\ 0 & 1 \\ 0 & 0 \end{bmatrix}, \quad V = \frac{1}{\sqrt{2}} \begin{bmatrix} 1 & -1 \\ 1 & 1 \end{bmatrix}.$$

Right Singular Vectors: Orthonormal Basis for the Domain Space

The right singular vectors (columns of V) form an orthonormal basis for the original 2D domain space. These basis vectors identify the directions in which the transformation A stretches or shrinks the unit circle most and least significantly.

Singular Values: Scaling Factors

The non-zero singular values indicate the scaling factors applied to the unit circle in the original 2D space along the two orthogonal directions defined by v_1 and v_2.

The largest singular value, $\sqrt{3}$, corresponds to the direction (v_1) where A stretches the unit circle the most, while the smaller singular value, 1, corresponds to the direction (v_2) where the stretch is minimal. Singular values therefore measure the "dominance" or "importance" of different directions in the transformation A.

Left Singular Vectors: Orthonormal Basis for the Transformed Space

The left singular vectors (columns of U) represent an orthonormal basis for the transformed space. These vectors are obtained by applying the transformation A to the right singular vectors (columns of V) and normalizing the resulting vectors.

Specifically, the first column of U, u_1, represents Av_1 normalized to unit length, and the second column, u_2, represents Av_2 normalized similarly. Together, u_1 and u_2 span the 2D subspace in the 3D range.

Space Embedding and Projection

In this example, Σ includes a third row without a diagonal element, reflecting the dimensionality increase from 2D to 3D. The transformed vectors occupy only the subspace spanned by u_1 and u_2 within the 3D space. The third singular vector, u_3, is orthogonal to this subspace and represents the part of 3D space not reached by the transformation. Thus, the transformed space is a 2D subspace embedded in 3D.

Taking the conjugate transpose of $A = U\Sigma V^\dagger$ in Eq. 13.15, we obtain:

$$A^\dagger = V\Sigma^T U^\dagger. \tag{13.22}$$

This shows that for $A^\dagger$, the roles of U and V are reversed, and Σ is transposed. In this case, $A^\dagger$ represents a transformation from 3D to 2D with the same singular values. However, Σ^T becomes a 2×3 matrix, with the third row omitted, leading to a reduction in dimensionality. Consequently, $A^\dagger$ projects the 3D space onto a 2D subspace.

Dimensionality reduction can also result from zero singular values. For instance, in Example 13.7, if the second singular value were zero, the transformed space would collapse to a line in 3D space, effectively becoming a 1D subspace embedded in 3D. ∎

13.2.3 ∗Constructing and Proving SVD

Having developed an intuitive understanding of SVD, we are now prepared to delve into the rigorous mathematics behind its construction and proof.

1   From SVD to $AA^\dagger$ and $A^\dagger A$

Theorem 13.3 If the SVD theorem holds, then U consists of the orthonormal eigenvectors of $AA^\dagger$, and V consists of the orthonormal eigenvectors of $A^\dagger A$.

Furthermore, the singular values of A (σ_i) are the square roots of the eigenvalues of $AA^\dagger$ and $A^\dagger A$ (λ_i):

$$\sigma_i = \sqrt{\lambda_i}. \tag{13.23}$$

Proof.   Starting from the SVD decomposition of A, we have:

$$AA^\dagger = (U\Sigma V^\dagger)(U\Sigma V^\dagger)^\dagger = U\Sigma V^\dagger V\Sigma^\dagger U^\dagger = U\Sigma\Sigma^\dagger U^\dagger,$$
$$A^\dagger A = (U\Sigma V^\dagger)^\dagger(U\Sigma V^\dagger) = V\Sigma^\dagger U^\dagger U\Sigma V^\dagger = V\Sigma^\dagger\Sigma V^\dagger.$$

Here, we used the unitarity of U and V.

From this, the eigenvalues of $AA^\dagger$ and $A^\dagger A$ are given by $\Sigma\Sigma^\dagger$ and $\Sigma^\dagger\Sigma$, respectively.

Since Σ is diagonal with its diagonal elements (the singular values) arranged in descending order, the non-zero diagonal elements of $\Sigma\Sigma^\dagger$ and $\Sigma^\dagger\Sigma$ are identical, even though these matrices may have different sizes.

This demonstrates that the singular values of A (σ_i) are the square roots of the eigenvalues of $AA^\dagger$ and $A^\dagger A$ (λ_i): $\sigma_i = \sqrt{\lambda_i}$.   □

2 Proof of the SVD Theorem by Constructing SVD from $AA^\dagger$ and $A^\dagger A$

Building on the intuition developed in Theorem 13.3, we can explicitly construct U, V, and Σ from $A^\dagger A$ and A. This construction process not only provides a practical method for obtaining the SVD but also serves as a proof of the SVD theorem (Theorem 13.2).

First, note that $A^\dagger A$ is Hermitian and thus has a spectral decomposition:

$$A^\dagger A = V\Lambda V^\dagger,$$

where V is a unitary matrix, and Λ is diagonal. Rearranging, we have:

$$\Lambda = V^\dagger A^\dagger AV = (AV)^\dagger AV.$$

The eigenvalues λ_i (diagonal elements of Λ) are non-negative, since:

$$\lambda_i = v_i^\dagger A^\dagger Av_i = \langle Av_i, Av_i \rangle \geq 0,$$

where v_i is the ith column of V. This confirms that $A^\dagger A$ is positive semidefinite.

Let r be the number of positive eigenvalues. Arranging the eigenvalues of Λ in descending order corresponds to a particular configuration of V. Define:

$$u_i = \frac{1}{\sigma_i} Av_i \quad \text{for } i = 1, 2, \ldots, r, \tag{13.24}$$

where $\sigma_i = \sqrt{\lambda_i}$. Then:

$$u_i^\dagger u_j = \left(\frac{Av_i}{\sigma_i}\right)^\dagger \left(\frac{Av_j}{\sigma_j}\right) = \frac{v_i^\dagger A^\dagger Av_j}{\sigma_i \sigma_j} = \frac{\lambda_j v_i^\dagger v_j}{\sigma_i \sigma_j} = \delta_{ij}.$$

Thus, $\{u_i\}$ forms an orthonormal set.

Let the size of A be $m \times n$. Then $A^\dagger A$ and V are $n \times n$ matrices, and the dimension of each u_i is m. If $r < m$, we can augment $\{u_i\}$ to a set of m orthonormal vectors using the Gram-Schmidt procedure. These additional vectors correspond to zero singular values and do not contribute to A. Ensuring their orthogonality to the existing vectors, we can construct a unitary matrix U from the augmented set $\{u_i\}$.

Finally, construct an $m \times n$ matrix Σ with σ_i as the first r diagonal elements and 0 elsewhere. These U, V, and Σ satisfy $A = U\Sigma V^\dagger$, completing the proof of the SVD theorem.

Exercise 13.6 Construct the SVD of A starting with U derived from $AA^\dagger$.

Exercise 13.7 Given a linear map $A : \mathbb{C}^n \to \mathbb{C}^m$, find two orthonormal bases $\{v_i\}$ for $\mathbb{C}^n$ and $\{u_i\}$ for $\mathbb{C}^m$, such that $\{v_i\}$ is transformed to $\{u_i\}$ under A.

13.2.4 $*$ SVD and Subspaces of Matrices

1 Four Fundamental Subspaces of a Matrix

Recall that a matrix $A \in \mathbb{C}^{m \times n}$ gives rise to four fundamental subspaces (introduced in § 8.5.3):

1. Row Space: The subspace spanned by the adjoint of the row vectors of A, i.e., $R(A) = \{A^\dagger x \mid x \in \mathbb{C}^m\}$. $R(A) \subseteq \mathbb{C}^n$.

2. Null Space (or Kernel): The set of all vectors x such that $Ax = 0$, i.e., $N(A) = \{x \in \mathbb{C}^n \mid Ax = 0\}$. $N(A) \subseteq \mathbb{C}^n$.

3. Column Space: The subspace spanned by the column vectors of A, i.e., $C(A) = \{Ax \mid x \in \mathbb{C}^n\}$. $C(A) \subseteq \mathbb{C}^m$.

4. Left Null Space: Defined as $\{x \in \mathbb{C}^m \mid x^\dagger A = 0\}$. Since $A^\dagger x = (x^\dagger A)^\dagger = 0$, the left null space of A is equivalent to the (right) null space of $A^\dagger$, i.e., $N(A^\dagger) = \{x \in \mathbb{C}^m \mid A^\dagger x = 0\}$. $N(A^\dagger) \subseteq \mathbb{C}^m$.

> ⓘ By convention, all four fundamental subspaces are treated as subspaces of column vectors. Hence, the adjoint in the definition of $R(A)$ and $N(A^\dagger)$.
>
> Note that in this convention, $R(A) = C(A^\dagger)$ instead of $R(A) = C(A^T)$.

2   Rank-Nullity Theorem

The row and column spaces share the same dimension, referred to as the rank of A. The dimensions of the null space and left null space are referred to as the nullity of A and $A^\dagger$, respectively. These dimensions are related by the following relationship:

Theorem 13.4 — Rank-Nullity Theorem.

$$\text{rank}(A) + \text{nullity}(A) = n,$$
$$\text{rank}(A) + \text{nullity}(A^\dagger) = m.$$

---

3   SVD Partition of the Subspaces

The SVD of a matrix A provides a complete decomposition of its domain and codomain into orthogonal subspaces, offering a comprehensive view of how A transforms vectors. In the singular value decomposition $A = U\Sigma V^\dagger$, the columns of V form an orthonormal basis for the domain space, and the columns of U form an orthonormal basis for the codomain space. Specifically:

Theorem 13.5 Given $A \in \mathbb{C}^{m \times n}$ with r non-zero singular values followed by zero singular values:

1. The number of non-zero singular values (r) equals the dimension of the row space and the column space of A, i.e., the rank of A.

2. The first r columns of V, $\{v_1, \ldots, v_r\}$, form an orthonormal basis for the row space $R(A)$.

3. The last $n - r$ columns of V, $\{v_{r+1}, \ldots, v_n\}$, form an orthonormal basis for the null space $N(A)$.

4. The first r columns of U, $\{u_1, \ldots, u_r\}$, form an orthonormal basis for the column space $C(A)$.

5. The last $m - r$ columns of U, $\{u_{r+1}, \ldots, u_m\}$, form an orthonormal basis for the left null space $N(A^\dagger)$.

Before delving into a rigorous proof, we first illustrate this theorem with an example.

■ **Example 13.8** SVD partitions the subspaces of two matrices (adjoint of each other) as shown:

$$A \quad = \quad U \qquad \Sigma \qquad V^\dagger$$

(a)

$\in C(A)$ $\in R(A)$

$$A = \begin{bmatrix} 1 & 0 & 1 \\ 0 & 1 & 1 \end{bmatrix} = \frac{1}{\sqrt{2}} \begin{bmatrix} 1 & -1 \\ 1 & 1 \end{bmatrix} \begin{bmatrix} \sqrt{3} & 0 & 0 \\ 0 & 1 & 0 \end{bmatrix} \frac{1}{\sqrt{6}} \begin{bmatrix} 1 & -\sqrt{3} & \sqrt{2} \\ 1 & \sqrt{3} & \sqrt{2} \\ 2 & 0 & -\sqrt{2} \end{bmatrix}$$

$N(A^\dagger) = \emptyset$ $\in N(A)$

(b)

$\in C(A)$ $\in R(A)$

$$A = \begin{bmatrix} 1 & 0 \\ 0 & 1 \\ 1 & 1 \end{bmatrix} = \frac{1}{\sqrt{6}} \begin{bmatrix} 1 & -\sqrt{3} & \sqrt{2} \\ 1 & \sqrt{3} & \sqrt{2} \\ 2 & 0 & -\sqrt{2} \end{bmatrix} \begin{bmatrix} \sqrt{3} & 0 \\ 0 & 1 \\ 0 & 0 \end{bmatrix} \frac{1}{\sqrt{2}} \begin{bmatrix} 1 & -1 \\ 1 & 1 \end{bmatrix}$$

$N(A^\dagger)$ $N(A) = \emptyset$

■

Proof.

1. Rank

Consider the SVD product:
$$A = U\Sigma V^\dagger.$$

Since U and V are unitary, they have full rank. Thus, the rank of A is determined by the rank of Σ. The matrix Σ is diagonal (with possibly rectangular shape), and its rank is the number of non-zero diagonal entries. Therefore, the number of non-zero singular values is precisely $r = \text{rank}(A)$.

2. Row Space

By definition:
$$R(A) = \{A^\dagger x \mid x \in \mathbb{C}^m\} \subseteq \mathbb{C}^n.$$

Using the SVD:
$$A^\dagger x = V\Sigma^\dagger U^\dagger x = (V\Sigma^\dagger)y,$$

where $y = U^\dagger x$.

Since U is unitary, there is a one-to-one mapping between x and y, both covering $\mathbb{C}^m$. Thus:
$$R(A) = \{(V\Sigma^\dagger)y \mid y \in \mathbb{C}^m\} = C(V\Sigma^\dagger).$$

$\Sigma^\dagger$ contains a leading real diagonal $r \times r$ block (corresponding to the r non-zero singular values of A), with zeros elsewhere. Therefore, $\Sigma^\dagger$ affects only the first r components of any vector, selecting the corresponding columns of V. Hence, $V\Sigma^\dagger$ effectively selects only the first r columns of V (see Example 13.8 for a visual illustration). Therefore:
$$R(A) = \text{span}\{v_1, \ldots, v_r\}. \qquad (13.25)$$

These vectors are orthonormal since V is unitary.

3. Null Space

By definition:

$$N(A) = \{x \in \mathbb{C}^n \mid Ax = 0\}.$$

Using the SVD:

$$Ax = U\Sigma V^\dagger x = U(\Sigma y) = 0,$$

where $y = V^\dagger x$.

Since U is unitary, we have $\Sigma y = 0$. For this to hold, the first r components of y must be zero. This means y lies in the subspace spanned by the last $n - r$ standard basis vectors in $\mathbb{C}^n$.

Translating y back to x using V:

$$x = Vy \in \text{span}\{v_{r+1}, \ldots, v_n\}.$$

Therefore:

$$N(A) = \text{span}\{v_{r+1}, \ldots, v_n\}. \tag{13.26}$$

4. Column Space

By definition:

$$C(A) = \{Ax \mid x \in \mathbb{C}^n\} \subseteq \mathbb{C}^m.$$

Using the SVD:

$$Ax = U\Sigma V^\dagger x = (U\Sigma)y,$$

where $y = V^\dagger x$.

Since U is unitary, there is a one-to-one mapping between x and y, both covering $\mathbb{C}^n$. The action of Σ in $U\Sigma$ effectively selects the first r columns of U. Thus, Ax must lie in the span of $\{u_1, \ldots, u_r\}$. Hence:

$$C(A) = \text{span}\{u_1, \ldots, u_r\}. \tag{13.27}$$

5. Left Null Space

Note in the SVD on $A^\dagger$:

$$A^\dagger = V\Sigma^\dagger U^\dagger,$$

the roles of U and V are switched compared to the SVD of A.

Applying the results from part (2) to $A^\dagger$ instead of A, we find:

$$N(A^\dagger) = \text{span}\{u_{r+1}, \ldots, u_m\}. \tag{13.28}$$

□

Exercise 13.8 If A is $m \times n$ with all singular values positive, what is the rank of A?

13.2.5 Uniqueness of Singular Values

The singular values of a matrix, being non-negative real numbers, are unique. This follows from the fact that the singular values of A are the square roots of the eigenvalues of $A^{\dagger}A$ (see § 13.2.3), and the eigenvalues of a matrix (and consequently their square roots) are uniquely determined, up to their order.

However, the singular vectors (i.e., the matrices U and V) are not unique. The columns of U (left singular vectors) and V (right singular vectors) corresponding to distinct singular values are unique only up to a phase factor—a complex scalar of unit modulus in the case of complex matrices, or a sign change in the case of real matrices. For repeated singular values, any orthonormal basis for the corresponding singular subspace is valid, leading to non-uniqueness.

13.2.6 Relationship to Eigenvalue Decomposition

1 Fundamental Differences

When comparing SVD to eigenvalue decomposition, as represented in Eq. 13.4, several key differences emerge. The SVD, written as $A = U\Sigma V^{\dagger}$, involves diagonalizing a matrix using two orthonormal bases, whereas eigenvalue decomposition, $A = X\Lambda X^{-1}$, uses a single basis. SVD is applicable to any $m \times n$ matrix, while eigenvalue decomposition is restricted to $n \times n$ square matrices with n linearly independent eigenvectors. Although SVD offers a more comprehensive analysis, it requires two sets of singular vectors.

2 Conditions for Coincidence

For SVD and eigenvalue decomposition to coincide, A must have orthonormal eigenvectors, enabling $X = U = V$. Additionally, the eigenvalues must be non-negative, allowing $\Lambda = \Sigma$. Thus, A must be a positive semidefinite Hermitian matrix for the decompositions to align perfectly, with $A = X\Lambda X^{-1}$ and $A = U\Lambda U^{\dagger}$ both matching $A = U\Sigma V^{\dagger}$.

For a general Hermitian matrix, even if not positive semidefinite, its SVD can still be derived from its eigendecomposition. If $A = U\Lambda U^{\dagger}$ and Λ contains negative elements, we can introduce a diagonal sign matrix T with entries 1 or -1 such that $T\Lambda$ has only non-negative elements. Then:

$$A = (UT)(T\Lambda)U^{\dagger},$$

which represents the SVD of A.

3 Positivity

A Hermitian matrix A is positive definite if, for all non-zero vectors $\boldsymbol{x}$, $\boldsymbol{x}^{\dagger}A\boldsymbol{x} > 0$. The eigenvalues of a positive definite matrix are all positive (see § 11.1.2.4). In this case, the SVD and eigenvalue decomposition coincide, meaning that a Hermitian matrix is positive definite if and only if all its singular values are strictly positive.

Exercise 13.9 Can we say that a Hermitian matrix is positive semidefinite if and only if all of its singular values are non-negative? Why?

4 Stability of Decompositions

Singular values are generally more stable than eigenvalues. Small perturbations in the elements of a matrix A can result in significant changes to its eigenvalues, while the singular values remain relatively stable. This distinction is illustrated in the following example.

■ **Example 13.9** Consider the matrix:

$$A = \begin{bmatrix} 0 & 1 & 0 & 0 \\ 0 & 0 & 2 & 0 \\ 0 & 0 & 0 & 3 \\ 0 & 0 & 0 & 0 \end{bmatrix}.$$

The SVD decomposition of A yields the matrices:

$$U = \begin{bmatrix} 0 & 0 & 1 & 0 \\ 0 & 1 & 0 & 0 \\ 1 & 0 & 0 & 0 \\ 0 & 0 & 0 & 1 \end{bmatrix}, \quad \Sigma = \begin{bmatrix} 3 & 0 & 0 & 0 \\ 0 & 2 & 0 & 0 \\ 0 & 0 & 1 & 0 \\ 0 & 0 & 0 & 0 \end{bmatrix}, \quad V = \begin{bmatrix} 0 & 0 & 0 & 1 \\ 0 & 0 & 1 & 0 \\ 0 & 1 & 0 & 0 \\ 1 & 0 & 0 & 0 \end{bmatrix}.$$

The SVD can also be expressed as a rank-1 decomposition:

$$A = U\Sigma V^{\dagger} = 3u_1 v_1^{\dagger} + 2u_2 v_2^{\dagger} + 1u_3 v_3^{\dagger}.$$

Note that all eigenvalues of A are zero. However, if we perturb one element slightly:

$$B = \begin{bmatrix} 0 & 1 & 0 & 0 \\ 0 & 0 & 2 & 0 \\ 0 & 0 & 0 & 3 \\ \frac{1}{60000} & 0 & 0 & 0 \end{bmatrix},$$

the characteristic equation $\lambda^4 - \frac{6}{60000} = 0$ yields eigenvalues $\frac{1}{10}\{1, i, -1, -i\}$, showing a dramatic change. Meanwhile, the SVD of B is adjusted by a minor additional term:

$$B = U\Sigma V^{\dagger} = 3u_1 v_1^{\dagger} + 2u_2 v_2^{\dagger} + 1u_3 v_3^{\dagger} + \frac{1}{60000} u_4 v_4^{\dagger}.$$

■

13.2.7 ✳ Applications

The Singular Value Decomposition (SVD) has numerous applications across various fields, including:

1. Image Compression: SVD approximates images by retaining only the largest singular values, reducing storage requirements while preserving essential features.

2. Principal Component Analysis (PCA): In PCA, SVD identifies the principal components of a dataset, corresponding to directions of maximum variance.

3. Regularization of Near-Singular Matrices: SVD stabilizes the inversion of near-singular matrices by truncating or damping small singular values, reducing sensitivity to noise.

4. Quantum Machine Learning: SVD enables dimensionality reduction, feature extraction, and classification in quantum machine learning algorithms, such as quantum support vector machines and quantum principal component analysis, leading to computational speedups.

5. Quantum Simulation: In quantum simulation algorithms, SVD is employed to represent and manipulate quantum systems efficiently. By decomposing the Hamiltonian (energy operator) into its singular values and vectors, quantum simulators can better study the system's dynamics and properties.

We illustrate the principles of these applications through examples of near-singular matrix regularization and image processing.

1 Regularization of Near-Singular Matrices Using SVD

When inverting near-singular (or ill-conditioned) matrices, small perturbations or numerical errors can lead to large changes in the computed inverse, making standard inversion methods unreliable. SVD regularization mitigates this issue by modifying the singular values of the matrix to stabilize the inversion process.

As an example, suppose we want to solve the linear system $Ax = b$, where A is:

$$A = \begin{bmatrix} 10 & 7 & 8.1 \\ 7.2 & 5 & 6.3 \\ 9.2 & 6.3 & 9 \end{bmatrix}.$$

Here, A may not accurately represent a physical problem; noise or other factors may have perturbed the matrix. In real-world problems, such matrices often have very large dimensions.

The solution can be computed from $x = A^{-1}b$. However, A is nearly singular, as one of its singular values is very small:

$$\Sigma = \text{diag}(23.14, 0.9617, 0.00162).$$

Standard inversion methods are unreliable because small perturbations or numerical errors can lead to large changes in the computed A^{-1}.

Using SVD, A^{-1} can be computed as:

$$A^{-1} = V\Sigma^{-1}U^\dagger,$$

where Σ^{-1} is the diagonal matrix with inverted singular values:

$$\Sigma^{-1} = \text{diag}\left(\frac{1}{23.14}, \frac{1}{0.9617}, \frac{1}{0.00162}\right) = \text{diag}(0.0432, 1.040, 618).$$

The extremely large value 618, corresponding to the smallest singular value 0.00162, indicates instability, as it amplifies small numerical errors in b. The goal is to approximate the physical problem while mitigating this instability.

Truncated SVD (TSVD)

With the TSVD method, we ignore small singular values and compute A^{-1} using only the terms corresponding to large singular values. For example, setting a threshold $\epsilon = 0.005$, we ignore the last element in Σ^{-1} corresponding to the singular value 0.00162:

$$\Sigma^{-1}_{\text{trunc}} = \text{diag}(0.0432, 1.040, 0).$$

The pseudo-inverse is then computed as:

$$A^{-1}_{\text{TSVD}} = V\Sigma^{-1}_{\text{trunc}}U^\dagger.$$

This method stabilizes the inversion process by eliminating the contribution of the smallest singular value.

Tikhonov Regularization

Instead of truncating small singular values, Tikhonov regularization modifies them by adding a small regularization parameter λ^2 (e.g., $\lambda = 0.005$):

$$\sigma_i^{\text{reg}} = \frac{\sigma_i}{\sigma_i^2 + \lambda^2}.$$

The regularized inverse is then computed as:

$$A_{\text{reg}}^{-1} = V\Sigma_{\text{reg}}^{-1}U^\dagger,$$

where:

$$\Sigma_{\text{reg}}^{-1} = \text{diag}(0.0432, 1.040, 58.57).$$

Tikhonov regularization retains some information from the smallest singular value while damping its impact, achieving a balance between stability and accuracy.

2 Noise Reduction and Image Compression Using SVD

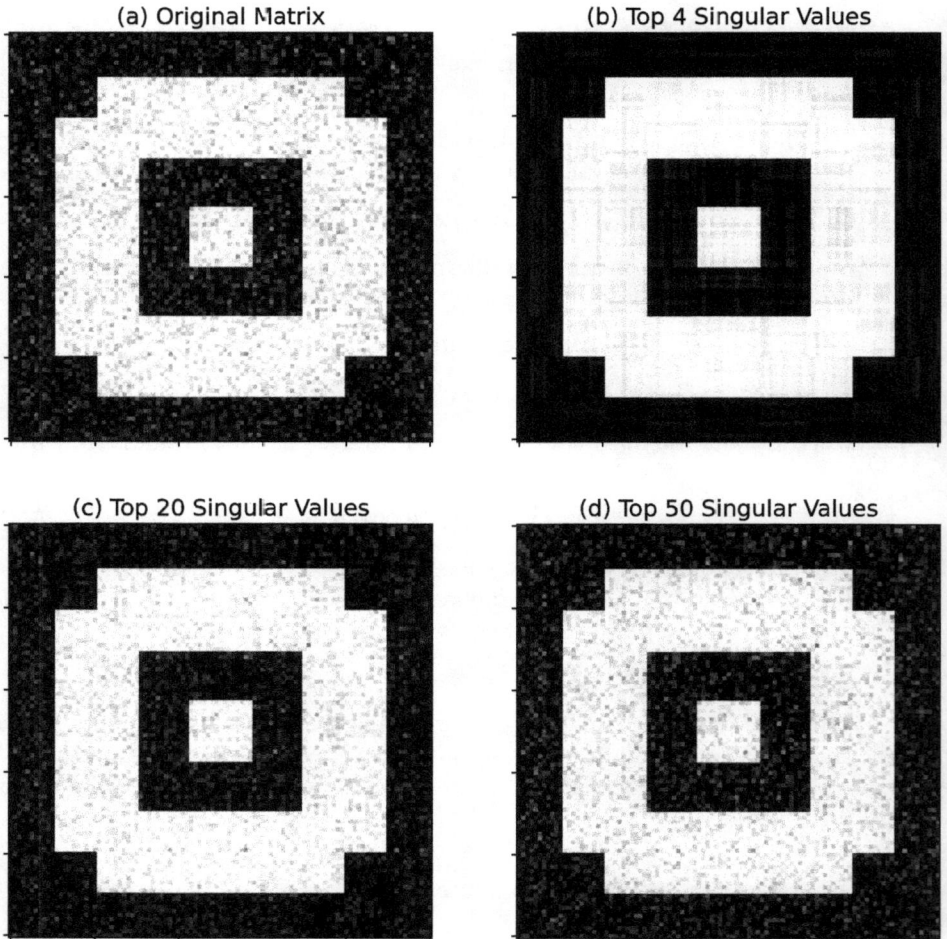

Figure 13.4: Example of Noise Reduction and Image Compression Using SVD

In Fig. 13.4, part (a) shows an image with 101×101 pixels. Each pixel is represented by a value from 0 to 1, where 0 denotes black, 1 denotes white, and intermediate values represent gray shades. The image is predominantly black-and-white, with some noise appearing as random grayish specks.

The image is represented by a 101×101 matrix A. The SVD of A yields the following singular values:

$$\sigma_1 = 60.5, \ \sigma_2 = 26.6, \ \sigma_3 = 10.3, \ \sigma_4 = 9.18, \ \sigma_5 = 2.96, \ldots,$$
$$\sigma_{10} = 2.59, \ \ldots, \ \sigma_{20} = 2.14, \ldots, \ \sigma_{50} = 1.25, \ldots, \ \sigma_{80} = 0.46, \ldots.$$

The singular values decrease progressively, reflecting the diminishing contributions of successive terms in the rank-1 representation:

$$A = \sigma_1 \boldsymbol{u}_1 \boldsymbol{v}_1^\dagger + \sigma_2 \boldsymbol{u}_2 \boldsymbol{v}_2^\dagger + \cdots + \sigma_r \boldsymbol{u}_r \boldsymbol{v}_r^\dagger.$$

By retaining only the leading terms and truncating the sum, we can obtain a meaningful approximation of the original matrix. This principle underpins many SVD applications.

In this example, retaining the first four terms, as shown in sub-figure (b), preserves the main features of the image while substantially reducing noise. This demonstrates noise reduction using SVD.

If the specks are considered fine details of the image that we aim to preserve, fewer terms are still sufficient; retaining 50 terms effectively maintains these features. This demonstrates image compression using SVD.

13.3 ✳ Composite Hilbert Spaces and Schmidt Decomposition

In this section, we extend the concept of singular value decompositions (SVDs) to tensor products of matrices and composite Hilbert spaces. A particularly useful result in this context is the Schmidt decomposition.

13.3.1 SVD of Matrix Tensor Products

1 SVD and Tensor Products

Theorem 13.6 Let A and B be matrices with the following singular value decompositions (SVDs):

$$A = U_A \Sigma_A V_A^\dagger \quad \text{and} \quad B = U_B \Sigma_B V_B^\dagger.$$

Then the SVD of $A \otimes B$ is:

$$A \otimes B = (U_A \otimes U_B)(\Sigma_A \otimes \Sigma_B)(V_A \otimes V_B)^\dagger, \tag{13.29}$$

where:

- $U_A \otimes U_B$ is a unitary matrix containing the left singular vectors of $A \otimes B$.

- $\Sigma_A \otimes \Sigma_B$ is a diagonal matrix containing the singular values of $A \otimes B$.

- $V_A \otimes V_B$ is a unitary matrix containing the right singular vectors of $A \otimes B$.

This result shows that taking the SVD of each matrix individually and then forming the tensor product of these decompositions is equivalent to first forming the tensor product of the matrices and then taking the SVD of the resulting matrix. In other words, the SVD and the tensor product operations "commute" in this context.

Proof.   Using the properties of the tensor product:

$$(A \otimes B)(C \otimes D) = (AC) \otimes (BD),$$
$$(A \otimes B)^\dagger = A^\dagger \otimes B^\dagger,$$

we derive the SVD of $A \otimes B$ as follows:

$$\begin{aligned} A \otimes B &= (U_A \Sigma_A V_A^\dagger) \otimes (U_B \Sigma_B V_B^\dagger) \\ &= (U_A \otimes U_B)(\Sigma_A \otimes \Sigma_B)(V_A^\dagger \otimes V_B^\dagger) \\ &= (U_A \otimes U_B)(\Sigma_A \otimes \Sigma_B)(V_A \otimes V_B)^\dagger. \end{aligned}$$

$\square$

However, an important consideration is that, by convention, singular values are arranged in descending order. This ordering may not be preserved in $\Sigma_A \otimes \Sigma_B$. Consequently, the singular values in the composite Σ may need to be reordered, requiring corresponding rearrangement of the rows and columns in the composite U and V. Despite this, many mathematical properties remain valid regardless of whether the reordering is performed.

The result in Eq. 13.29 can be extended to tensor products of multiple matrices:

$$\bigotimes_i A_i = \bigotimes_i U_i \bigotimes_i \Sigma_i \bigotimes_i V_i^\dagger. \tag{13.30}$$

■ **Example 13.10** Consider the SVDs of the matrices from Example 13.6 and Example 13.7:

$$A = \frac{1}{3}\begin{bmatrix} 3 & 0 \\ 4 & 5 \end{bmatrix}, \quad B = \begin{bmatrix} 1 & 0 \\ 0 & 1 \\ 1 & 1 \end{bmatrix}.$$

$$U_A = \frac{1}{\sqrt{10}}\begin{bmatrix} 1 & -3 \\ 3 & -1 \end{bmatrix}, \quad \Sigma_A = \frac{\sqrt{5}}{3}\begin{bmatrix} 3 & 0 \\ 0 & 1 \end{bmatrix}, \quad V_A = \frac{1}{\sqrt{2}}\begin{bmatrix} 1 & -1 \\ 1 & 1 \end{bmatrix}.$$

$$U_B = \frac{1}{\sqrt{6}}\begin{bmatrix} 1 & -\sqrt{3} & \sqrt{2} \\ 1 & \sqrt{3} & \sqrt{2} \\ 2 & 0 & -\sqrt{2} \end{bmatrix}, \quad \Sigma_B = \begin{bmatrix} \sqrt{3} & 0 \\ 0 & 1 \\ 0 & 0 \end{bmatrix}, \quad V_B = \frac{1}{\sqrt{2}}\begin{bmatrix} 1 & -1 \\ 1 & 1 \end{bmatrix}.$$

The singular values of $A \otimes B$ are given by:

$$\Sigma = \Sigma_A \otimes \Sigma_B = \frac{\sqrt{5}}{3}\begin{bmatrix} 3\Sigma_B & 0 \\ 0 & \Sigma_B \end{bmatrix} = \frac{\sqrt{5}}{3}\begin{bmatrix} 3\sqrt{3} & 0 & 0 & 0 \\ 0 & 3 & 0 & 0 \\ 0 & 0 & 0 & 0 \\ 0 & 0 & \sqrt{3} & 0 \\ 0 & 0 & 0 & 1 \\ 0 & 0 & 0 & 0 \end{bmatrix}.$$

After reordering the singular values to close the row "gaps," we obtain:

$$\Sigma' = \frac{\sqrt{5}}{3} \begin{bmatrix} 3\sqrt{3} & 0 & 0 & 0 \\ 0 & 3 & 0 & 0 \\ 0 & 0 & \sqrt{3} & 0 \\ 0 & 0 & 0 & 1 \\ 0 & 0 & 0 & 0 \\ 0 & 0 & 0 & 0 \end{bmatrix}.$$

The corresponding V is:

$$V = \frac{1}{2} \begin{bmatrix} 1 & -1 & -1 & 1 \\ 1 & 1 & -1 & -1 \\ 1 & -1 & 1 & -1 \\ 1 & 1 & 1 & 1 \end{bmatrix}.$$

The matrix U is 9×9, and its explicit form is omitted for brevity. ∎

2 Dirac Notation

In Dirac notation (see Eq. 13.18), if:

$$A = \sum_{i=1}^{m} \sigma_{Ai} \, |u_i\rangle_A \, \langle v_i|_A \quad \text{and} \quad B = \sum_{j=1}^{n} \sigma_{Bj} \, |u_j\rangle_B \, \langle v_j|_B,$$

then:

$$A \otimes B = \sum_{i=1}^{m} \sum_{j=1}^{n} \sigma_{Ai}\sigma_{Bj}(|u_i\rangle_A \otimes |u_j\rangle_B)(\langle v_i|_A \otimes \langle v_j|_B). \tag{13.31}$$

In the quantum literature, when the context is clear and unambiguous, subscripts A and B are often omitted, and shorthand notation $|u_i\rangle \otimes |u_j\rangle \equiv |u_i u_j\rangle$ is used. Hence:

$$A \otimes B = \sum_{i=1}^{m} \sum_{j=1}^{n} \sigma_i\sigma_j \, |u_i u_j\rangle \, \langle v_i v_j|. \tag{13.32}$$

ⓘ In the shorthand notation $|u_i u_j\rangle$, $|u_i\rangle$ in the first position and $|u_j\rangle$ in the second position are vectors in different spaces; they may even have different dimensions.

3 Parallel for Spectral Decomposition

Similar to Eq. 13.32, if A and B have the spectral decompositions:

$$A = \sum_{i=1}^{m} \lambda_{Ai} \, |\psi_i\rangle_A \, \langle \psi_i|_A \quad \text{and} \quad B = \sum_{j=1}^{n} \lambda_{Bj} \, |\psi_j\rangle_B \, \langle \psi_j|_B,$$

then the tensor product $A \otimes B$ in terms of their spectral decompositions is:

$$A \otimes B = \sum_{i=1}^{m} \sum_{j=1}^{n} \lambda_{Ai}\lambda_{Bj}(|\psi_i\rangle_A \otimes |\psi_j\rangle_B)(\langle \psi_i|_A \otimes \langle \psi_j|_B). \tag{13.33}$$

In shorthand notation:

$$A \otimes B = \sum_{i=1}^{m} \sum_{j=1}^{n} \lambda_i\lambda_j \, |\psi_i \psi_j\rangle \, \langle \psi_i \psi_j|. \tag{13.34}$$

13.3.2 Composite Hilbert Spaces (Refresher)

In quantum mechanics and quantum computing, composite Hilbert spaces are essential for analyzing systems composed of multiple subsystems (e.g., multiple qubits). If two quantum systems are described by the Hilbert spaces $\mathcal{H}_A$ and $\mathcal{H}_B$, their composite system is represented by the tensor product of these spaces, denoted $\mathcal{H}_A \otimes \mathcal{H}_B$.

The tensor product of vector spaces was discussed in detail in Chapter 10. This subsection revisits key concepts necessary for the Schmidt decomposition in the next subsection.

1 Definition

A composite Hilbert space $\mathcal{H}_A \otimes \mathcal{H}_B$ is constructed as follows:

- If $\mathcal{H}_A$ is an m-dimensional Hilbert space and $\mathcal{H}_B$ is an n-dimensional Hilbert space, then $\mathcal{H}_A \otimes \mathcal{H}_B$ is an mn-dimensional Hilbert space.

- The basis vectors of $\mathcal{H}_A \otimes \mathcal{H}_B$ are formed by taking the tensor product of the basis vectors of $\mathcal{H}_A$ and $\mathcal{H}_B$.

Formally, given two Hilbert spaces $\mathcal{H}_A$ and $\mathcal{H}_B$ with orthonormal bases $\{|a_i\rangle\}$ for $\mathcal{H}_A$ and $\{|b_j\rangle\}$ for $\mathcal{H}_B$, the composite Hilbert space $\mathcal{H}_A \otimes \mathcal{H}_B$ is defined as:

$$\mathcal{H}_A \otimes \mathcal{H}_B = \text{span}\{|a_i\rangle \otimes |b_j\rangle \ \mid \ |a_i\rangle \in \mathcal{H}_A, |b_j\rangle \in \mathcal{H}_B\}. \tag{13.35}$$

■ **Example 13.11 — Two-Qubit Systems.** Consider two qubits, each described by the Hilbert space $\mathbb{C}^2$. The composite Hilbert space for the two-qubit system is $\mathbb{C}^2 \otimes \mathbb{C}^2$.

For a single qubit, the basis vectors are $|0\rangle$ and $|1\rangle$. The basis vectors for the composite space are $\{|00\rangle, |01\rangle, |10\rangle, |11\rangle\}$, where:

$$|00\rangle \equiv |0\rangle \otimes |0\rangle, \quad |01\rangle \equiv |0\rangle \otimes |1\rangle, \quad |10\rangle \equiv |1\rangle \otimes |0\rangle, \quad |11\rangle \equiv |1\rangle \otimes |1\rangle. \tag{13.36}$$

The matrix representation of these basis vectors is:

$$|00\rangle = \begin{bmatrix} 1 \\ 0 \\ 0 \\ 0 \end{bmatrix}, \quad |01\rangle = \begin{bmatrix} 0 \\ 1 \\ 0 \\ 0 \end{bmatrix}, \quad |10\rangle = \begin{bmatrix} 0 \\ 0 \\ 1 \\ 0 \end{bmatrix}, \quad |11\rangle = \begin{bmatrix} 0 \\ 0 \\ 0 \\ 1 \end{bmatrix}. \tag{13.37}$$

These four basis vectors form a complete orthonormal set. Therefore, any vector in the composite Hilbert space can be represented as:

$$|\psi\rangle = \sum_{i,j\in\{0,1\}} c_{ij} |ij\rangle = c_{00} |00\rangle + c_{01} |01\rangle + c_{10} |10\rangle + c_{11} |11\rangle. \tag{13.38}$$

This space is 4-dimensional, as each qubit contributes 2 dimensions, resulting in $2^2 = 4$. ■

> ⓘ $\mathbb{C}^2 \otimes \mathbb{C}^2$, $\mathbb{C}^{2\times2}$, and $\mathbb{C}^4$ represent different structures (tensor products, matrices, and vectors, respectively) but are isomorphic as complex vector spaces. This means there is a one-to-one correspondence between elements of these spaces,

| allowing them to be used interchangeably in many contexts.

2   Bipartite Composite Systems

Building on Example 13.11, we generalize the two-qubit composite system to systems with arbitrary dimensions, referred to as general bipartite composite systems.

For general Hilbert spaces $\mathcal{H}_A = \mathbb{C}^m$ and $\mathcal{H}_B = \mathbb{C}^n$, the composite Hilbert space $\mathcal{H}_A \otimes \mathcal{H}_B$ is $\mathbb{C}^{mn}$. The basis vectors for $\mathcal{H}_A \otimes \mathcal{H}_B$ are:

$$\{|a_i b_j\rangle \equiv |a_i\rangle \otimes |b_j\rangle \quad | \quad 1 \leq i \leq m, \ 1 \leq j \leq n\}, \tag{13.39}$$

which form an orthonormal basis for the composite Hilbert space $\mathbb{C}^{mn}$.

A general vector in the composite space has the form:

$$|\psi\rangle = \sum_{i=1}^{m} \sum_{j=1}^{n} c_{ij} |a_i b_j\rangle, \tag{13.40}$$

where c_{ij} are complex coefficients.

3   Multi-Qubit Composite Systems

We generalize the concept of a two-qubit system (Example 13.11) to systems with more qubits.

For an n-qubit system, the composite space is described by the tensor product of the individual qubit spaces. Such a system is represented by a 2^n-dimensional Hilbert space.

The basis for an n-qubit system is a set of orthonormal vectors that span the Hilbert space of the composite system. The basis vectors are:

$$|0\cdots 00\rangle, \ |0\cdots 01\rangle, \ |0\cdots 10\rangle, \ \cdots, \ |1\cdots 11\rangle, \tag{13.41}$$

where each binary string inside $|\ \rangle$ (e.g., $010110\cdots 1$, of length n) uniquely identifies a basis vector.

Each basis vector is a column vector of dimension 2^n, with a '1' at the position indexed by the corresponding binary number. For example:

$$|0\cdots 00\rangle = \begin{bmatrix} 1 & 0 & 0 & \cdots & 0 & 0 \end{bmatrix}^T,$$
$$|0\cdots 01\rangle = \begin{bmatrix} 0 & 1 & 0 & \cdots & 0 & 0 \end{bmatrix}^T,$$
$$|0\cdots 10\rangle = \begin{bmatrix} 0 & 0 & 1 & \cdots & 0 & 0 \end{bmatrix}^T,$$
$$\cdots$$
$$|1\cdots 11\rangle = \begin{bmatrix} 0 & 0 & 0 & \cdots & 0 & 1 \end{bmatrix}^T.$$

These basis vectors are orthonormal because each basis state has exactly one '1' at a unique position.

A general n-qubit state can be expressed as:

$$|\psi\rangle = \sum_{x_1, x_2, \cdots, x_n \in \{0,1\}} c_{x_1 x_2 \cdots x_n} |x_1 x_2 \cdots x_n\rangle, \tag{13.42}$$

or more compactly:

$$|\psi\rangle = \sum_{x \in \{0,1\}^n} c_x |x\rangle, \tag{13.43}$$

where $x \in \{0,1\}^n$ represents an n-bit string (see Example 4.9).

Since the composite Hilbert space is 2^n-dimensional, describing the state vector $|\psi\rangle$ of an entangled system requires up to 2^n complex coefficients. Operators on this space are represented by $2^n \times 2^n$ complex matrices. The growth of 2^n highlights the scalability challenges in quantum computing: for $n = 300$, $2^{300} \approx 10^{90}$, which surpasses the estimated number of atoms in the observable universe.

4 Product Vectors vs. Non-Product Vectors

Although the basis vectors in the composite Hilbert space are tensor products of the basis vectors of the individual spaces (see Eqs. 13.39 and 13.41), not all vectors in the composite Hilbert space are expressible as tensor products of vectors from the individual spaces.

A *tensor-product vector* has the form:

$$|\psi\rangle = |\psi\rangle_A \otimes |\psi\rangle_B \otimes |\psi\rangle_C \otimes \dots. \tag{13.44}$$

A *non-tensor-product vector* cannot be expressed in this form, by definition.

A well-known example of a non-product vector is the Bell state:

$$|\psi\rangle = |00\rangle + |11\rangle,$$

which cannot be written as a tensor-product vector.

> *(i)* In quantum mechanics, a quantum state is described by a normalized vector in the Hilbert space. Therefore, the terms 'state', 'state vector', 'vector', and 'ket' are used interchangeably in the quantum literature.
>
> In this context, a tensor-product vector corresponds to the state of independent (or separable) systems, while a non-tensor-product vector corresponds to the state of dependent (or entangled) systems.

This concept of product vs. non-product properties extends naturally to operators (matrices) in composite Hilbert spaces. For tensor-product operators, matrix operations such as adjoint, inverse, spectral decomposition, and SVD can be distributed to the tensor product of matrices subjected to these operations individually, as shown in § 13.3.1. For non-product operators, the situation becomes far more complex. However, there is an elegant result for general bipartite composite vectors, namely, the Schmidt decomposition, which we discuss next.

13.3.3 Schmidt Decomposition

The Schmidt decomposition provides a way to express a vector in the composite space of two Hilbert spaces. It is a specific form of singular value decomposition (SVD) achieved by reshaping the coordinates of the vector into a matrix.

1 Description

> **Theorem 13.7 — Schmidt Decomposition.** Let $\mathcal{H}_A$ and $\mathcal{H}_B$ be Hilbert spaces of dimensions m and n, respectively. Assume $m \geq n$ without loss of generality. The Schmidt decomposition states that for any vector $|\psi\rangle$ in the composite Hilbert space $\mathcal{H}_A \otimes \mathcal{H}_B$:
>
> $$|\psi\rangle = \sum_{i=1}^{m} \sum_{j=1}^{n} c_{ij} |a_i\rangle_A \otimes |b_j\rangle_B, \tag{13.45}$$
>
> where c_{ij} are complex coefficients, there exist orthonormal bases $\{|u_i\rangle_A\}$ in $\mathcal{H}_A$ and $\{|v_i\rangle_B\}$ in $\mathcal{H}_B$ such that:
>
> $$|\psi\rangle = \sum_{i=1}^{n} \sigma_i |u_i\rangle_A \otimes |v_i\rangle_B, \tag{13.46}$$
>
> where:
>
> - $\{\sigma_i\}$ are non-negative real numbers called the Schmidt coefficients.
>
> - The number of non-zero σ_i is called the Schmidt rank.
>
> - The bases $\{|u_i\rangle_A\}$ and $\{|v_i\rangle_B\}$ are called the Schmidt bases, which are generally different from the original bases $\{|a_i\rangle_A\}$ and $\{|b_i\rangle_B\}$.

To illustrate the Schmidt decomposition, we consider a simple example before presenting the formal proof.

■ **Example 13.12** Consider the two-qubit state vector:

$$|\psi\rangle = \frac{1}{2}(|00\rangle + |01\rangle + |10\rangle + |11\rangle).$$

The expansion coefficients are $c_{00} = c_{01} = c_{10} = c_{11} = \frac{1}{2}$. These coefficients can be arranged into a 2×2 matrix:

$$C = \frac{1}{2} \begin{bmatrix} 1 & 1 \\ 1 & 1 \end{bmatrix}.$$

Here, the rows correspond to the basis states of the first qubit, and the columns correspond to the basis states of the second qubit.

The SVD of C is:

$$C = \frac{1}{\sqrt{2}} \begin{bmatrix} 1 & 1 \\ 1 & -1 \end{bmatrix} \begin{bmatrix} 1 & 0 \\ 0 & 0 \end{bmatrix} \frac{1}{\sqrt{2}} \begin{bmatrix} 1 & 1 \\ 1 & -1 \end{bmatrix}.$$

Or, in rank-1 decomposition form:

$$C = \sigma_1 |u_1\rangle \langle v_1| + \sigma_2 |u_2\rangle \langle v_2|,$$

where $\sigma_1 = 1$, $\sigma_2 = 0$, $|u_1\rangle = |v_1\rangle = \frac{1}{\sqrt{2}} \begin{bmatrix} 1 \\ 1 \end{bmatrix}$, and $|u_2\rangle = |v_2\rangle = \frac{1}{\sqrt{2}} \begin{bmatrix} 1 \\ -1 \end{bmatrix}$.

Converting back to the 4-vector form, we obtain:

$$|\psi\rangle = \sigma_1 |u_1\rangle \otimes |v_1\rangle + \sigma_2 |u_2\rangle \otimes |v_2\rangle,$$

which is the Schmidt decomposition of $|\psi\rangle$. ∎

2   Procedure of Schmidt Decomposition

Building on Example 13.12, we outline a general procedure for constructing the Schmidt decomposition, which also serves as a proof.

Construct the Matrix Representation

Given:

$$|\psi\rangle = \sum_{i=1}^{m}\sum_{j=1}^{n} c_{ij} |a_i\rangle_A \otimes |b_j\rangle_B,$$

construct an $m \times n$ matrix C with elements c_{ij}. This mapping is achieved by associating the basis states $|a_i\rangle_A \otimes |b_j\rangle_B$ with the outer product $|a_i\rangle\langle b_j|$:

$$C = \sum_{i=1}^{m}\sum_{j=1}^{n} c_{ij} |a_i\rangle\langle b_j|.$$

Compute the SVD of C

Perform singular value decomposition (SVD) of C:

$$C = \sum_{k=1}^{r} \sigma_k |u_k\rangle\langle v_k|,$$

where σ_k are the singular values of C, and r is the rank of C.

Map Back to the Tensor Product Space

Use the reverse mapping $|u_k\rangle\langle v_k| \mapsto |u_k\rangle_A \otimes |v_k\rangle_B$:

$$|\psi\rangle = \sum_{k=1}^{r} \sigma_k |u_k\rangle_A \otimes |v_k\rangle_B,$$

which is the Schmidt decomposition.

Verify Orthonormality

Verify that $\{|u_k\rangle_A\}$ and $\{|v_k\rangle_B\}$ form orthonormal sets. This follows from the fact that $|u_k\rangle$ and $|v_k\rangle$ are columns of the unitary matrices U and $V^\dagger$ in the SVD of C.

Exercise 13.10 Find the Schmidt decomposition for $|\psi\rangle = \frac{1}{\sqrt{2}}(|01\rangle + |10\rangle)$.

3   Multipartite Systems

For bipartite systems, Schmidt decomposition allows us to express a composite vector as a sum of tensor-product vectors between the two subsystems, each weighted by a Schmidt coefficient. However, the Schmidt decomposition does not always extend cleanly to multipartite systems.

One approach is to group the subsystems into two parts and treat the two groups as a bipartite system, allowing the application of the Schmidt decomposition. However, such grouping is not unique, and the resulting decomposition may have multiple forms, as illustrated by the following example.

■ **Example 13.13** Consider a composite space consisting of three individual spaces, A, B, and C, each with dimension two and basis vectors $\{|0\rangle, |1\rangle\}$. A composite vector is given by:

$$|\psi\rangle = |0\rangle_A |0\rangle_B |0\rangle_C + 2 |1\rangle_A |0\rangle_B |1\rangle_C + 3 |1\rangle_A |1\rangle_B |0\rangle_C + 4 |1\rangle_A |1\rangle_B |1\rangle_C .$$

To apply the Schmidt decomposition, we represent this vector as a matrix by grouping the subsystems. For example, grouping subsystems B and C together and treating subsystem A separately gives a bipartite split between A and BC. We form a matrix where the rows correspond to the basis states of A, and the columns correspond to the basis states of BC:

$$\psi_{(A)(BC)} = \begin{bmatrix} 1 & 0 & 0 & 0 \\ 0 & 2 & 3 & 4 \end{bmatrix}.$$

The SVD of this matrix yields two non-zero Schmidt coefficients, $\sqrt{29}$ and 1.

Similarly, if we group the subsystems as C and AB, the vector is represented as:

$$\psi_{(C)(AB)} = \begin{bmatrix} 1 & 0 & 0 & 3 \\ 0 & 0 & 2 & 4 \end{bmatrix}.$$

The SVD of this matrix gives two different non-zero Schmidt coefficients, $2\sqrt{7}$ and $\sqrt{2}$, along with distinct Schmidt bases.

■

Exercise 13.11 Complete Example 13.13 by finding the Schmidt coefficients for the grouping B and AC.

This example introduces the concept of tensors. While matrices arrange mathematical objects in two dimensions (rows and columns), tensors extend this concept to multiple dimensions. For instance, a 3D tensor has rows, columns, and sheets. A tripartite system like the one described in Example 13.13 can be naturally represented as a 3D tensor.

_____

The Schmidt decomposition has numerous applications in quantum information theory, including entanglement characterization and quantum state purification. With your mastery of linear algebra and matrix analysis, you are now well-prepared to delve into these fascinating topics.

13.3.4 ＊Other Matrix Decompositions

Matrix decomposition refers to breaking down a matrix into a product of several special matrices. In addition to widely used methods such as spectral decomposition, Jordan decomposition, SVD, and Schmidt decomposition, there are several others that, while less commonly employed in quantum computing, are important in specific contexts. Below is an introduction to these decompositions.

1 Positive Semidefinite Decomposition .

Let A be a positive semidefinite matrix with spectral decomposition $A = U\Lambda U^\dagger$, where U is a unitary matrix and Λ is a diagonal matrix containing the eigenvalues of A. Since A is positive semidefinite, $\Lambda^{1/2}$ is well-defined. Using this, A can be expressed as the product of a matrix and its adjoint:

$$A = SS^\dagger, \tag{13.47}$$

where $S = U\Lambda^{1/2}$.

2   Cholesky Decomposition

Cholesky decomposition applies to positive semidefinite Hermitian matrices A:

$$A = LL^\dagger, \tag{13.48}$$

where L is a lower triangular matrix with all entries above the main diagonal equal to zero. This method is particularly advantageous in optimization problems and certain numerical methods relevant to quantum simulation.

3   LU Decomposition

LU decomposition factors a square matrix A into the product of a lower triangular matrix L and an upper triangular matrix U:

$$A = LU. \tag{13.49}$$

Here, L has ones along its diagonal. LU decomposition is useful for solving systems of linear equations, computing matrix inverses, and evaluating determinants. It is especially relevant in numerical methods for quantum simulation and other linear algebra problems.

4   QR Decomposition

QR decomposition factors a matrix $A \in \mathbb{C}^{m \times n}$ into the product of a unitary (orthogonal) matrix Q and an upper triangular matrix R:

$$A = QR. \tag{13.50}$$

This decomposition is a key tool in iterative eigenvalue algorithms, such as the QR algorithm, and is widely used in least-squares fitting problems. QR decomposition also has applications in quantum algorithms involving orthogonalization or state preparation.

5   Polar Decomposition

Polar decomposition expresses any square complex matrix A as:

$$A = UP, \tag{13.51}$$

where U is a unitary matrix and P is a positive semidefinite Hermitian matrix. If A is normal, U and P commute in this decomposition (Problem 13.15). Polar decomposition is particularly useful in quantum process tomography and in analyzing quantum gates and operations, as it separates unitary transformations from positive transformations.

6   Schur Decomposition

Schur decomposition states that any square complex matrix can be transformed into the form:

$$A = UTU^\dagger, \tag{13.52}$$

where U is a unitary matrix and T is an upper triangular matrix. This decomposition is primarily used for theoretical analysis.

These decompositions involve advanced linear algebra techniques and mathematical concepts that extend beyond the scope of this text. Readers interested in exploring these topics further are encouraged to consult specialized texts on advanced linear algebra and matrix analysis.

Problem Set 13

13.1 Construct a matrix whose eigenvalues are $e^{ik\frac{2\pi}{N}}$, where $k = 0, 1, 2, \ldots, N-1$, and the first eigenvector is $v_1 = \frac{1}{\sqrt{N}}[1\ 1\ \ldots\ 1]^T$. Start with $N = 2$ and $N = 3$, and then develop a general formula.

13.2 Consider the matrix
$$A = \begin{bmatrix} 1 & 0 & 0 \\ 1 & 2 & -1 \\ 1 & 1 & 0 \end{bmatrix}.$$

(a) Find the eigenvalues and eigenvectors of A. Show that A is not diagonalizable.

(b) Find the Jordan form decomposition of A: $A = PJP^{-1}$.

13.3 Consider the 2×2 Jordan block:
$$J = \begin{bmatrix} \lambda & 1 \\ 0 & \lambda \end{bmatrix}.$$

(a) Derive a general formula for J^k, where k is a positive integer.

(b) Compute the matrix exponential e^J.

13.4 Compute the SVD of the following matrices:
$$A = \begin{bmatrix} 0 & -1 \\ 1 & 0 \end{bmatrix}, \quad B = \begin{bmatrix} 1 & -1 \\ 0 & 1 \\ 1 & 0 \end{bmatrix}, \quad C = \begin{bmatrix} 1 & 1 & 1 \\ -1 & 0 & -2 \\ 1 & 2 & 0 \end{bmatrix}.$$

13.5 For the matrices A, B, and C defined in Problem 13.4:

(a) Find the SVD of $A \otimes B$.

(b) Find the singular values of $A \otimes B \otimes C$.

13.6 Two square matrices A and B are similar if there exists an invertible matrix P such that $A = P^{-1}BP$. Show that if A is a square matrix (potentially complex), then $AA^\dagger$ and $A^\dagger A$ are similar.

13.7 Prove that the SVD of a square matrix is invariant under unitary transformations. Namely, given that A is a square matrix and U is a unitary matrix, demonstrate that $UAU^\dagger$ has the same singular values as A.

13.8 (a) Find a complex square matrix A such that $A \neq 0$ but all its eigenvalues are 0.

(b) Show that for any complex matrix A, $A = 0$ if and only if all its singular values are 0.

13.9 Let A be a positive semidefinite matrix with spectral decomposition $A = U\Lambda U^\dagger$, where U is unitary and Λ is diagonal with the eigenvalues of A. Using $S = U\Lambda^{1/2}$, the decomposition $A = SS^\dagger$ holds (see § 13.3.4.1).

Express $A = SS^\dagger$ in bra-ket notation and verify that it matches the spectral decomposition $A = \sum_{i=1}^{n} \lambda_i |\phi_i\rangle \langle\phi_i|$.

13.10 If A has non-zero singular values $\sigma_1, \sigma_2, \ldots, \sigma_r$, what are the singular values of:

(a) $A^\dagger$,

(b) tA, where $t > 0$ is real,

(c) A^{-1}, assuming A is invertible?

13.11 If G and H are positive definite matrices of the same size, show that $G + H$ is positive definite. Furthermore, if G and H are square, determine whether GH is positive definite.

13.12 Consider the composite Hilbert space for a three-qubit system.

(a) Define the composite space.

(b) Formulate the basis vectors.

(c) Demonstrate that the basis vectors form a complete, orthonormal basis.

(d) Express a general vector as a linear combination of the above basis vectors.

13.13 Consider the composite Hilbert space for a two-qubit system.

(a) Show that both:

$$|\psi_1\rangle = \frac{1}{2}(|00\rangle + |01\rangle + |10\rangle + |11\rangle),$$

$$|\psi_2\rangle = \frac{1}{2}(|00\rangle - |01\rangle - |10\rangle + |11\rangle),$$

can be expressed as tensor products of single-qubit vectors of the form $|\psi\rangle_A \otimes |\psi\rangle_B$.

(b) Show that *neither* of:

$$|\phi_1\rangle = \frac{1}{\sqrt{2}}(|00\rangle + |11\rangle),$$

$$|\phi_2\rangle = \frac{1}{\sqrt{2}}(|01\rangle - |10\rangle),$$

can be expressed as tensor products of the form $|\phi\rangle_A \otimes |\phi\rangle_B$.

(c) Find the Schmidt decompositions of $|\phi_1\rangle$ and $|\phi_2\rangle$.

13.14 $*$ For a real $m \times n$ matrix A with positive singular values $\sigma_1, \sigma_2, \ldots, \sigma_r$, define:

$$s(x) = (x - \sigma_1)(x - \sigma_2) \cdots (x - \sigma_r).$$

For a square matrix B, define:

$$c_B(\lambda) = \det(\lambda I - B),$$

where I is the identity matrix of the same dimension as B.

(a) Show that $c_{A^\dagger A}(x) = s(x)x^{n-r}$ and $c_{AA^\dagger}(x) = s(x)x^{m-r}$.

(b) If $m \le n$, show that $c_{A^\dagger A}(x) = s(x)x^{n-m}$.

13.15 Prove that a square matrix $A \in \mathbb{C}^{n\times n}$ is a normal matrix if and only if the unitary matrix U and the positive semidefinite Hermitian matrix P in its polar decomposition $A = UP$ commute. You do not need to prove the polar decomposition itself.

IV

A Probability Primer for Quantum Computing

Quantum mechanics, the foundation of quantum computing, can be viewed as a specialized extension of probability theory. Both quantum information theory and quantum computing—spanning areas such as quantum Monte Carlo simulations—are deeply rooted in probabilistic principles. A robust understanding of probability theory is therefore essential for mastering these fields.

In the chapters of Part IV, we will first explore the core concepts of probability theory and stochastic processes to establish a strong foundation for more advanced topics.

We then explore Markov chains and Monte Carlo simulations, chosen for two key reasons:

1. Markov chains and Monte Carlo simulations are vital for studying quantum algorithms, reflecting the intrinsic randomness and probabilistic framework of quantum computing.

2. They exemplify the practical applications of linear algebra, demonstrating how abstract mathematical tools underpin powerful computational techniques.

Our focus will be on discrete probability, which is particularly relevant to quantum computing, emphasizing discrete sample spaces and random variables. The term "discrete" refers to sets that are finite or countably infinite, making them representable through vectors and matrices. This approach not only aligns with the mathematical foundations of quantum computing but also deepens our understanding of linear algebra.

14. Fundamentals of Probability

Contents

This chapter provides the foundation by introducing fundamental probabilistic concepts such as random experiments, sample spaces, and events. These elements form the essential building blocks for the more advanced discussions on stochastic processes that follow.

14.1 Basic Concepts

14.1.1 Sample Space, and Events

> **Definition 14.1 — Random Experiments.** A random experiment or trial is a process that yields a definite set of possible outcomes, none of which can be precisely predicted in advance. Such experiments can be repeated under consistent conditions, with each repetition being independent of the others.

Definition 14.2 — Sample Space. The sample space, denoted by $\Omega = \{s_1, s_2, \ldots, s_n\}$, includes all possible outcomes of an experiment, where each s_i represents a distinct outcome. This space may be finite or countably infinite in discrete cases, and must satisfy two criteria:

1. Exhaustiveness: It must cover all potential outcomes of the experiment.

2. Mutual Exclusivity: No two outcomes can occur simultaneously in a single execution of the experiment.

Definition 14.3 — Events. An event, denoted as E, is any subset of the sample space, $E \subseteq \Omega$. An event E occurs if the outcome of the random experiment is contained within E.

■ **Example 14.1** An event may be an elementary event (consisting of a single outcome) or a compound event (comprising multiple outcomes).

Consider the roll of a six-sided die as a random experiment, where the sample space is $\Omega = \{1, 2, 3, 4, 5, 6\}$.

The event of rolling an even number is denoted as $E = \{2, 4, 6\}$, while the event of rolling an odd number is denoted as $O = \{1, 3, 5\}$. Events E and O are exhaustive and mutually exclusive.   ■

14.1.2 Random Variables and Probability Distributions

Definition 14.4 — Random Variable. A random variable is a *function* that assigns a numerical value (or other mathematical construct) to each outcome in the sample space of a random experiment, effectively mapping from the sample space to a set of possible values.

■ **Example 14.2** Consider the process of rolling a fair six-sided die.

Conventionally, the random variable X takes values in the set $\{1, 2, 3, 4, 5, 6\}$, each corresponding to the face-up number after the roll.

Now, let Y be another random variable defined to represent the parity of the die's outcome. For example, Y could take values in the set $\{o, e, o, e, o, e\}$, where 'o' denotes an odd outcome and 'e' denotes an even outcome, reflecting whether the result is odd or even.

Each distinct mapping, such as X or Y, constitutes a unique random variable. ■

Definition 14.5 — Probability Distribution. A probability distribution specifies a probability for each possible outcome of an experiment, adhering to the following principles:

1. Non-Negativity: Each probability is non-negative.

2. Normalization: The sum of probabilities for all possible outcomes equals 1.

3. Additivity: For any two mutually exclusive events, the probability of their union equals the sum of their individual probabilities.

While probability distribution originates from the concept of a sample space, it can also be defined in the context of random variables. In the latter case, a probability distribution describes how probabilities are distributed over the different possible values that the random variable can take.

■ **Example 14.3 — Probability Distribution Over Sample Space.** For a fair die, the probability distribution across Ω is uniform:

$$p_i = p(\{i\}) = \frac{1}{6}, \quad \text{for each } i \in \Omega. \tag{14.1}$$

Consequently, the probability of event E (rolling an even number) is calculated as:

$$p(E) = p(\{2\}) + p(\{4\}) + p(\{6\}) = \frac{1}{6} + \frac{1}{6} + \frac{1}{6} = \frac{1}{2}. \tag{14.2}$$

■

(*i*) In this text, we use the lowercase p to represent probability or probability distributions. This follows the notation often used in quantum information literature, as opposed to the uppercase P commonly found in classical probability texts.

Additionally, instead of distinguishing between "probability mass function" for discrete random variables and "probability density function" for continuous ones, we will use the term "probability distribution" to refer to both cases.

■ **Example 14.4 — Probability Distribution for Random Variables.** If X denotes the outcome of a fair die roll, with possible values $\{1, 2, 3, 4, 5, 6\}$, its probability distribution is uniform, given by $p(X = x) = \frac{1}{6}$ for $x = 1, 2, \ldots, 6$.

When a random variable maps multiple outcomes to a single value (a many-to-one assignment), it aggregates those outcomes, potentially altering the probability distribution. For instance, if X is defined to indicate whether a die roll is even (e) or odd (o), with e for $\{2, 4, 6\}$ and o for $\{1, 3, 5\}$, then $p(X = e) = p(X = o) = \frac{1}{2}$. ■

Events can also be characterized using random variables. For instance, the event $E = \{X \leq 3\}$ corresponds to rolling a die and obtaining a value of 3 or less.

(*i*) A discrete random variable X can be represented as a vector $\boldsymbol{x}$, which contains all possible values of X. Similarly, the probability distribution of X can be expressed as a vector $\boldsymbol{p}$, where each element corresponds to the probability of a respective value in $\boldsymbol{x}$. This vector notation simplifies mathematical operations in some scenarios. However, we will initially focus on fundamental probability concepts and principles before delving into these representations.

14.2 Interactions of Multiple Random Variables

Analyzing the interactions between multiple random variables is essential for understanding the composite behavior they exhibit and how these interactions influence their associated probabilities. Initially, we will explore the case involving two random variables, with the intent to later expand these principles to encompass multiple variables.

14.2.1 Probabilities of Two Random Variables

Consider two random variables, X and Y, defined over the same sample space Ω. We examine specific events associated with these variables, represented by the outcomes x for X and y for Y. The probabilities linked to these events are outlined as follows:

Definition 14.6 — Joint Probability. The joint probability of the events $X = x$ and $Y = y$ occurring simultaneously is denoted by $p(X = x, Y = y)$, or more succinctly as $p(x, y)$ when context allows.

The joint probability distribution, represented as $p(X, Y)$, accounts for $p(x, y)$ across all conceivable combinations of outcomes x and y.

Joint probability is symmetric, i.e., $p(x, y) = p(y, x)$ and $p(X, Y) = p(Y, X)$.

Alternatively, in some contexts, the joint probability $p(x, y)$ might be denoted as $p(x \cap y)$, highlighting the simultaneous occurrence of events x and y. Similarly, $p(X \cap Y) \equiv p(X, Y)$.

Definition 14.7 — Conditional Probability. Conditional probability represents the probability of event $X = x$ occurring, given that event $Y = y$ has already occurred, denoted as $p(x|y)$.

This concept is further extended to the conditional probability distribution $p(X|Y)$, indicating the conditional probability of all combinations of the outcomes of X given Y.

Note that $p(x|y)$ may not be equivalent to $p(y|x)$.

■ **Example 14.5** Let Ω represent people in the US. Let A denote a person's age, and S denote the state the person resides in, with 1 for New York, 2 for New Jersey, etc. Then:

$$p(A = 18) = \frac{\# \text{ of people in the US who are 18}}{\# \text{ of people in the US}},$$

$$p(A = 18|S = 1) = \frac{\# \text{ of people in New York who are 18}}{\# \text{ of people in New York}}.$$

The condition $S = 1$ restricts the population (sample space) to New York. This intuition is often useful in understanding complex conditional probability relations.

■

Definition 14.8 — Marginal Probability. The marginal probability of an event $X = x$ refers to the likelihood of $X = x$ occurring, irrespective of the outcomes of another variable Y. It is denoted as $p(X = x)$ or simply $p(x)$, and is determined by summing over all possible outcomes of Y:

$$p(X = x) = \sum_{y \in \Omega} p(X = x, Y = y). \tag{14.3}$$

This concept can be extended to a probability distribution, where the marginal probability distribution of X is given by:

$$p(X) = \sum_{y \in \Omega} p(X, Y = y), \tag{14.4}$$

---

summarizing the probabilities of X over its entire range of outcomes, irrespective of Y.

Through this marginalization process, we focus on the likelihood of one specific event within a more complex system of variables, thereby isolating the probability of that event in the broader context.

■ **Example 14.6** Consider the scenario involving two variables: the weather (with possible outcomes being rainy or sunny) and your choice of ice cream (chocolate or vanilla). The marginal probability of choosing vanilla ice cream, denoted as $p(\text{vanilla})$, is calculated by summing the probabilities of choosing vanilla for all possible weather conditions: $p(\text{vanilla}) = p(\text{vanilla, rainy}) + p(\text{vanilla, sunny})$. This calculation effectively disregards the weather's potential influence on the ice cream choice. ■

(*i*) In a strict mathematical sense, symbols such as $p(X)$, $p(Y)$, $p(X,Y)$, $p(X|Y)$, and $p(Y|X)$ denote distinct functions and ideally should be represented by unique symbols. However, for simplicity and to minimize notation clutter, as in common probability literature, we use the same symbol p across these expressions. It is crucial to understand that p signifies "the probability of" in this context.

Furthermore, we use $p(x)$ to represent $p(X = x)$, $p(x, y)$ for $p(X = x, Y = y)$, and so on, whenever there is no ambiguity. Here, $p(X)$ can be thought of as a function that accepts a set of events as input and returns a probability distribution, effectively acting as a vector function. Conversely, $p(x)$ refers to the probability of X assuming the specific value x.

Definition 14.9 — Probability of the Union. The probability that at least one of the events x or y occurs is denoted as $p(x \cup y)$. This is calculated by considering the probability of either event happening, taking into account their possible overlap.

Exercise 14.1 A fair coin is tossed, and independently, a fair die is thrown. Let x be the event that a head is tossed, and y be the event that an odd number is thrown. Calculate $p(x \cap y)$ and $p(x \cup y)$.

14.2.2 Independent vs. Dependent Random Variables

1 Independent Random Variables

Definition 14.10 — Independent Events. Two events are independent if the occurrence of one event does not influence the occurrence of the other.

For instance, the outcomes of two rolls of a die are independent. The probabilities for independent events satisfy:

$$p(x, y) = p(x) \cdot p(y), \quad p(x|y) = p(x), \quad p(y|x) = p(y). \tag{14.5}$$

This concept is similarly applicable to random variables. Random variables X and Y are independent if the probability distribution of one is not affected by the realization of the other. Their joint and conditional probability distributions satisfy:

$$p(X, Y) = p(X) \cdot p(Y), \quad p(X|Y) = p(X), \quad p(Y|X) = p(Y). \tag{14.6}$$

2   Dependent Random Variables

Events, or random variables, are dependent if the outcome of one affects the likelihood of the other, such as in drawing cards from a deck without replacement. Their probabilities are described by the chain rule:

$$p(x, y) = p(x|y) \cdot p(y) = p(y|x) \cdot p(x). \tag{14.7}$$

Similarly, for random variables X and Y, the joint probability can be expressed as:

$$p(X, Y) = p(X|Y) \cdot p(Y) = p(Y|X) \cdot p(X). \tag{14.8}$$

Exercise 14.2 Consider that random variables X and Y are perfectly correlated, such that X completely determines Y, for example, $Y = 3X$. Analyze the implications of this relationship on $p(X, Y)$, $p(X|Y)$, $p(X)$, and $p(Y)$.

14.2.3   Mutually Exclusive vs. Overlapping Events

1   Mutually Exclusive Events

> **Definition 14.11 — Mutually Exclusive Events.** Mutually exclusive (or disjoint) events are events that cannot occur simultaneity.

Mutually exclusive events lead to the following probability relations:

$$p(x, y) = 0, \quad p(x \cup y) = p(x) + p(y). \tag{14.9}$$

2   Overlapping Events

Overlapping events can occur simultaneously ($p(x, y) \neq 0$) and adhere to the **inclusion-exclusion principle**:

$$p(x \cup y) = p(x) + p(y) - p(x, y). \tag{14.10}$$

This relationship is often illustrated using a Venn diagram, as shown in Figure 14.1.

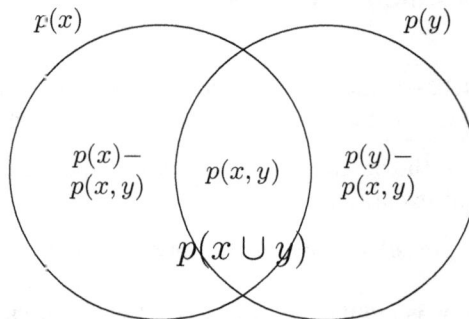

Figure 14.1: Venn Diagram for Probability Relationships

Exercise 14.3 Is $p(x) - p(x, y) = p(x|\bar{y})$ true, where $p(\bar{y})$ represents the probabilty that y has not happened?

■ **Example 14.7** Consider two random variables, X and Y, each taking values in $\Omega = \{1, 2, \ldots, 6\}$ for fair dice.

Given two events: x is rolling a 1 on die one, and y is rolling a 3 on die two. The joint probability $p(x, y)$, the probability of both x and y occurring together, is given by the product of the probabilities of x and y, as the dice rolls are independent:

$$p(x, y) = p(x) \cdot p(y) = \frac{1}{6} \cdot \frac{1}{6} = \frac{1}{36}. \tag{14.11}$$

The probability of the union of events x and y, $p(x \cup y)$, representing either event occurring, is given by:

$$p(x \cup y) = p(x) + p(y) - p(x, y) = \frac{1}{6} + \frac{1}{6} - \frac{1}{36} = \frac{11}{36}. \tag{14.12}$$

Now, we impose the condition that $X + Y$ is even. In this case, X and Y are no longer independent. The joint probability distribution $p(X, Y)$ becomes uniform over the subset of $\Omega \times \Omega$ where the sum of the dice is even, including pairs such as $\{(1, 1), (1, 3), (2, 2), (2, 4), \ldots\}$, each with a probability of $\frac{1}{18}$. However, $p(X, Y)$ is not uniform over $\Omega \times \Omega$, as $p(X = x, Y = y) = 0$ if $x + y$ is odd.

If Alice "holds" X and Bob "holds" Y, using the marginalization formula for Alice's marginal distribution

$$p(X) = \sum_{y \in \Omega} p(X, Y = y), \tag{14.13}$$

we discover that $p(X)$ is still uniform over Ω, each outcome having a probability of $\frac{1}{6}$. Similarly, for Bob's $p(Y)$.

This shows that if X and Y are dependent, the marginal distributions can be uniform over Ω, despite their joint distribution $p(X, Y)$ not being uniform over $\Omega \times \Omega$.
■

Exercise 14.4 Two events x and y are such that $p(x) = 0.5$, $p(y) = 0.4$ and $p(x \cap y) = 0.2$. Calculate

(a) $p(x|y)$;

(b) $p(y|x)$;

(c) $p(x \cup y)$;

(d) $p(x \cap y|x \cup y)$.

14.2.4 Key Relations

1 Normalization Conditions

For any random variable X, the probabilities of all possible outcomes must sum to 1, reflecting the certainty that one of the outcomes in the sample space will occur. This normalization condition applies to marginal, joint, and conditional probabilities:

$$\sum_{x \in \Omega_X} p(x) = 1, \tag{14.14a}$$

$$\sum_{x \in \Omega_X} \sum_{y \in \Omega_Y} p(x, y) = 1, \tag{14.14b}$$

$$\sum_{x \in \Omega_X} p(x|y) = 1, \quad \text{for any given } y \in \Omega_Y \text{ where } p(y) > 0. \tag{14.14c}$$

Here, Ω_X and Ω_Y are the sample spaces of X and Y, respectively.

2 Law of Total Probability

The Law of Total Probability, also known as the Partition Theorem, provides a method for calculating the probability of an event based on a partition of the sample space Ω into k disjoint subsets $\Omega_1, \Omega_2, \ldots, \Omega_k$:

$$\Omega_i \cap \Omega_j = \emptyset \quad \text{for all } i, j \text{ with } i \neq j, \tag{14.15}$$

and

$$\bigcup_{i=1}^{k} \Omega_i = \Omega. \tag{14.16}$$

Accordingly,

$$p(x) = \sum_{i=1}^{k} p(x, \Omega_i) = \sum_{i=1}^{k} p(x|\Omega_i)p(\Omega_i). \tag{14.17}$$

For example, considering a specific event y and its complement $\bar{y}$,

$$p(x) = p(x|y)p(y) + p(x|\bar{y})p(\bar{y}). \tag{14.18}$$

3 Chain Rule

The chain rule, which describes how to compute the joint probability of a sequence of random variables, is given for two variables by Eq. 14.8.

This can be extended to more variables as follows:

$$p(X_1, X_2, \ldots, X_n) = p(X_1) \cdot p(X_2|X_1) \cdot p(X_3|X_1, X_2) \cdots p(X_n|X_1, \ldots, X_{n-1}). \tag{14.19}$$

4 Bayes' Theorem

Bayes' Theorem allows us to "invert" conditional probabilities, i.e., express $p(x|y)$ in terms of $p(y|x)$. It is a direct consequence of the chain rule in Eq. 14.7:

$$p(x|y) = \frac{p(y|x) \cdot p(x)}{p(y)}. \tag{14.20}$$

More generally, in terms of the probability distributions of multiple random variables, Bayes' theorem can be expressed as:

$$p(X_1|X_2, X_3, \ldots, X_n) = \frac{p(X_2, X_3, \ldots, X_n|X_1) \cdot p(X_1)}{p(X_2, X_3, \ldots, X_n)}. \tag{14.21}$$

Bayes' Theorem relates conditional probabilities in two directions, often described as "cause and effect" or "prior and posterior" beliefs. The left-hand side, $p(x|y)$, represents how likely x is given information about y. The right-hand side reverses the conditioning, adjusting with scaling factors from the marginal and joint distributions. This relationship forms the foundation of Bayesian inference.

Bayesian Inference

Bayes' Theorem is a fundamental tool in statistics and probability for updating our beliefs about the likelihood of hypotheses (or events) in light of new evidence. Here's how the components of Bayes' Theorem represent prior and new information:

- Prior Probability $p(x)$: This is the probability of the event $X = x$ before considering any information about Y. It reflects our initial belief about the event.

- Likelihood $p(y|x)$: This is the probability of observing the new evidence $Y = y$ given that $X = x$ is true. It is how we expect the evidence to be distributed if our hypothesis $X = x$ were true.

- Marginal Likelihood or Evidence $p(y)$: This is the probability of observing the evidence $Y = y$ under all possible outcomes of X. It normalizes the posterior to ensure that the probabilities sum to one.

- Posterior Probability $p(x|y)$: This is the updated probability of $X = x$ after taking into account the new evidence $Y = y$. It combines our prior belief with the likelihood of the new evidence to give us a revised belief.

When we apply Bayes' Theorem, we are performing a Bayesian update from the prior to the posterior. The prior represents what we know about X before seeing Y, and the posterior represents what we know after observing Y. If we receive more evidence Z, we could then take our posterior $p(x|y)$ as the new prior and update it with the likelihood of Z to get a new posterior, and so on. This process is at the heart of Bayesian inference, where we iteratively update our beliefs as new data becomes available.

■ **Example 14.8 — Bayesian Inference.** In screening for a certain disease, the probability that a healthy person wrongly gets a positive result is 0.05. The probability that a diseased person wrongly gets a negative result is 0.02. The overall rate of the disease in the population being screened is 1%. If the test gives a positive result, what is the probability of actually having the disease?

Solution. Let

$$D = \{\text{have disease}\}, \qquad P = \{\text{postive test}\},$$
$$\overline{D} = \{\text{don't have disease}\}, \quad \overline{P} = \{\text{negative test}\}.$$

Then

$$p(P|\overline{D}) = 0.05,$$
$$p(\overline{P}|D) = 0.02, \quad \Rightarrow p(P|D) = 1 - p(\overline{P}|D) = 0.98,$$
$$p(D) = 0.01.$$

From the Law of Total Probability,

$$p(P) = p(P|D)p(D) + p(P|\overline{D})p(\overline{D})$$
$$= 0.98 \times 0.01 + 0.05 \times (1 - 0.01) \approx 0.0593.$$

Finally, using Bayes' Theorem,

$$p(D|P) = \frac{p(P|D) \cdot p(D)}{p(P)}$$

$$= \frac{0.98 \times 0.01}{0.0593} \approx 0.165,$$

indicating the probability of actually having the disease is about 16.5%. ∎

14.3 Expectation and Variance

14.3.1 Expectation

Definition 14.12 — Expectation. The expectation, expected value, or mean, of a random variable X, denoted $E[X]$, quantifies the statistical average of its possible values, each weighted by its probability:

$$E[X] = \sum_x x \cdot p(X = x). \tag{14.22}$$

This formula applies to discrete random variables; for continuous random variables, integration over the variable's range replaces the summation.

ⓘ Square brackets are preferred over parentheses in the notation for expectation, variance, and other statistical operators, highlighting their role as operators on random variables, rather than conventional functions.

For example, the expected value of a roll of a fair six-sided die is:

$$E[X] = \sum_{x=1}^{6} x \cdot p(X = x) = 3.5, \tag{14.23}$$

assuming a uniform probability distribution where each outcome is equally likely.

If a discrete random variable X is represented as a vector $\boldsymbol{x}$ and its probability distribution as a vector $\boldsymbol{p}$, then $E[X]$ can also be expressed as the dot product of $\boldsymbol{x}$ and $\boldsymbol{p}$:

$$E[X] = \boldsymbol{x} \cdot \boldsymbol{p}. \tag{14.24}$$

The expectation is a central measure of a random variable's distribution. For a large number N of observations of X, the average of these N values approaches $E[X]$ as N becomes infinitely large.

Definition 14.13 — Expectation of $f(X)$. For a function $f(x)$,

$$E[f(X)] = \sum_x f(x) \cdot p(X = x), \tag{14.25}$$

applies to any transformation of X, where $f(x)$ is applied to each value of X and weighted by its probability.

In particular, the expectation of X^2 as $E[X^2] = \sum_x x^2 \cdot p(X = x)$.

Definition 14.14 — Expectation of $f(X, Y)$. For two random variables X and Y, and a function $f(x, y)$,

$$E[f(X, Y)] = \sum_{x,y} f(x, y) \cdot p(X = x, Y = y). \tag{14.26}$$

In particular,

$$E[X + Y] = \sum_{x,y} (x + y) \cdot p(X = x, Y = y), \tag{14.27a}$$

$$E[XY] = \sum_{x,y} xy \cdot p(X = x, Y = y). \tag{14.27b}$$

This definition allows computing the expectation of sums, products, and other functions involving multiple random variables.

Properties of Expectation

For functions g and h, and constants a and b, for any random variables X and Y,

$$E[aX + bY] = aE[X] + bE[Y], \tag{14.28a}$$
$$E[af(X) + bg(Y)] = aE[f(X)] + bE[g(Y)], \tag{14.28b}$$

illustrating the linearity of expectation.

For *independent* random variables X and Y,

$$E[XY] = E[X] \cdot E[Y], \tag{14.29a}$$
$$E[f(X)g(Y)] = E[f(X)] \cdot E[g(Y)], \tag{14.29b}$$

highlighting that the expectation of the product is the product of the expectations only when X and Y are independent.

Exercise 14.5 For two standard dice all 36 outcomes of a throw are equally likely. Find $p(X + Y = j)$ for all j and calculate $E(X + Y)$. Confirm that $E(X) + E(Y) = E(X + Y)$.

14.3.2 Variance

Definition 14.15 — Variance. The variance of a random variable X, denoted Var$[X]$, measures the mean squared deviation of X from its expected value, indicating the variability of X:

$$\text{Var}[X] = E\left[(X - E[X])^2\right] = E[X^2] - (E[X])^2. \tag{14.30}$$

Definition 14.16 — Standard Deviation. The standard deviation of a random variable X, denoted ΔX, is the square root of its variance, quantifying the spread

of X's values:
$$\Delta X = \sqrt{\text{Var}[X]} = \sqrt{E[X^2] - (E[X])^2}. \qquad (14.31)$$

Definition 14.17 — Covariance. The covariance between two random variables X and Y, denoted $\text{Cov}[X, Y]$, reflects their joint variability:

$$\text{Cov}[X, Y] = E\left[(X - E[X])(Y - E[Y])\right] = E[XY] - E[X]E[Y]. \qquad (14.32)$$

Proof. To prove the final result of Eq. 14.32, we begin with the definition:
$$\text{Cov}[X, Y] = E\left[(X - E[X])(Y - E[Y])\right].$$

Expanding the product inside the expectation:
$$\text{Cov}[X, Y] = E\left[XY - XE[Y] - E[X]Y + E[X]E[Y]\right].$$

Now, apply the linearity of expectation:
$$\text{Cov}[X, Y] = E[XY] - E[X]E[Y] - E[X]E[Y] + E[X]E[Y].$$

Simplifying:
$$\text{Cov}[X, Y] = E[XY] - E[X]E[Y].$$

$\square$

Covariance indicates the direction of the linear relationship between X and Y. Positive covariance implies that X and Y tend to move in the same direction, while negative covariance suggests they move in opposite directions.

Note that $\text{Cov}[X, Y] = 0$ suggests no linear dependency but does not necessarily imply independence, as illustrated by the example below.

■ **Example 14.9** Consider the random variable X that takes values from the set $\{-1, 0, 1\}$ with probabilities:

$$p(X = -1) = \frac{1}{4}, \quad p(X = 0) = \frac{1}{2}, \quad p(X = 1) = \frac{1}{4}.$$

Define another random variable Y as:

$$Y = \begin{cases} 1 & \text{if } X = -1 \text{ or } X = 1, \\ 0 & \text{if } X = 0. \end{cases}$$

Clearly, Y depends on X, so they are not independent. We now show that $\text{Cov}[X, Y] = 0$.

First, compute $E[X]$:

$$E[X] = (-1) \cdot \frac{1}{4} + 0 \cdot \frac{1}{2} + 1 \cdot \frac{1}{4} = 0.$$

Next, compute $E[Y]$:

$$E[Y] = 1 \cdot (p(X = -1) + p(X = 1)) = 1 \cdot \left(\frac{1}{4} + \frac{1}{4}\right) = \frac{1}{2}.$$

Now compute $E[XY]$:

$$E[XY] = (-1) \cdot 1 \cdot p(X = -1) + 1 \cdot 1 \cdot p(X = 1) = 0.$$

Thus:

$$\mathrm{Cov}[X, Y] = E[XY] - E[X]E[Y] = 0 - 0 \cdot \frac{1}{2} = 0.$$

∎

Definition 14.18 — Correlation. The correlation coefficient between X and Y, denoted $\mathrm{Corr}[X, Y]$, normalizes the covariance to measure the strength and direction of their linear relationship:

$$\mathrm{Corr}[X, Y] = \frac{\mathrm{Cov}[X, Y]}{\sqrt{\mathrm{Var}[X] \cdot \mathrm{Var}[Y]}}. \tag{14.33}$$

The correlation coefficient ranges from -1 to $+1$, with ± 1 indicating a perfect linear relationship. A correlation of ± 1 occurs if and only if $Y = aX + b$ for some constants a and b, with $+1$ for $a > 0$ and -1 for $a < 0$, provided that $\mathrm{Var}[X] > 0$ and $\mathrm{Var}[Y] > 0$.

Properties of Variance

For functions g and h, and constants a and b, the variance has the following properties for any random variables X and Y:

$$\mathrm{Var}[af(X) + b] = a^2 \, \mathrm{Var}[f(X)], \tag{14.34a}$$
$$\mathrm{Var}[X + Y] = \mathrm{Var}[X] + \mathrm{Var}[Y] + 2 \, \mathrm{Cov}[X, Y]. \tag{14.34b}$$

These properties highlight the influence of scaling and shifting operations on variance, as well as how the covariance between X and Y affects the variance of their sum.

Exercise 14.6 Consider a fair six-sided die with four green faces and two red faces. Let the random variable X represent the score received from a single roll: 3 points for green and 2 point for red. Determine X's probability distribution and calculate its expectation and variance.

14.3.3 Sampling Error of the Mean

1 Theoretical Expectation vs. Empirical Measurement

In many scientific and engineering fields, a key task is estimating a quantity of interest through measurements or observations. This could mean measuring signal intensity, calculating the ground state energy with a quantum variational eigenvalue solver, or evaluating other measurable phenomena.

The goal is to get as close as possible to the theoretical mean value. This is the ideal value that reflects the expected outcome under perfect conditions.

While theory tells us that the empirical average of an infinite number of measurements will converge to the theoretical mean, practical limitations often restrict us to a finite number of observations. As a result, we encounter sampling error—a difference between the empirical average and the theoretical mean.

This raises an important question: How many measurements do we need to keep the error within a desired limit, such as 1%?

2 Random Sampling and Measurements

Before discussing sampling error, it's important to understand random sampling in measurements. In quantum computing and physical experiments, random sampling means selecting measurements based on the probability distribution of the random variable X.

This is different from simple random sampling, where each sample is chosen with equal probability, regardless of the underlying distribution.

When we measure a physical quantity influenced by quantum or thermal fluctuations, instrumental limitations, or other uncertainties, each measurement of the random variable X is a random sample from its distribution. Repeated measurements of X give us multiple independent samples from this distribution.

By doing this, we can estimate X's theoretical mean (or expectation). This is the average result we would expect if we could measure X an infinite number of times under the same conditions.

3 Expectation and Variance of the Empirical Mean

Let random variable X be the quantity we will be measuring. Let random variables $X_1, X_2, \ldots, X_n$ represent n independent and identically distributed (i.i.d.) random samples drawn from the distribution of X. The **empirical mean**, or **sample mean**, is the average of n measurements:

$$\bar{X} = \frac{1}{n}(X_1 + X_2 + \cdots + X_n). \tag{14.35}$$

The expectation of $\bar{X}$ is:

$$E[\bar{X}] = E\left[\frac{1}{n}(X_1 + X_2 + \cdots + X_n)\right] \tag{14.36a}$$

$$= \frac{1}{n}(E[X_1] + E[X_2] + \cdots + E[X_n]) \tag{14.36b}$$

$$= \frac{1}{n}(nE[X]) \tag{14.36c}$$

$$= E[X], \tag{14.36d}$$

where $E[X]$ is the theoretical mean of X (and each X_i). Thus, the expectation of the sample mean is the theoretical mean, as expected. We have used the linear operator property of expectation (Eq. 14.28).

Similarly, the **variance of the sample mean** is obtained using Eq. 14.34:

$$\mathrm{Var}[\bar{X}] = \mathrm{Var}\left[\frac{1}{n}(X_1 + X_2 + \cdots + X_n)\right] \tag{14.37a}$$

$$= \frac{1}{n^2}(\mathrm{Var}[X_1] + \mathrm{Var}[X_2] + \cdots + \mathrm{Var}[X_n]) \tag{14.37b}$$

$$= \frac{n\,\mathrm{Var}[X]}{n^2} \tag{14.37c}$$

$$= \frac{\mathrm{Var}[X]}{n}, \tag{14.37d}$$

where $\mathrm{Var}[X]$ is the **theoretical variance** of X.

Or, in terms of standard deviation, referred to as the standard error of the mean (SEM) in this context,

$$\text{SEM} \equiv \Delta \bar{X} = \frac{\Delta X}{\sqrt{n}}. \tag{14.38}$$

The SEM can be regarded as the typical level of sampling error. This result indicates that as the sample size n increases, SEM decreases as $\frac{1}{\sqrt{n}}$. As $n \to \infty$, SEM approaches zero, indicating that the empirical mean $E[\bar{X}]$ indeed converges to the true expected value. This observation is a direct illustration of the **Law of Large Numbers**, which states that the sample mean converges to the expected value as the number of samples increases.

Note that SEM is proportional to ΔX, the standard deviation of X. If $\Delta X = 0$, indicating the measurement process is non-stochastic, then $\Delta \bar{X} = 0$, i.e., $E[X]$ can be determined accurately even with a single measurement.

(i) The two key results, the expectation of the sample mean equals the theoretical mean (Eq. 14.36), and the standard deviation of the sample mean decreases as $\frac{1}{\sqrt{n}}$ (Eq. 14.38), are independent of the underlying distribution of X.

At this point, we can already answer the question of how many measurements are necessary, at least roughly. For example, assume $\Delta X = 1$, and we want the typical measurement error SEM to be within 0.01, then n has to be 10^4. However, if we want to be more precise than the "typical measurement error", we need to introduce the Central Limit Theorem and the concept of confidence level, as we will discuss next.

4   Distribution of the Empirical Mean: The Central Limit Theorem

Now that we know the expectation of the sample mean (or the true population mean, $E[X]$ given by Eq. 14.36) and its standard deviation (SEM in Eq. 14.38), our next inquiry is: what is the probability distribution of the sample mean? This is a different concept from the original distribution of X. It is governed by the Central Limit Theorem (CLT) for large sample sizes (n).

The CLT is one of the most powerful and widely used tools in statistics. We omit the proof of CLT here, as the complicated mathematics is less relevant to our discussion. This theorem justifies the use of the normal distribution as a model in many situations and underpins the validity of inferential statistics techniques, such as confidence intervals and hypothesis testing.

The CLT states that the distribution of the average of a large number of independent, identically distributed variables ($\bar{X}$ in Eq. 14.35) approaches a normal distribution (which we will detail in § 15.4), regardless of the original distribution of the variables. Specifically, $\bar{X}$ will have a distribution that is approximately normal with mean $\mu = E[X]$ and standard deviation $\sigma = \text{SEM}$ for large n.

To understand how this distribution helps us estimate the population mean, we introduce the concepts of margin of error and confidence intervals.

Margin of Error (MOE): The margin of error provides a measure of how much the sample mean is likely to differ from the true population mean. It is calculated as:

$$\text{MOE} = \tilde{z} \cdot \text{SEM}, \tag{14.39}$$

where $\tilde{z}$ is the critical value corresponding to a specific confidence level. The critical value reflects the desired level of certainty in our estimate. For example, a 95% confidence level has a critical value of approximately 1.96. This number is related to the z-score discussed below.

Confidence Level and Interval: A confidence level quantifies the likelihood that the true population mean lies within a confidence interval constructed from repeated sampling. For example, a 95% confidence level implies that if we were to construct confidence intervals from repeated samples, 95% of these intervals would contain the true mean, while 5% would not.

A confidence interval represents the range of values within which the true population mean is expected to lie, given a specified confidence level. It is calculated as:

$$\text{Confidence Interval} = \bar{X} \pm \text{MOE} = \bar{X} \pm \tilde{z} \cdot \text{SEM}. \qquad (14.40)$$

z-score: The z-score (or standard score) measures how many standard errors a particular value is away from the mean. In the context of the sampling distribution of the mean, it is calculated as:

$$z = \frac{\bar{X} - \mu}{\text{SEM}}. \qquad (14.41)$$

Since $\bar{X}$ follows a normal distribution with standard deviation $\sigma = \text{SEM}$ for large n (as per the Central Limit Theorem), the z-score follows a standard normal distribution $N(0, 1)$. Consequently, each z-score corresponds to a specific percentage of data within an interval around the mean, commonly known as the 68-95-99.7 rule, as illustrated in Fig. 14.2:

- $z = 1$ corresponds to 68.27% of the data falling within one standard deviation from the mean ($\sigma = \text{SEM}$).

- $z = 2$ corresponds to 95.45% of the data falling within $\mu \pm 2\sigma$.

- $z = 3$ corresponds to 99.73% of the data falling within $\mu \pm 3\sigma$.

Or equivalently, we can state this rule in terms of the confidence level (c):

- $c = 68\%$ corresponds to $z \approx 1$, meaning the margin of error (MOE) equals the SEM.

- For $c = 95\%$, $z \approx 2$, and SEM must be compressed to $\frac{1}{2}$ MOE by increasing n by 4 times.

- For $c = 99.7\%$, $z \approx 3$, and n must be about 9 times as large as required for $c = 68\%$.

■ **Example 14.10** The height of individuals in a large population is known to follow a normal distribution. Consider the average height of adult men in a population being 70 inches with a standard deviation of 3 inches. An adult man whose height is 73 inches would have a z-score of 1. This implies that approximately 68% of men in this population have heights between 67 inches (one standard deviation below the mean) and 73 inches (one standard deviation above the mean).   ■

■ **Example 14.11** A coin, which may be biased, is tossed n times, resulting in m heads. We estimate the theoretical probability of heads p using the sample mean

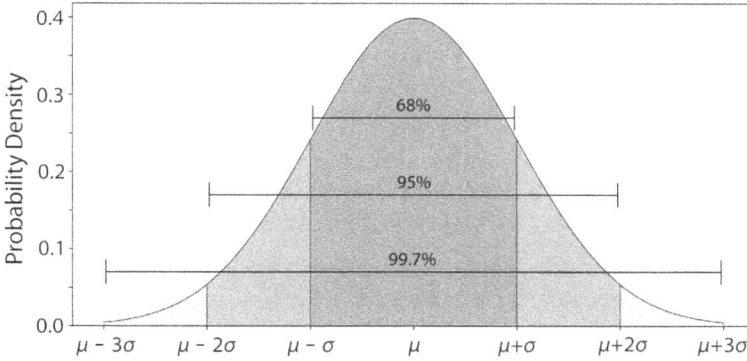

Figure 14.2: Normal Distribution and the 68-95-99.7 Rule

$\bar{p} = \frac{m}{n}$. To parallel Eq. 14.35, we introduce a random variable X, with $X = 1$ for heads and $X = 0$ for tails. Then

$$\bar{X} = \frac{1}{n} \sum_{i=1}^{n} X_i. \tag{14.42}$$

Apparently, $\bar{p} = \bar{X}$, and $p = E[X]$.

Assuming that through measurement with a reasonable n, we have determined p to be approximately 0.3. The question is: how large must n be to ensure our measurement of p is within an accuracy of 99% with a 95% confidence level?

To determine n, we first need to estimate the standard deviation of X. In this case, since tossing coins is a Bernoulli process and X follows a binomial distribution, the standard deviation of X is given by $\Delta X = \sqrt{p(1-p)}$. (We will delve into this in § 15.2.2.) With $p \approx 0.3$, we find that $\Delta X \approx 0.46$, which gives:

$$\text{SEM} \approx \frac{0.46}{\sqrt{n}}. \tag{14.43}$$

Alternatively, we can also estimate ΔX through empirical measurements:

$$\Delta X \approx \sqrt{\frac{1}{n} \sum_{i=1}^{n} (X_i - p)^2}. \tag{14.44}$$

To measure p with an accuracy of 99%, the margin of error is $\text{MOE} = 0.01p \approx 0.003$. Setting $\text{SEM} \approx \frac{0.46}{\sqrt{n}} = 0.003$ gives us:

$$n \approx 23000. \tag{14.45}$$

That n, calculated with $\text{MOE} = \text{SEM}$, provides a confidence level of only 68%. For a 95% confidence level, n must be increased by a factor of 4, giving approximately:

$$n \approx 93000. \tag{14.46}$$

■

Problem Set 14

14.1 Suppose three six-sided fair dice are thrown, each with faces numbered from 1 to 6. Let X_1, X_2 and X_3 denote the scores obtained from each die, respectively. Calculate the probability $p(X_1 + X_2 < X_3^2)$.

14.2 Two coins are tossed, and each can land showing either Heads (H) or Tails (T). Describe the sample space for this experiment. Then, list all possible events. Hint: There are 16 such events in total. The empty set, representing the impossible event, is also considered an event.

14.3 Count the number of distinct ways of putting m balls into n boxes when:

(a) all boxes and balls are distinguishable;

(b) the boxes are different but the balls are identical;

(c) the balls are identical, the boxes are different but each can hold at most a single ball. (In this case, $n \geq m$.)

14.4 Suppose an exam consists of n multiple-choice questions, each with five possible answers. For each question, a score of 1 is awarded for a correct answer and -0.2 for a wrong answer.

(a) If a test taker randomly guesses the answers to all questions, what is the expected total score for the entire exam?

(b) If a test taker has a 75% likelihood of answering any given question correctly and randomly guesses the answers to the rest, what is the expected total score for the entire exam?

14.5 Suppose we have a set of 10 coins: 8 of them are ordinary with equal chances of landing on Heads or Tails when tossed, one of them is a double-headed coin (two Heads), and one is a double-tailed coin (two Tails).

(a) If you randomly select one coin, toss it, and it lands on Heads, what is the probability that the selected coin is the double-headed coin?

(b) If you randomly select another coin (without replacing the first one) to toss and it lands on Tails, what is the probability that it is one of the ordinary coins?

14.6 ✳ The probability distribution of a random variable X is given as:

$$p(X = x) = \begin{cases} kx^2 & \text{for } x = -1, -2, -3 \\ 2kx & \text{for } x = 1, 2, 3 \\ 0 & \text{otherwise} \end{cases}$$

where k is a constant. Determine k and calculate

(a) $p(X \geq 0)$

(b) $E[X]$

(c) $\text{Var}[X]$

14.7 Consider two fair six-sided dice with faces numbered from 1 to 6. Let X and Y represent the outcomes of rolling these two dice, respectively. Define the following variables based on the results of the rolls:

o $S = X + Y$: The sum of the two dice rolls.

o $D = X - Y$: The difference between the two dice rolls.

o $U = \min(X, Y)$: The smaller outcome of the two dice rolls.

o $V = \max(X, Y)$: The larger outcome of the two dice rolls.

Determine the covariance and correlation for each of the following pairs of these variables:

(a) (X, Y) (c) (X, D) (e) (U, V) (g) (V, D)

(b) (X, S) (d) (X, U) (f) (U, S)

You may find it beneficial to use a programming language like Python or R to facilitate the calculations for this problem.

14.8 $*$ Consider a busy intersection where the number of cars passing through is known to follow a Poisson distribution. You know that roughly 20 cars pass through the intersection every minute.

You are tasked with measuring the average number of cars per minute that pass through this intersection, aiming to ensure your estimate is within a 95% confidence level with a measurement error margin of less than 5%.

(a) Let X represent the number of cars passing through the intersection per minute. Calculate the expected value $E[X]$ and the standard deviation ΔX.

(b) Given that you observe the traffic at this intersection for n minutes, express the sample mean $\bar{X}$ of the number of cars observed per minute. Use the Central Limit Theorem to describe the distribution of $\bar{X}$ when n is large.

(c) Calculate the minimum number of minutes n you need to observe the intersection to estimate the average traffic flow within the specified accuracy and confidence level. Recall that the standard error of the mean is SEM $= \frac{\Delta X}{\sqrt{n}}$, and the z-score for a 95% confidence level is approximately 2.

14.9 Charlie prepares many pairs of coins and sends one coin from each pair to Alice and the other to Bob. The coins land randomly as Heads (H) or Tails (T), but each pair always maintains perfect correlation, i.e., both coins show either HH or TT.

Alice and Bob each have three landing pads for the coins, labeled A_1, A_2, A_3 for Alice and B_1, B_2, B_3 for Bob. The pairs land on these pads in a rotating sequence: $A_1\text{-}B_1, A_2\text{-}B_2, A_3\text{-}B_3, A_1\text{-}B_1, \ldots$.

To determine the correlation of their coin pairs, Alice and Bob compute the following quantity based on many observations:

$$C = P(A_1 = B_2) + P(A_2 = B_3) + P(A_3 = B_1),$$

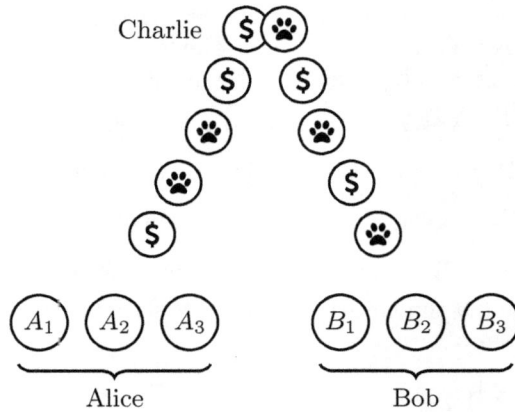

where $P(A_i = B_j)$ is the probability that Alice's coin on pad A_i and Bob's coin on pad B_j show the same face (i.e., both H or both T).

Prove the following:

(a) If all pairs of coins are perfectly correlated (e.g., HH, HH, HH, ...), then $C = 3$.

(b) If the pairs alternate perfectly between HH and TT, then $C = 1$.

(c) In the general case, C satisfies $C \geq 1$.

Background: The inequality $C \geq 1$ is a specific form of Bell's inequality within a classical probability framework.

However, in quantum mechanics, where entangled particle pairs are involved, this classical constraint no longer holds. Quantum measurements can produce correlations that violate Bell's inequality, achieving values as low as $C = 0.75$. This highlights the fundamentally non-classical nature of quantum entanglement.

15. Stochastic Processes

Contents

15.1 Introduction to Stochastic Processes

A random or stochastic process is essentially a collection of random variables, each indexed by time or space, that describe the evolution of systems in a stochastic, or probabilistic, manner. This concept is central in understanding systems that evolve unpredictably over time or space. Here are some common examples:

- Fundamental Stochastic Processes:

  - Bernoulli Process is a sequence of independent binary (success/failure) experiments, leading to the Binomial distribution when tallying the number of successes.

  - Poisson Process models the occurrence of independent events over time at a constant rate, giving rise to the Poisson distribution for the count of events in fixed intervals.

  - Gaussian Process consists of random variables with joint Gaussian distributions, used extensively in modeling continuous outcomes and in fields like quantum mechanics for state estimation.

  These processes are foundational in the study of stochastic systems and contribute to our understanding of random phenomena, with each process corresponding to a distinct probability distribution that quantifies uncertainty and variability in various contexts.

- Markov Chains: A Markov chain is a stochastic process where the probability of moving to the next state depends only on the current state and not on the sequence of events that preceded it. Markov chains are used in a variety of fields, including quantum algorithms, economics, game theory, and biology.

- Random Walk: Random walk is a special case of Markov chain where each step is taken randomly and independently, typically with equal probability, in a specific state space like a lattice or a grid. Random walks are used to model various phenomena, including stock market fluctuations and particle movements in liquids.

- Queueing Models: Used extensively in operations research, these models study the behavior of queues (or lines). They help in understanding and predicting queue lengths and waiting times, important in designing and managing facilities like call centers, hospitals, and manufacturing plants.

In this chapter, we will focus on the Bernoulli, Poisson, and Gaussian Processes to lay the foundation for understanding basic concepts of stochastic processes. These models illustrate how randomness can be systematically described and analyzed. Subsequently, we will delve into more complex models such as Markov chains and explore their applications in various fields, including Monte Carlo methods, which are pivotal in quantum algorithms, computational physics and finance for simulating random processes and for optimization problems.

15.2   Bernoulli Process and Binomial Distribution

15.2.1   Bernoulli Process

A Bernoulli process is a sequence of independent random experiments, each with two possible outcomes, commonly termed success and failure. When the number of successes is counted over a fixed number of trials, this leads to various important stochastic processes. This concept is fundamental to probability theory and underlies many other stochastic processes. Here are a few examples:

- Coin Flipping: The most classic example of a Bernoulli process is flipping a fair coin. Each flip is independent of the others, and there are two possible outcomes: heads (success) or tails (failure).

- Quality Control: In a manufacturing process, the inspection of items as either defective (failure) or non-defective (success) is a Bernoulli process. Each inspection is independent, and there are two possible outcomes for each item inspected.

- Medical Trials: In a clinical trial for a new drug, each patient's reaction can be modeled as a Bernoulli process, with outcomes being effective treatment (success) or not effective (failure).

- Survey Responses: In surveys or opinion polls, each response to a yes/no question is a Bernoulli trial. For instance, agreeing (success) or not agreeing (failure) with a statement.

- Network Packet Transmission: In computer networks, the transmission of a packet can be seen as a Bernoulli trial, where each packet is either successfully transmitted (success) or lost (failure).

- Biology and Ecology: The germination of seeds can be treated as a Bernoulli process, where each seed either germinates (success) or does not (failure).

- Finance and Investing: The performance of an investment can be modeled as a Bernoulli process, with each period (like a day) yielding a profit (success) or a loss (failure).

15.2.2 Binomial Distribution

The resultant distribution of successes is the binomial distribution. Formally, we can use a random variable X to represent number of successes in n independent trials, each with probability p of success. The probability distribution is given by:

$$p(X = x) = \binom{n}{x} p^x (1 - p)^{n-x} \tag{15.1}$$

where $\binom{n}{x}$ is the number of ways to choose x successes from n trials. Key properties of binomial distribution is summarized in Table 15.1.

Property	Description or Formula
Notation	$X \sim \text{Binomial}(n, p)$
Description	Number of successes in n independent trials, each with probability p of success.
Probability	$p(X = x) = \binom{n}{x} p^x (1 - p)^{n-x}$ for $x = 0, 1, 2, \ldots, n$
Mean	$E(X) = np$
Variance	$\text{Var}(X) = np(1 - p) = npq$, where $q = 1 - p$
Sum	If $X \sim \text{Binomial}(n, p)$, $Y \sim \text{Binomial}(m, p)$, and X and Y are independent, then $X + Y \sim \text{Binomial}(n + m, p)$.

Table 15.1: Key Properties of Binomial Distribution

Exercise 15.1 Eight fair coins are tossed simultaneously. Find the probability of getting

   (a) exactly 6 Heads,

   (b) at most 6 Heads, and

   (c) at least 6 Heads.

15.2.3 Related Probability Distributions

Several probability distributions are related to the concept of Bernoulli trial and extend it in different ways to model a wide variety of scenarios that can be encountered in quantum algorithms and stochastic processes. These include:

1 Geometric Distribution

The geometric distribution reflects the number of failures before the *first* success in a sequence of independent Bernoulli trials, each with probability p of success. This distribution can be viewed as the waiting time for a success in a Bernoulli process. Its key properties are summarized in Table 15.6 at the end of this chapter.

2 Negative Binomial Distribution

The negative binomial distribution represents the number of failures before the kth success in a sequence of independent Bernoulli trials. It generalizes the Geometric distribution by allowing for more than one success. It can be thought of as the sum of k independent geometric distributions. Its key properties are summarized in Table 15.7.

3 Hypergeometric Distribution

The hypergeometric distribution describes the number of 'special' objects drawn without replacement from a finite population consisting of two types of objects. It differs from the binomial distribution in that the trials are not independent; each draw changes the composition of the population. Its key properties are summarized in Table 15.9.

4 Multinomial Distribution

The multinomial distribution generalizes the binomial distribution to scenarios with more than two possible outcomes for each trial. If a Bernoulli trial is a coin toss, the multinomial distribution is analogous to rolling a die with k sides. It gives the joint probability of a vector of outcomes $(X_1, \ldots, X_k)$ from n independent trials, where each trial results in one of k categories with respective probabilities $(p_1, \ldots, p_k)$. Its key properties are summarized in Table 15.8.

15.3 Poisson Process and Poisson Distribution

15.3.1 Poisson Process

A Poisson process models situations where events occur randomly in time (or space), with some average rate. (For the sake of brevity, we only consider Poisson processes in time here.) The key property of a Poisson process is that the number of events occurring in any interval of time only depends on the length of the interval, not on when the interval starts. Here are a few examples:

- Telecommunications: The number of phone calls arriving at a call center within a given time frame can often be modeled as a Poisson process. This model helps in understanding call frequency and is crucial for staffing and capacity planning.

- Healthcare: The arrival of patients at an emergency department can be modeled as a Poisson process, which is critical for resource allocation, including staffing and equipment availability.

- Retail and Commerce: The number of customers entering a store or website hits over time can often be modeled as a Poisson process. This information is used for staffing, inventory management, and server capacity planning.

- Physics and Engineering: The emission of particles from a radioactive source can be modeled as a Poisson process, where the counts of emitted particles per unit time follow a Poisson distribution. This is crucial for understanding radioactive decay and for designing detectors and safety measures.

- Finance and Insurance: The occurrence of certain insurance claim events, such as accidents or natural disasters (assuming they are rare and independent), can be modeled as a Poisson process. This aids in pricing insurance products and assessing risk.

15.3.2 Poisson Distribution

Formally, we use a random variable X to represent the number of events occurring within a fixed interval of time. The process is characterized by a rate parameter λ, which represents the average number of events per unit time. The resultant probability distribution of X is the Poisson distribution, whose properties are summarized in Table 15.2.

Property	Description or Formula
Notation	$X \sim \text{Poisson}(\lambda)$
Description	Number of events occurring within a fixed interval of time, given average rate λ.
Probability	$p(X = x) = \dfrac{\lambda^x e^{-\lambda}}{x!}$ for $x = 0, 1, 2, \ldots$
Mean	$E(X) = \lambda$
Variance	$\text{Var}(X) = \lambda$
Sum	If $X \sim \text{Poisson}(\lambda)$, $Y \sim \text{Poisson}(\mu)$, and X and Y are independent, then $X + Y \sim \text{Poisson}(\lambda + \mu)$.

Table 15.2: Key Properties of Poisson Distribution

15.3.3 From Binomial to Poisson Distribution

The Poisson distribution can be viewed as a limiting case of the binomial distribution (Eq. 15.1) when the number of trials goes to infinity while the expected number of successes remains constant, $np = \lambda$. That is,

$$p(X = x) = \binom{n}{x} p^x (1-p)^{n-x} \approx \frac{\lambda^x}{x!} e^{-\lambda} \qquad \text{given } n \gg x \text{ and } np = \lambda. \qquad (15.2)$$

The derivation is as follows:

$$p(X = x) = \binom{n}{x} p^x (1-p)^{n-x}$$

$$= \frac{n!}{x!(n-x)!} p^x (1-p)^{n-x}$$

$$\approx \frac{n^x}{x!} p^x (1-p)^{n-x} \qquad \because n \gg x,\ n-x \approx n,\ \frac{n!}{(n-x)!} \approx n^x$$

$$\approx \frac{\lambda^x}{x!} \left(1 - \frac{\lambda}{n}\right)^n \qquad \because np = \lambda,\ p = \frac{\lambda}{n}$$

$$\approx \frac{\lambda^x}{x!} e^{-\lambda} \qquad \because \lim_{n\to\infty} \left(1 + \frac{\lambda}{n}\right)^n = e^{-\lambda}.$$

> **Exercise 15.2** The number of customers arriving at a store can be modeled by a Poisson process with a mean rate $\lambda = 100$ customers per hour.
>
> (a) Find the probability that there are 50 customers between 10:00 and 10:30.
>
> (b) Find the probability that there are 50 customers between 10:00 and 10:20 and 50 customers between 10:20 and 11:00.

15.3.4 Related Distributions

Two key probability distributions related to the Poisson distribution extend it in various ways to model a broad range of scenarios encountered in statistical simulations and quantum algorithms.

1 Exponential Distribution

The time between events in a Poisson process follows an exponential distribution. This is often used to model waiting times between continuous and memoryless stochastic events, such as the arrival of emails or bus arrivals, assuming the events are equally likely to occur at any time. Its key properties are summarized in Table 15.10 at the end of this chapter.

Derivation

This derivation exemplifies the internal connections between different types of probability distributions. Therefore, we present it here.

Let T be the random variable representing the time until the next event occurs. We will first find the probability that T is greater than some time t, denoted $p(T > t)$.

If we consider an interval of length t, we can divide it into n small subintervals of length $\tau = \frac{t}{n}$. As $n \to \infty$, τ becomes infinitesimally small.

By the definition of the Poisson process, the probability of exactly one event occurring in a small interval τ is $\lambda\tau$. Therefore, the probability of no event occurring in that small interval is $1 - \lambda\tau$.

The probability of no event occurring in the entire interval t is the product of the probabilities of no event in each small interval, given that each small interval is independent:

$$p(\text{no event in } t) = p(\text{no event in } \tau)^n = (1 - \lambda\tau)^n.$$

Taking the limit as $n \to \infty$, we get:

$$\lim_{n \to \infty} \left(1 - \frac{\lambda t}{n}\right)^n = e^{-\lambda t}.$$

Hence, the probability that no event occurs by time t (which is the same as the time until the first event being greater than t) is:

$$p(T > t) = e^{-\lambda t}.$$

To find the probability density function (PDF) of T, we take the derivative of the cumulative distribution function (CDF), $p(T \leq t) = 1 - p(T > t)$, with respect to t:

$$f_T(t) = \frac{d}{dt}(1 - e^{-\lambda t}) = \lambda e^{-\lambda t}.$$

This is the exponential distribution with parameter λ, which describes the time between events in a Poisson process.

Discrete vs. Continuous Probability Distributions

Note that the exponential distribution is a continuous distribution: the sample space of the random variable is a continuous range of real numbers, whereas the Poisson distribution is a discrete distribution. The key differences are summarized in Table 15.3.

> The relationship between continuous and discrete probability distributions can be illustrated by considering the digitization of a continuous random variable.
>
> For example, consider a continuous random variable X uniformly distributed over $[0, 1]$. This can be approximated by a discrete variable Y taking values in $\{0, 0.01, 0.02, \ldots, 1\}$. The interval $[0, 1]$ is divided into 100 equal parts, and each discrete value of Y corresponds to the midpoint of one of these intervals.
>
> In this case, the probability $p(Y = 0.01)$ approximates the probability of X falling within $[0.005, 0.015]$.

2 Gamma Distribution

The Gamma distribution generalizes the exponential distribution and is associated with the sum of several independent exponentially distributed random variables. It is used to model waiting times for the kth event in a Poisson process, where k is a positive integer. While its derivation is beyond the scope of this text, we summarize its key properties in Table 15.11 at the end of this chapter.

In another context, the Gamma distribution is used in Bayesian statistics (see § 14.2.4.4) as a prior distribution for rates or proportions that must be strictly positive. In Bayesian inference, the choice of a prior distribution is crucial because it

	Discrete	Continuous
Description	Sample space is finite or countably infinite, comprising separate points.	Sample space consists of an interval (or intervals) of real numbers.
Example	Poisson distribution $$p(x) = \frac{\lambda^x e^{-\lambda}}{x!}$$ for $x = 0, 1, 2, \ldots$	Exponential distribution $p(x) = \lambda e^{-\lambda x}$ for $0 \leq x < \infty$
Function Type	Probability mass function (PMF)	Probability density function (PDF)
Cumulative Distribution	$\mathrm{CDF}(X < \tilde{x}) = \sum_{x < \tilde{x}} p(x)$	$\mathrm{CDF}(X < \tilde{x}) = \int_{-\infty}^{\tilde{x}} p(x)\, dx$
Normalization	$\sum_x p(x) = 1$	$\int_{-\infty}^{\infty} p(x)\, dx = 1$

Table 15.3: Discrete vs. Continuous Probability Distributions

informs the posterior distribution, which melds our prior beliefs with the empirical data. The Gamma distribution, with its adaptable shape and compatibility with Poisson and exponential models, frequently serves as a mathematically convenient and conceptually clear prior for parameters that represent rates or proportions in Bayesian frameworks.

Exercise 15.3 Let $N(t)$ be a Poisson process with rate $\lambda = 2$ per unit time, and let T be the random variable representing the time interval between two successive events, which follows an exponential distribution.

(a) Find the probability that the first event occurs after $t = 1$, i.e., $p(t > 1)$.

(b) Given that the third event occurred at $t = 2$, find the probability that the fourth event occurs after $t = 4$.

(c) Given that no event has occured before $t = 2$, find the probability that the first event occurs before $t = 4$.

15.4 Gaussian Process and Normal Distribution

15.4.1 Single-Variable Normal Distribution

The normal (or Gaussian) distribution is characterized by its bell-shaped curve. It serves as an approximation to the binomial distribution when the number of trials (n) is large. The limit is reached when the probability of success (p) is neither too close to 0 nor 1, and $np(1 - p)$ is sufficiently large.

(Related to this, the Poisson distribution can also be seen as a limiting case of the binomial distribution as $n \to \infty$. In this case, the limit occurs when $p \to 0$, while the expected number of successes ($np = \lambda$) remains constant. See § 15.3.3.)

Furthermore, the Central Limit Theorem (see § 14.3.3.4) states that the distribution of the sum (or average) of a large number of independent, identically distributed variables approaches a normal distribution, regardless of the original distribution of the variables. This theorem justifies the use of the normal distribution as a model for many stochastic scenarios.

For example, human height is determined by a large number of factors, both genetic and environmental, which are additive in their effects. Thus, it follows a normal distribution.

The normal distribution is defined by two parameters: mean μ and standard deviation σ (or variance σ^2). Its key properties are summarized in Table 15.4.

Property	Description or Formula
Notation	$X \sim \text{Normal}(\mu, \sigma^2)$
Description	A distribution characterized by its mean μ and variance σ^2, producing the well-known bell-shaped curve.
Probability	$p(X = x) = \dfrac{1}{\sqrt{2\pi\sigma^2}} e^{-\frac{(x-\mu)^2}{2\sigma^2}}$ for $-\infty < x < \infty$
Cumulative	No simple closed form, but tabulated and available through statistical software.
Mean	$E(X) = \mu$
Variance	$\text{Var}(X) = \sigma^2$
Sum	If $X_1, X_2, \ldots, X_n$ are independent, and each $X_i \sim \text{Normal}(\mu_i, \sigma_i^2)$, then $X_1 + X_2 + \ldots + X_n \sim \text{Normal}\left(\sum_{i=1}^{n} \mu_i, \sum_{i=1}^{n} \sigma_i^2\right)$.

Table 15.4: Key Properties of Normal (Gaussian) Distribution

Exercise 15.4 In a normal distribution with mean 20 and standard deviation 2, find the probability for the interval of $[14, 26]$.

15.4.2 Multivariate Normal Distribution

The Multivariate Normal Distribution extends the concept of the normal distribution to multiple dimensions, describing the joint distribution of a vector of random variables. Each variable follows a normal distribution, and any linear combination of these variables also follows a normal distribution.

As summarized in Table 15.5, this distribution is characterized by a mean vector and a covariance matrix (also known as a kernel), where the mean vector defines the expected value of each variable, and the covariance matrix defines the variance of each variable along with the covariance between every pair of variables.

15.4.3 Gaussian Process

A Gaussian process is a stochastic process where any finite collection of random variables has a multivariate normal distribution. It's a powerful and flexible tool for modeling functions in a probabilistic framework. Unlike the Poisson Process, which deals with discrete events in time, a Gaussian Process can model continuous functions.

Property	Description or Formula
Notation	$X \sim \text{Normal}(\mu, \Sigma)$
Description	A vector of n random variables with a joint normal distribution.
Probability	$p(X = x) = \dfrac{1}{\sqrt{(2\pi)^n \det(\Sigma)}} e^{-\frac{1}{2}(x-\mu)^\top \Sigma^{-1}(x-\mu)}$
Mean Vector	$\mathbb{E}(X) = \mu$
Covariance Matrix	$\text{Cov}(X) = \Sigma$
Marginals	Each variable X_i is normally distributed, $X_i \sim \text{Normal}(\mu_i, \sigma_{ii})$.

Table 15.5: Key Properties of Multivariate Normal Distribution

Gaussian processes are widely used in modeling stochastic processes. Here are a few examples:

- Spatial Data Analysis: Gaussian processes model the elevation of a landscape at different points, where the elevation values at locations close to each other are more likely to be similar. This application is common in geostatistics and environmental science.

- Finance: Modeling stock prices or interest rates over time, where the future values are assumed to have a normal distribution centered around a mean that evolves over time.

- Machine Learning: In Gaussian Process Regression (GPR), the function that maps input features to output predictions is modeled as a Gaussian process. This allows for the estimation of uncertainty in predictions, making GPR particularly useful for tasks where it's required to quantify confidence in the predictions.

15.5 ∗Generating Probability Distributions

In stochastic process modeling, creating a random variable X with a specific probability distribution often begins with generating random numbers using a pseudorandom number generator (RAND) that produces a uniform distribution. These uniformly distributed numbers are then transformed into the desired probability distribution.

In this section, we explore various methods for transforming uniformly distributed random numbers into random variables that follow specific distributions. These techniques are essential in simulation and provide practical tools for generating random variables, enabling the study and implementation of complex processes across a wide range of applications.

15.5.1 Uniform Distribution

A uniform distribution is characterized by an equal probability for all outcomes within a specified range. Formally, a continuous uniform distribution over an interval

$[a, b]$ has a constant probability density function (pdf) over a to b and zero elsewhere, denoted as Uniform(a, b).

The standard uniform distribution Uniform$(0, 1)$, with pdf constant over 0 to 1, is frequently used as the starting point to generate other distributions. On a computer, this distribution can be simulated using a pseudorandom number generator that outputs values in the range $[0, 1]$.

To convert a variable $U \sim$ Uniform$(0, 1)$ to $V \sim$ Uniform(a, b), and vice versa, the following formulas are used:

$$V = a + (b - a)U, \quad U = \frac{V - a}{b - a}. \tag{15.3}$$

This transformation linearly scales and shifts U to fit within the interval $[a, b]$ for V, maintaining uniformity due to the linearity of the transformation.

15.5.2 Uniform to Binomial

As a simple example, to simulate a biased coin—say with a probability of 0.6 for heads and 0.4 for tails—we first generate random numbers to construct the random variable $U \sim$ Uniform$(0, 1)$. For each trial, we consider it a 'heads' if $U < 0.6$, and a 'tails' otherwise.

To simulate a random variable $X \sim$ Binomial(n, p) from Uniform$(0, 1)$, we perform n independent trials of $U_i \sim$ Uniform$(0, 1)$, where $i = 1, 2, \ldots, n$. Each trial is counted as a success if $U_i < p$, and a failure otherwise. The total number of successes gives us X:

$$X = \sum_{i=1}^{n} I(U_i < p), \tag{15.4}$$

where I is the indicator function that equals 1 if $U_i < p$ and 0 otherwise.

This counting method is a form of the Monte Carlo simulation but can be inefficient for large n. For such cases, more sophisticated methods are typically used.

15.5.3 Uniform to Exponential

To generate a random variable X with an Exponential(λ) distribution from a Uniform$(0, 1)$ distribution, we can employ the Inverse Transform Sampling (ITS) method. Let's first use an example to demostrate how ITS works.

■ **Example 15.1** Suppose we aim to generate a random variable X that follows an exponential distribution with a rate parameter $\lambda = 2$. The cumulative distribution function (CDF) for this exponential distribution is $F(x; 2) = 1 - e^{-2x}$.

Steps:

1. Generate Uniform Number: Generate a random number U from the Uniform(0,1) distribution. Assume $U = 0.35$ is obtained.

2. Invert the CDF: To find the corresponding x value, solve the equation $F(x; 2) = U$: $1 - e^{-2x} = 0.35$. By solving for x, we find $x \approx 0.21$.

3. Output: The value $x \approx 0.21$ is the sample from the Exponential(2) distribution we sought.

Why It Works:

The CDF maps values from the target distribution to their cumulative probabilities. By generating a uniform random number U, we effectively choose a random cumulative probability. ITS finds the value in the target distribution's domain whose cumulative probability equals U. The monotonic nature of the CDF ensures this value is unique.   ∎

For a broader application, the CDF of the exponential distribution is generally given by:
$$F(x; \lambda) = 1 - e^{-\lambda x} \text{ for } x \geq 0.$$

Given a random sample U from Uniform(0,1), we invert the CDF of the exponential distribution to find the corresponding x:
$$U = 1 - e^{-\lambda x}$$

Solving for x gives us:
$$x = -\frac{\ln(1 - U)}{\lambda}.$$

Since U is uniformly distributed, $1 - U$ has the same distribution. Thus, we simplify the inversion formula to:

$$X = -\frac{\ln(U)}{\lambda}, \tag{15.5}$$

where X is the generated random variable with an Exponential(λ) distribution.

ITS plays a crucial role in statistical simulation by converting uniformly distributed samples into variables with a specific desired distribution, not limited to the exponential distribution.

Exercise 15.5 Given $X \sim$ Uniform$(0, 1)$, calculate the probability distribution of $Y = 1/X$.

15.5.4 Uniform and Exponential to Poisson

ITS can also be used to generate a $k \sim$ Poisson(λ) random variable from a uniform distribution $U \sim$ Uniform$(0, 1)$. The process is as follows:

First, compute the cumulative probability $F(k; \lambda) = p(X \leq k)$ for the Poisson distribution up to some k where $F(k; \lambda) \approx 1$.

Then, find the smallest integer k such that $U \leq F(k; \lambda)$. This k is the generated Poisson-distributed random variable.

Using the inherent property of the Poisson process that waiting times between events follow an exponential distribution, we can also generate $k \sim$ Poisson(λ) from Exponential(λ) as follows:

Keep generating Exponentials, adding them to a variable t, until t exceeds the fixed time interval we are interested in (often taken as 1 unit of time for convenience). Then count the events: The number of events we needed to sum before exceeding the interval, k, is the random variable drawn from $k \sim$ Poisson(λ).

Exercise 15.6 Write a Python program to generate a sequence of 1000 numbers following a Poisson distribution with a given $\lambda = 5$ using the Inverse Transform Sampling (ITS) technique. Provide a plot or histogram of the generated numbers to visualize their distribution. Ensure to include comments in your code to explain each step of the process.

You may use libraries like 'numpy' for random number generation and 'scipy.stats' for distribution functions, but the focus should be on applying the ITS manually.

15.5.5   Uniform to Gaussian

The Box-Muller transform is a popular method for generating normally distributed random numbers from uniformly distributed random numbers. It cleverly uses trigonometric functions and the natural logarithm to convert two uniform random variables into two independent standard normal variables.

Given $U_1, U_2 \sim \text{Uniform}(0, 1)$, two standard normal random variables Z_1 and Z_2 are generated as follows:

$$Z_1 = \sqrt{-2 \ln U_1} \cos(2\pi U_2), \quad Z_2 = \sqrt{-2 \ln U_1} \sin(2\pi U_2) \qquad (15.6)$$

Then, $X = \mu + \sigma Z_1$ (or σZ_2) will be distributed according to $\text{Normal}(\mu, \sigma^2)$.

To generate a multivariate normal random variable $\boldsymbol{X} \sim \text{Normal}(\boldsymbol{\mu}, \boldsymbol{\Sigma})$ from $\text{Uniform}(0, 1)$, one approach is to first generate a set of independent standard normal variables $\boldsymbol{Z}$ using the Box-Muller transform or other methods. Then, transform $\boldsymbol{Z}$ using the Cholesky decomposition of $\boldsymbol{\Sigma}$, $\boldsymbol{\Sigma} = \boldsymbol{L}\boldsymbol{L}^\top$, where $\boldsymbol{L}$ is a lower triangular matrix:

$$\boldsymbol{X} = \boldsymbol{\mu} + \boldsymbol{L}\boldsymbol{Z} \qquad (15.7)$$

This method relies on the linear transformation of standard normal variables to achieve the desired mean vector $\boldsymbol{\mu}$ and covariance matrix $\boldsymbol{\Sigma}$.

Exercise 15.7 Write a Python program to generate a sequence of 1000 numbers following the normal distribution (with $\mu = 100$ and $\sigma = 5$) using the Box-Muller transform. Provide a plot or histogram of the generated numbers to visualize their distribution. Ensure to include comments in your code to explain each step of the process.

Problem Set 15

15.1 A box contains 20 red balls, 30 green balls, and 50 blue balls. If 10 balls are selected at random without replacement, find the probability that:

(a) exactly 3 are red, 3 are green, and 4 are blue;

(b) at least 3 are red;

(c) at most 3 are red.

15.2 Consider a factory that produces items with a 0.1% probability of being defective.

(a) Determine the maximum number of items that can be packed into a box so that the probability of containing one or more defective items does not exceed 1%.

(b) Calculate the expected value and standard deviation of the number of defective items in a box with this maximum number of items.

15.3 Suppose that 1% of items of a product are randomly defective. The items are packaged in boxes of 100.

(a) What is probability distribution that a given box contains n defectives?

(b) Suppose you have m boxes of this product. What is the distribution of the number of boxes with no defective items? What is the expected number of boxes with no defective items?

15.4 Suppose the number of typos a person makes has an average rate of 0.2 typos per page, and the typos are independent of each other. Find the probability that:

(a) a 5-page document contains no typos;

(b) a 10-page document contains at most 2 typos.

15.5 The annual heating oil consumption for a building is known from past statistics to average 1000 gallons, with a standard deviation of 200 gallons. Assuming the consumption follows a normal distribution:

(a) If 1200 gallons are stocked, what is the probability that this amount will be sufficient for the year's needs? (This is sometimes referred to as the "coverage probability.")

(b) What stock level is required to ensure with 95% confidence that the amount will be sufficient for the year's consumption?

15.6 Suppose we're analyzing the performance of two stocks, X and Y. We model their joint daily returns using a multivariate normal distribution with the following parameters:

Mean vector: $\mu = \begin{bmatrix} 0.01 & 0.015 \end{bmatrix}$

Covariance vector: $\Sigma = \begin{bmatrix} 0.005 & 0.002 \\ 0.002 & 0.008 \end{bmatrix}$

(a) Calculate the correlation coefficient between the returns of stock X and Y. Interpret its meaning.

(b) You invest 60% of your capital in stock X and 40% in stock Y. Find the expected daily return of this portfolio. Calculate the variance of the portfolio's daily return.

(c) Given that the return of stock X was 2% today, what is the probability that the return of stock Y was greater than 1%?

15.7 ∗ Consider a sequence of n coin flips where each flip results in heads with probability p and tails with probability $1 - p$. Let h denote the number of heads and t the number of tails obtained.

(a) Find the probability distribution of the random variable X defined as $X = h - t$.

(b) Find the probability distribution of $Y = 2h - t$.

(c) Discuss the behavior of the distributions of X and Y for large n. (Hint: use the central limit theorem.)

Appendix: Properties of Additional Probability Distributions

Property	Description or Formula
Notation	$X \sim \text{Geometric}(p)$
Description	Number of failures before the *first* success in a sequence of independent trials, each with probability p of success.
Probability	$p(X = x) = (1 - p)^x p \quad \text{for} \quad x = 0, 1, 2, \ldots$
Mean	$E(X) = (1 - p)/p$
Variance	$\text{Var}(X) = (1 - p)/p^2$
Sum	If $X_1, X_2, \ldots, X_k$ are independent, and each $X_i \sim \text{Geometric}(p)$, then $X_1 + X_2 + \ldots + X_k \sim \text{NegBinomial}(k, p)$.

Table 15.6: Key Properties of Geometric Distribution

Property	Description or Formula
Notation	$X \sim \text{NegBinomial}(n, p)$
Description	Number of failures before the kth success in a sequence of independent trials, each with probability p of success.
Probability	$p(X = x) = \binom{k+x-1}{x} p^k (1 - p)^x \quad \text{for} \quad x = 0, 1, 2, \ldots$
Mean	$E(X) = k(1 - p)/p$
Variance	$\text{Var}(X) = k(1 - p)/p^2$
Sum	If $X \sim \text{NegBinomial}(k, p)$, $Y \sim \text{NegBinomial}(m, p)$, and X and Y are independent, then $X + Y \sim \text{NegBinomial}(k + m, p)$.

Table 15.7: Key Properties of Negative Binomial Distribution

Property	Description or Formula
Notation	$(X_1, X_2, \ldots, X_k) \sim \text{Multinomial}(n; p_1, p_2, \ldots, p_k)$
Description	For n independent trials with k outcomes, X_i counts trials resulting in outcome i, where p_i is the probability of outcome i, for $i = 1, 2, \ldots, k$.
Probability	$p(X_1 = x_1, \ldots, X_k = x_k) = \dfrac{n!}{x_1! \ldots x_k!} p_1^{x_1} p_2^{x_2} \ldots p_k^{x_k}$, with $\sum_{i=1}^{k} x_i = n$ and $\sum_{i=1}^{k} p_i = 1$.
Marginal P	$X_i \sim \text{Binomial}(n, p_i)$
Mean	$E(X_i) = np_i$
Variance	$\text{Var}(X_i) = np_i(1 - p_i)$
Coariance	$\text{Cov}(X_i, X_j) = -np_i p_j \quad \text{for all} \quad i \neq j$

Table 15.8: Key Properties of Multinomial Distribution

Property	Description or Formula
Notation	$X \sim \text{Hypergeometric}(N, M, n)$
Description	Given N objects, of which M are 'special'. Draw n objects *without replacement*. X is the number of the n objects that are 'special'.
Probability	$p(X = x) = \binom{M}{x}\binom{N-M}{n-x}/\binom{N}{n}$ for $\quad x = \max(0, n + M - N) \quad$ to $\quad \min(n, M)$
Mean	$E(X) = np$, where $p = M/N$
Variance	$\text{Var}(X) = np(1-p)\frac{N-n}{N-1}$, where $p = \frac{M}{N}$

Table 15.9: Key Properties of Hypergeometric Distribution

Property	Description or Formula
Notation	$X \sim \text{Exponential}(\lambda)$
Description	The time between two events in a Poisson process with average rate λ.
Probability	$p(X = x) = \lambda e^{-\lambda x} \quad$ for $0 \le x < \infty$
Cumulative	$p(X \le x) = 1 - e^{-\lambda x} \quad$ for $0 \le x < \infty$
Mean	$E(X) = 1/\lambda$
Variance	$\text{Var}(X) = 1/\lambda^2$
Sum	If $X_1, X_2, \ldots, X_k$ are independent, and each $X_i \sim \text{Exponential}(\lambda)$, then $X_1 + X_2 + \ldots + X_k \sim \text{Gamma}(k, \lambda)$.

Table 15.10: Key Properties of Exponential Distribution

Property	Description or Formula
Notation	$X \sim \text{Gamma}(k, \lambda)$
Description	Waiting time for the kth event in a Poisson process with average rate λ.
Probability	$p(X = x) = \dfrac{\lambda^k x^{k-1} e^{-\lambda x}}{\Gamma(k)} \quad$ for $0 \le x < \infty$ where $\Gamma(k) = \int_0^\infty y^{k-1} e^{-y} \, dy$ (the Gamma function).
Mean	$E(X) = k/\lambda$
Variance	$\text{Var}(X) = k/\lambda^2$
Sum	If $X_1, X_2, \ldots, X_n$ are independent, and each $X_i \sim \text{Gamma}(k_i, \lambda)$, then $X_1 + X_2 + \ldots + X_n \sim \text{Gamma}\left(\sum_{i=1}^n k_i, \lambda\right)$.

Table 15.11: Key Properties of Gamma Distribution

16. Markov Chains

Contents

(i) Although this topic is not essential for studying the second book, *Quantum Computing and Information*, it will become important in the third book, *Quantum Algorithms and Applications*. Readers may defer studying it until encountering the relevant quantum algorithms.

16.1 Introduction

Markov Chains are mathematical models for analyzing systems transitioning between states within a finite or countably infinite set. Their applicability spans across diverse fields such as quantum algorithms, economics, game theory, and biology. Understanding these foundational concepts is crucial for delving into their quantum computing applications, including stochastic modeling, Monte Carlo simulations, and the development of quantum algorithms.

Consider a simplistic model of weather in a city, which can be categorized as either clear or rainy. One might start with a Bernoulli process for day-to-day weather prediction, where there is a fixed probability of rain or clear skies on any given day. For instance, assume a random variable X represents the weather condition with a 10% chance of rain $(X = R)$ and a 90% chance of clear skies $(X = C)$.

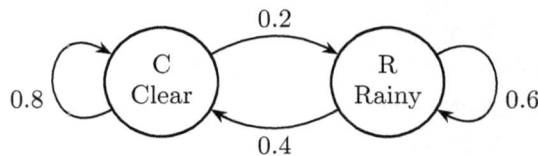

Figure 16.1: A Markov Chain Model for Weather

Such a Bernoulli process does not, however, capture the tendency of weather conditions to persist—clear days often follow clear days, and rainy days may come in streaks. To incorporate this behavior, we can extend the model to a two-state Markov process, as shown in Fig. 16.1.

In the Markov model (Fig. 16.1), if today is clear, the probability of tomorrow being clear remains high at 80%, while the chance of it turning rainy is 20%. Conversely, if today is rainy, the probability is split at 60% rainy and 40% clear for the next day's weather. This introduces a more nuanced understanding of weather patterns, allowing us to ask questions about long-term weather probabilities.

A fundamental assumption of Markov processes is the Markov property of memorylessness, meaning the system's next state depends only on its current state and not on the sequence of events that preceded it. For the weather model, this implies that the prediction for tomorrow's weather relies solely on today's weather, without consideration for past conditions.

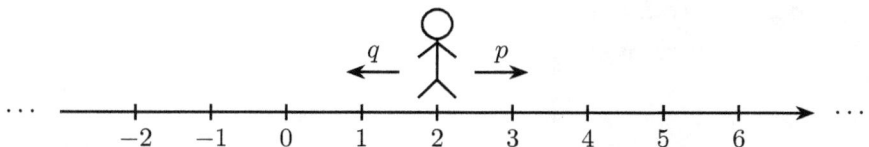

Figure 16.2: One-Dimensional Random Walk

A random walk on a line (or 1D random walk), depicted in Fig. 16.2, is another classic example of a Markov chain. Here, a man takes steps along a line, moving forward with probability p and backward with probability q. (For a symmetric random walk, $p = q = 0.5$.) If the man starts at position $x = 0$ and takes n steps, we might ask about the mean position $\bar{x}$ after those steps, the variance of his position, and the probability of reaching a cliff at position $x = m$. Given a sufficiently large n, we can also ponder whether reaching the cliff is an inevitability.

16.2 Definitions

To fully analyze Markov chains and make predictions, we need a precise mathematical framework. Let's now formalize the definitions that underpin Markov chains.

Recall that a stochastic process (§ 15.1) is a collection of random variables, $X_0, X_1, X_2, \ldots$, each indexed by time or space, that models the evolution of systems in a probabilistic manner. In the following, we will use time t to index the random variables unless otherwise stated.

> **Definition 16.1 — Markov Chain.** A Markov chain is a stochastic process X_0, X_1, X_2, ... in which each X_t represents the state of the system at time t. This process is characterized by the Markov property, which asserts that the probability of transitioning to any particular state is dependent solely on the current state and not on the sequence of events that preceded it.

For illustration, consider a board game where a die roll determines your next move. This game is memoryless because the result of the next roll depends only on the current position on the board and the outcome of the roll, without any dependence on the sequence of previous rolls. In contrast, memory plays a crucial role in strategic card games like bridge, where players must remember past plays to inform their strategy. The former is an example of a Markov process, while the latter is not.

> **Definition 16.2 — State.** The state of a Markov chain at time t is the value of X_t.

> **Definition 16.3 — State Space.** The state space of a Markov chain, denoted by S, is the set of all possible values that each X_t can assume.

For example, in the two-state weather model (Fig. 16.1), the state space is {C, R}, representing {Clear, Rainy}. Here, X_t indicates the weather on day t.

In the random walk on a number line example (Fig. 16.2), the state space is $\mathbb{Z} = \{\ldots, -2, -1, 0, 1, 2, \ldots\}$, and X_n represents the walker's position, x, after n steps.

> **Definition 16.4 — Trajectory.** A trajectory of a Markov chain is a specific sequence of states $X_0, X_1, X_2, \ldots$ that the process passes through over time.

In the Markov weather model (Fig. 16.1), a particular trajectory could be C, C, C, R, C, R, R, C, C, ... For the 1D random walk (Fig. 16.2), a particular trajectory might be $0, 1, 0, -1, -2, -1, 0, 1, 0, 1, 2, \ldots$.

The Markov property, a foundational aspect of Markov chains, can be formally defined as follows:

> **Definition 16.5 — Markov Property.**
>
> $$p(X_{n+1} = x \mid X_n = x_n, \ldots, X_0 = x_0) = p(X_{n+1} = x \mid X_n = x_n), \qquad (16.1)$$
>
> for all possible states of $x, x_n, x_{n-1}, \ldots, x_0$ and for all $n \geq 0$. Here, p represents the transition probability, indicating the likelihood of moving from one state to another.

This property allows Markov chains to be represented using transition diagrams, as illustrated in Figs. 16.1 and 16.2, and enables the computation of their evolutions through transition matrices, as we will delve into next.

16.3 Time Evolution

Now that we have the language to describe the states and transitions within a Markov chain, let's explore how these chains evolve over time and how we can predict their behavior.

16.3.1 Transition Matrices

A Markov transition diagram, such as in Fig. 16.1, depicts the transitions between different states in a Markov chain. We can present the same information with a transition matrix which is a stochastic matrix.

> **Definition 16.6 — Stochastic Matrix.** A stochastic matrix is a real square matrix P with the properties:
>
> - Non-negativity: Each entry P_{ij} is non-negative, i.e., $P_{ij} \geq 0$ for all i, j.
>
> - Row Sum to One: The sum of elements in each row is one, i.e., $\sum_{j=1}^{N} P_{ij} = 1$ for all i.

> **Definition 16.7 — Transition Matrix.** A transition matrix P for a Markov chain with N states is an $N \times N$ stochastic matrix where the entry P_{ij} is the probability of transitioning from state i to state j.

In other words, in a transition matrix P,

- the rows represent *now* or "from X_t"
- the columns represent *next* or "to X_{t+1}"
- the element P_{ij} is the conditional probability $p(X_{t+1} = j \mid X_t = i)$
- each row sums up to 1 (but each column may not).

In principle, P can be time dependent. However, in this text, we will only explore *homogeneous* Markov chains where P does not change over time.

For example, the transition matrix for the Markov weather model (Fig. 16.1) is given by:

$$P = X_t \left\{ \begin{array}{c} C \\ R \end{array} \right. \overbrace{\left[\begin{array}{cc} C & R \\ 0.8 & 0.2 \\ 0.4 & 0.6 \end{array} \right]}^{X_{t+1}} \tag{16.2}$$

> ⓘ The property that each row in a transition matrix sums to 1 corresponds to the fact that the total exit probabilities from any state in a transition diagram must sum to 1.

16.3.2 Probability Vectors

When considering a Markov chain, it's common to focus on its state at a specific time point, denoted as $X_t = i$. However, to fully capture the behavior of the Markov chain, it's insightful to consider all possible states, or the enssemble of states, it could be in at any given time. This broader perspective introduces us to the concept of a probability vector.

Definition 16.8 — Probability Vector. The probability vector of a Markov chain with N states at time t, $\boldsymbol{p}(t)$, is a row vector whose elements represent the probabilities of the chain being in each of the N states at that time:

$$\boldsymbol{p}(t) = [p_1(t), p_2(t), \ldots, p_N(t)] \tag{16.3}$$

where each element $p_i(t)$ satisfies $0 \le p_i(t) \le 1$, and $\sum_{i=1}^{N} p_i(t) = 1$.

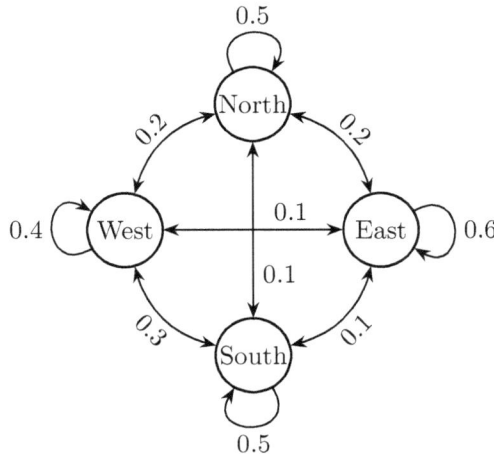

Figure 16.3: A Markov Chain Model for Bike Sites

The concept of a probability vector is illustrated by the bike-sharing model shown in Fig. 16.3. In this model, bikes are rented from and returned to any of four stations in a city. The transition probabilities represent the likelihood of bikes being relocated between stations. For a total of M bikes, the vector $M\boldsymbol{p} = [Mp_1, Mp_2, Mp_3, Mp_4]$ reflects the expected number of bikes at each station at a given time.

> ⓘ Stating that a Markov chain is in state 1 is equivalent to defining its probability
>   vector as $p = [1, 0, \ldots, 0]$, where the chain is certain to be in state 1 and has a
>   zero probability of being in any other state.

In essence, the vector $p(t)$ embodies the entire probability distribution of X_t. Each component $p_i(t)$ within $p(t)$ signifies the probability $p(X_t = i)$, concisely conveying the likelihood of the Markov chain being in each state at the indicated time. This form of representation efficiently encapsulates the state distribution of the chain.

Exercise 16.1 Express the Markov chain model for bike sites depicted in Fig. 16.3 as a transition matrix.

16.3.3 Time Evolution

The Markov transition matrix P fully characterizes a Markov chain and its evolution over time through the equation:

$$p(t + 1) = p(t)P. \tag{16.4}$$

This relationship embodies the Markov property by demonstrating that the state probability vector at time $t + 1$, $p(t + 1)$, is determined solely by its immediate predecessor $p(t)$ and the transition matrix P, independent of any prior states. Thus, the future state of the Markov chain is conditionally independent of past states, given the present state.

By iteration, we also have:

$$p(t) = p(0)P^t, \tag{16.5}$$

where P^t represents the tth power of P.

The evolution of $p(t)$ mirrors the state transitions in the Markov chain, as represented by the sequence X_t. This evolution serves as a generalization of the trajectory concept outlined in Def. 16.4. It is important to note that each computation of $p(t)$ accounts for all possible state transitions, thus capturing the entire spectrum of potential states at each step.

Exercise 16.2 Consider the bike-sharing model depicted in Fig. 16.3. If the initial probability vector is $p(0) = [0.25, 0.25, 0.25, 0.25]$, determine the probability vector at the next time step, $p(1)$.

16.3.4 Predicting the Weather

Let's apply the principles in this section to predict the weather using the Markov model shown in Fig. 16.1 with the transition matrix given in Eq. 16.2.

1 Probability for a Trajectory

A trajectory, defined as a specific sequence of states $X_0, X_1, X_2, \ldots$, represents one possible path the Markov chain might follow. The probability for the trajectory C, C, C, R, C, R, R, given the initial weather is clear, is given by

$$p = p_C(0) \cdot p_{CC}(1) \cdot p_{CC}(2) \cdot p_{CR}(3) \cdot p_{RC}(4) \cdot p_{CR}(5) \cdot p_{RR}(6)$$

$$= 1 \cdot 0.8 \cdot 0.8 \cdot 0.8 \cdot 0.2 \cdot 0.4 \cdot 0.6 \approx 0.0154.$$

Here, $p_C(0)$ is the initial probability of being in a clear state, p_{CC} is the probability of transitioning from clear to clear, p_{CR} from clear to rainy, and so on. These probabilities are directly taken from the transition diagram or matrix.

2 Weather for a Week

Assume today is clear, which means $X_0 = C$ and $p(0) = [1, 0]$. Then the probability vector for tomorrow's weather is:

$$p(1) = p(0)P = [0.8, 0.2],$$

indicating an 80% probability of clear weather and a 20% probability of rain.

For the next few days, the probability vectors can be computed iteratively:

$$p(2) = p(1)P = [0.72, 0.28],$$
$$p(3) = p(2)P = [0.688, 0.312],$$
$$p(4) = p(3)P \approx [0.675, 0.325],$$
$$p(5) = p(4)P \approx [0.670, 0.330].$$

We can also calculate $p(5)$ from $p(0)$ directly using the power of the transition matrix P:

$$p(5) = p(0)P^5.$$

3 Stationary Distribution

If we continue the process, we find the probabilities converge to a stationary distribution (or stationary state) $\tilde{p} \approx [0.667, 0.333]$. This vector is unaffected by further applications of P. It is also reflected in the powers of P:

$$\tilde{P} = \lim_{n \to \infty} P^n \approx \begin{bmatrix} 0.667 & 0.333 \\ 0.667 & 0.333 \end{bmatrix},$$

where each row is the stationary vector $\tilde{p}$. Thus, $\tilde{P}$ transforms any probability vector into $\tilde{p}$, independent of the initial weather. ($\tilde{p}$ is often denoted as π in some literature.)

> **Definition 16.9 — Stationary Distribution.** A stationary distribution $\tilde{p}$ is a probability vector that is unchanged by the operation of transition matrix P: $\tilde{p}P = \tilde{p}$.

We can calculate the stationary distribution $\tilde{p}$ directly from $\tilde{p}P = \tilde{p}$:

$$\begin{bmatrix} \tilde{p}_1 & \tilde{p}_2 \end{bmatrix} \begin{bmatrix} 0.8 & 0.2 \\ 0.4 & 0.6 \end{bmatrix} = \begin{bmatrix} \tilde{p}_1 & \tilde{p}_2 \end{bmatrix},$$

which yields $\tilde{p}_1 = 2\tilde{p}_2$. Together with $\tilde{p}_1 + \tilde{p}_2 = 1$, we obtain $\tilde{p}_1 = \frac{2}{3}$ and $\tilde{p}_2 = \frac{1}{3}$.

In this simple weather model, predictions for the weather on more distant days change less and less with each subsequent day and tend towards a steady state. The stationary vector $\tilde{p}$ represents the long-term probabilities of clear and rainy weather.

The stationary state in this model indicates the limitations of predictions far into the future, as the system tends towards equilibrium. However, in some Markov chains, the stationary state is a key feature that can provide meaningful insights.

4   Expected Duration of Clear Weather Between Rainy Days

In our exploration of weather prediction using Markov models, an interesting metric is the average number of clear days we can expect before a rainy day occurs again.

Given that a day can either remain clear with probability p_{CC} or transition to rainy with probability p_{CR}, the process of staying clear can be viewed as a geometric distribution (see Table 15.6) with the probability of "success" being the transition to rain.

Thus, the expected number of clear days before a rainy day can be expressed as:

$$E[C \to R] = \frac{1 - p_{CR}}{p_{CR}} = \frac{1 - 0.2}{0.2} = 4.$$

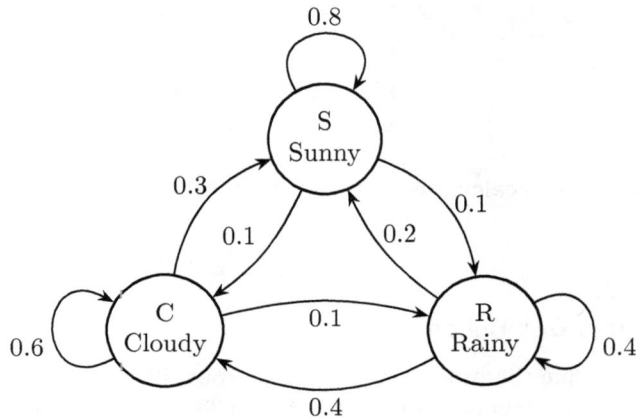

Figure 16.4: A Three-State Weather Model

Exercise 16.3 To improve the weather model, we expand the state space to include Sunny (C), Rainy (R), and Cloudy (C), as depicted in Fig. 16.4.

(a) Construct the transition matrix for this Markov model.

(b) Calculate the probability for the trajectory S, S, R, C, R, S, C, S.

(c) Assume we start with a sunny day. Predict the weather for the next five days.

(d) Calculate the probability distribution for the stationary state.

(e) ✳ Determine the mean break time between two rainy days.

16.4 Stationary Distribution and Detailed Balance

In § 16.3, we used a simple two-state weather example to illustrate the basic features of Markov chains. In particular, we introduced the stationary distribution, and observed the remarkable result that a Markov chain automatically finds its way to an equilibrium distribution as the chain evolves over time. It turns out such an equilibrium exists for many Markov chains, but not all.

In this section, we will explore the conditions required for a chain to converge to an equilibrium distribution. Our investigation will reveal several fundamental properties of Markov chains.

16.4.1 Stationary Distribution

A stationary distribution for a Markov Chain, denoted $\tilde{p}$, is a probability vector unchanged by the application of the transition matrix P: $\tilde{p}P = \tilde{p}$. Once a Markov chain reaches a stationary distribution, it remains in this state indefinitely, achieving equilibrium.

The stationary distribution $\tilde{p}$ is determined by solving the equation $\tilde{p}P = \tilde{p}$ alongside the condition $\sum_{i=1}^{N} \tilde{p}_i = 1$, as illustrated in § 16.3.4.3.

Equilibrium implies that the distribution of X_{t+1} is identical to that of X_t. If $X_t \sim \tilde{p}$ for any t, then $X_{t+r} \sim \tilde{p}$ for all $r \geq 0$.

Exercise 16.4 Consider the stochastic matrix P given by

$$P = \begin{bmatrix} 0.3 & 0.4 & 0.3 \\ 0.3 & 0.5 & 0.2 \\ 0.4 & 0.1 & 0.5 \end{bmatrix}.$$

(a) Draw a transition diagram for the Markov chain represented by P.

(b) Find the stationary distribution $\tilde{p}$ for this Markov chain.

16.4.2 Equilibrium and Detailed Balance

The stationary distribution of a Markov chain is related to a property called the detailed balance condition. This concept is particularly useful in the context of Markov Chain Monte Carlo (MCMC) simulations, which we will study in Chapter 17.

1 Definition and Implications

> **Definition 16.10 — Detailed Balance.** Let P be the transition matrix of a Markov chain. A probability distribution p on the chain satisfies the detailed balance condition if
> $$p_i P_{ij} = p_j P_{ji}, \quad \forall i, j. \tag{16.6}$$

As an illustration, imagine a network of interconnected rooms where each room represents a state in a Markov chain and the passages between rooms are transitions between states. If the network is in a steady state, the number of individuals entering any given room must equal the number exiting it.

This concept is formalized through the detailed balance condition, which essentially states that for every pair of states, the rate of flow from state i to state j is exactly balanced by the rate of flow from state j to state i. This mutual balancing act ensures that the overall system remains in equilibrium over time, without any net flux between any two states.

The detailed balance condition not only characterizes equilibrium but also provides a crucial criterion for the reversibility of the Markov chain. When a Markov chain satisfies this condition, it implies that observing the chain's transitions forwards in time is statistically indistinguishable from observing them in reverse, a property that can significantly simplify the analysis and computation of steady-state probabilities.

2   Sufficient Condition for Stationary Distribution

Detailed balance directly implies the existence of a stationary distribution:

> **Theorem 16.1** If a probability distribution p on a Markov chain satisfies the detailed balance condition, then p is a stationary distribution of the chain. That is, $p = \tilde{p}$, where $\tilde{p}P = \tilde{p}$.

Proof.

$$(pP)_j = \sum_i p_i P_{ij} \quad \text{(matrix multiplication)}$$

$$= \sum_i p_j P_{ji} \quad \text{(by the detailed balance condition)}$$

$$= p_j \sum_i P_{ji} = p_j \quad (\because \sum_i P_{ji} = 1).$$

$\square$

Because of this theorem, some literature describes a Markov chain with a detailed-balanced distribution as being 'detailed balanced.'

A special case of detailed balance is:

> **Theorem 16.2** If the transition matrix P of a Markov chain is symmetric, a uniform distribution over all states satisfies the detailed balance condition and is thus a stationary distribution.

3   Sufficient but Not Necessary

The converse of Theorem 16.1 is untrue: not all stationary distributions satisfy the detailed balance condition. As an example, consider

$$P = \begin{bmatrix} 0 & p & 1-p \\ 1-p & 0 & p \\ p & 1-p & 0 \end{bmatrix},$$

which represents a three-node random walk with cyclic boundaries. A stationary distribution is $\tilde{p} = \left[\frac{1}{3}, \frac{1}{3}, \frac{1}{3}\right]$. Apparently, unless $p = 0.5$, it is not detailed balanced:

$$\tilde{p}_1 P_{12} = \frac{1}{3}p \quad \neq \quad \tilde{p}_2 P_{21} = \frac{1}{3}(1-p).$$

16.5   Class Structure, Ergodicity, and Convergence

While detailed balance is a sufficient condition for the existence of a stationary distribution, it does not inherently guarantee the convergence of a Markov chain to this distribution, as such convergence also requires the chain to be ergodic. Consider the matrix $P = \begin{bmatrix} 0 & 1 \\ 1 & 0 \end{bmatrix}$, which satisfies the detailed balance condition with $p = [0.5, 0.5]$. Despite this, the matrix properties $P^2 = I$, $P^3 = P$, and so forth, indicate that starting from an initial state $[1, 0]$, the chain will perpetually oscillate between two states. This oscillation demonstrates that without ergodicity, the system fails to converge to a long-term, stable distribution, even if the chain has a detailed-balance distribution. Therefore, while detailed balance ensures stationarity under certain conditions, it is neither necessary nor sufficient to guarantee *convergence*, which fundamentally depends on the ergodic nature of the process.

16.5.1 Class Structure

The class structure of Markov chains is a central concept that aids in understanding the dynamics of these mathematical systems over time. It allows us to categorize states into meaningful groups based on their interactions. This structure is crucial for determining the behavior of a system over time, in particular, whether it can reach a steady state with a stationary distribution.

1 Classes of States

Markov chain states can be grouped into classes, which may consist of a single state or multiple states, based on the reachability from one state to another. The main types of classes include:

- Communicating Classes: A class is communicating if every state in the class can be reached from every other state in the class, possibly over multiple steps. Formally, state i communicates with state j, denoted as $i \leftrightarrow j$, if there exists some $n \geq 0$ such that the probability of transitioning from i to j in n steps is greater than zero.

  Mathematically, the communicating relation $\leftrightarrow$ is an equivalence relation, which means that it partitions the state space S into non-overlapping equivalence classes. Therefore, every state is a member of exactly one communicating class.

- Transient and Recurrent Classes: States within communicating classes can be further classified as transient or recurrent. A state is *transient* if there's a nonzero probability that the system will not return to it once it has left. Conversely, a state is *recurrent* (or persistent) if the system is certain to return to it eventually. A class containing at least one recurrent state is recurrent; otherwise, it is transient.

- Absorbing States: An absorbing state is a specific type of recurrent state from which, once entered, the system cannot exit. A Markov chain with at least one absorbing state, where every state can reach an absorbing state (possibly over multiple steps), is termed an *absorbing Markov chain*.

In the example shown in Fig. 16.5, state 1 leads to state 2, but no state leads back to state 1, making 1 a transient state. If the chain enters state 6, it remains there indefinitely, designating it as an absorbing state. State 6 is accessible from any other state, classifying this chain as an absorbing chain. This chain comprises four communicating classes: $\{1\}$, $\{2, 3\}$, $\{4, 5\}$, and $\{6\}$.

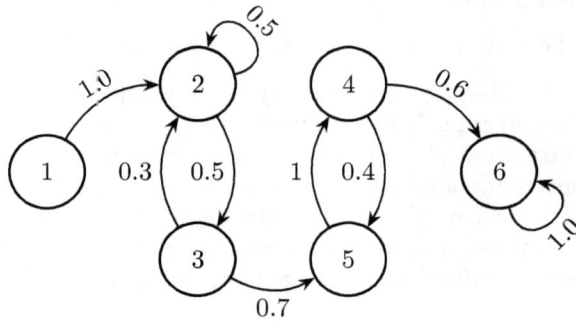

Figure 16.5: Example of a Reducible Markov Chain

Exercise 16.5 Consider the Markov chain depicted in Fig. 16.5. Identify the communicating classes for the following modifications to the transition probabilities between nodes:

(a) There is a positive transition probability from node 5 to 3;

(b) There is a positive transition probability from node 6 to 1;

(c) There is a positive transition probability from node 1 to 6.

Assume all other transition probabilities remain positive.

2 Reducibility

A Markov chain or transition matrix P is *irreducible* if its state space constitutes a single communicating class, implying the feasibility of transitioning from any state to any other state in a finite number of steps. This indicates a connected state space without isolated subsets of states.

In a finite-state Markov chain, irreducibility suggests that all states are recurrent, as the ability to reach any state from any other ensures all states must be recurrent if at least one state is recurrent.

The Markov chain depicted in Fig. 16.5 is reducible, whereas those in Figs. 16.1 and 16.4 are irreducible.

3 Periodicity

One of the key properties of a Markov chain is whether it is *aperiodic* or *periodic*. This property is related to the regularity with which the chain revisits a state. To understand this concept, let's consider an example:

Consider a neighborhood where each house represents a state in your Markov chain, and your movements to different houses simulate the chain's transitions. In an aperiodic Markov chain, your return to any house isn't constrained by a fixed pattern; you could come back at any time, be it after a brief or extended sequence of visits to other houses. Conversely, a periodic Markov chain would see you returning to each house at regular, fixed intervals, like following a strict visitation schedule.

Formally, the period of a state is the greatest common divisor (gcd) of the lengths of all possible cycles from that state back to itself. If the gcd is 1, the state (and its class) is aperiodic; otherwise, it is periodic. A Markov chain is aperiodic if all its states are aperiodic.

For instance, in a 1D random walk (Fig. 16.2), every state has a period of 2, as $2n$ steps, n forward followed by n backward, return to the original position. Conversely, the weather models (Figs. 16.1 and 16.4) have aperiodic states, as returns to the same state are possible in 1, 2, or more steps.

If a Markov chain is irreducible and contains at least one aperiodic state, then all its states are aperiodic. This is because the irreducibility ensures a path from an aperiodic state i to any state j, and the aperiodicity of i implies that j also cannot have a fixed period.

4   Ergodicity

A Markov chain is *ergodic* if it possesses two key properties: aperiodicity and recurrence. Aperiodicity ensures that the chain can visit any state at any time without being confined to a fixed cycle. Recurrence guarantees that every state will be revisited indefinitely. Imagine an ideal explorer navigating the landscape of the target distribution—this chain moves freely (aperiodicity) and consistently revisits all important areas, reflecting their underlying significance (recurrence).

Over time, the chain's path becomes independent of its starting point. The distribution of time spent in different states converges to accurately represent the true probabilities within the target distribution. More technically, an ergodic Markov chain eventually settles into a unique stationary distribution, which captures the steady-state probabilities of all states.

16.5.2   Convergence

The convergence of a Markov chain indicates that, irrespective of the initial state, the probability distribution over states tends towards a unique stationary distribution $\tilde{p}$ as the number of transitions increases.

Ergodicity, defined as the condition of being both irreducible and aperiodic in finite-state chains, is necessary for convergence. This principle is encapsulated by the following theorem, which we will revisit in § 16.6.3.

Theorem 16.3 — Perron-Frobenius Theorem. If a finite Markov chain is ergodic, it possesses a unique stationary distribution $\tilde{p}$. Moreover, P^n converges to a rank-one matrix where each row is the stationary distribution $\tilde{p}$ as n approaches infinity:

$$\lim_{n \to \infty} P^n = \mathbf{1}\tilde{p} \qquad (16.7)$$

where $\mathbf{1}$ is a column vector with all entries equal to 1.

Exercise 16.6 Show that for a stochastic matrix $\tilde{P}$ where each row is $\tilde{p}$, $v\tilde{P} = \tilde{p}$ for any probability vector v.

Irreducibility is crucial for ensuring that a Markov chain converges to a unique equilibrium, independent of the initial state. In reducible chains, where states are partitioned into isolated classes, the initial state can significantly influence the eventual distribution, preventing the chain from 'forgetting' its start and converging to a single, global equilibrium.

The importance of aperiodicity is highlighted by considering a Markov chain with a transition matrix $P = \begin{bmatrix} 0 & 1 \\ 1 & 0 \end{bmatrix}$, starting in state $X_0 = 1$. This chain alternates between states, with $X_n = 1$ for even n and $X_n = 2$ for odd n, demonstrating periodic behavior with a period of 2. Such chains do not reach a stable equilibrium as they are trapped in a cycle.

16.6 *Stochastic Matrices

Transition matrices in Markov chains are a specific type of matrix known as stochastic matrices. These are real, square matrices with non-negative elements, where each row sums to 1 (Def. 16.6).

Stochastic matrices are fundamental not only in the analysis of Markov chains but also in the broader study of stochastic processes. Understanding their properties is essential for analyzing and predicting the behavior of systems that evolve probabilistically.

There are two conventions for stochastic matrices: row-based, where each row represents a probability distribution over possible next states given the current state, and column-based. In Markov chain analysis, row-based stochastic matrices are more common, and this will be the convention we follow.

16.6.1 Preservation of Probability Vectors

> **Theorem 16.4** If p is a probability vector and P a row-based stochastic matrix, then $q = pP$ is also a probability vector.

Proof.

Non-negativity: By the definition of matrix multiplication,

$$q_j = \sum_{i=1}^{N} p_i P_{ij}$$

Since $p_i \geq 0$ for all i, and $P_{ij} \geq 0$ for all i, j by the properties of probability vectors and stochastic matrices, respectively, it follows that each term in the sum is non-negative. Therefore, $q_j \geq 0$ for all j.

Elements sum to one: The sum of all elements of q is:

$$\sum_{j=1}^{N} q_j = \sum_{j=1}^{N} \left(\sum_{i=1}^{N} p_i P_{ij} \right).$$

By changing the order of summation, we get:

$$\sum_{j=1}^{N} q_j = \sum_{i=1}^{N} p_i \left(\sum_{j=1}^{N} P_{ij} \right).$$

Since $\sum\limits_{j=1}^{N} P_{ij} = 1$ for all i (each row of P sums to one), we have:

$$\sum_{j=1}^{N} q_j = \sum_{i=1}^{N} p_i = 1. \tag{16.8}$$

□

16.6.2 Maximum Eigenvalue 1

> **Theorem 16.5** The largest eigenvalue (both left and right) of a stochastic matrix P is 1. That is, there exist a column vector v such that $Pv = v$, and a row vector u such that $uP = u$.

Proof.

Let v be a regular (i.e., right) eigenvector of P with eigenvalue λ, such that $Pv = \lambda v$. Consider an element v_i in v with the maximum absolute value. Without loss of generality, assume $v_i \geq 0$; otherwise, we could consider $-v$, maintaining the same argument.

By the definition of matrix multiplication, the ith element of Pv is:

$$(Pv)_i = \sum_{j=1}^{N} P_{ij}v_j \tag{16.9}$$

Given $\sum\limits_{j=1}^{N} P_{ij} = 1$ for all i, and $P_{ij} \geq 0$, and considering $v_j \leq v_i$ for all j by the choice of v_i, it follows:

$$(Pv)_i = \lambda v_i \leq \sum_{j=1}^{N} P_{ij}v_i = v_i \tag{16.10}$$

This implies $\lambda \leq 1$. To demonstrate that $\lambda = 1$ is an eigenvalue, consider the vector $e = [1, 1, \ldots, 1]^{\top}$. Clearly, $Pe = e$, showing that 1 is indeed an eigenvalue of P.

The transpose P^T shares the same set of eigenvalues as P. Hence, the largest eigenvalue of P^T, or the largest left eigenvalue of P, is also 1. □

16.6.3 Perron-Frobenius Theorem

A standard stochastic matrix can have 0 entries. This indicates some states are inaccessible from others within a single step. However, it is possible that, after some number of steps, it is feasible to transition from any state to any other state with positive probability. This has motivated the following definition:

> **Definition 16.11 — Regular Stochastic Matrix.** A stochastic matrix P is called regular if there exists a positive integer k such that all entries of P^k are strictly

positive, meaning $P^k > 0$ for some $k \in \mathbb{N}$.

A Markov chain with a regular stochastic transition matrix is ergodic, as the regularity ensures both irreducibility and aperiodicity, and for finite state spaces, also implies positive recurrence of all states. Thus, the Perron-Frobenius theorem introduced in § 16.5.2 can be stated as follows:

Theorem 16.6 A regular stochastic matrix P possesses a unique stationary distribution $\tilde{p}$. Furthermore, P^n converges to a rank-one matrix where each row is the stationary distribution $\tilde{p}$ as n approaches infinity:

$$\lim_{n \to \infty} P^n = \mathbf{1}\tilde{p}$$

where $\mathbf{1}$ is a column vector with all entries equal to 1.

Proof.

A rigorous proof of this theorem is beyond the scope of this text. Below is an outline of the proof.

Existence of a Stationary Distribution: Let v be an arbitrary probability distribution. Consider the sequence $vP, vP^2, vP^3, \ldots$. Due to the regularity of P, there exists k such that P^k has all positive entries, making the chain irreducible and ensuring the sequence becomes strictly positive, indicating the chain is communicating. The set of probability distributions is compact, thus this sequence has a convergent subsequence by the Bolzano-Weierstrass theorem.

Convergence to Stationary Distribution: The limit of any convergent subsequence of vP^n as $n \to \infty$ is a stationary distribution $\tilde{p}$, satisfying $\tilde{p}P = \tilde{p}$, due to the continuity of matrix multiplication and the preservation of total probability.

Uniqueness: Suppose there exist two distinct stationary distributions, $\tilde{p}$ and $\tilde{q}$. Consider their average $\tilde{r} = \frac{1}{2}(\tilde{p} + \tilde{q})$. This distribution $\tilde{r}$ is also stationary because

$$\tilde{r}P = \frac{1}{2}(\tilde{p}P + \tilde{q}P) = \frac{1}{2}(\tilde{p} + \tilde{q}) = \tilde{r}.$$

Given the regularity of P, $\tilde{r}P^k$ would be in the interior of the probability simplex after k steps, indicating strictly positive entries due to the convex combination of $\tilde{p}$ and $\tilde{q}$ under P^k. This interior point cannot be the average of two distinct points unless $\tilde{p} = \tilde{q}$, hence proving the uniqueness of the stationary distribution.

Summary: By demonstrating that any sequence of distributions derived from repeatedly applying P to any initial distribution converges to a stationary distribution, and arguing the uniqueness of this distribution, we conclude that a regular stochastic matrix P has a unique stationary distribution $\tilde{p}$.   □

16.6.4   Doubly Stochastic Matrix

Definition 16.12 — Doubly Stochastic Matrix. A doubly stochastic matrix is a stochastic matrix where both the rows and columns sum to 1.

Note that a doubly stochastic matrix may not be a regular stochastic matrix. An example is $P = \begin{bmatrix} 0 & 1 \\ 1 & 0 \end{bmatrix}$, where $P^2 = I$, $P^3 = P$, and so on.

Theorem 16.7 A doubly stochastic matrix has at least one stationary distribution, which is the uniform distribution.

Proof.

Consider a doubly stochastic matrix P of size $N \times N$. By definition, for all i,
$$\sum_{j=1}^{N} P_{ij} = 1 \text{ (each row sums to 1) and } \sum_{i=1}^{N} P_{ij} = 1 \text{ (each column sums to 1)}.$$

Let's assume $\tilde{p}$ is a uniform distribution, meaning $\tilde{p}_i = \frac{1}{N}$ for all i. We need to show that this choice of $\tilde{p}$ satisfies $\tilde{p}P = \tilde{p}$.

The ith element of the product $\tilde{p}P$ is given by:

$$(\tilde{p}P)_i = \sum_{j=1}^{N} \tilde{p}_j P_{ji} = \sum_{j=1}^{N} \frac{1}{N} P_{ji} = \frac{1}{N} \sum_{j=1}^{N} P_{ji}$$

Since P is doubly stochastic, $\sum_{j=1}^{N} P_{ji} = 1$ for all i. Therefore,

$$(\tilde{p}P)_i = \frac{1}{N} \cdot 1 = \frac{1}{N} = \tilde{p}_i$$

This shows that $\tilde{p}P = \tilde{p}$, and hence $\tilde{p}$ is indeed a stationary distribution for P.

Thus, a doubly stochastic matrix has at least one stationary distribution, which is the uniform distribution, proving the theorem.   ☐

A symmetric stochastic matrix is a doubly stochastic matrix. But the converse is untrue.

Exercise 16.7 Show that the doubly stochastic matrix $P = \begin{bmatrix} 0 & 1 \\ 1 & 0 \end{bmatrix}$ has a stationary distribution even though P is not a regular stochastic matrix.

16.7 ✳ Random Walks

In this section, we explore random walks as key applications of Markov chains. Random walks are fundamental models in both theoretical and applied mathematics, offering valuable insights into a wide range of phenomena in physics, economics, biology, computer science, and quantum computing. They provide a simplified framework for studying complex random processes.

Random walks are essential for understanding diffusion processes, stock market fluctuations, population genetics, and even algorithms in computer networks. In quantum computing, they play a crucial role in the development of quantum algorithms, enabling more efficient exploration of computational spaces compared to classical methods.

16.7.1 One-Dimensional Random Walk

The one-dimensional random walk is a simple, illustrative model of random behavior. It can be visualized as walking along a line, as depicted in Fig. 16.2, with each step being of uniform length. Starting at the origin, the direction of each step is determined by flipping a coin: a heads (with probability p) results in a step forward, and tails (with probability q) results in a step backward.

1 Symmetric Random Walk

We will consider $p = q = 0.5$ first. This problem essentially models the net effect of flipping a *fair* coin and keeping track of the number of heads minus the number of tails. Statistically, this is akin to summing up a series of independent and identically distributed random variables, each taking the value $+1$ or -1 with equal probability of 0.5.

The problem of interest is determining the probability of being at a specific point d after n steps, given that the starting point is 0. We denote this probability as $f_n(d)$. Since the walker must be at some position after n steps, the sum of these probabilities for all possible positions d must equal 1, i.e., $\sum_d f_n(d) = 1$.

The pattern in these probabilities becomes clearer when factoring out $\frac{1}{2^n}$:

d	-5	-4	-3	-2	-1	0	1	2	3	4	5
$f_0(d)$						1					
$2f_1(d)$					1		1				
$2^2 f_2(d)$				1		2		1			
$2^3 f_3(d)$			1		3		3		1		
$2^4 f_4(d)$		1		4		6		4		1	
$2^5 f_5(d)$	1		5		10		10		5		1

To better understand these probabilities, one can enumerate the sequences of coin flips leading to each position. For instance, in a three-step walk ($n = 3$), the sequence HHH will end at $d = 3$, while HHT, HTH, and THH will end at $d = 1$. The negative positions correspond to the reverse combinations of H and T. With $2^3 = 8$ possible three-step walks, the probabilities for the different positions are: $f_3(-3) = 1/8$, $f_3(-1) = 3/8$, $f_3(1) = 3/8$, $f_3(3) = 1/8$.

These probabilities, after scaling by 2^n, correspond to the coefficients in Pascal's Triangle, and they are the same as those in the binomial expansion of $(a + b)^n$. Therefore, the probability distribution is given by:

$$f_n(d) = \frac{1}{2^n}\binom{n}{\frac{n+d}{2}} = \frac{n!}{2^n\left(\frac{n+d}{2}\right)!\left(\frac{n-d}{2}\right)!},\qquad(16.11)$$

where d is an integer such that $-n \le d \le n$ and both $n + d$ and $n - d$ are even. In fact, $\frac{n+d}{2}$ and $\frac{n-d}{2}$ are the number of forward and backward steps, respectively.

The probability distribution is graphed in Fig. 16.6 for $n = 50$. Apparently, the graph resembles a bell-shaped curve centered around $d = 0$. Here is why:

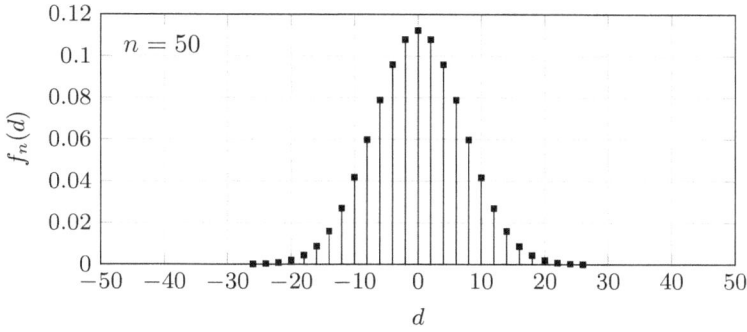

Figure 16.6: Probability Distribution of 1D Random Walk

Let $X \in \{-1, 1\}$ be the random variable representing each step of the random walk. The expectation, variance, and standard deviation of X are given by:

$$E[X] = 0, \tag{16.12a}$$
$$\text{Var}[X] = 1^2 p + (-1)^2 q = 1, \tag{16.12b}$$
$$\Delta X = 1. \tag{16.12c}$$

The position after n steps is given by the random variable $d = X_1 + X_2 + \ldots + X_n$. We have:

$$E[d] = nE[X] = 0, \tag{16.13a}$$
$$\text{Var}[d] = n\,\text{Var}[X] = n, \tag{16.13b}$$
$$\Delta d = \sqrt{n}. \tag{16.13c}$$

According to the central limit theorem (see § 14.3.3), d approximates a normal distribution for large n with $\mu = E[d] = 0$ and $\sigma = \Delta d = \sqrt{n}$. This implies that while the walker's average displacement from the origin remains zero, the typical distance grows as the square root of the number of steps taken. More precisely, the walker's position is within $\sqrt{n}$ of the origin with 68% of probability, and $3\sqrt{n}$ with 99.7% of probability.

2   Asymmetric Random Walk

Now let's extend our investigation to asymmetric random walks where $p \neq q$. We define $\delta = p - q$. Similar to Eq. 16.11, the probability distribution for this scenario becomes:

$$f_n(d) = \binom{n}{\frac{n+d}{2}} p^{\frac{n+d}{2}} q^{\frac{n-d}{2}}. \tag{16.14}$$

This equation represents the binomial distribution discussed in § 15.2.2, except with $d = 0$ labeling the center.

For a single step, the expected value, variance, and standard deviation are given by:

$$E[X] = 1p + (-1)q = \delta, \tag{16.15a}$$
$$\text{Var}[X] = 1^2 p + (-1)^2 q - \delta^2 = 1 - \delta^2 = 4pq, \tag{16.15b}$$

$$\Delta X = \sqrt{1 - \delta^2} = 2\sqrt{pq}. \tag{16.15c}$$

For n steps, the expected value and standard deviation evolve as:

$$E[d] = n\delta, \tag{16.16a}$$
$$\text{Var}[d] = 4npq, \tag{16.16b}$$
$$\Delta d = 2\sqrt{npq}. \tag{16.16c}$$

Similar to the symmetric random walk, for large n, d follows a normal distribution with mean $\mu = E[d] = n\delta$ and standard deviation $\sigma = \Delta d = 2\sqrt{npq}$. However, the center of the bell curve moves according to the mean, indicating a linear walkoff on average, skewed by the asymmetry in p and q.

3   General Case

Now let's consider the most general case, where at each step, the walker has probability p to move forward, q to move backward, and r to stay put. Of course, $p + q + r = 1$. This situation is depicted by the Markov chain in Fig. 16.7.

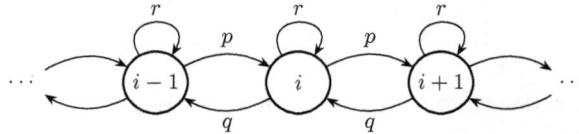

Figure 16.7: Markov Chain Model of 1D Random Walk

We still define $\delta = p - q$. In this case, the expected value, variance, and standard deviation are calculated as follows:

$$E[X] = 1p + (-1)q + 0r = \delta, \tag{16.17a}$$
$$\text{Var}[X] = 1^2 p + (-1)^2 q + 0^2 r - \delta^2 = 4pq + r(1 - r), \tag{16.17b}$$
$$\Delta X = \sqrt{4pq + r(1 - r)}, \tag{16.17c}$$

The expected value and standard deviation for n steps are:

$$E[d] = n\delta, \tag{16.18a}$$
$$\Delta d = \sqrt{n(4pq + r(1 - r))}. \tag{16.18b}$$

The probability for f forward, b backward, and $c = n - f - b$ circling steps, forms a trinomial distribution. However, the total displacement $d = f - b$ depends only on f and b. Therefore, the probability function of d is a marginal distribution of the trinomial distribution, which is too complicated to detail here.

Nonetheless, for large n, we can still glimpse the general behavior of d owing to the central limit theorem: d approximates a normal distribution for large n with $\mu = E[d] = n\delta$ and $\sigma \propto \sqrt{n}$.

4   Finite Random Walk with Boundaries

In the preceding cases, the random walk has no boundaries. The corresponding Markov chain has no stationary distribution because the probability distribution of d is ever evolving with n. The Markov chain is considered infinite because the transition matrix has an infinite dimension:

$$P = \begin{bmatrix} \ddots & \vdots & \vdots & \vdots & \vdots & \vdots & \\ \cdots & r & p & 0 & 0 & 0 & \cdots \\ \cdots & q & r & p & 0 & 0 & \cdots \\ \cdots & 0 & q & r & p & 0 & \cdots \\ \cdots & 0 & 0 & q & r & p & \cdots \\ \cdots & 0 & 0 & 0 & q & r & \cdots \\ & \vdots & \vdots & \vdots & \vdots & \vdots & \ddots \end{bmatrix} \tag{16.19}$$

Exercise 16.8 Explain why P in Eq. 16.19 is a tridiagonal matrix, i.e., having nonzero elements only on the main diagonal, and the diagonals above and below the main diagonal.

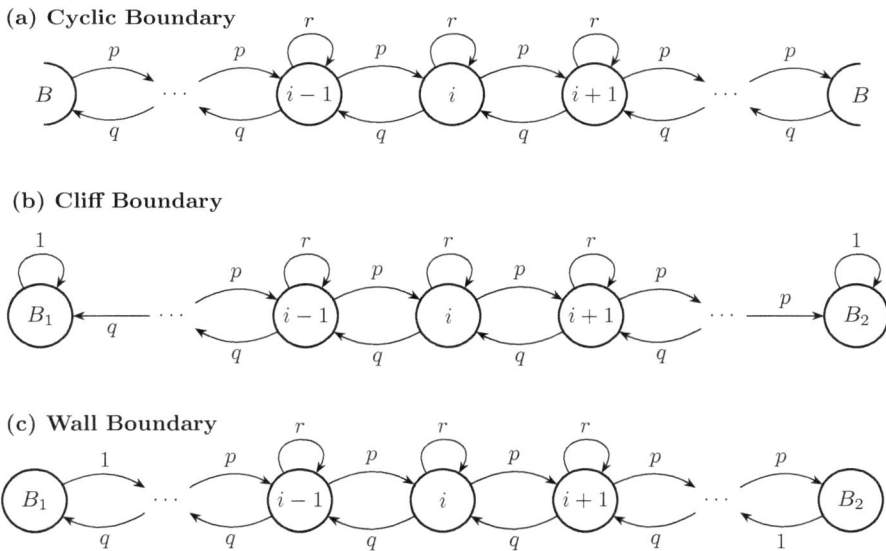

(a) Cyclic Boundary

(b) Cliff Boundary

(c) Wall Boundary

Figure 16.8: Examples of 1D Random Walk with Boundaries

Random walks can also be bounded. Figure 16.8 shows three types of boundaries as examples. Sub-figure (a) depicts a cyclic boundary: when the walker steps off the right boundary, he enters from the left boundary, and vice versa. So in the figure, both edge nodes are labelled B, indicating they are the same node. Since in this scenario all nodes are equivalent, the Markov chain has a stationary distribution - a uniform distribution. The transition matrix is given by:

$$
P = \begin{bmatrix}
q & r & p & 0 & 0 & 0 & \cdots \\
0 & q & r & p & 0 & 0 & \cdots \\
0 & 0 & q & r & p & 0 & \cdots \\
& & & \cdots & & & \\
\cdots & 0 & q & r & p & 0 & 0 \\
\cdots & 0 & 0 & q & r & p & 0 \\
\cdots & 0 & 0 & 0 & q & r & p
\end{bmatrix}
\tag{16.20}
$$

Sub-figure (b) depicts a cliff boundary: when the walker reaches either the left or right boundary, he falls off a cliff and stops walking. So in the figure, both edge nodes, labelled B_1 and B_2, are absorbing nodes. Because of the absorbing nodes, this Markov chain is reducible. So it has no stationary distribution. Interestingly, even when $p = q = 0.5$, the walker will inevitably fall off a cliff no matter how far away the cliff is, because the spread of the displacement increases with $\sqrt{n}$ without bound.

Sub-figure (c) depicts a wall boundary: when the walker reaches a boundary, he just turns around. So the edge nodes B_1 and B_2 both have exit probability of 1. This Markov chain is ergodic and has a stationary distribution, but its analytic solution is beyond the scope of this text.

Exercise 16.9 Formulate the transition matrices for the random walk with cliff boundary and wall boundary, respectively.

16.7.2 ✴ Random Walk in Higher Dimensions

Higher-dimensional random walks exhibit behaviors that are distinct from one-dimensional random walks. These differences stem from the increased complexity of the system and the greater freedom of movement. In various scientific fields, the dimensionality of the random walk significantly affects how models are constructed and interpreted. Here, we explain the key differences and core concepts without delving into the mathematical intricacies.

1   Recurrence vs. Transience

Consider a random walker embarking on a journey. Here's how recurrence and transience define its eventual fate in different dimensions:

A random walk is considered recurrent if the walker, starting from any point, is guaranteed to revisit its starting position infinitely often during its journey. This is analogous to a forgetful explorer who, despite getting lost, eventually stumbles back home. One-dimensional and two-dimensional random walks are examples of such recurrent walks.

Conversely, a random walk is transient if there's a nonzero probability that the walker will venture indefinitely without returning to its starting point. Picture an intrepid explorer venturing in a random direction, moving perpetually outward. This behavior characterizes random walks in three or more dimensions.

The key to this difference lies in the number of available paths. In one dimension, the walker alternates between left and right, with a finite "width" to explore, leading to recurrence. In two dimensions, despite more freedom, the walker is still likely to return due to the bounded area of exploration.

In contrast, three-dimensional space expands the number of paths exponentially. With this vastness, the likelihood of the walker returning to the starting point decreases significantly, allowing for the possibility of never returning, hence the transient nature.

To visualize, consider a drunkard navigating a one-dimensional path: they must eventually return to their starting point. Now, place them in a two-dimensional cityscape: while freer to roam, they may still, by chance, return home. However, in the boundless expanse of a three-dimensional universe, they could wander endlessly without returning.

Understanding recurrence and transience is vital for predicting the long-term behavior of random walks, which is crucial for applications such as search algorithms and modeling particle behavior in gases.

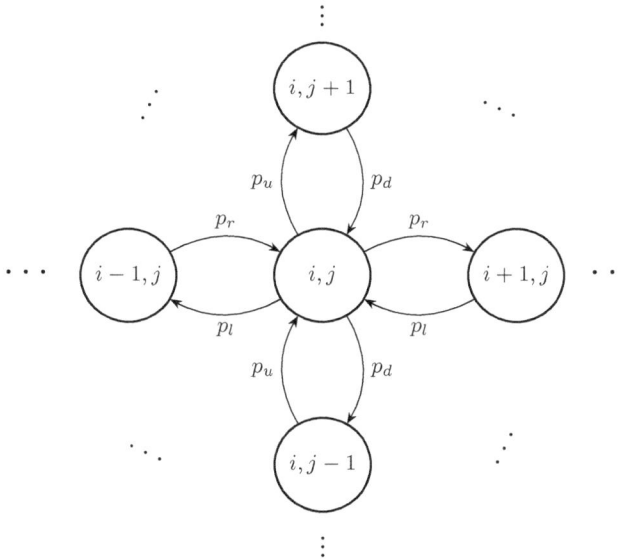

Figure 16.9: Markov Chain Model of 2D Random Walk

Figure 16.9 depicts the transition diagram for a two-dimensional (2D) random walk, where the movement can occur in four directions—left, right, up, and down—each with a specified probability. This diagram should be compared to the one-dimensional (1D) random walk shown in Fig. 16.7. While it is possible to conceptualize similar diagrams for three-dimensional (3D) walks, envisioning them beyond 3D becomes increasingly challenging. Even for a 2D random walk, the transition structure is no longer a regular matrix; it becomes a 3D tensor.

2 Diffusion Behavior

The diffusion process in random walks describes how a walker's position becomes increasingly spread out over time. The spread rate differs across dimensions. Higher dimensions, offering more directions for movement, result in a more rapid dispersal.

The Central Limit Theorem (CLT) states that the sum of a large number of independent random variables, like the steps in a random walk, will converge to a normal distribution, regardless of the distributions of the individual steps.

The spread of the distribution, characterized by its standard deviation, grows in proportion to the square root of the step count. However, this growth factor is greater in higher dimensions, indicating a faster diffusion rate and more spread-out distribution.

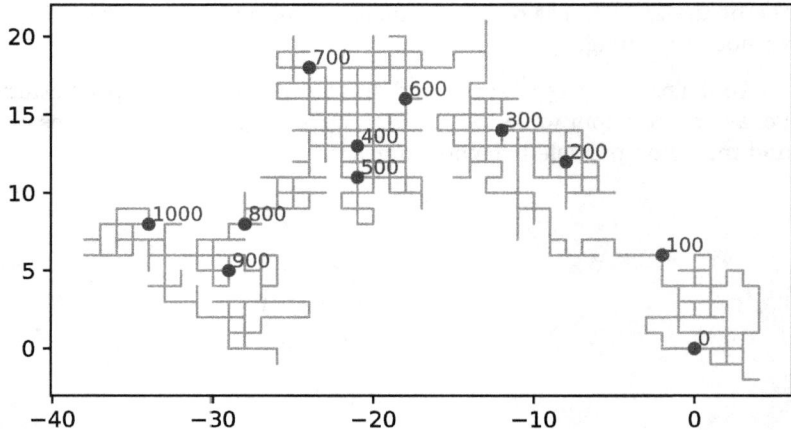

Figure 16.10: Simulation of a 2D Random Walk

Figure 16.10 depicts a 2D random walk simulation. Notable observations include the 2D walk's additional degree of freedom and its ability to cover more area, potentially circling back to previously visited points, which is not possible in 1D without retracing steps.

3 Escape Probability

Escape probability quantifies the likelihood of a random walker leaving a region and never returning. This metric is useful for assessing the transience of random walks and is relevant in various complex models.

In one dimension, this probability is zero, indicating that the walker is guaranteed to return to its starting point infinitely often. In two dimensions, it is greater than zero but finite. In three dimensions and beyond, the vastness of space makes it overwhelmingly likely for the walker to drift away, resulting in an escape probability approaching one.

4 Random Walk on Graphs

Graphs are mathematical structures consisting of vertices (or nodes) connected by edges. They are instrumental in visualizing relationships and processes, such as higher-dimensional Markov chains, which can often be represented as random walks on these graphs. Graph theory offers a rich set of tools to analyze such walks, aiding in the study of properties like recurrence, transience, and mixing times.

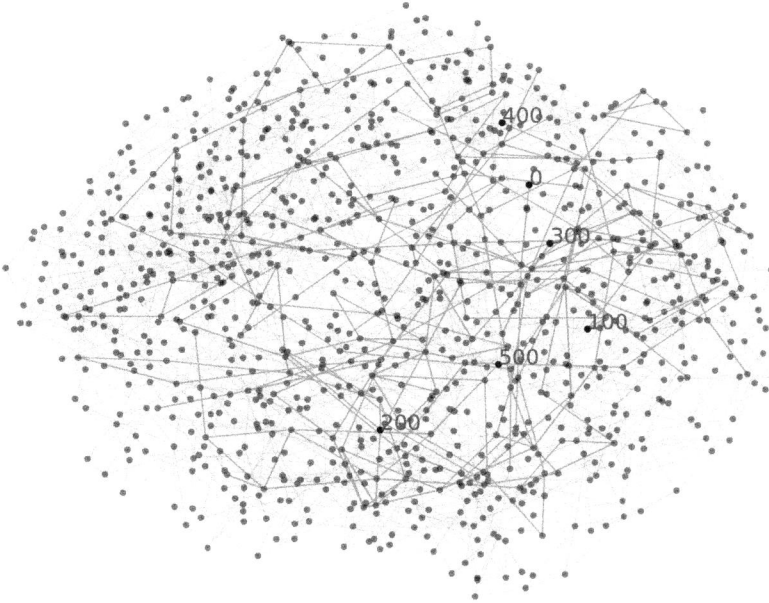

Figure 16.11: Simulation of a Random Walk on A Graph

Figure 16.11 depicts a simulation of a random walk on an interconnected graph of 1000 nodes. The path of the random walk is emphasized, with specific nodes labeled at steps 0, 100, 200, ..., 500. This graph models a network of 1000 users, connected in a manner that suggests potential email exchanges among them. Each step in the graph simulates the transmission of a single hypothetical email carrying a virus from one user to another in the network.

16.8 ✳ Additional Examples

Markov chains provide a powerful framework for modeling a variety of real-world systems where outcomes are dependent only on the current state and not on how that state was reached. This section explores several practical applications of Markov chains across different fields, demonstrating their versatility and effectiveness in solving complex problems.

1 Modeling Web Traffic

Scenario: Imagine a user navigating through a website. The pages of the website can be represented as the states of a Markov chain. Transitions between pages are determined by link probabilities (if the user clicks a link from page A to page B, that's a transition).

Transition Matrix: The transition matrix P would have entries where P_{ij} represents the probability of moving from page i to page j.

The Stationary State: The stationary distribution in this scenario reveals the long-term proportion of time a user is expected to spend on each page of the website.

Insight: Pages with higher probability in the stationary distribution are the ones users naturally gravitate towards. This information can be used for optimizing website layout, ad placement, and identifying potential bottlenecks in navigation.

2　Search Engine Ranking

The above web-traffic model can be extended to the PageRank search engine ranking algorithm.

Scenario: Model the web as a massive graph. Webpages are the states of a Markov chain, and links between them represent transitions. A "random surfer" follows links with probabilities determined by how pages are connected.

Transition Matrix: The transition matrix P is constructed with entries P_{ij} representing the probability of going from page i to page j. Importantly, this matrix is very sparse as most pages don't link to each other.

The Stationary State: The stationary distribution represents the long-term probability of the random surfer being on each page. This probability distribution forms the core of the PageRank algorithm.

Insight: Pages with higher stationary probability are considered more important or authoritative within the overall web structure. Search engines use this information (alongside other factors) to rank results.

3　Finance: Credit Rating Transitions

Scenario: In the financial industry, credit ratings assigned to bonds or other securities indicate their credit quality and are crucial for investment decisions. These ratings can be modeled as states in a Markov chain, where transitions represent rating upgrades or downgrades.

Transition Matrix: The transition matrix P in this context contains entries P_{ij} that indicate the likelihood of a bond's rating moving from grade i to grade j over a specified time frame.

The Stationary State: The stationary distribution provides insights into the long-term stability of different credit ratings and the expected lifetime of securities within various credit grades.

Insight: Understanding the transition probabilities helps investors assess the risk of credit changes and make more informed decisions regarding their portfolios. It also aids in pricing bonds according to their risk of downgrade or default.

4　Language Processing: Word Sequence Prediction

Scenario: Predict the next word in a sequence. Words are the states of a Markov chain, and transitions represent how likely a word is to follow another in natural language.

Transition Matrix: The transition matrix P has entries P_{ij} representing the probability of word j immediately following word i, estimated from a large text corpus (collection of text).

The Stationary State: While less directly interpretable in this context, the stationary distribution provides a sense of the overall frequency distribution of words within the text corpus.

Insight: This model can be used for:

- Next Word Prediction: Given a few starting words, the next most likely word can be suggested according to the highest transition probability.

- Text Generation: Sample from the model to generate realistic-looking sequences of words.

- Sentence plausibility: Calculate the probability of a whole sentence (i.e., a chain trajectory) by multiplying transition probabilities from the model, helping distinguish nonsensical word orders.

5   Large Language Models

Large Language Models (LLMs) can be regarded as massively complex and sophisticated versions of the word-sequence Markov chain idea. They retain the core of sequential probability modeling but overcome the limitations with vastly expanded memory, richer representations, and the ability to learn from gigantic amounts of language data.

Core Similarity: Sequence Modeling

Basic Markov Chain: Predicting the next word based on the immediate preceding words is inherently a sequential modeling task. Markov chains excel at this, capturing short-term dependencies within language.

LLMs: At their core, LLMs are also sequence modeling powerhouses. They're trained on massive text datasets to learn the statistical patterns governing how words and sentences follow each other. This training allows them to predict the next most likely word or phrase in a sequence, based on both the immediate and more extended textual context.

Key Differences: Complexity and Context

Markov Chain Limitations:

- Short Memory: Simple Markov chains for word prediction usually only look at a few words back, leading to limited contextual understanding. This restricts their ability to produce contextually appropriate responses over longer text spans.

- Sparsity: Relying on exact word sequences can result in many unseen transitions in real text, making the model stumble. "Unseen transitions" refer to word pairs or sequences that do not appear in the training data, leading to situations where the Markov chain has no guidance on how to proceed. This sparsity issue limits the applicability of traditional Markov chains in dynamic or varied linguistic environments.

LLM Advancements:

- Transformers: LLMs use the Transformer architecture, allowing them to attend to long-range dependencies in text, far beyond just the previous few words. This architecture uses mechanisms like attention to weigh the importance of different words in a sentence or passage, regardless of their position, enhancing understanding and response accuracy.

- Embeddings: Rather than treating words as discrete units, words are represented by rich numerical vectors (embeddings) capturing semantic relationships. This helps generalize and handle unseen words and phrases more effectively, enabling more flexible and nuanced language generation.

- Context Window: LLMs leverage a large context window to analyze and generate text. The context window refers to the span of text the model can consider at once. A larger context window allows the model to understand and generate more coherent and contextually relevant text by considering broader linguistic context.

- Massive Scale: LLMs are trained on enormous datasets and possess billions of parameters, enabling them to capture nuances and produce more coherent, human-like text. Their vast training corpus includes a wide array of language uses and styles, providing a robust framework for generating text that can adapt to various contexts and requirements.

These enhancements have enabled LLMs to far surpass traditional Markov models in performance and applicability, particularly in tasks requiring deep understanding and generation of natural language. LLMs have become foundational tools in a wide range of applications, from automated writing assistants to advanced conversational agents, reflecting their significance in the evolving landscape of artificial intelligence.

6 Further Exploration

For a more in-depth exploration of stochastic processes, Markov chains, and random walks, readers are encouraged to consult *Probability and Statistics* by Morris H. DeGroot and Mark J. Schervish [4].

Problem Set 16

16.1 A Markov chain model for bike rental is depicted in Fig. 16.3. In this model, individuals rent bikes from any of four sites within a city. The transition probabilities, as shown in the figure, indicate the likelihood of bikes being returned to each site from any other site.

 (a) Based on the transition probabilities given in Fig. 16.3, formulate the transition matrix P.

 (b) Calculate the stationary probability vector $\tilde{p}$. Utilize the method discussed in § 16.3.4.3 for finding stationary distributions.

 (c) Given the city has a total of 1 million bikes, determine the expected number of bikes at each site in the equilibrium state.

 (d) Verify that as n becomes large, P^n approaches a matrix $\tilde{P}$ where each column is equal to $\tilde{p}$. Interpret the significance of this result in the context of the bike rental model.

16.2 For a general two-state Markov chain, consider the diagram below:

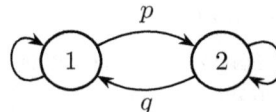

Given the transition probabilities p and q:

(a) What condition must p and q satisfy for the chain to be irreducible?

(b) What condition must p and q satisfy for the chain to be aperiodic?

(c) Assuming it exists, derive the stationary distribution of the chain.

16.3 You have $10 and play a game where in each round you can win $1 with probability 0.6 or lose $1 with probability 0.4. The game stops when you either reach $0 or $20. Let X_n represent the amount of money you have after the nth round (with $X_0 = \$10$ at the start of the game).

(a) Construct the transition matrix of this Markov chain.

(b) Compute $p(X_1 = 9)$.

(c) Compute $p(X_2 = 10)$.

(d) Compute $p(X_6 = 10$ given $X_4 = 10)$.

(e) Compute $p(X_6 = 10$ and $X_4 = 10$ and $X_2 = 10)$.

(f) Find the probability of ending the game with $20.

(g) Determine the period of each state in the chain.

(h) Identify all communicating classes within the Markov chain.

(i) Indicate which of the communicating classes are transient and which are recurrent.

(j) Calculate the expected number of rounds that the game will last.

(k) Suppose you take a snack break each time your money totals $19 during the game. How many such breaks do you expect to take?

(l) Map this problem to a random walk.

16.4 Consider a random walk where at each step, the walker moves forward two units with probability $\frac{1}{3}$ or moves one unit backward with probability $\frac{2}{3}$.

(a) Derive the theoretical probability distribution of the walker's position d after n steps, similar to Eq. 16.14. Assume that the walker starts at position $d = 0$.

(b) Using the central limit theorem, discuss the expected distribution and behavior of d for large n.

(c) ✳ Confirm that the result from part (a) is approximated by the normal distribution predicted in part (b) when n is large.

16.5 ✳ The stationary distribution of a Markov chain can be studied through simulation on a computer. The underlying principle is as follows: let P be the transition matrix, and v_0 the initial probability vector. If a stationary distribution exists, then the sequence $v_1 = v_0 P$, $v_2 = v_1 P$, $v_3 = v_2 P$, ..., or equivalently, $v_0 P^i$ for $i = 1, 2, 3, \ldots$, should converge to the stationary distribution, potentially independent of the choice of v_0.

Simulate this process for the three boundary cases of the random walk shown in Fig. 16.8: (a) cyclic, (b) cliff, and (c) wall. Investigate whether a stationary distribution exists, how it depends on the choice of v_0 in each case, and discuss the ergodicity of each scenario.

16.6 ✳

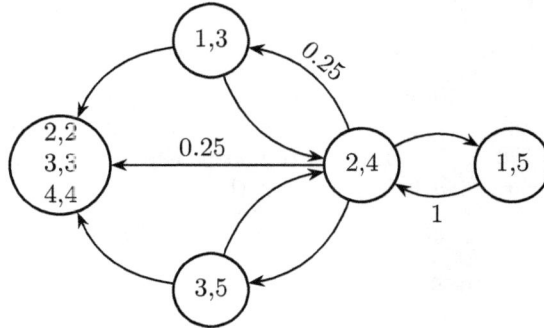

Figure 16.12: Cat and Mouse Game

Consider a game involving five sequential boxes, labeled from one to five. Initially, a cat starts in box one and a mouse in box five. Each midnight, they independently jump to an adjacent box. (Assume they are blind and oblivious to each other's presence.) The game ends tragically for the mouse if they ever land in the same box.

This scenario can be modeled using a Markov chain, where the states are the position tuples of (cat, mouse). Although there are $5 \times 5 = 25$ possible combinations, many are unattainable; the mouse cannot overtake and then move beyond the cat, and the indices' sum always retains even parity. Furthermore, the states leading to the mouse's demise—(2, 2), (3, 3), and (4, 4)—are amalgamated into a single 'terminal' state in our model.

The transition diagram of the Markov chain, reflecting these rules, is depicted in Fig. 16.12.

(a) Construct a transition matrix for the Markov chain model.

(b) Determine the probability that the mouse is caught within 8 turns of the game.

(c) Calculate the expected lifetime of the mouse.

You can address the last two questions either theoretically or using computer simulations.

Contents

---

i | Although this topic is not essential for studying the second book, *Quantum Computing and Information*, it will become important in the third book, *Quantum Algorithms and Applications*. Readers may defer studying it until encountering the relevant quantum algorithms.

17.1 Monte Carlo Simulations

Monte Carlo simulations are a fundamental technique in the field of computational physics and other disciplines, born out of the integration of mathematical stochastic processes, physics, and the veil of secrecy surrounding mid-twentieth-century wartime research. The methodology derives its name from the Monte Carlo Casino in Monaco, reflecting the element of randomness and chance intrinsic to both the simulations and the games of chance held at the casino.

17.1.1 Historical Context

American mathematicians Stanislaw Ulam and John von Neumann are credited with the development of Monte Carlo methods. In the 1940s, while engaged in the Manhattan Project, Ulam conceived the idea during a period of illness. Pondering over the probabilities of solitaire outcomes led him to consider the use of random sampling to address complex mathematical problems that were analytically intractable.

Upon discussing the concept with von Neumann, they recognized its potential for nuclear physics applications. They subsequently utilized Monte Carlo methods for critical calculations of neutron diffusion in fissionable materials, providing a novel approach to simulate the stochastic behavior of particles in complex systems where deterministic solutions were unattainable.

17.1.2   Basic Procedures

Two examples will illustrate the application of Monte Carlo simulation.

■ **Example 17.1 — Estimating π.** One of the simplest yet effective uses of Monte Carlo methods is estimating π. The estimation involves randomly placing points within a square enclosing a quarter circle. The proportion of points within the circle to the total points gives an approximation of π. Below is a Python program demonstrating this method:

```python
import random

# Estimates pi using Monte Carlo
def estimate_pi(num_points):
    inside = 0
    for _ in range(num_points):
        # Random point between -1 and 1
        x, y = random.uniform(-1, 1), random.uniform(-1, 1)
        # Inside quarter circle?
        if x**2 + y**2 <= 1:
            inside += 1
    # Pi is approximately 4 times ratio
    return 4 * inside / num_points

# Simulation with specified points
num_points = 10000
pi_estimate = estimate_pi(num_points)
print(f"Estimated pi: {pi_estimate}")
```

The simulation yields a result approximating 3.14.

This technique showcases the utility of random sampling for problems where analytical solutions are difficult or impossible.   ■

■ **Example 17.2 — Solitaire Probability.** To illustrate Ulam's insight that led to the Monte Carlo method, consider estimating the probability of a specific solitaire outcome—finding at least one King and one Queen adjacent or one card apart. This problem's analytical complexity stems from the vast number of card permutations and decision trees, making brute force solutions impractical.

Below is a Python program that uses Monte Carlo simulation to estimate this probability:

```python
import random

# Checks for adjacent or near K and Q
def is_king_queen_next(deck):
    for i in range(len(deck) - 1):
        # Check if the current and next cards are K and Q, or Q and K
        if (deck[i][0] == 'K' and deck[i + 1][0] == 'Q') or \
           (deck[i][0] == 'Q' and deck[i + 1][0] == 'K'):
```

```
            return True
    # Check one card away if there's space
    if i < len(deck) - 2:
        if (deck[i][0] == 'K' and deck[i + 2][0] == 'Q') or \
           (deck[i][0] == 'Q' and deck[i + 2][0] == 'K'):
            return True
    return False

# Monte Carlo simulation
def monte_carlo(num_trials):
    suits = ['Hearts', 'Diamonds', 'Clubs', 'Spades']
    ranks = ['2', '3', '4', '5', '6', '7', '8', '9', 'T', \
             'J', 'Q', 'K', 'A']
    deck = [r + ' of ' + s for s in suits for r in ranks]
    success = 0
    for _ in range(num_trials):
        random.shuffle(deck)
        if is_king_queen_next(deck):
            success += 1
    return success / num_trials

# Run the simulation
trials = 10000
prob = monte_carlo(trials)
print(f"Probability is approximately {prob:.3f}")
```

The simulation result is approximately 0.73.

The Monte Carlo simulation follows these steps:

- Set up a computer program to simulate the card game with established rules.

- Use random sampling for each simulation, representing different possible shuffles of the 52-card deck.

- Execute the game.

- Repeat the simulation a significant number of times to ensure statistical relevance.

- Estimate the probability by dividing the number of successful outcomes by the total simulations.

■

Exercise 17.1

(a) Perform the Monte Carlo estimation of π using 10,000 shots (num_points) per simulation, as detailed in Example 17.1. Compute the 95% confidence error limit for the estimated value of π. Refer to § 14.3.3 for methods of estimating sampling error using the Central Limit Theorem.

(b) Confirm the 95% confidence error limit obtained in part (a) using an additional Monte Carlo simulation: Repeat the estimation 100 times. For each run, compute the absolute error between the estimated value of π and its true value (e.g., using math.pi in Python). Verify that the absolute error is within the 95% confidence error limit in approximately 95 out of the 100 simulation runs.

17.1.3   Definition and Principle

Monte Carlo Simulation is a computational technique that utilizes random sampling to approximate complex mathematical or physical systems. Given a domain of possible inputs, it involves:

1. Random Sampling: Generating values (samples) from a suitable probability distribution over the problem domain where the probability of obtaining any particular value follows that distribution. Each sample generated can be considered an outcome of an independent random variable that follows the given distribution. This distribution is not limited to uniform and can be any distribution that reflects the underlying process of the problem, such as normal, exponential, binomial, etc.

2. Computational Evaluation: Performing deterministic computations on the samples as inputs. These computations could involve evaluating a mathematical function, simulating a physical process, or any operation that yields a measurable output.

3. Results Aggregation: Aggregating the results of the individual computations to produce a final result, which serves as an approximation to the true solution of the problem. The output of a Monte Carlo Simulation is not limited to counting occurrences; it can encompass a wide range of statistical estimations including means, variances, confidence intervals, and other descriptors of probability distributions.

The accuracy of a Monte Carlo Simulation typically improves as the number of trials increases, converging to the true value.

Monte Carlo simulations are effective due to the statistical principle of the *law of large numbers*. This law asserts that as the number of trials in a random process increases, the average of the results obtained from the random samples will converge to the expected value, which is the mean of the population. In the context of Monte Carlo simulations, this translates to the following key elements:

- Fidelity in Sampling: The sampling must accurately reflect the true probabilistic nature of the system being modeled, often requiring samples to be drawn from distributions that closely match those of the actual system.

- Comprehensive Scenario Coverage: Monte Carlo simulations encompass a broad range of possible outcomes, capturing the variability and uncertainties inherent in real-world systems. This comprehensive coverage allows the simulation to provide detailed insights into the system's behavior under various conditions.

- Convergence: With a sufficient number of samples, the aggregate results from a Monte Carlo simulation tend to approximate the actual distribution of outcomes. This convergence provides reliable estimations of the true characteristics of the system, including its mean, variance, and other statistical properties.

Thus, the effectiveness of a Monte Carlo simulation hinges not just on the number of trials, but also on the fidelity of the sampling method to the true system characteristics and the statistical interpretation of the simulation results.

Monte Carlo methods are particularly useful for systems where direct analytical solutions are infeasible due to complexity or an inherent randomness in the system itself. They have widespread applications in fields such as finance, physics, and engineering, for problems ranging from numerical integration to the simulation of physical processes. In quantum computing, Quantum Monte Carlo simulations are employed to study the properties of quantum systems where the complexity of quantum behaviors precludes classical approximations.

17.1.4 ＊Variance Reduction Techniques

In Monte Carlo simulations, variance reduction refers to strategies designed to improve the efficiency and accuracy of estimates without increasing the number of simulations. These techniques are essential for obtaining precise results within a limited computational budget. They are especially valuable when simulations exhibit high variability or when the outcome of interest involves rare events.

When applied correctly, variance reduction techniques can significantly enhance the reliability of simulation results, often requiring fewer runs than traditional random sampling methods. To demonstrate their impact, let us consider an example in the context of financial risk assessment.

■ Example 17.3 — Importance Sampling in Financial Risk Assessment. Importance Sampling is a variance reduction technique particularly useful in scenarios where certain outcomes are critical but have a low probability of occurrence. It is widely used in financial risk management to assess the risk of extreme losses, often referred to as tail risks.

Problem Context

Consider a financial institution that needs to estimate the probability of experiencing a loss that exceeds a certain threshold due to market fluctuations. These extreme loss events are rare but can be devastating. Using a normal simulation approach with the actual distribution of market returns might not generate sufficient samples from the tail of the distribution, where these losses occur, leading to high variance and inaccurate estimates.

Simulation Setup

1. Model the Market: Assume the daily returns of a market index follow a normal distribution $f(x)$, with mean $\mu = 0\%$ and standard deviation $\sigma = 1\%$.

2. Determine the Threshold: Define a significant loss as any return less than -3%. In a standard normal distribution, this is a 3σ event, which is statistically rare.

3. Traditional Sampling: Normally, simulating returns from the historical distribution would rarely yield events as extreme as -3%, thus making it challenging to estimate their probability accurately.

4. Apply Importance Sampling:

   • Modify the Distribution: Use a normal distribution $g(x)$ with the same mean but a higher standard deviation ($\sigma = 3\%$), making the threshold of -3% only 1σ, thus more frequently sampled.

   • Weight Adjustment: Compute the weight $w(x)$ for each return x, using the ratio of the probability densities of the original distribution $f(x)$ to the modified distribution $g(x)$:

$$w(x) = \frac{f(x)}{g(x)}. \tag{17.1}$$

Conducting the Simulation

- Generate a large number of daily returns (x_i), where each x_i is a sample drawn from the modified distribution $g(x)$.

- For each return x_i, calculate its weight $(w(x_i))$ according to Eq. 17.1.

- Check if the return x_i falls below the loss threshold of -3%. Retain only those samples that qualify.

- To compute the weighted average of these extreme loss occurrences, sum the products of each qualified return's value and its weight, then divide by the sum of the weights:

$$\text{Weighted Average} = \frac{\sum_i w(x_i)\, x_i}{\sum_i w(x_i)}, \tag{17.2}$$

where the summation includes only qualified samples, i.e., $x_i < -3\%$.

Why It Works

Let X be the random variable representing the daily returns. Each x_i is considered an independent realization of X. The weighted average in Eq. 17.2 approximates the following expected value for qualified returns:

$$E[X] = \sum_x g(x)w(x)x. \tag{17.3}$$

Here, the summation iterates over all qualified values $(x < -3\%)$ of the random variable X. Substituting $w(x)$, we obtain:

$$E[X] = \sum_x g(x)\frac{f(x)}{g(x)}x = \sum_x f(x)x, \tag{17.4}$$

which restores the expectation under the original distribution $f(x)$ for the tail events. Thus, the weighting effectively adjusts each return's impact on the final estimate based on its altered probability due to the change in distribution, ensuring that the estimate remains unbiased for the occurrence of rare events.

Benefits of Importance Sampling

This technique enhances the precision in estimating the probability of rare, significant losses, thus aiding financial analysts in making better-informed risk management decisions. Importance Sampling directly addresses the challenges posed by the rarity of critical events and the need for their accurate quantification in risk models. ∎

Exercise 17.2 Conduct a Monte Carlo simulation to assess financial risk as described in Example 17.3. Perform two sets of simulations:

(a) A standard Monte Carlo simulation using the original distribution of market returns.

> (b) An importance sampling Monte Carlo simulation using a modified distribution to enhance the occurrence of significant loss events.
>
> Compare the variance of the estimated means from both simulations. Discuss how importance sampling affects the precision and reliability of the risk assessment.

The utilization of importance sampling in the preceding example demonstrates one approach to variance reduction. However, it is just one of several techniques available for this purpose. The list below introduces additional methods, which are further illustrated in Problems 17.4 to 17.6.

- Importance Sampling: This approach changes the probability distribution from which random variables are sampled to give more weight to the important regions of the domain, which contribute more significantly to the integral or sum being estimated.

- Antithetic Variates: This technique involves using pairs of negatively correlated variables to cancel out some of their variability. By simulating both a variable and its antithetic counterpart, one can reduce the variance of the estimator.

- Control Variates: By identifying variables with known expected values (control variates), one can adjust the simulation outcomes using the difference between the known and observed means of the control variates to reduce the variance.

- Stratified Sampling: This method divides the domain of inputs into distinct strata and ensures that samples are taken from each stratum. Stratification can improve the representativeness of the samples and consequently reduce the variance.

- Latin Hypercube Sampling: An extension of stratified sampling to multiple dimensions, Latin Hypercube Sampling ensures that each variable is sampled uniformly across its range, thereby improving the sample diversity and reducing variance.

The examples in this section highlight the utility of Monte Carlo simulations using common distributions like uniform, normal, or exponential. However, real-world problems are often more complex. The distribution of random variables in such cases can be intricate and not easily described by standard forms.

When the target distribution is not easily accessible, or we need to explore high-dimensional spaces, more advanced methods are required. This is where Markov Chain Monte Carlo (MCMC) comes in. MCMC techniques use the properties of Markov chains to sample from complex probability distributions, expanding the power of Monte Carlo simulations to a wider range of practical applications. We will explore these techniques in the next section.

17.2 ✷ Markov Chain Monte Carlo (MCMC) Simulations

Markov Chain Monte Carlo (MCMC) methods are a class of algorithms essential for sampling from complex probability distributions that are difficult to analyze directly. MCMC generates a Markov Chain, where the states evolve over time into samples from the target distribution. This process can be visualized as a guided random walk that gradually homes in on the high-probability regions of the distribution. After an initial "warm-up" period, called the burn-in phase, the samples produced start to reflect the characteristics of the target distribution more accurately.

MCMC methods are particularly valuable when dealing with high-dimensional problems, where distributions involve many variables. These methods are indispensable for solving such problems efficiently. Additionally, MCMC is a cornerstone of Bayesian statistical inference, enabling the exploration of posterior distributions of parameters based on observed data.

Quantum Monte Carlo (QMC) encompasses a set of techniques that leverage quantum computational resources to enhance Monte Carlo simulations. Quantum computers have the potential to speed up certain aspects of MCMC algorithms, allowing faster convergence to the target distribution. QMC techniques are especially useful for simulating quantum systems, which naturally involve complex distributions that are difficult to sample.

Bosonic Sampling on Quantum Computers

One of the most notable demonstrations of quantum algorithms' potential to solve classically hard problems is Google's 2019 achievement of quantum supremacy using Bosonic sampling. Bosonic sampling involves generating samples from a specific quantum system, where the particles follow Bose-Einstein statistics. This problem becomes computationally intractable for classical computers because the required resources grow exponentially with the number of particles.

Using the Sycamore quantum computer, Google performed a sampling task exponentially faster than the best-known classical algorithms, marking a significant milestone in quantum computing. This breakthrough demonstrates the potential of quantum algorithms to revolutionize fields that rely on complex sampling, such as statistical physics, materials science, and even classical MCMC applications. The achievement underscores that quantum computers can efficiently solve certain sampling problems that are prohibitively difficult for classical methods.

To establish a foundation for studying QMC and advanced topics such as Bosonic sampling, this section will delve into MCMC by examining two prevalent algorithms:

1. Metropolis-Hastings Algorithm: A cornerstone of MCMC methods, this algorithm utilizes a proposal mechanism to suggest new states for the Markov Chain and an acceptance criterion to determine whether these states should be accepted or rejected.

2. Gibbs Sampling: A specialized form of MCMC that simplifies sampling by updating one variable at a time, conditioned on the current values of all other variables in the system.

17.2.1   Metropolis-Hastings (MH) Algorithm

The Metropolis-Hastings (MH) algorithm is a versatile MCMC sampling technique used to draw samples from complex probability distributions that are difficult to sample from directly. It operates by proposing candidate moves from a proposal distribution and accepting or rejecting these moves based on an acceptance ratio. This mechanism systematically favors proposals in regions of higher probability within the target distribution, ensuring flexibility and broad applicability.

To illustrate these core concepts, let's start with a simple example.

1   An Illustrative Example

■ **Example 17.4 — Sampling from a Bimodal Distribution.** Consider a bimodal distribution defined by two Gaussian (normal) distributions, a classic test case for the MH algorithm due to the distinct peaks which simple random sampling might not effectively explore. The distribution is defined as:

$$f(x) = 0.5\frac{1}{\sqrt{2\pi\sigma_1^2}}e^{-\frac{(x-\mu_1)^2}{2\sigma_1^2}} + 0.5\frac{1}{\sqrt{2\pi\sigma_2^2}}e^{-\frac{(x-\mu_2)^2}{2\sigma_2^2}}, \qquad (17.5)$$

where each Gaussian has an equal weight of 0.5, and $\mu_1 = -2$, $\sigma_1 = 1$, $\mu_2 = 2$, and $\sigma_2 = 1$.

Below is a Python program that demonstrates the use of the MH algorithm to sample from our bimodal distribution. The program, while amazingly only 20 lines long, employs sophisticated and subtle methods. We will start by visualizing the intuition behind proposal acceptance and rejection, and then explore the intricate details of the accept-reject mechanism.

```
import numpy as np
import matplotlib.pyplot as plt
from scipy.stats import norm

# Define the target distribution: a mixture of two Gaussians
def target_dist(x):
    # Weights are equal, with centers at -2 and 2, and std dev 1
    return 0.5 * norm.pdf(x, -2, 1) + 0.5 * norm.pdf(x, 2, 1)

# MH algorithm to sample from the target distribution
def metropolis_hastings(num_samples, prop_std):
    samples = np.zeros(num_samples)
    # Start from a random place within a typical normal range
    current = np.random.randn()

    for i in range(num_samples):
        # Propose a new state
        proposal = current + np.random.normal(0, prop_std)
        # Compute the acceptance ratio
        r_accept = target_dist(proposal) / target_dist(current)
        # Accept the proposal with the probability r_accept
        if np.random.rand() < r_accept:
            current = proposal
        samples[i] = current   # Collect this sample

    return samples

# Run the simulation
num_samples = 12000 # Total samples excluding burn-in
burn_in = 2000    # Number of samples to discard as burn-in
prop_std = 1.0    # Standard deviation of the proposal distribution
samples = metropolis_hastings(num_samples, prop_std)
# Keep samples after removing burn-in
samples = samples[burn_in:]

# Visualization code would go here (omitted for brevity)
```

The output of the MH algorithm, as depicted in Fig. 17.1, illustrates the sampler's proficiency in covering both peaks of the bimodal target distribution. The histogram displays both the accepted and rejected samples after the burn-in period. The density of rejected samples highlights regions of lower probability density, demonstrating the sampler's tendency to accept proposals in regions of higher probability density. ■

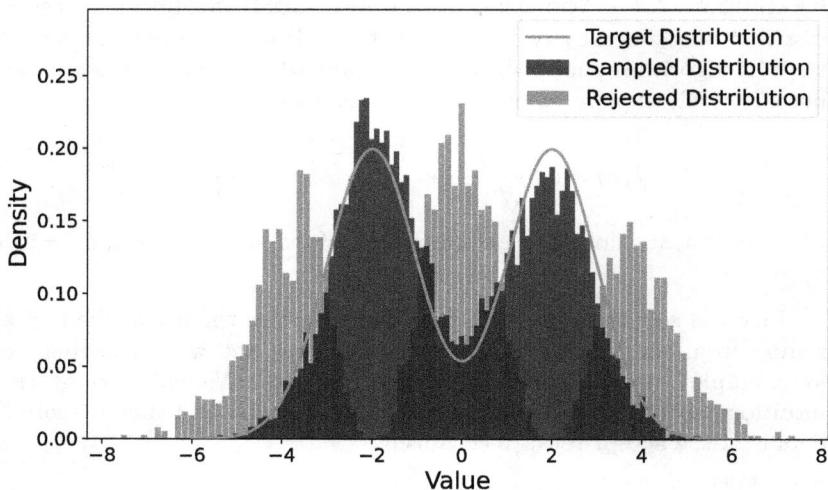

Figure 17.1: Metropolis-Hastings Sampling from a Bimodal Distribution

2   Core Procedure

The Metropolis-Hastings (MH) algorithm iteratively constructs a Markov chain that converges to the target distribution, denoted $p(x)$. The procedure is outlined as follows:

1. Initialization: Start with an initial position, x_0, representing the first state of the Markov chain.

2. Iterations: Perform a large number of iterations. For each step t:

    a. Proposal Generation: Generate a proposed next position, x'_{t+1}, from a proposal distribution. The conditional probability distribution for the proposed new position given the current position x_t is denoted $q(x'_{t+1}|x_t)$. This proposal distribution is commonly symmetric and centered at x_t, such as the normal distribution used in Example 17.4.

    b. Acceptance Ratio: Compute the acceptance ratio r using the formula:

$$r = \frac{p(x'_{t+1})q(x_t|x'_{t+1})}{p(x_t)q(x'_{t+1}|x_t)}. \tag{17.6}$$

    For symmetric proposals, as in Example 17.4, this simplifies to:

$$r = \frac{p(x'_{t+1})}{p(x_t)}. \tag{17.7}$$

    c. Accept-Reject Sampling: Generate a uniform random number u from the interval $[0, 1]$. If $u \leq r$, then accept x'_{t+1} as the new current state: $x_{t+1} = x'_{t+1}$; otherwise, retain the existing state: $x_{t+1} = x_t$.

3. Burn-in Removal: After the iterations, discard the initial samples from the period known as 'burn-in'. The number of samples to discard should be

experimentally determined to allow the Markov chain to reach equilibrium, thus ensuring that the remaining samples are more representative of the target distribution.

Acceptance Probability

The Accept-Reject Sampling step in the above procedure is equivalent to accepting a proposal with probability α and rejecting it with $1 - \alpha$, where α is given by:

$$\alpha(x_t \rightarrow x'_{t+1}) = \min(1, r) = \min\left(1, \ \frac{p(x'_{t+1})q(x_t|x'_{t+1})}{p(x_t)q(x'_{t+1}|x_t)}\right). \tag{17.8}$$

Exercise 17.3 To internalize the above procedure, consider the task of sampling from a simple joint distribution $p(X, Y)$ where $X, Y \in \{0, 1\}$ using the Metropolis-Hastings algorithm. The joint probabilities $p(X = i, Y = j)$ are given by the elements of the matrix P_{ij}:

$$P = \begin{bmatrix} 0.1 & 0.2 \\ 0.3 & 0.4 \end{bmatrix}.$$

(a) Implement the Metropolis-Hastings algorithm to sample from this distribution. Start with the initial state $(X, Y) = (0, 0)$ and propose to flip each variable independently with probability 0.5. Perform 5,000 iterations and compute the empirical distribution of (X, Y).

(b) Compare the empirical distribution obtained from the Metropolis-Hastings simulation with the theoretical joint distribution $p(X, Y)$.

(c) You are encouraged to vary the initial state and the number of iterations to observe their effect on the convergence of the algorithm. Use visual comparisons such as heat maps to illustrate the differences in empirical distributions under different settings.

3 ✳ Markov Chain and Convergence

At the heart of MCMC is a dynamically constructed Markov chain for which the MCMC target distribution satisfies the detailed balance condition (refer to § 16.4.2). Owing to Theorem 16.1, the MCMC target distribution acts as a stationary distribution of the Markov chain. If the chain is also designed to be ergodic, it will converge to the MCMC target distribution, thereby ensuring convergence to the desired target distribution. This subsection aims to elucidate these points.

Transition Diagram and Probabilities

Figure 17.2 illustrates the transitions of an MH MCMC Markov chain, conceptualized here in terms of iteration steps (temporal) and state transitions (spatial). While the chain might inherently possess multidimensional and continuous properties, for illustrative purposes, it is represented as one-dimensional and discrete, with x values placed on a finite grid depicted as thick dots on the number line.

In each MCMC iteration, the node x_t—the current position—can potentially transition to multiple proposed locations x'_{t+1}, labeled by x'^j_{t+1}, each with a corresponding transition probability s_j, where $j = 0, \pm1, \pm2, \ldots$. However, during the accept-reject phase, only one of these proposed locations is accepted, and the rest remain unrealized. For example, if the proposal at x'^3_{t+1} is accepted, the chain moves to this new state as x_{t+1}. If no proposal is accepted, the chain remains at x_t, represented by the loop back to x_t in the figure.

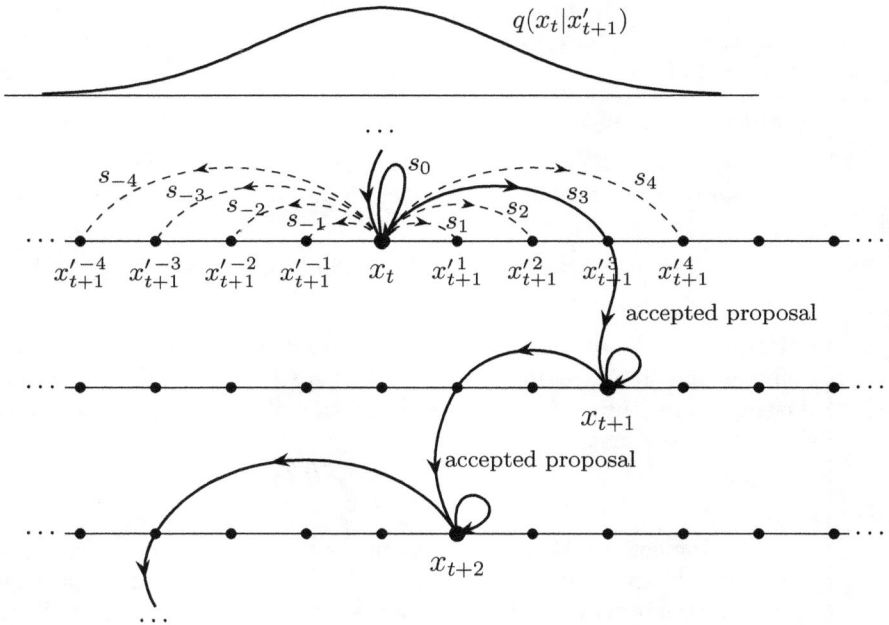

Figure 17.2: Transition Diagram of Markov Chain Monte Carlo (MCMC)

After the accept-reject step, the unrealized locations become irrelevant to the main Markov chain. Consequently, for brevity, we will drop the index j from subsequent discussions, using x'_{t+1} to refer to the accepted location, exemplified as x'^3_{t+1} in the figure.

The probability of proposing any x'_{t+1} is governed by the proposal distribution $q(x'_{t+1}|x_t)$, which is represented in the figure as a Gaussian curve centered at x_t.

The acceptance probability for any proposal x'_{t+1} is $\alpha(x_t \rightarrow x'_{t+1})$, as defined in Eq. 17.8.

The transition probability from x_t to x'_{t+1} is given by the product of these probabilities, reflecting the influence of two independent stochastic processes:

$$s(x_t \rightarrow x'_{t+1}) = \alpha(x_t \rightarrow x'_{t+1}) \cdot q(x'_{t+1}|x_t). \tag{17.9}$$

The looping probability at x_t, denoted as s_0, encapsulates the likelihood that x_t remains unchanged in the subsequent iteration, accounting for both the rejection of the proposed transition and the scenario where no effective proposal change occurs. It is quantified as:

$$s_0 = 1 - s(x_t \rightarrow x'_{t+1}). \tag{17.10}$$

This ensures the total transition probability from x_t sums to one.

Random Walk MCMC

Random walk MCMC algorithms are a subset of Markov Chain Monte Carlo (MCMC) methods characterized by their reliance on generating successive sample states that depend on the current state through a random step, often determined by

> a stochastic proposal mechanism. These steps typically have no direct awareness of the gradient or structure of the target distribution; rather, they "wander" through the state space, hence the term "random walk." MH MCMC a prime example of random walk MCMC.

Detailed Balance

To ensure that the long-term transition probabilities lead the chain to a state where it accurately represents the target distribution, the acceptance criterion must be engineered to establish detailed balance with the target distribution. (Refer to § 16.4.2 for details of this concept.) For MCMC, the detailed balance condition for the target distribution $p(x)$ demands:

$$p(x_t) \cdot s(x_t \to x'_{t+1}) = p(x'_{t+1}) \cdot s(x'_{t+1} \to x_t), \quad \text{for any } x_t \text{ and } x'_{t+1}. \quad (17.11)$$

Next, we show that this condition holds for MH MCMC. The left-hand side of the above equation can be transformed as:

$$\begin{aligned} \text{LHS} &= p(x_t) \cdot s(x_t \to x'_{t+1}) \\ &= p(x_t) \cdot \alpha(x_t \to x'_{t+1}) \cdot q(x'_{t+1}|x_t) \\ &= p(x_t) \cdot \min\left(1, \frac{p(x'_{t+1})q(x_t|x'_{t+1})}{p(x_t)q(x'_{t+1}|x_t)}\right) \cdot q(x'_{t+1}|x_t). \end{aligned}$$

Since $p(x_t)$ and $q(x'_{t+1}|x_t)$ are positive, we can move them inside the min function without affecting the results. So:

$$\text{LHS} = \min\left(p(x_t)q(x'_{t+1}|x_t),\ p(x'_{t+1})q(x_t|x'_{t+1})\right).$$

The denominator from α gets cancelled, and we are left with a symmetric expression. If we start from the right-hand side, we arrive at the same expression. This proves the detailed balance condition for MH MCMC.

Ergodicity

While the detailed balance condition ensures that our target distribution is the stationary distribution of the Markov chain, ergodicity is required to guarantee convergence to this distribution.

To ensure the ergodic condition in the MH algorithm, it is crucial to select a proposal distribution that supports irreducibility, allowing transitions between all possible states in the support of the target distribution. A common approach is to use symmetric distributions such as a Gaussian centered at the current state, which ensures that the proposal covers the entire state space. Aperiodicity can be enhanced by allowing a non-zero probability of staying in the current state. Fine-tuning the proposal's parameters, such as the variance in a Gaussian proposal, helps maintain a reasonable acceptance rate (typically between 20% to 50%), balancing the breadth and frequency of state exploration.

Practical implementation also involves validating ergodicity through both theoretical considerations and empirical diagnostics. Initiating the Markov chain from various initial states and discarding an initial set of samples as burn-in can mitigate the impact of the starting position on the chain's convergence to the stationary distribution. Monitoring convergence through trace plots, autocorrelation plots, and specific diagnostics like the Gelman-Rubin statistic ensures that the chain behaves as expected and converges reliably. These practices are helpful for implementing the MH algorithm effectively, guaranteeing that the generated chain is ergodic and the sampling robustly reflects the target distribution.

4 Advantages and Caveats

Advantages

- Normalization Tolerance: A key advantage of the MH algorithm over methods like Inverse Transform Sampling (ITS, see § 15.5.3), is its ability to sample from distributions where the normalization constant is unknown or computationally infeasible to calculate. This is particularly valuable in high-dimensional spaces and applications that rely only on the unnormalized probability density.

  The MH algorithm reduces the computational burden by using the following acceptance ratio, where the normalizing constant of the target distribution $p(x)$ cancels out. Assume $p(x) = A \cdot u(x)$, then Eq. 17.8 simplifies to

$$\alpha(x_t \to x'_{t+1}) = \min\left(1, \; \frac{u(x'_{t+1})q(x_t|x'_{t+1})}{u(x_t)q(x'_{t+1}|x_t)}\right), \qquad (17.12)$$

  independent of A.

- Flexibility: MH can sample from complex, non-standard probability distributions where direct sampling and other techniques are inadequate.

  The choice of the proposal distribution $q(x'_{t+1}|x_t)$ influences the efficiency of sampling but does not compromise the algorithm's ability to produce statistically correct samples, provided there is sufficient overlap with the target distribution and adequate mixing occurs during the sampling process.

- High-Dimensional Capability: MH handles high-dimensional spaces effectively, offering advantages over other sampling methods like native Monte Carlo or ITS.

- Bayesian Inference: MH serves as a powerful computational tool in navigating complex posterior probability distributions that arise in Bayesian statistical inference.

Caveats

- Proposal Distribution Dependence: Despite its robustness, the choice of proposal distribution in MH can lead to inefficiencies if it does not adequately cover all modes of a multimodal target distribution or uniformly covers the target leading to low acceptance rates in sparse regions. This underscores the importance of thoughtful proposal distribution design, particularly in complex or high-dimensional spaces where a poor choice can lead to significant inefficiencies and biases in the sampling process.

  The optimal acceptance rate depends on the target distribution; however, theoretical studies have shown that the ideal acceptance rate for a one-dimensional Gaussian target distribution is about 50%, which decreases to about 20% for a high-dimensional Gaussian target distribution.

- Correlation: In MCMC simulations, each sample is generated based on the previous one, introducing autocorrelation between the samples. This correlation means that consecutive samples are not independent, potentially leading to an overestimation of the precision of estimators derived from the chain. Hence, there arises the concept of effective sample size (ESS), which represents the number of independent samples that would be equivalent to the autocorrelated samples obtained from the MCMC simulation.

- Convergence Assessment: There is no foolproof method to definitively know if the Markov chain has converged. Diagnostics are commonly used but they do not guarantee convergence.

- Computational Cost: MH can be computationally demanding for complex distributions or when a large number of samples are required.

17.2.2 Gibbs Sampling

While the classic Metropolis-Hastings (MH) algorithm, as described earlier, can be employed to sample across multiple dimensions, this approach often becomes challenging in high-dimensional spaces. The difficulty primarily arises from the need to tune the proposal distribution: a process that must effectively balance the exploration of the space without leading to excessive rejection rates. This balance is harder to achieve as the dimensionality increases, because each additional dimension substantially complicates the tuning process.

Gibbs sampling offers an alternative that is often more effective in such high-dimensional settings. This method simplifies the task by sequentially updating each dimension separately, rather than attempting to generate a new sample for all dimensions simultaneously. By reducing the high-dimensional sampling problem to a series of one-dimensional or low-dimensional problems, Gibbs sampling can efficiently navigate the complexities associated with each dimension.

This technique is particularly advantageous when direct sampling from the joint distribution is impractical. However, it necessitates that the conditional probabilities of each dimension—or several dimensions grouped together as 'blocks'—be well-defined and computationally accessible, at least in their functional forms without the need for normalization constants. Gibbs Sampling is thus especially useful in scenarios where these conditional distributions are known and can be straightforwardly sampled, even if only up to a proportionality constant.

1 An Illustrative Example

To provide a program that illustrates Gibbs Sampling, let's consider a simple example involving two conditional distributions.

■ **Example 17.5 — Bivariate Normal Distribution.** Suppose X and Y follow a bivariate normal distribution, denoted as Normal($\boldsymbol{\mu}, \boldsymbol{\Sigma}$) (see § 15.4.2). The mean vector $\boldsymbol{\mu}$ is given by $\begin{bmatrix} \mu_X & \mu_Y \end{bmatrix}$, and the covariance matrix $\boldsymbol{\Sigma}$ is

$$\boldsymbol{\Sigma} = \begin{bmatrix} \sigma_X^2 & \rho\sigma_X\sigma_Y \\ \rho\sigma_X\sigma_Y & \sigma_Y^2 \end{bmatrix}.$$

In this scenario, the conditional distributions are as follows:

$$X|Y \sim \text{Normal}\left(\mu_X + \rho\sigma_X\frac{(Y-\mu_Y)}{\sigma_Y}, \sigma_X^2(1-\rho^2)\right),$$

and similarly, $Y|X$. This setup allows us to demonstrate Gibbs sampling in a straightforward manner using the following Python program:

```python
import numpy as np
import matplotlib.pyplot as plt

# Set the parameters for the bivariate normal distribution
mu_X, mu_Y = 0, 0  # Means
sigma_X, sigma_Y = 2, 1  # Standard deviations
rho = 0.5  # Correlation coefficient

# Function to sample from conditional distributions
def sample_X_given_Y(y):
    mean_X_given_Y = mu_X + rho * sigma_X / sigma_Y * (y - mu_Y)
    var_X_given_Y = sigma_X**2 * (1 - rho**2)
    return np.random.normal(mean_X_given_Y, np.sqrt(var_X_given_Y))

def sample_Y_given_X(x):
    mean_Y_given_X = mu_Y + rho * sigma_Y / sigma_X * (x - mu_X)
    var_Y_given_X = sigma_Y**2 * (1 - rho**2)
    return np.random.normal(mean_Y_given_X, np.sqrt(var_Y_given_X))

# Gibbs Sampling function
def gibbs_sampling(iterations):
    samples = np.zeros((iterations, 2))
    x, y = 0, 0  # Initial values
    for i in range(iterations):
        x = sample_X_given_Y(y)
        y = sample_Y_given_X(x)
        samples[i] = [x, y]
    return samples

# Run the Gibbs Sampling
iterations = 5000 # Total number of samples
burn_in = 2000  # Number of samples to discard as burn-in
samples = gibbs_sampling(iterations)
samples = samples[burn_in:]

# Visualization code would go here (omitted for brevity)
```

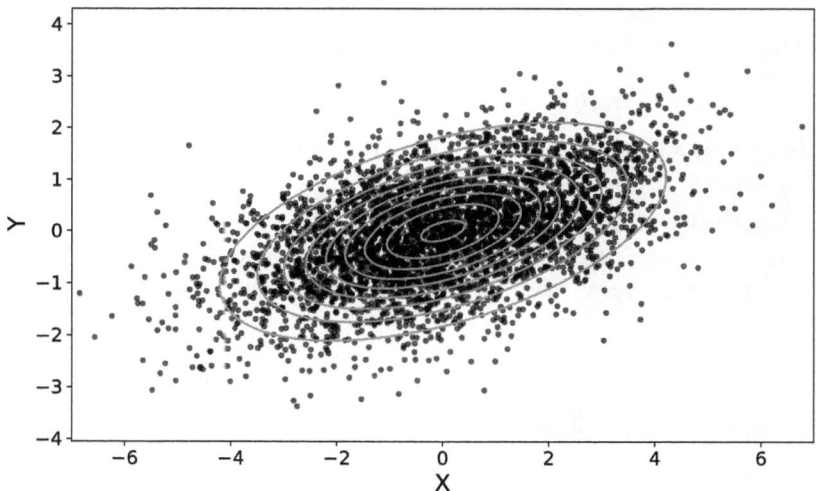

Figure 17.3: Gibbs Sampling from a Bivariate Normal Distribution

Figure 17.3 illustrates the output of Gibbs sampling applied to a bivariate normal distribution. The diagram displays 3,000 samples in the $X - Y$ plane, each representing a draw from the distribution.

Overlaid contour lines represent the density levels of the distribution, with each successive contour moving outward from the center indicating a lower density region. The dense clustering of samples near the central contours visually confirms the algorithm's effectiveness in capturing the high-density areas of the distribution. This pattern also suggests a potential correlation between the variables X and Y, highlighting the linear relationship typical of bivariate normal distributions. ■

2 Core Procedure

The procedure for Gibbs sampling is outlined as follows:

1. Initialization: Start with an initial set of values for the variables $x_0 = (x_{1,0}, x_{2,0}, \ldots, x_{n,0})$, representing the first state of the Markov chain. In practice, these might come from prior distributions or other estimates.

2. Iterations: Perform a large number of iterations. For each iteration t:

   a. Sample Each Conditional: For each variable x_i in $x_t = (x_1, x_2, \ldots, x_n)$, sample a new x_i' for $t + 1$ from the conditional distribution:

   $$x_i' \sim p(x_i' \mid x_1', \ldots, x_{i-1}', x_{i+1}, \ldots, x_n), \qquad (17.13)$$

   where $x_1', \ldots, x_{i-1}'$ denote the already updated x values.

   b. Update All Variables: After all variables are updated sequentially, set $x_{t+1} = (x_1', x_2', \ldots, x_n')$, for the next iteration. In practical programing, this and the previous steps are combined using in-place variable update.

   The order in which variables x_i are updated can follow a fixed sequence or be chosen randomly, depending on the specific implementation and convergence properties.

3. Burn-in Removal: After completing the necessary iterations, discard the initial samples from the period known as the 'burn-in' phase. The length of this phase should be determined experimentally to ensure that the Markov chain has reached equilibrium. The samples remaining after the burn-in period are considered to be representative of the target distribution.

The effectiveness of Gibbs sampling relies on the assumption that the conditional distributions are simpler to sample from than the joint distribution, making it a powerful tool for high-dimensional problems where direct sampling from the joint distribution is computationally infeasible.

3 Detailed Balance Condition

The detailed balance condition (refer to § 16.4.2 for details) is crucial for ensuring that the target distribution that we aim to sample from is the stationary distribution of the Markov chain. Mathematically, for any two states x and x' in the multidimensional state space, the condition is expressed as:

$$p(x)s(x \to x') = p(x')s(x' \to x), \qquad (17.14)$$

where $p(x)$ is the target distribution, and $s(x \to x')$ is the transition probability from state x to state x'.

For Gibbs sampling, here is how the detailed balance is maintained. In each iteration, for each variable x_i, the update involves sampling x_i' from its conditional distribution given the other variables, in such a way that it depends only on the most recently updated values of the other variables:

$$x_i' \sim p(x_i' \mid x_1', \ldots, x_{i-1}', x_{i+1}, \ldots, x_n). \tag{17.15}$$

The transition probability for moving from state $\boldsymbol{x} = (x_1, \ldots, x_n)$ to state $\boldsymbol{x}' = (x_1', \ldots, x_n')$ in one full cycle of updates (where each variable is updated once, independently) can be decomposed as:

$$s(\boldsymbol{x} \to \boldsymbol{x}') = \prod_{i=1}^{n} p(x_i' \mid x_1', \ldots, x_{i-1}', x_{i+1}, \ldots, x_n). \tag{17.16}$$

The detailed balance condition can be shown by noting the symmetry in the sequence of updates where each variable x_i' or x_i is conditioned on a mix of old and new values. This symmetry ensures that the conditionals are balanced in both the forward and reverse transitions. Consequently, the product of conditional probabilities (from the current state to the next) equals the product in the reverse order (from the next state back to the current), under the target distribution $p(x)$:

$$p(\boldsymbol{x}) \prod_{i=1}^{n} p(x_i' \mid x_1', \ldots, x_{i-1}', x_{i+1}, \ldots, x_n) =$$
$$p(\boldsymbol{x}') \prod_{i=1}^{n} p(x_i \mid x_1, \ldots, x_{i-1}, x_{i+1}', \ldots, x_n'). \tag{17.17}$$

4   Normalization Constant Independence

In Gibbs sampling, it is sufficient to know the proportional function of the conditional distributions for each dimension without needing the exact normalization constants. This is because in Eq. 17.17, such constants all cancel out.

This is a significant advantage of Gibbs sampling, especially in Bayesian computation where the full conditional distributions often arise from complex models with difficult-to-compute normalization constants. In this case, each conditional distribution can be derived from the joint distribution. If the joint distribution is $p(x_1, x_2, \ldots, x_n)$, then:

$$p(x_i \mid x_{-i}) = \frac{p(x_1, x_2, \ldots, x_n)}{p(x_{-i})} \propto p(x_1, x_2, \ldots, x_n), \tag{17.18}$$

Here, $p(x_{-i})$ is a marginal distribution of all variables except x_i. A 'constant' in $p(x_i \mid x_{-i})$ means a quantity that does not depend on x_i, and $p(x_{-i})$ is such a constant.

This characteristic of Gibbs Sampling allows for significant computational savings, especially in complex Bayesian models where normalization constants are often computationally expensive to calculate.

5 Ergodicity and Convergence

For Gibbs Sampling to be ergodic and converge to the correct distribution, each variable's conditional distribution must be properly defined and cover the entire support of the target distribution. This ensures irreducibility, meaning the algorithm can transition between all possible states of the joint distribution. Aperiodicity is usually not an issue in Gibbs Sampling, as the process of sampling from the conditionals avoids the fixed cycles that can occur in more restricted Markov chains.

When implementing Gibbs Sampling, it's important to carefully design how each variable is updated. Typically, variables are updated one at a time in a fixed order, though random or adaptive ordering can sometimes improve mixing and reduce the correlation between samples. It's essential to monitor the convergence of the chain using tools like trace plots and autocorrelation plots to ensure the samples represent the target distribution well.

To further ensure accuracy, it's often helpful to start the chain from multiple initial points and discard some initial samples as burn-in. These practices reduce the influence of the starting conditions and help guarantee that the Gibbs Sampling process produces an ergodic chain that captures the target distribution reliably.

6 Combining Gibbs and Metropolis-Hastings (MH)

In some complex models, especially those with high-dimensional or highly correlated parameters, it can be difficult or impossible to sample directly from some conditional distributions. This is particularly true in cases with mixed-type data or models where parameters interact in complex ways that are not easily captured by standard distributions.

In such situations, combining Gibbs Sampling with Metropolis-Hastings (MH) can be useful. This hybrid approach uses a Metropolis-Hastings step for variables where direct sampling is challenging. The MH step proposes a new value for a variable using a proposal distribution, and then accepts or rejects this proposal based on an acceptance ratio, just like in standard MH algorithms.

> **Exercise 17.4** Suppose we want to sample from a joint distribution $p(X, Y)$ where $X, Y \in \{1, 2\}$ and each joint probability $p(X = i, Y = j)$ is given by the element P_{ij} of the following matrix:
> $$P = \begin{bmatrix} 0.1 & 0.2 \\ 0.3 & 0.4 \end{bmatrix}.$$
>
> (a) Calculate the marginal probabilities $p(X)$ and $p(Y)$, and list the conditional probabilities $p(X|Y)$ and $p(Y|X)$.
>
> (b) Conduct a Gibbs sampling simulation with an initial state $(X, Y) = (1, 1)$. Compare the empirical distribution obtained from your samples with the original joint distribution $p(X, Y)$.

17.2.3 Advanced Topics

This subsection introduces some advanced topics related to Markov Chain Monte Carlo (MCMC) methods, highlighting their sophisticated applications and potential integration with quantum computing technologies.

1   Hamiltonian Monte Carlo (HMC)

Hamiltonian Monte Carlo (HMC) represents an evolution in MCMC techniques, utilizing gradient information to make informed, efficient moves across the state space. This method is particularly effective in sampling from complex, high-dimensional distributions. Unlike the Metropolis-Hastings (MH) algorithm, which performs a random walk through the state space, HMC simulates the physical dynamics of a particle moving through a potential energy landscape. This approach leverages both position and momentum to navigate the distribution, leading to faster convergence and more thorough exploration.

The Hamiltonian, central to this method, is defined as:

$$H(\boldsymbol{q}, \boldsymbol{p}) = U(\boldsymbol{q}) + K(\boldsymbol{p}), \tag{17.19}$$

where $\boldsymbol{q}$ denotes the position coordinates (parameters of interest), $\boldsymbol{p}$ the corresponding momenta, $U(\boldsymbol{q})$ the potential energy (linked to the probability density), and $K(\boldsymbol{p})$ the kinetic energy, usually expressed as $\frac{1}{2}\boldsymbol{p}^T M^{-1} \boldsymbol{p}$ with M being the mass matrix. These dynamics facilitate efficient navigation through the parameter space.

Advantages and Limitations

HMC excels in handling complex, high-dimensional distributions by using gradient information to guide proposals, making large, effective steps across the state space and avoiding the inefficiencies of MH and Gibbs sampling in correlated dimensions. Additionally, the simultaneous update of all parameters and conservation of Hamiltonian energy contribute to high acceptance rates and rapid convergence.

However, the computational expense of calculating gradients, especially in models where derivatives are complex, limits HMC's applicability. Furthermore, HMC's effectiveness is highly dependent on the tuning of hyperparameters such as step size and leapfrog steps, requiring extensive experimentation and a deeper understanding for optimal performance.

2   Convergence Diagnostics

Ensuring that an MCMC algorithm has converged to the target distribution is crucial for the validity of the samples it generates. Several diagnostic tools are employed to assess and confirm convergence:

Trace Plots

Trace plots display the values of samples across iterations, offering a visual method to inspect chain mixing and stability. A well-mixed chain appears as a "fuzzy caterpillar," indicating random sampling without visible trends or patterns, which would suggest non-convergence and potential issues in chain mobility.

Autocorrelation

This diagnostic measures the correlation of the chain with itself at different lags. A quick decline in autocorrelation is indicative of effective mixing, with autocorrelation approaching zero suggesting near independence of samples, a hallmark of proper convergence in MCMC sampling.

Potential Scale Reduction Factor (PSRF)

Also known as the Gelman-Rubin statistic, PSRF evaluates convergence by comparing the variance between multiple parallel chains to the variance within each chain. A

PSRF value close to 1 (typically less than 1.1) indicates that the chains are likely exploring the distribution adequately and have converged.

These diagnostic tools are essential for a robust assessment of MCMC performance, often used in combination to provide a comprehensive view of convergence. Adjustments to the model or algorithm might be necessary if these diagnostics indicate issues, ensuring the generation of representative samples.

Further Exploration

Due to the breadth of content covered in this text, we will not explore these topics in further detail. Readers who are keen to delve deeper into the advanced aspects of Markov Chain Monte Carlo methods are encouraged to consult:

Handbook of Markov Chain Monte Carlo, edited by Steve Brooks, Andrew Gelman, Galin Jones, and Xiao-Li Meng [5].

This comprehensive reference also includes an extensive discussion on Hamiltonian Monte Carlo (HMC) among other advanced topics, providing valuable insights for those aspiring to expand their knowledge in this field.

3 Quantum Mechanics: A Different Kind of Probability

Quantum mechanics introduces a fundamentally different approach to understanding probability, where the concept of probability amplitude is key. Unlike classical probability, which deals with certainties and direct probabilities, quantum theory handles probabilities in a way that is inherently probabilistic and often counterintuitive. As the great physicist Richard Feynman noted in 1951:

> "The new theory [quantum mechanics] asserts that there are experiments for which the exact outcome is fundamentally unpredictable, and that in these cases one has to be satisfied with computing probabilities of various outcomes. But far more fundamental was the discovery that in nature the laws of combining probabilities were not those of the classical probability theory of Laplace. *What is changed, and changed radically, [from classical mechanics to quantum mechanics] is the method of calculating probabilities.*"

(Here Laplace probability refers to the classical view where probabilities are calculated by counting equally likely outcomes, assuming certainty and predictability when enough information is available.)

This shift highlights a central aspect of quantum mechanics: the probabilities of different outcomes are determined not directly, but through the squares of probability amplitudes. These amplitudes can interfere with each other, leading to phenomena such as superposition and entanglement, which have no counterparts in classical physics. Our introduction of probability and statistics in this text aims to bridge the foundational mathematical principles with the advanced concepts in quantum mechanics and quantum computing.

4 Applications in Quantum Computing

Quantum computing provides a unique context for applying MCMC methods, which can be adapted to exploit quantum properties for enhanced performance.

Quantum Annealing as MCMC

Quantum annealing can be seen as a quantum analog to classical simulated annealing, an MCMC method. This technique uses quantum fluctuations to transcend local minima, improving the exploration of complex energy landscapes. Ongoing research examines how closely quantum annealing aligns with MCMC properties like detailed balance, which is critical for confirming its theoretical validity and practical utility.

Quantum-Enhanced Sampling

Innovations in quantum computing have spurred interest in designing MCMC algorithms that leverage quantum capabilities to manipulate and explore complex probability distributions more efficiently than classical methods. This includes faster computation of proposal distributions and acceptance criteria, with significant implications for studying quantum systems and materials.

Applications of MCMC in Quantum Settings

MCMC methods find applications in several quantum computing tasks, such as quantum state preparation, Bayesian inference for quantum systems, and quantum machine learning. These applications illustrate potential synergies between classical statistical methods and quantum computational advances, paving the way for new hybrid techniques.

These advanced topics not only illustrate the breadth of MCMC applications but also highlight the intersections with emerging quantum technologies, offering pathways for significant advancements in both fields.

Further Exploration

With the robust mathematical foundation you have established through this text, you are well-prepared to tackle these advanced topics in quantum computing [1, 6, 8, 9]. We hope you continue your journey of exploration and discovery with our other publications on quantum computing, each designed to further enrich your understanding and expertise in this exciting field.

Problem Set 17

17.1 Consider the game of Texas Hold'em Poker, a popular variant of poker. In this game, each player is dealt two private cards (known as 'hole cards') that belong to them alone. Five community cards are then dealt face-up on the 'board'. All players in the game use these shared community cards in conjunction with their own hole cards to each make their best possible five-card poker hand.

To illustrate the use of Monte Carlo methods, estimate the probability of getting a 'flush' on the flop. A flush in poker is a hand where all cards are of the same suit. Assume a player holds two cards of the same suit, and you are to estimate the probability that the next three community cards (the 'flop') are also of the same suit, completing the flush.

Simulation steps:

(a) Simulate dealing two hole cards of the same suit to a player. Assume the deck is a standard 52-card deck.

(b) Simulate the flop by randomly selecting three more cards from the remaining 50 cards in the deck.

(c) Record whether these three cards match the suit of the hole cards to complete a flush.

(d) Repeat this simulation several thousand times (e.g., 10,000 times) to estimate the probability of getting a flush on the flop.

17.2 Assume a simplified SIR (susceptible, infected, recovered) model to simulate the spread of an infectious disease in a small community. In this model, the population is classified into three compartments:

- Susceptible (S): Individuals who have not yet contracted the disease but are vulnerable to infection.

- Infected (I): Individuals currently infected and capable of transmitting the disease.

- Recovered (R): Individuals who have recovered from the disease and are assumed to be immune for the duration of the simulation.

The dynamics of the disease are governed by several key parameters:

- Basic Reproduction Number (R_0): The average number of new infections caused by an infected individual in a fully susceptible population.

- Transmission Rate (β): The rate at which the disease is transmitted from infected to susceptible individuals, calculated as $\beta = R_0 \cdot \gamma$.

- Recovery Rate (γ): The rate at which infected individuals recover and gain immunity.

Use Monte Carlo methods to estimate the peak number of infections over the course of the disease's spread within the community.

The parameters for the simulation are as follows:

- Total population $N = 2000$.

- Initial number of infected individuals $I_0 = 10$.

- $R_0 = 2.5$.

- $\gamma = 1/14$ days^{-1}, corresponding to an average infectious period of 14 days.

Simulation steps:

(a) Initialize the population with $S = N - I_0$, $I = I_0$, and $R = 0$.

(b) Perform daily updates where each infected individual has the potential to:

- Transmit the disease to a susceptible individual with a probability of $\beta \cdot \frac{S}{N}$.

- Recover based on the recovery rate γ, affecting their transition from I to R.

(c) Update the compartments daily by adjusting S, I, and R based on new infections and recoveries.

(d) Record the number of individuals in each compartment at each time step to identify the peak of the infected compartment.

(e) Repeat the simulation multiple times (e.g., 1000 simulations) to estimate the expected peak number of infections and its variability.

17.3 Consider a simplified financial model where a stock price follows geometric Brownian motion, a common method for modeling stock prices in financial mathematics. This exercise uses Monte Carlo simulation to estimate the price of a European call option, which is a type of financial contract that gives the holder the right, but not the obligation, to buy a stock at a predetermined price (the strike price) on or before a specific date (the expiration date).

Background: A European call option allows the buyer to purchase shares of a stock at a set price (the strike price) on a specific future date (the expiration date). The value of this option depends on the probability that the stock's price will exceed the strike price by the expiration date, making the option profitable to exercise.

Parameters for the simulation:

- Initial stock price $S_0 = \$100$
- Strike price of the option $K = \$105$
- Risk-free rate $r = 5\%$ annually (the theoretical rate of return on an investment with zero risk)
- Volatility $\sigma = 20\%$ annually (a measure of the stock's price fluctuations)
- Time to expiration $T = 1$ year

Simulation steps:

(a) Simulate the stock price at expiration using the formula:

$$S_T = S_0 e^{(r - \frac{1}{2}\sigma^2)T + \sigma\sqrt{T}Z}$$

where Z is a standard normal random variable, $Z \sim \text{Normal}(0, 1)$. This represents the random component of stock price movement due to market volatility.

(b) Calculate the payoff of the call option for each simulation using:

$$\text{Payoff} = \max(S_T - K, 0)$$

This formula reflects the option's value only if the stock price S_T exceeds the strike price K, otherwise, the value is zero because the option would not be exercised.

(c) Repeat the simulation 10,000 times to estimate the average payoff. Discount this average back to its present value using the formula:

$$\text{Present Value} = \text{Average Payoff} \times e^{-rT}$$

This step accounts for the time value of money, reflecting the idea that money available today is worth more than the same amount in the future due to its potential earning capacity.

17.4 ✳ We will estimate the value of π using Monte Carlo simulations with variance reduction via the *Antithetic Variates* technique. It is advisable to use a large sample size, such as $N = 10000$, for clearer observation of the variance reduction effect.

(a) Standard Monte Carlo Estimation

- Generate N random points (X_i, Y_i) within the unit square $[0, 1] \times [0, 1]$.

- Count the number of points C inside the quarter circle $(X_i^2 + Y_i^2 \leq 1)$.

- Estimate π as $\bar{\pi} = 4 \cdot \frac{C}{N}$.

- Compute the variance of $\bar{\pi}$.

(b) Antithetic Variates

- Generate $\frac{N}{2}$ random points (X_i, Y_i) within the unit square.

- For each point (X_i, Y_i), generate its antithetic counterpart $(1 - X_i, 1 - Y_i)$.

- Count the number of points from both the original and antithetic sets that fall inside the quarter circle.

- Estimate π using the combined set of points.

- Compute the variance of the combined estimate $\bar{\pi}$.

(c) Comparison

- Compare the variance of π estimations from the standard method and the Antithetic Variates technique.

- Discuss the reduction in variance when using Antithetic Variates for estimating π.

17.5 ✳ Estimating π using *Stratified Sampling*.

(a) Divide the unit square $[0, 1] \times [0, 1]$ into $4 \times 4 = 16$ equally-sized squares (strata).

(b) Within each stratum, generate an equal number of samples. This approach ensures that all areas of the sampling space are uniformly explored, reducing the risk of over- or under-sampling certain regions that might occur with simple random sampling. This method can potentially reduce the variance of the estimator because each stratum will likely contain both points inside and outside the quarter-circle, balancing the overall sample.

(c) For each sample (X, Y) in a stratum, determine if it falls inside the quarter-circle using:
$$X^2 + Y^2 \leq 1$$

(d) Calculate the proportion of points that fall within the quarter-circle across all strata. In this exercise, since each stratum contributes an equal number of points, simply average the results from each stratum to estimate π as:

$$\pi \approx 4 \times \text{Average Proportion of Points within the Quarter-Circle}$$

(e) Compare the variance of π estimations from the standard method and the Stratified Sampling technique. This comparison should be done with a significant number of total samples (e.g., 10,000) to effectively demonstrate the variance reduction capabilities of Stratified Sampling.

17.6 ✳ Estimating π using a *Control Variate*.

(a) Simulate a set of points within the unit square $[0, 1] \times [0, 1]$ and calculate whether each point falls inside the quarter-circle using the formula:

$$X^2 + Y^2 \leq 1$$

(b) For each point (X, Y), also calculate if it falls within a half-square control, for instance, the lower half defined by $Y \leq 0.5$. This area is 0.5, and it provides a variance that is non-zero since not all points will fall in this half-square, making it useful for variance reduction.

(c) Compute the covariance between the quarter-circle indicator and the half-square indicator. Adjust the estimate for the area of the quarter-circle using the formula:

$$\text{Adjusted Estimate} = \text{Empirical Estimate of Quarter-Circle}$$
$$- b(\text{Empirical Estimate of Half-Square} - 0.5)$$

where $b = \frac{\text{Cov}(I_{\text{circle}}, I_{\text{half-square}})}{\text{Var}(I_{\text{half-square}})}$, and I_{circle}, $I_{\text{half-square}}$ are indicators for being inside the quarter-circle and half-square, respectively.

(d) Estimate π by multiplying the adjusted average indicator by 4.

(e) Compare the variance of π estimations from the standard method and the Control Variate technique with a significant number of total samples (e.g., 10,000).

17.7 ✳ Implement the MH MCMC algorithm to sample from a probability distribution that combines both normal and exponential behaviors. The target distribution is defined as a mixture of a truncated normal distribution and an exponential distribution, with the following probability density function:

$$f(x) = \alpha \frac{1}{\sqrt{2\pi\sigma^2}} e^{-\frac{(x-\mu)^2}{2\sigma^2}} + 0.4\lambda e^{-\lambda x}, \quad x \geq 0,$$

where α is a normalization factor that ensures $\int_0^\infty f(x)\,dx = 1$. For this exercise, set $\mu = -2$, $\sigma = 1$, and $\lambda = 0.2$.

Tasks:

(a) Write a Python function `target_dist(x)` that computes the value of $f(x)$ for a given $x \geq 0$. Determine the normalization factor α numerically.

(b) Implement the MH MCMC algorithm to sample from $f(x)$. Start from an initial point $x_0 = 1$ and use a normal distribution centered at the current sample as the proposal distribution, with a standard deviation of 1.

(c) Run the algorithm for 10,000 iterations and plot a histogram of the sampled values. Overlay the target distribution $f(x)$ on the histogram for comparison.

(d) Discuss the performance of the MH MCMC algorithm. How well does the algorithm explore the target distribution? Comment on the transitions between the normal and exponential components and identify any potential challenges in sampling.

Hint: You may reject all proposals that result in $x < 0$.

V

Supporting Materials

Key Formulas and Concepts

A Complex Numbers

Basic Relations

Imaginary unit: $i \equiv \sqrt{-1}$. $i^2 = -1$, $i^3 = -i$, $i^4 = 1$, $i^5 = i$, ...

	Cartesian Form	Exponential Form
	$z = x + iy$	$z = re^{i\theta}$
Conjugate	$z^* = x - iy$	$z^* = re^{-i\theta}$
Modulus	$|z| = \sqrt{zz^*} = \sqrt{x^2 + y^2}$	$|z| = r$
Conversion	$x = r\cos\theta$	$r = \sqrt{x^2 + y^2}$
	$y = r\sin\theta$	$\theta = \arctan2(y, x)$

Basic Operations

Given

$z_1 = x_1 + iy_1 = r_1 e^{i\theta_1}$, $z_2 = x_2 + iy_2 = r_2 e^{i\theta_2}$:

$$(z_1 \cdot z_2)^* = z_1^* \cdot z_2^* \qquad |z_1 \cdot z_2| = |z_1| \cdot |z_2|$$

$$z_1 \cdot z_2 = r_1 e^{i\theta_1} \cdot r_2 e^{i\theta_2} = r_1 r_2 e^{i(\theta_1 + \theta_2)}$$

$$\frac{z_1}{z_2} = \frac{r_1 e^{i\theta_1}}{r_2 e^{i\theta_2}} = \frac{r_1}{r_2} e^{i(\theta_1 - \theta_2)}$$

$$z_1 \cdot z_2 = x_1 x_2 - y_1 y_2 + i(x_1 y_2 + x_2 y_1)$$

$$\frac{z_1}{z_2} = \frac{z_1 z_2^*}{z_2 z_2^*} = \frac{x_1 x_2 + y_1 y_2 - i(x_1 y_2 - x_2 y_1)}{x_2^2 + y_2^2}$$

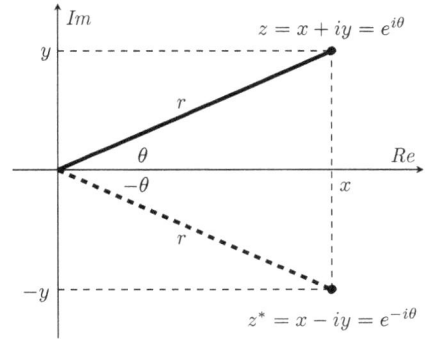

Useful Formulas

Euler's formula: $e^{i\theta} = \cos\theta + i\sin\theta$

De Moivre's formula: $(\cos\theta + i\sin\theta)^n = \cos(n\theta) + i\sin(n\theta)$

Roots of unity ($\omega^n = 1$): $\omega_k = e^{ik\frac{2\pi}{n}}$, where $k = 0, 1, \ldots, n-1$. $\displaystyle\sum_{k=0}^{n-1} \omega_k = 0$.

B Trigonometry

Definitions and Basic Properties

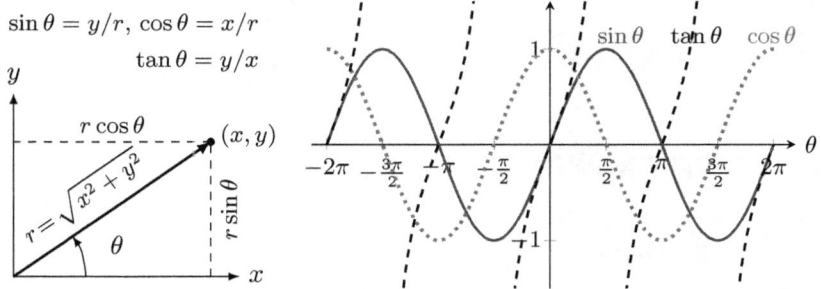

	$\sin\theta$	$\cos\theta$	$\tan\theta$	$\csc\theta$	$\sec\theta$	$\cot\theta$
Definition	y/r	x/r	y/x	r/y	r/x	x/y
Period	2π	2π	π	2π	2π	π
Range	$[-1,1]$	$[-1,1]$	$(-\infty,\infty)$	$(-\infty,-1]\cup[1,\infty)$	$(-\infty,\infty)$
Zeros	$n\pi$	$(n+\frac{1}{2})\pi$	$n\pi$			$(n+\frac{1}{2})\pi$
Poles			$(n+\frac{1}{2})\pi$	$n\pi$	$(n+\frac{1}{2})\pi$	$n\pi$
Inv Range	$[-\frac{\pi}{2},\frac{\pi}{2}]$	$[0,\pi]$	$(-\frac{\pi}{2},\frac{\pi}{2})$	$[-\frac{\pi}{2},\frac{\pi}{2}]\setminus\{0\}$	$[0,\pi]\setminus\{\frac{\pi}{2}\}$	$(-\frac{\pi}{2},\frac{\pi}{2})$

Note: n is an integer. $\sin\theta$ is also denoted as $\sin(\theta)$. The inverse of $x = \sin\theta$, $\theta = \arcsin x$, is also written as $\theta = \sin^{-1} x$. The extended inverse $\text{arcsin2}(y,x)$, or $\text{asin2}(y,x)$, extends the range to $[-\pi,\pi]$ by considering the signs of x and y. Similarly, for other trig functions.

Special Values

θ	0	$\frac{\pi}{6},30°$	$\frac{\pi}{4},45°$	$\frac{\pi}{3},60°$	$\frac{\pi}{2},90°$	$\frac{3\pi}{4},135°$	$\pi,180°$	$\frac{5\pi}{4},225°$
$\sin\theta$	0	$\frac{1}{2}$	$\frac{\sqrt{2}}{2}$	$\frac{\sqrt{3}}{2}$	1	$\frac{\sqrt{2}}{2}$	0	$-\frac{\sqrt{2}}{2}$
$\cos\theta$	1	$\frac{\sqrt{3}}{2}$	$\frac{\sqrt{2}}{2}$	$\frac{1}{2}$	0	$-\frac{\sqrt{2}}{2}$	-1	$-\frac{\sqrt{2}}{2}$
$\tan\theta$	0	$\frac{\sqrt{3}}{3}$	1	$\sqrt{3}$	∞	-1	0	1

Interrelations

$$\tan\theta = \frac{\sin\theta}{\cos\theta} \qquad \cot\theta = \frac{1}{\tan\theta} \qquad \sec\theta = \frac{1}{\cos\theta} \qquad \csc\theta = \frac{1}{\sin\theta}$$

Cofunction Formulas

$$\sin\left(\tfrac{\pi}{2} - \theta\right) = \cos\theta \qquad \cos\left(\tfrac{\pi}{2} - \theta\right) = \sin\theta \qquad \cot\left(\tfrac{\pi}{2} - \theta\right) = \tan\theta$$

Pythagorean Identities

$$\sin^2\theta + \cos^2\theta = 1 \qquad \tan^2\theta + 1 = \sec^2\theta \qquad 1 + \cot^2\theta = \csc^2\theta$$

Symmetry Properties

$$\sin(-\theta) = -\sin\theta \qquad \sin(\pi - \theta) = \sin\theta \qquad \sin(\pi + \theta) = -\sin\theta$$
$$\cos(-\theta) = \cos\theta \qquad \cos(\pi - \theta) = -\cos\theta \qquad \cos(\pi + \theta) = -\cos\theta$$
$$\tan(-\theta) = -\tan\theta \qquad \tan(\pi - \theta) = -\tan\theta \qquad \tan(\pi + \theta) = \tan\theta$$

Double Angle Formulas

$$\sin 2\theta = 2\sin\theta\cos\theta$$
$$\cos 2\theta = \cos^2\theta - \sin^2\theta = 2\cos^2\theta - 1 = 1 - 2\sin^2\theta$$
$$\tan 2\theta = \frac{2\tan\theta}{1 - \tan^2\theta}$$

Half Angle Formulas

$$\sin^2\frac{\theta}{2} = \frac{1 - \cos\theta}{2} \qquad \cos^2\frac{\theta}{2} = \frac{1 + \cos\theta}{2} \qquad \tan\frac{\theta}{2} = \frac{\sin\theta}{1 + \cos\theta} = \frac{1 - \cos\theta}{\sin\theta}$$

Sum and Difference Formulas

$$\sin(\alpha \pm \beta) = \sin\alpha\cos\beta \pm \cos\alpha\sin\beta \qquad \cos(\alpha \pm \beta) = \cos\alpha\cos\beta \mp \sin\alpha\sin\beta$$
$$\tan(\alpha \pm \beta) = \frac{\tan\alpha \pm \tan\beta}{1 \mp \tan\alpha\tan\beta}$$

Product-to-Sum Formulas

$$2\sin\alpha\sin\beta = \cos(\alpha - \beta) - \cos(\alpha + \beta) \qquad 2\cos\alpha\cos\beta = \cos(\alpha - \beta) + \cos(\alpha + \beta)$$
$$2\sin\alpha\cos\beta = \sin(\alpha + \beta) + \sin(\alpha - \beta) \qquad 2\cos\alpha\sin\beta = \sin(\alpha + \beta) - \sin(\alpha - \beta)$$

Sum-to-Product Formulas

$$\sin\alpha + \sin\beta = 2\sin\frac{\alpha + \beta}{2}\cos\frac{\alpha - \beta}{2} \qquad \sin\alpha - \sin\beta = 2\cos\frac{\alpha + \beta}{2}\sin\frac{\alpha - \beta}{2}$$
$$\cos\alpha + \cos\beta = 2\cos\frac{\alpha + \beta}{2}\cos\frac{\alpha - \beta}{2} \qquad \cos\alpha - \cos\beta = -2\sin\frac{\alpha + \beta}{2}\sin\frac{\alpha - \beta}{2}$$

Laws of Sines and Cosines

$$\frac{\sin A}{a} = \frac{\sin B}{b} = \frac{\sin C}{c} \qquad a^2 = b^2 + c^2 - 2bc\cos A$$

Spherical Coordinate System

Useful Taylor Expansions

$$e^x = \sum_{n=0}^{\infty} \frac{x^n}{n!} = 1 + x + \frac{x^2}{2!} + \frac{x^3}{3!} + \cdots$$

$$\sin x = \sum_{n=0}^{\infty} (-1)^n \frac{x^{2n+1}}{(2n + 1)!} = x - \frac{x^3}{3!} + \frac{x^5}{5!} - \cdots$$

$$\cos x = \sum_{n=0}^{\infty} (-1)^n \frac{x^{2n}}{(2n)!} = 1 - \frac{x^2}{2!} + \frac{x^4}{4!} - \cdots$$

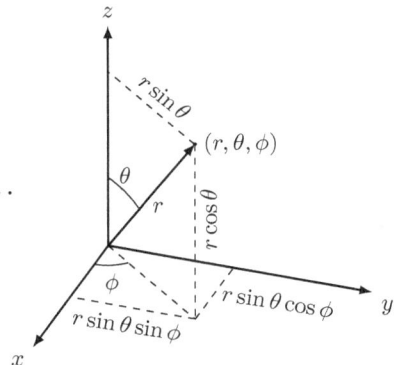

C Linear Algebra for QCI

Arrays in 0D, 1D, 2D, ...

0D: scalar, c, in $\mathbb{C}$

1D: vector, $|a\rangle = \begin{bmatrix} a_1 \\ a_2 \\ \vdots \\ a_n \end{bmatrix}$, in $\mathbb{C}^n$

2D: matrix, $A = \begin{bmatrix} a_{11} & a_{12} & \cdots \\ a_{21} & a_{22} & \ddots \\ \vdots & & \end{bmatrix}$, in $\mathbb{C}^{n \times n}$

3D, 4D, ...: tensor, in $\mathbb{C}^{n \times n \times \cdots \times n}$ (k times), or $\mathbb{C}^{n^k}$

Matrix Definitions

A^*: complex conjugate of A

A^T: transpose of A

$A^\dagger$: Hermitian conjugate or adjoint of A, $A^\dagger = \left(A^T\right)^*$

Hermitian (self-adjoint) matrix: $H^\dagger = H$

Unitary matrix: $U^\dagger = U^{-1}$ or $U^\dagger U = I$

Normal matrix: $AA^\dagger = A^\dagger A$ (includes Hermitian matrix and unitary matrix)

Trace: $\operatorname{tr} A \equiv \sum_i a_{ii}$

Determinant: $\det A = |A|$

Commutator: $[A, B] \equiv AB - BA$

Anti-commutator: $\{A, B\} \equiv AB + BA$

Direct Sum: $A \oplus B$ (block diagonal matrix of A and B)

Ket, Bra, and Braket

Ket: $|v\rangle \equiv \vec{v} = \begin{bmatrix} v_1 \\ v_2 \\ v_3 \end{bmatrix}$

Bra: $\langle v| \equiv |v\rangle^\dagger = \begin{bmatrix} v_1^* & v_2^* & v_2^* \end{bmatrix}$

Braket (inner product):

$$\langle u|v \rangle \equiv \langle u| \cdot |v \rangle = \begin{bmatrix} u_1^* & u_2^* & u_3^* \end{bmatrix} \begin{bmatrix} v_1 \\ v_2 \\ v_3 \end{bmatrix} = u_1^* v_1 + u_2^* v_2 + u_3^* v_3$$

$$\langle u|v \rangle = \langle v|u \rangle^*$$

Norm: $\|v\| = \sqrt{\langle v|v \rangle} = \sqrt{v_1^* v_1 + v_2^* v_2 + v_3^* v_3}$

Normalization: $|v \rangle \to |\hat{v} \rangle = \dfrac{|v \rangle}{\|v\|}$ so that $\|\hat{v}\| = 1$

Vector Matrix Product

$$A |v \rangle = \begin{bmatrix} a_{11} & a_{12} & a_{13} \\ a_{21} & a_{22} & a_{23} \\ a_{31} & a_{32} & a_{33} \end{bmatrix} \begin{bmatrix} v_1 \\ v_2 \\ v_3 \end{bmatrix}$$

$$(A |v \rangle)^\dagger \equiv \langle Av| = \langle v| A^\dagger$$

$$\langle u|A|v \rangle = \begin{bmatrix} u_1^* & u_2^* & u_3^* \end{bmatrix} \begin{bmatrix} a_{11} & a_{12} & a_{13} \\ a_{21} & a_{22} & a_{23} \\ a_{31} & a_{32} & a_{33} \end{bmatrix} \begin{bmatrix} v_1 \\ v_2 \\ v_3 \end{bmatrix}$$

$$\langle u|A|v \rangle = \langle u|Av \rangle = \langle uA|v \rangle = \langle A^\dagger u|v \rangle$$

Eigenvalues and Eigenvectors

$$H |\phi_i \rangle = \lambda_i |\phi_i \rangle$$

A normal matrix allows for spectral decomposition. Its eigenvalues are complex, whereas those of Hermitian matrices are real, and skew-Hermitian matrices imaginary. The eigenvectors of these matrices form complete, orthonormal bases.

Orthonormal property: $\langle \phi_i|\phi_j \rangle = \delta_{ij}$

Completeness: $\sum_i |\phi_i \rangle\langle \phi_i| = I$

Spectral decomposition of H: $H = \sum_i \lambda_i |\phi_i \rangle\langle \phi_i|$

Statistics

Observables in QM (A and B) are Hermitian matrices.

Expected value: $\langle A \rangle_\psi = \langle \psi|A|\psi \rangle$

Standard deviation: $\Delta A_\psi = \sqrt{\left\langle \left(A - \langle A \rangle_\psi \right)^2 \right\rangle_\psi} = \sqrt{\langle A^2 \rangle_\psi - \langle A \rangle_\psi^2}$

Cauchy-Schwarz inequality: $|\langle u|v \rangle|^2 \leq \langle u|u \rangle \langle v|v \rangle$

Uncertainty theorem: $\Delta A_\psi \cdot \Delta B_\psi \geq \frac{1}{2}|\langle AB - BA \rangle_\psi|$

Outer Product

$$|v\rangle\langle u| = \begin{bmatrix} v_1 \\ v_2 \\ v_3 \end{bmatrix} \begin{bmatrix} u_1^* & u_2^* & u_3^* \end{bmatrix} = \begin{bmatrix} v_1 u_1^* & v_1 u_2^* & v_1 u_3^* \\ v_2 u_1^* & v_2 u_2^* & v_2 u_3^* \\ v_3 u_1^* & v_3 u_2^* & v_3 u_3^* \end{bmatrix}$$

$$|0\rangle\langle 0| = \begin{bmatrix} 1 & 0 \\ 0 & 0 \end{bmatrix}, \quad |1\rangle\langle 1| = \begin{bmatrix} 0 & 0 \\ 0 & 1 \end{bmatrix}, \quad |0\rangle\langle 1| = \begin{bmatrix} 0 & 1 \\ 0 & 0 \end{bmatrix}, \quad |1\rangle\langle 0| = \begin{bmatrix} 0 & 0 \\ 1 & 0 \end{bmatrix}$$

$$\begin{bmatrix} a_{00} & a_{01} \\ a_{10} & a_{11} \end{bmatrix} = a_{00}|0\rangle\langle 0| + a_{01}|0\rangle\langle 1| + a_{10}|1\rangle\langle 0| + a_{11}|1\rangle\langle 1|$$

Projection

Projector onto $|u\rangle$: $P_u = |u\rangle\langle u|$

Projection of $|v\rangle$ onto $|u\rangle$: $P_u|v\rangle = |u\rangle\langle u|v\rangle = \langle u|v\rangle|u\rangle$

Projectors are idempotent: $P_u^2 = P_u$

Unitary Matrices

U is unitary: $U^\dagger U = I$ or $U^\dagger = U^{-1}$

The columns (or rows) of a unitary matrix form an orthonormal basis:

$$(U|i\rangle)^\dagger(U|j\rangle) = \langle i|U^\dagger U|j\rangle = \langle i|j\rangle = \delta_{ij}$$

Unitary transformation preserves inner product:

$$\langle U\phi|U\psi\rangle = (U|\phi\rangle)^\dagger(U|\psi\rangle) = \langle\phi|U^\dagger U|\psi\rangle = \langle\phi|\psi\rangle$$

$\{|\phi_i\rangle\}$ are orthonormal and $|\lambda_i| = 1$ $\Leftrightarrow$ $\sum_i \lambda_i|\phi_i\rangle\langle\phi_i|$ is unitary.

If U and V are unitary, so is $U \otimes V$.

$|\det U| = 1$

Standard (or Computational) Basis

Example: $|0\rangle = \begin{bmatrix} 1 \\ 0 \\ 0 \end{bmatrix}, \quad |1\rangle = \begin{bmatrix} 0 \\ 1 \\ 0 \end{bmatrix}, \quad |2\rangle = \begin{bmatrix} 0 \\ 0 \\ 1 \end{bmatrix}$

Orthonormality and completeness: $\langle i|j\rangle = \delta_{ij}, \quad \sum_i |i\rangle\langle i| = I$

Vector decomposition: $|v\rangle = \sum_i v_i|i\rangle, \quad v_i = \langle i|v\rangle$

Matrix decomposition: $A = \sum_{i,j} a_{ij}|i\rangle\langle j|, \quad a_{ij} = \langle i|A|j\rangle, \quad A|j\rangle = \sum_i a_{ij}|i\rangle$

Change of Basis

Let $\{|b_i\rangle\}$ and $\{|b_i'\rangle\}$ be complete and orthonormal bases:

$$\langle b_i|b_j\rangle = \langle b_i'|b_j'\rangle = \delta_{ij}, \quad \sum_i |b_i\rangle\langle b_i| = \sum_i |b_i'\rangle\langle b_i'| = I$$

Change of basis from $\{|b_i\rangle\}$ to $\{|b_i'\rangle\}$ via unitary operator U:

U is defined by $|b_i'\rangle = U|b_i\rangle$ and given by

$$U = \sum_i |b_i'\rangle\langle b_i| = \sum_{i,j} \langle b_i|b_j'\rangle\, |b_i'\rangle\langle b_j'| = \sum_{i,j} \langle b_i|b_j'\rangle\, |b_i\rangle\langle b_j|$$

Vector and Matrix Representations under Change of Basis

Suppose $|\psi\rangle = \sum_j c_j |b_j\rangle = \sum_j c_j' |b_j'\rangle$. Then $c_i' = \sum_j c_j \langle b_i'|b_j\rangle, \quad c_i = \sum_j c_j' \langle b_i|b_j'\rangle.$

Below, $|v\rangle$ is the vector formed by (c_i), and $|v'\rangle$ by (c_i'). Similarly, for matrices.

$$|v'\rangle = U^\dagger |v\rangle \qquad A' = U^\dagger A U \qquad\qquad \mathrm{tr}\, A' = \mathrm{tr}\, A$$
$$\langle u'|v'\rangle = \langle u|v\rangle \quad \langle u'|A'|v'\rangle = \langle u|A|v\rangle$$
$$H|\psi_i\rangle = \lambda_i |\psi_i\rangle \quad \Rightarrow \quad H'|\psi_i'\rangle = \lambda_i |\psi_i'\rangle$$

Change from Standard Basis to Eigenvector Basis

Given H is Hermitian and $H|\phi_i\rangle = \lambda_i |\phi_i\rangle$, to change basis from $\{|i\rangle\}$ to $\{|\phi_i\rangle\}$:

$$|\phi_i\rangle = U|i\rangle, \quad U = \sum_i |\phi_i\rangle\langle i| = \sum_{i,j} \langle i|\phi_j\rangle\, |i\rangle\langle j| = \begin{bmatrix} \langle 0|\phi_0\rangle & \langle 0|\phi_1\rangle & \cdots \\ \langle 1|\phi_0\rangle & \langle 1|\phi_1\rangle & \\ \vdots & & \ddots \end{bmatrix}$$

H is diagonal in $\{|\phi_i\rangle\}$: $H = \sum_i \lambda_i |\phi_i\rangle\langle\phi_i|$

$U^\dagger H U$ is diagonal in $\{|i\rangle\}$: $U^\dagger H U = \sum_i \lambda_i |i\rangle\langle i|$

Tensor Product

$$|u\rangle \otimes |v\rangle \equiv |u\rangle |v\rangle \equiv |uv\rangle$$

$$|uv\rangle = \begin{bmatrix} u_1 \\ u_2 \end{bmatrix} \otimes \begin{bmatrix} v_1 \\ v_2 \end{bmatrix} = \begin{bmatrix} u_1 |v\rangle \\ u_2 |v\rangle \end{bmatrix} = \begin{bmatrix} u_1 v_1 \\ u_1 v_2 \\ u_2 v_1 \\ u_2 v_2 \end{bmatrix}$$

$A \otimes B =$

$$\begin{bmatrix} a_{11} & a_{12} \\ a_{21} & a_{22} \end{bmatrix} \otimes \begin{bmatrix} b_{11} & b_{12} \\ b_{21} & b_{22} \end{bmatrix} = \begin{bmatrix} a_{11}B & a_{12}B \\ a_{21}B & a_{22}B \end{bmatrix} = \begin{bmatrix} a_{11}b_{11} & a_{11}b_{12} & a_{12}b_{11} & a_{12}b_{11} \\ a_{11}b_{21} & a_{11}b_{22} & a_{12}b_{21} & a_{12}b_{21} \\ a_{21}b_{11} & a_{21}b_{12} & a_{22}b_{11} & a_{22}b_{11} \\ a_{21}b_{21} & a_{21}b_{22} & a_{22}b_{21} & a_{22}b_{21} \end{bmatrix}$$

In general, $(A \otimes B) \neq (B \otimes A)$

$(A \otimes B)^\dagger = A^\dagger \otimes B^\dagger \quad (A \otimes B)^{-1} = A^{-1} \otimes B^{-1}$

$(A \otimes B)(C \otimes D) = AC \otimes BD$

$(A \otimes B)(|u\rangle \otimes |v\rangle) \equiv (A \otimes B)|uv\rangle = |Au\rangle \otimes |Bv\rangle$

$A \otimes B = (A \otimes I)(I \otimes B)$

$(A_1 \otimes B_1)(A_2 \otimes B_2) \cdots (A_n \otimes B_n) = (A_1 A_2 \cdots A_n) \otimes (B_1 B_2 \cdots B_n)$

$(A_1 \otimes A_2 \cdots \otimes A_n)(B_1 \otimes B_2 \cdots \otimes B_n) = (A_1 B_1) \otimes (A_2 B_2) \cdots \otimes (A_n B_n)$

Tenser product and inner product: $\langle \phi_1 \otimes \phi_2 | \psi_1 \otimes \psi_2 \rangle = \langle \phi_1 | \psi_1 \rangle \langle \phi_2 | \psi_2 \rangle$

Tenser product and outer product:

$(|i\rangle_1 \otimes |j\rangle_2)(\langle k|_1 \otimes \langle l|_2) = |i\rangle_1 \langle k|_1 \otimes |j\rangle_2 \langle l|_2$

Shorthand notation: $ij\rangle\langle kl| = |i\rangle\langle k| \otimes |j\rangle\langle l|$

Exponent notation for $\otimes$: $|x\rangle^{\otimes n} \equiv \bigotimes_{i=1}^{n} |x\rangle \equiv |xx \cdots x\rangle$

Basis expansion in linear space $\mathbb{C}^{2^n}$ (i.e., general state vector of an n-qubit system):

$$|\psi\rangle = \sum_{\substack{x_i \in \{0,1\} \\ i \in \{1,2,\cdots,n\}}} c_{x_1 x_2 \cdots x_n} |x_1 x_2 \cdots x_n\rangle \equiv \sum_{k=0}^{2^n-1} c_k |k\rangle$$

Functions of Matrices

If A is Hermitian with real eigenvalues λ_i and eigenvectors $|\phi_i\rangle$, then

$$f(A) = \sum_i f(\lambda_i) |\phi_i\rangle\langle\phi_i|$$

$$e^A = \sum_i e^{\lambda_i} |\phi_i\rangle\langle\phi_i|$$

$$\log A = \sum_i \log \lambda_i |\phi_i\rangle\langle\phi_i| \text{ (for positive } A\text{)}$$

With orthonormal basis $\{|\phi_i\rangle\}$ and $m \in \mathbb{Z}^+$,

$$(|\phi_i\rangle\langle\phi_i|)^m = |\phi_i\rangle\langle\phi_i|, \quad \left(\sum_i \lambda_i |\phi_i\rangle\langle\phi_i|\right)^m = \sum_i \lambda_i^m |\phi_i\rangle\langle\phi_i|$$

Alternative definition: $\exp(A) \equiv e^A \equiv \sum_{n=0}^{\infty} \frac{1}{n!} A^n$

If A is normal, then e^A is also normal, and the eigenvalues of e^A are the exponentials of the eigenvalues of A, with the same eigenvectors.

If H is Hermitian, then e^{iH} is unitary.

If U is unitary, $e^{UAU^\dagger} = Ue^AU^\dagger$

Generalized Euler formula: If γ is real and $A^2 = I$, then $e^{i\gamma A} = \cos\gamma I + i\sin\gamma A$

In general, $e^{A+B} \neq e^Ae^B$, unless A and B commute.

BCH formula: $e^{A+B} = e^Ae^Be^{-\frac{1}{2}[A,B]+\cdots}$

In general, $e^{A\otimes B} \neq e^A \otimes e^B$, but $e^{A\otimes I+I\otimes B} = e^A \otimes e^B$.

Partial Product Notations

$$U_A\,|x_1x_2\rangle \equiv U\,|x_1\rangle \otimes |x_2\rangle \qquad U_B\,|x_1x_2\rangle \equiv |x_1\rangle \otimes U\,|x_2\rangle$$

$$\langle a^{(1)}|x_1x_2\rangle \equiv \langle a|x_1\rangle\,|x_2\rangle \qquad \langle a^{(2)}|x_1x_2\rangle \equiv \langle a|x_2\rangle\,|x_1\rangle$$

$$\langle a^{(j)}|x_1x_2\cdots x_j\cdots x_n\rangle \equiv \langle a|x_j\rangle\,|x_1x_2\cdots x_{j-1}x_{j+1}\cdots x_n\rangle$$
$$\equiv (I_{j-1}\otimes\langle a|\otimes I_{n-j})\,|x_1x_2\cdots x_j\cdots x_n\rangle$$

$$\langle a_1^{(1)}a_2^{(2)}\cdots a_m^{(m)}|x_1x_2\cdots x_m\cdots x_n\rangle \equiv \langle a_1a_2\cdots a_m|x_1x_2\cdots x_m\rangle\,|x_{m+1}x_{m+2}\cdots x_n\rangle$$
$$\equiv (\langle a_1a_2\cdots a_m|\otimes I_{n-m})\,|x_1x_2\cdots x_m\cdots x_n\rangle$$

$$\text{Given } |\psi\rangle = \sum_{x_1,x_2,\cdots,x_n\in\{0,1\}} c_{x_1x_2\cdots x_n}\,|x_1x_2\cdots x_n\rangle,$$

$$\langle a_1^{(1)}a_2^{(2)}\cdots a_m^{(m)}|\psi\rangle = \sum_{x_{m+1},\cdots,x_n\in\{0,1\}} c_{a_1a_2\cdots a_mx_{m+1}\cdots x_n}\,|x_{m+1}\cdots x_n\rangle$$

Trace

$$\text{tr }A \equiv \text{tr}(A) \equiv \sum_{i=1}^{n} a_{ii}$$

$\text{tr}\left(A^T\right) = \text{tr}(A) \quad \text{tr}\left(A^\dagger\right) = \text{tr}\left(A^*\right) = (\text{tr }A)^*$

$\text{tr}(cA) = c\,\text{tr }A \quad \text{tr}(A+B) = \text{tr }A + \text{tr }B$

$\text{tr}(AB) = \text{tr}(BA) \quad$ (Note in general, $AB \neq BA$, and $\text{tr}(AB) \neq \text{tr}(A)\,\text{tr}(B)$)

$\text{tr}([A,B]) = 0$

For Pauli matrices: $\text{tr }\sigma_j = 0, \quad \text{tr}(\sigma_j\sigma_k) = 2\delta_{jk}$

Similarity invariance: $\text{tr}\left(PAP^{-1}\right) = \text{tr}\,A$

As sum of eigenvalues: $\text{tr}(A) = \sum_{i=1}^{n} \lambda_i, \quad \text{tr}(A^k) = \sum_{i=1}^{n} \lambda_i^k$

Tensor product property: $\text{tr}(A \otimes B) = \text{tr}(A)\,\text{tr}(B)$

Outer product property: $\langle v|u\rangle = \text{tr}(|u\rangle\langle v|), \quad \langle v|A|u\rangle = \text{tr}(A\,|u\rangle\langle v|)$

Cyclic product property: $\text{tr}(ABC) = \text{tr}(BCA) = \text{tr}(CAB) \quad (\neq \text{tr}(BAC))$

For orthonormal basis $\{|\phi_i\rangle\}$, $\text{tr}\left(\sum_{i=1}^{n} |\phi_i\rangle\langle\phi_i|\right) = n, \quad \text{tr}\,A = \sum_{i=1}^{n} \langle\phi_i|A|\phi_i\rangle$

A useful identity: $\langle\phi_i|A|\phi_i\rangle = \text{tr}(\langle\phi_i|A|\phi_i\rangle) = \text{tr}(A\,|\phi_i\rangle\langle\phi_i|)$

Cauchy-Schwarz inequality: $0 \leq [\text{tr}(AB)]^2 \leq \text{tr}(A^2)\,\text{tr}(B^2) \leq [\text{tr}(A)]^2[\text{tr}(B)]^2$

For normal matrix: $\text{tr}(A) = \log\left(\det(e^A)\right)$

Determinant

$\det\left(A^T\right) = \det(A) \quad \det\left(A^\dagger\right) = \det\left(A^*\right) = \det(A)^*$

$\det(A^{-1}) = \det(A)^{-1}$

$\det(AB) = \det(A)\,\det(B)$

$\det(A \otimes B) = (\det A)^n (\det B)^m \quad$ (where m is the dimension of A, and n of B.)

For unitary matrix: $|\det(U)| = 1$

For normal matrix: $\det(A) = \prod_i \lambda_i \quad \det(e^A) = e^{\text{tr}\,A}$

Vector Space of Matrices

Inner product of two matrices:

$$\langle A, B\rangle \equiv A \cdot B \equiv \sum_{i=1}^{m}\sum_{j=1}^{n} a_{ij}^* b_{ij} = \text{tr}(A^\dagger B) = \text{tr}(BA^\dagger)$$

Decomposition of $A \in \mathbb{C}^{2^n \times 2^n}$ in Pauli basis:

$$P_i \in \{I, X, Y, Z\}^{\otimes n}, \quad P_i \cdot P_j = 2^n \delta_{ij}, \quad \text{with } i, j = 1, 2, ..., 4^n$$

$$A = \sum_{i=1}^{4^n} a_i P_i, \quad a_i = \frac{1}{2^n} P_i \cdot A \equiv \frac{1}{2^n}\,\text{tr}(AP_i^\dagger) = \frac{1}{2^n}\,\text{tr}(AP_i)$$

D Pauli Matrices

Definitions

$$\sigma_0 \quad = I = |0\rangle\langle0| + |1\rangle\langle1| \qquad = \begin{bmatrix} 1 & 0 \\ 0 & 1 \end{bmatrix}$$

$$\sigma_1 = \sigma_x = X = |0\rangle\langle1| + |1\rangle\langle0| \qquad = \begin{bmatrix} 0 & 1 \\ 1 & 0 \end{bmatrix}$$

$$\sigma_2 = \sigma_y = Y = |0\rangle\langle1|(-i) + |1\rangle\langle0|i \quad = \begin{bmatrix} 0 & -i \\ i & 0 \end{bmatrix}$$

$$\sigma_3 = \sigma_z = Z = |0\rangle\langle0| - |1\rangle\langle1| \qquad = \begin{bmatrix} 1 & 0 \\ 0 & -1 \end{bmatrix}$$

Eigenvalues and Vectors

Pauli Matrix	Eigenvalue	Eigenvector
$\sigma_z \equiv Z$	1	$|0\rangle$
	-1	$|1\rangle$
$\sigma_x \equiv X$	1	$|+\rangle = \frac{1}{\sqrt{2}}(|0\rangle + |1\rangle)$
	-1	$|-\rangle = \frac{1}{\sqrt{2}}(|0\rangle - |1\rangle)$
$\sigma_y \equiv Y$	1	$|+_i\rangle = \frac{1}{\sqrt{2}}(|0\rangle + i\,|1\rangle)$
	-1	$|-_i\rangle = \frac{1}{\sqrt{2}}(|0\rangle - i\,|1\rangle)$

Properties

For $i, j, k \in \{1, 2, 3\}$,

Unitary and Hermitian: $\sigma_j^2 = \sigma_0, \quad \sigma_j^\dagger = \sigma_j$

Commutation relation: $[\sigma_j, \sigma_k] = 2i\varepsilon_{jkl}\sigma_l$, or $\sigma_1\sigma_2 = \sigma_3$, etc.

Anti-commutation relation: $\{\sigma_j, \sigma_k\} = 2\delta_{jk}\sigma_0$, or $\sigma_j\sigma_k = -\sigma_k\sigma_j$ for $j \neq k$

Orthogonality: $\langle\sigma_j, \sigma_k\rangle \equiv \mathrm{tr}(\sigma_j^\dagger\sigma_k) = \mathrm{tr}(\sigma_j\sigma_k) = 2\delta_{jk}$

In addition, $\mathrm{tr}\,\sigma_j = 0$, $\quad \det\sigma_j = -1, \quad \sigma_j\sigma_k = i\varepsilon_{jkl}\sigma_l + \delta_{jk}\sigma_0$

Basis for 2×2 Matrices

Any matrix $A \in \mathbb{C}^{2\times2}$ can be expanded over the set $\{\sigma_0, \sigma_1, \sigma_2, \sigma_3\}$:

$$A = \sum_{j=0}^{3} a_j\sigma_j, \text{ where } a_0 = \tfrac{1}{2}\mathrm{tr}\,A, \text{ and } a_j = \tfrac{1}{2}\mathrm{tr}(\sigma_j A).$$

Representing Spins in Any Direction

For any real unit vector $\boldsymbol{u} = (u_x, u_y, u_z)$ where $u_x^2 + u_y^2 + u_z^2 = 1$,

define $\sigma_u = \boldsymbol{u} \cdot \boldsymbol{\sigma} = u_x \sigma_x + u_y \sigma_y + u_z \sigma_z$.

For $\boldsymbol{u} = (\sin\theta\cos\phi,\ \sin\theta\sin\phi,\ \cos\theta)$,

$$\sigma_u = \sin\theta\cos\phi\,\sigma_x + \sin\theta\sin\phi\,\sigma_y + \cos\theta\,\sigma_z \quad = \begin{bmatrix} \cos\theta & \sin\theta e^{-i\phi} \\ \sin\theta e^{i\phi} & -\cos\theta \end{bmatrix}$$

Eigenvectors:

$$|+_u\rangle = \cos\frac{\theta}{2}|0\rangle + \sin\frac{\theta}{2}e^{i\phi}|1\rangle \quad = \begin{bmatrix} \cos\frac{\theta}{2} \\ \sin\frac{\theta}{2}e^{i\phi} \end{bmatrix}$$

$$|-_u\rangle = -\sin\frac{\theta}{2}|0\rangle + \cos\frac{\theta}{2}e^{i\phi}|1\rangle \quad = \begin{bmatrix} -\sin\frac{\theta}{2} \\ \cos\frac{\theta}{2}e^{i\phi} \end{bmatrix}$$

Expected values:

$$\langle +_u|\sigma_x|+_u\rangle = \sin\theta\cos\phi$$
$$\langle +_u|\sigma_y|+_u\rangle = \sin\theta\sin\phi$$
$$\langle +_u|\sigma_z|+_u\rangle = \cos\theta$$

Exponentiation

For any real γ, $e^{i\gamma\sigma_j} = \cos\gamma\sigma_0 + i\sin\gamma\sigma_j$

Also, $e^{i\gamma\sigma_u} = \cos\gamma\sigma_0 + i\sin\gamma\sigma_u$

Rotation Operators

$$R_j(\gamma) \equiv e^{-i\frac{\gamma}{2}\sigma_j} = \sigma_0\cos\frac{\gamma}{2} - i\sigma_j\sin\frac{\gamma}{2}$$

Explicitly,

$$R_1(\gamma) \equiv R_x(\gamma) = \begin{bmatrix} \cos\frac{\gamma}{2} & -i\sin\frac{\gamma}{2} \\ -i\sin\frac{\gamma}{2} & \cos\frac{\gamma}{2} \end{bmatrix}$$

$$R_2(\gamma) \equiv R_y(\gamma) = \begin{bmatrix} \cos\frac{\gamma}{2} & -\sin\frac{\gamma}{2} \\ \sin\frac{\gamma}{2} & \cos\frac{\gamma}{2} \end{bmatrix}$$

$$R_3(\gamma) \equiv R_z(\gamma) = \begin{bmatrix} e^{-i\frac{\gamma}{2}} & 0 \\ 0 & e^{i\frac{\gamma}{2}} \end{bmatrix}$$

Rotation about axis $\boldsymbol{u} = (n_x, n_y, n_z)$:

$$R_u(\gamma) = e^{-i\frac{\gamma}{2}\sigma_u} = \sigma_0\cos\frac{\gamma}{2} - i\sigma_u\sin\frac{\gamma}{2}$$

Rotation from $|0\rangle$ to $|+_u\rangle$ and $|-_u\rangle$:

$$|+_u\rangle = R_z(\phi)R_y(\theta)|0\rangle$$
$$|-_u\rangle = R_z(\phi)R_y(-\theta)|0\rangle$$

Other Forms

$$\sigma_+ \equiv \tfrac{1}{2}(\sigma_1 + i\sigma_2) = \begin{bmatrix} 0 & 1 \\ 0 & 0 \end{bmatrix} \qquad \sigma_- \equiv \tfrac{1}{2}(\sigma_1 - i\sigma_2) = \begin{bmatrix} 0 & 0 \\ 1 & 0 \end{bmatrix}$$

$$\sigma_h \equiv \tfrac{1}{2}(\sigma_0 + \sigma_3) = \begin{bmatrix} 1 & 0 \\ 0 & 0 \end{bmatrix} \qquad \sigma_v \equiv \tfrac{1}{2}(\sigma_0 - \sigma_3) = \begin{bmatrix} 0 & 0 \\ 0 & 1 \end{bmatrix}$$

Gate Relations

$$XY = -YX \quad = iZ$$
$$YZ = -ZY \quad = iX$$
$$ZX = -XZ \quad = iY$$

$$XXX = X, \quad XYX = -Y, \quad XZX = -Z,$$
$$YXY = -X, \quad YYY = Y, \quad YZY = -Z,$$
$$ZXZ = -X, \quad ZYZ = -Y, \quad ZZZ = Z$$

$$HXH = Z, \text{ where } H = \frac{1}{\sqrt{2}} \begin{bmatrix} 1 & 1 \\ 1 & -1 \end{bmatrix}$$

$$HYH = -Y$$
$$HZH = X$$

$$SXS^\dagger = Y, \text{ where } S = \begin{bmatrix} 1 & 0 \\ 0 & i \end{bmatrix}$$

$$SYS^\dagger = -X$$
$$SZS^\dagger = Z$$

Pauli Strings

Definition: $P = A_1 \otimes \cdots \otimes A_n$ with $A_i \in \{I, X, Y, Z\}$, or $P \in \{I, X, Y, Z\}^{\otimes n}$

Orthogonality: $P_i \cdot P_j \equiv \operatorname{tr}(P_i^\dagger P_j) = 2^n \delta_{ij}$, with $i, j = 1, 2, ..., 4^n$.

Two Pauli strings P_i and P_j commute if they have an even number of qubits where their corresponding Pauli matrices are different and anticommute (e.g., one is X and the other is Y or Z, but not I).

Bibliography

Books

[1] Peter Y. Lee, Huiwen Ji, and Yan Cheng. *Quantum Computing and Information: A Scaffolding Approach*. 2nd edition. Polaris QCI Publishing, 2024. ISBN: 978-1-961880-06-1 (cited on pages 89, 92, 120, 167, 178, 267, 283, 298, 356, 378, 512, 552).

[2] Lokenath Debnath and Piotr Mikusinski. *Introduction to Hilbert Spaces with Applications*. 3rd. Burlington, MA: Elsevier Academic Press, 2005. ISBN: 978-0-12-208438-6 (cited on page 114).

[3] Gilbert Strang. *Introduction to Linear Algebra*. English. 6th edition. Wellesley-Cambridge Press, Apr. 2023. ISBN: 978-1733146678 (cited on page 170).

[4] Morris H. DeGroot and Mark J. Schervish. *Probability and Statistics*. 4th. Boston, MA: Pearson, 2012. ISBN: 978-0-321-50046-5 (cited on page 488).

[5] Steve Brooks et al., editors. *Handbook of Markov Chain Monte Carlo*. Boca Raton, FL: CRC Press, 2011. ISBN: 978-1-4200-7941-8 (cited on page 511).

[6] Michael A Nielsen and Isaac L Chuang. *Quantum Computation and Quantum Information: 10th Anniversary Edition*. Cambridge University Press, 2010. ISBN: 978-1-107-00217-3 (cited on pages 512, 552).

Articles

[7] Paul A. M. Dirac. A New Notation for Quantum Mechanics. *Mathematical Proceedings of the Cambridge Philosophical Society* 35.3 (1939) (cited on page 90).

[8] Peter Nimbe, Benjamin Asubam Weyori, and Adebayo Felix Adekoya. Models in quantum computing: a systematic review. *Quantum Information Processing* 20.2 (Feb. 2021). ISSN: 1573-1332 (cited on pages 512, 552).

[9] Laszlo Gyongyosi and Sandor Imre. A Survey on quantum computing technology. *Computer Science Review* 31 (2019). ISSN: 1574-0137 (cited on pages 512, 552).

[10] Alexander M. Dalzell et al. Quantum algorithms: A survey of applications and end-to-end complexities. *Preprint* arXiv:2310.03011 (2023) (cited on page 552).

[11] Victor Montenegro et al. Review: Quantum Metrology and Sensing with Many-Body Systems. *Preprint* arXiv:2408.15323 (2024) (cited on page 552).

List of Figures

List of Tables

Index

Z

Journey Forward

A Journey Well Traveled

Congratulations to all who have completed this book, especially those who engaged with the exercises and problems throughout. You've successfully established a strong foundation in the mathematical concepts underpinning quantum computing, equipping yourselves for further exploration in this rapidly advancing field.

The Mathematical Horizon

As quantum computing continues to evolve, so too does its mathematical framework. New discoveries in areas such as quantum algorithms, cryptography, and error correction challenge us to deepen our mathematical understanding and extend it into new realms. The foundational topics you have mastered—linear algebra, Dirac notation, matrix analysis, and probability—will remain central as quantum technology progresses.

But there are still many exciting mathematical challenges ahead. As the quest for quantum algorithms and scalable, fault-tolerant quantum computing continues, new theoretical tools will be required. Quantum computing presents unique mathematical demands, and the journey ahead will offer opportunities to expand upon the knowledge you have gained here.

Looking Forward

As you continue your journey, consider the following next steps:

- Study the fundamentals of quantum computing and information [6, 8, 9], particularly by exploring the second book in this series, *Quantum Computing and Information: A Scaffolding Approach* [1].

- Delve into quantum algorithms after establishing a solid understanding of quantum computing, as these offer solutions across a wide range of applications [10].

- Explore other fields within quantum science, such as quantum sensing, where new frontiers in technology and mathematics are rapidly emerging [11].

- Stay updated with the latest advancements in quantum computing, including hardware innovations, error correction techniques, and emerging algorithms.

With the knowledge gained from this book, you are well-positioned to further explore the fascinating realm of quantum computing and related fields. Keep learning, stay curious, and continue to expand your understanding of this ever-evolving subject.

www.ingramcontent.com/pod-product-compliance
Lightning Source LLC
Chambersburg PA
CBHW051748200326
41597CB00025B/4481